The Organometallic Chemistry
of N-heterocyclic Carbenes

Inorganic Chemistry

A Wiley Series of Advanced Textbooks
ISSN: 1939-5175

Editorial Board

Previously Published Books in this Series

Bioinorganic Chemistry – Inorganic Elements in the Chemistry of Life: An Introduction and Guide, 2nd Edition
Wolfgang Kaim, Brigitte Schwederski & Axel Klein; ISBN: 978-0-470-97523-7

Structural Methods in Molecular Inorganic Chemistry
David W. H. Rankin, Norbert W. Mitzel & Carole A. Morrison; ISBN: 978-0-470-97278-6

Introduction to Coordination Chemistry
Geoffrey Alan Lawrance; ISBN: 978-0-470-51931-8

Chirality in Transition Metal Chemistry
Hani Amouri & Michel Gruselle; ISBN: 978-0-470-06054-4

Bioinorganic Vanadium Chemistry
Dieter Rehder; ISBN: 978-0-470-06516-7

Inorganic Structural Chemistry 2nd Edition
Ulrich Müller; ISBN: 978-0-470-01865-1

Lanthanide and Actinide Chemistry
Simon Cotton; ISBN: 978-0-470-01006-8

Mass Spectrometry of Inorganic and Organometallic Compounds: Tools-Techniques-Tips
William Henderson & J. Scott McIndoe; ISBN: 978-0-470-85016-9

Main Group Chemistry, Second Edition
A.G. Massey; ISBN: 978-0-471-19039-5

Synthesis of Organometallic Compounds: A Practical Guide
Sanshiro Komiya; ISBN: 978-0-471-97195-5

Chemical Bonds: A Dialog
Jeremy Burdett; ISBN: 978-0-471-97130-6

The Molecular Chemistry of the Transition Elements: An Introductory Course
Francois Mathey & Alain Sevin; ISBN: 978-0-471-95687-7

Stereochemistry of Coordination Compounds
Alexander von Zelewsky; ISBN: 978-0-471-95599-3

For more information on this series see: www.wiley.com/go/inorganic

The Organometallic Chemistry of N-heterocyclic Carbenes

Han Vinh Huynh

Department of Chemistry
National University of Singapore
Republic of Singapore

WILEY

This edition first published 2017
© 2017 John Wiley & Sons Ltd

The right of Han Vinh Huynh to be identified as the author of this work has been asserted in accordance with law.

Registered Offices
John Wiley & Sons, Inc., 111 River Street, Hoboken, NJ 07030, USA
John Wiley & Sons Ltd, The Atrium, Southern Gate, Chichester, West Sussex, PO19 8SQ, UK

Editorial Office
The Atrium, Southern Gate, Chichester, West Sussex, PO19 8SQ, UK

For details of our global editorial offices, customer services, and more information about Wiley products visit us at www.wiley.com.

Wiley also publishes its books in a variety of electronic formats and by print-on-demand. Some content that appears in standard print versions of this book may not be available in other formats.

Library of Congress Cataloging-in-Publication Data

Names: Huynh, Han Vinh, 1972–
Title: The organometallic chemistry of N-heterocyclic carbenes / Han Vinh Huynh, Department of Chemistry, National University of Singapore.
Description: First edition. | Hoboken, NJ : John Wiley & Sons, Inc., 2017. | Series: Inorganic chemistry : a textbook series | Includes bibliographical references and index.
Identifiers: LCCN 2016046161 (print) | LCCN 2016046392 (ebook) | ISBN 9781118593776 (cloth) | ISBN 9781118698808 (pdf) | ISBN 9781118698792 (epub)
Subjects: LCSH: Organometallic chemistry. | Heterocyclic chemistry. | Carbenes (Methylene compounds)
Classification: LCC QD411 .H89 2017 (print) | LCC QD411 (ebook) | DDC 547/.412–dc23
LC record available at https://lccn.loc.gov/2016046161

Cover image: Courtesy of author
Cover design by Wiley

Set in 10/12pt Times by SPi Global, Pondicherry, India
Printed in Singapore by C.O.S. Printers Pte Ltd

10 9 8 7 6 5 4 3 2 1

Contents

Foreword

Rarely over the last 50 years has a new type of ligand captured the imaginations of chemists as was the case with the N-heterocyclic carbenes (NHCs). The report by Arduengo *et al.* in 1991 on stable, "bottle-able" N-heterocyclic carbenes derived from imidazolium salts initiated an intensive stream of research on these compounds and their metal complexes which has now lasted unabated for 25 years. While NHC complexes have been known since the work of Öfele *et al.* and Wanzlick *et al.* in 1969, the parent free carbenes remained elusive up until the studies by Arduengo, validating earlier proposals by Wanzlick *et al.*

Meanwhile, a large number of cyclic diaminocarbenes derived from smaller or larger rings than the imidazolium salts used originally have been produced. In addition, the heteroatoms in the carbene ring have been varied to give P-heterocyclic carbenes or even the cyclic alkyl-amino carbenes (CAACs) featuring only one nitrogen atom in the heterocyclic carbene ring. From transient laboratory curiosities, NHCs have developed into an important class of carbon(II) donor ligands next to carbonyls and isocyanides. They have found multiple applications as sterically unique and demanding spectator ligands for the preparation of a various catalytically active metal complexes, culminating in the development of ruthenium-NHC complexes catalyzing various types of olefin metathesis. Other NHC complexes have found applications as organometallic metallodrugs, and poly-NHCs have even found use for the construction of discrete metallosupramolecular assemblies.

Knowing the very substantial current interest in NHCs and their metal complexes and the maturity of the research field in general, the usefulness of a text appealing to the novice in carbene chemistry and to serve as a reference for the advanced researcher becomes apparent. Good progress in this area has been achieved with this book by H. V. Huynh. Written by an experienced researcher in the field, the text discusses in nine chapters general aspects of NHC chemistry, as well as advanced topics such as mesoionic carbenes and expanded ring NHCs. Special emphasis is placed on NHC-metal complexes of groups 10 and 11, Ruthenium, Rhodium, and Iridium. The book illustrates clearly various methods for the preparation of NHCs and their complexes thereby demonstrating their vast potential in modern organometallic chemistry. In order not to discourage the reader by too voluminous a book, the author has placed emphasis on general concepts and methods rather than listing all known NHCs or their metal complexes.

Given the current state of research on NHCs, a general text on the subject is overdue. The author has found a middle way between a comprehensive, exhaustive description of the subject and mere conceptual highlighting of important topics. It is hoped that this text appeals to the diverse readership it was written for.

F. Ekkehardt Hahn
Münster, September 2016

Preface

Since its humble beginning, which started with curiosity-driven works of Wanzlick and Öfele in 1968, the chemistry of N-heterocyclic carbenes (NHCs) has matured considerably. The isolation of the first free NHC by Arduengo reported in 1991 has reignited interest in an almost forgotten area, and early catalytic applications of NHC complexes in C–C couplings reported by Herrmann further fueled the development of the field. Nowadays, NHCs are ubiquitous in mainstream and modern chemistry, where they are most commonly used as ligands and organocatalysts. Other areas for their application are just beginning to unfold. Given their prominent standing chemistry, it is surprising to note that there are currently no general textbooks dedicated to NHCs. The several books on NHCs available in the scientific literature are rather specific or mainly concentrate on their use in catalysis. These texts are suitable and valuable for experienced scientists. However, for university students at the undergraduate or graduate level and for any newcomer to the field, they may be too advanced, sometimes lacking basic, but crucial explanations for a better understanding of the subject.

This textbook attempts to bridge the gap between the novice and the expert. As such it is targeting the advanced undergraduate and graduate student and anyone wishing to explore NHC chemistry. It includes an historical account highlighting important milestones in the development of this chemistry from its early beginnings. Key properties of NHCs, their precursors and complexes are addressed, and a critical discussion on the nomenclature of NHCs is included. Moreover, common synthetic methods are explained in terms of outcome and driving force. Four out of nine chapters are dedicated to the NHC chemistry of the most common transition metals in this field, while the last two chapters provide an overview and outlook on developments "beyond" the field of classical NHCs.

Many examples given in this book were selected from works of early pioneers to pay tribute to their contributions. Others were chosen to highlight a specific property, reactivity, or behavior of NHCs or their complexes. The purpose of this textbook is not to provide a review on the extensive and most recent organometallic chemistry of NHCs in general. Therefore, I wish to apologize to all researchers in the field whose contributions could not be included, or have unknowingly been left out. A principal task of science is detecting mistakes and errors in a system for rectification and improvement. This textbook should certainly not be exempted from such scrutiny, and any constructive suggestions from colleagues all around the world will be greatly valued.

Being a sole author, it was a tremendous task of more than four years for me to complete this textbook in addition to my usual duties as a scientist and lecturer. I am grateful to all my current and former coworkers for their dedicated work in NHC chemistry, and wish to particularly mention Dr. Qiaoqiao Teng, Dr. Van Ha Nguyen, Mr Xiaoke Xie (Ph.D. cand.), Dr. Guo Shuai and Dr. Jan C. Bernhammer, who helped me in one way or another in the realization of this textbook. Moreover, I wish to thank my Wiley contacts Sarah Hall, Sarah Tilley Keegan, Nithya Sechin, Hari Sridharan, Chris Cartwright and Emma Strickland for their input and patience. I am indebted to my friend, colleague, and former mentor Professor F. Ekkehardt Hahn for providing a Foreword to this book. The depictions in the textbook were created using site-licensed *ChemDraw Prime* and the freeware *Mercury 3.8* using CIF files from the CCDC websites.

Finally, I hope that students, lecturers, and active researchers in this field will find this textbook useful and stimulating. The chemistry of NHCs is fun, and indeed not that difficult at all.

Han Vinh Huynh
Singapore, September 2016

List of Abbreviations and Definitions

Abbreviation/term	Definition
$\%V_{Bur}$	Percentage of buried volume
γ	Gyromagnetic ratio
$+I$ effect	Positive inductive effect
$-I$ effect	Negative inductive effect
$+M$ effect	Positive mesomeric effect
$-M$ effect	Negative mesomeric effect
°C	Degree Celsius
5-NHC	Five-membered N-heterocyclic carbene
6-DAC	Six-membered diamidocarbene
6-NHC	Six-membered N-heterocyclic carbene
7-DAC	Seven-membered diamidocarbene
7-NHC	Seven-membered N-heterocyclic carbene
8-NHC	Eight-membered N-heterocyclic carbene
Å	Ångström
Ad	Adamantyl
aka	Also known as
All_2-bimy	1,3-diallylbenzimidazolin-2-ylidene
Ar	Aryl
BDE	Bond dissociation energy
Bh_2-bimy	1,3-dibenzhydrylbenzimidazolin-2-ylidene
bimy	Benzimidazolin-2-ylidene
BINAP	2,2′-bis(diphenylphosphino)-1,1′-binaphthyl
BMIM	1-butyl,3-methylimidazolin-2-ylidene
Bn_2-bimy	1,3-dibenzylbenzimidazolin-2-ylidene
Bn_2-tazy	1,4-dibenzyl-1,2,4-triazolin-5-ylidene
Bn-btzy	3-benzylbenzothiazolin-2-ylidene
BQ	p-benzoquinone
bzoxNHC	Benzoxazole-functionalized NHC
CAAC	Cyclic alkyl amino carbene
COD	1,5-cyclooctadiene
Cp	Cyclopentadienyl anion
Cp*	Pentamethylcyclopentadienyl anion
CuAAC	Copper-catalyzed azide-alkyne cycloaddition
Cy	Cyclohexyl
DAC	Diamidocarbene
DAE	Diallylether
dba	Dibenzylideneacetone
de	Diastereomeric excess
Dipp	2,6-diisopropylphenyl
ditz	1,2,4-triazolidine-3,5-diylidene
DMFU	Dimethylfumarate
DMSO	Dimethyl sulfoxide

dppd	1,3-diphenyl-1,3-propanedionate
DVDS	1,1,3,3-tetramethyl-1,3-divinyldisiloxane
EMIM	1-ethyl,3-methylimidazolin-2-ylidene
equiv	Equivalent(s)
ESI	Electrospray ionization
Et	Ethyl
eV	Electron volt
FPyr	1,2,3,4,6,7,8,9-octahydropridazino[1,2-*a*]-indazolin-11-ylidene
h	Hour
HB	Conjugate acid of general Brønsted base
HMBC	Heteronuclear multiple-bond correlation
HX	General Brønsted acid
Hz	Hertz
IAd	1,3-diadamantylimidazolin-2-ylidene
IBh	1,3-dibenzhydrylimidazolin-2-ylidene
^{i}Bu$_2$-bimy	1,3-diisobutylbenzimidazolin-2-ylidene
ICy	1,3-dicyclohexylimidazolin-2-ylidene
IFc	1,3-diferrocenylimidazolin-2-ylidene
I^{i}Pr$_{Me2}$	1,3-diisopropyl-4,5-dimethylimidazolin-2-ylidene
IMe	1,3-dimethylimidazolin-2-ylidene
IMe$_{Me2}$	1,3,4,5-tetramethylimidazolin-2-ylidene
IMes	1,3-dimesitylimidazolin-2-ylidene
imy	Imidazolin-2-ylidene
in situ	"In the reaction mixture", without isolation
INap	1,3-di(1-naphtyl)imidazolin-2-ylidene
Ind	Indenyl anion
Indy	Indazolin-3-ylidene
IPh	1,3-diphenylimidazolin-2-ylidene
IPr	1,3-(2,6-diisopropylphenyl)imidazolin-2-ylidene
^{i}Pr	Isopropyl
^{i}Pr,Bn-bimy	1-isopropyl,3-benzylbenzimidazolin-2-ylidene
^{i}Pr$_2$-bimy	1,3-diisopropylbenzimidazolin-2-ylidene
IPy	1,3-di(2-pyridyl)imidazolin-2-ylidene
IR	Infrared
I^{t}Bu	1,3-di(tert-butyl)imidazolin-2-ylidene
IUPAC	International Union of Pure and Applied Chemistry
kJ	Kilojoule
M	Molar
MAH	Maleic anhydride
Me	Methyl
Me$_2$-bimy	1,3-dimethylbenzimidazolin-2-ylidene
Mes	Mesityl
MHMDS	metal hexamethyldisilazide
MS	Molecular sieves
MVS	Metal vapor synthesis
NCA	Weakly or non-coordinating anion
NHC	N-heterocyclic carbene
NMR	Nuclear Magnetic Resonance
NOHC	N,O-heterocyclic carbene
Np	Neopentyl
Np$_2$-bimy	1,3-(2,2-dimethylpropyl)benzimidazolin-2-ylidene
NQ	1,4-naphthoquinone

NSHC	N,S-heterocyclic carbene
NuH	General nucleophile
p-cymene	1-Methyl-4-(1-methylethyl)benzene
PEPPSI	Pyridine-enhanced precatalyst preparation stabilization and initiation
Ph	Phenyl
Ph$_3$tria	1,3,4-triphenyl-1,2,4-triazolin-5-ylidene
ppm	Parts per million
Pr$_2$-bimy	1,3-di(*n*-propyl)benzimidazolin-2-ylidene
pta	1,3,5-triaza-7-phosphaadamantane
Pyry	Pyrazolin-3-ylidene
R	Usually alkyl group
RCM	Ring-closing metathesis
Ref	Reference
ROMP	Ring-opening metathesis polymerization
RT	Room temperature
SIBn	1,3-dibenzylimidazolidin2-ylidene
SIEt	1,3-diethylimidazolidin-2-ylidene
SIMe	1,3-dimethylimidazolidin-2-ylidene
SIMes	1,3-(2,4,6-trimethylphenyl)imidazolidin-2-ylidene
SIPh	1,3-diphenylimidazolidin-2-ylidene
SIPr	1,3-(2,6-diisopropylphenyl)imidazolidin-2-ylidene
SI^{t}Bu	1,3-di(tert-butyl)imidazolidin-2-ylidene
t	Time
T	Temperature
tazy	1,2,4-triazolin-5-ylidene
^{t}Bu	Tertiary butyl
TCNE	Tetracyanoethylene
TEP	Tolman's Electronic Parameter
THF	Tetrahydrofurane
THT	Tetrahydrothiophene
timy	1,3,4,5-tetramethylimidazolin-2-ylidene
tmeda	N,N,N′N′-tetramethylethylenediamine
TMSCl	Chlorotrimethylsilane
tol	Tolyl
tria	Triazolin-5-ylidene
UV-vis	Ultraviolet–visible
via	by using
*a*NHC	*Abnormal* N-heterocyclic carbene
MIC	*Mesoionic* carbene, here *mesoionic* 1,2,3-triazolin-5-ylidene
*r*NHC	*Remote* N-heterocyclic carbene
trz	*Normal* 1,2,3-triazolin-5-ylidene
Pyry	Pyrazolin-3/5-ylidene
D	Debye (unit for dipole moment)
Indy	Indazolin-3-ylidene

1 General Introduction

1.1 Definition of Carbenes

According to the International Union of Pure and Applied Chemistry (IUPAC) a carbene [1] is "*the electrically neutral species H_2C: and its derivatives, in which the carbon is covalently bonded to two univalent groups of any kind or a divalent group and bears two nonbonding electrons, which may be spin-paired (singlet state) or spin-non-paired (triplet state)*." In general terms, carbenes are therefore neutral compounds R_2C: derived from the parent methylene (H_2C:) that feature a divalent carbon atom with only six valence electrons, which result from four bonding electrons in the two R–C bonds and two non-bonding electrons remaining at the carbene center. The geometry at the carbene carbon can be either linear or bent, depending on the degree of hybridization [2]. The linear geometry is based on an *sp*-hybridized carbene center with two nonbonding, energetically degenerate *p* orbitals (p_x and p_y). On the other hand, the bent geometry is adopted when the carbene carbon atom is sp^2-hybridized. On transition from the *sp*- to sp^2-hybridization, the energy of one *p* orbital, usually called p_π, remains almost unchanged, while the newly formed sp^2-hybrid orbital, normally called σ, is energetically stabilized as it acquires partial *s* character (Figure 1.1). However, the linear geometry is rarely observed, and most carbenes contain a sp^2-hybridized carbene center and are therefore bent.

For the simple linear case and without considering π contributions of the R-substituents, only the $p_x^1 p_y^1$ electronic configuration is feasible according to Hund's first rule [3], due to the degeneracy of the p_x and p_y orbitals. The two unpaired electrons are both "spin up" ($m_s = \frac{1}{2}$) giving rise to a total spin of $S = 1$, which in turn results in a spin multiplicity of $M = 3$ (Equation 1.1). Therefore, linear carbenes are generally in a triplet state. On the other hand, two common electronic configurations are possible for the carbene carbon in bent species. The two nonbonding electrons can singly occupy the two different σ and p_π orbitals with parallel spins ($\sigma^1 p_\pi^1$), which also leads to a triplet ground state ($^3B_1$). Alternatively, the two nonbonding electrons can also be spin-paired in the energetically more favorable σ orbital ($\sigma^2 p_\pi^0$) leading to a singlet ground state ($^1A_1$).

In addition to the two ground states in bent carbenes, two less favorable electronic configurations are conceivable (not depicted) that give rise to singlet states. The first has two spin-paired electrons in the p_π orbital ($\sigma^0 p_\pi^2$, $^1A_1$), and the second has two electrons singly occupying the σ and p_π orbitals, but with opposite spins ($\sigma^1 p_\pi^1$), giving rise to an excited singlet state ($^1B_1$) [4]. The latter two electronic configurations and their states have little significance for the discussion in this work.

The properties and reactivities of bent carbenes are primarily determined by their ground state spin multiplicity [5]. The two singly occupied orbitals in triplet carbenes are unsaturated (open-shell) and can accommodate one more electrons of opposite spin each. Thus, it

The Organometallic Chemistry of N-heterocyclic Carbenes, First Edition. Han Vinh Huynh.
© 2017 John Wiley & Sons Ltd. Published 2017 by John Wiley & Sons Ltd.

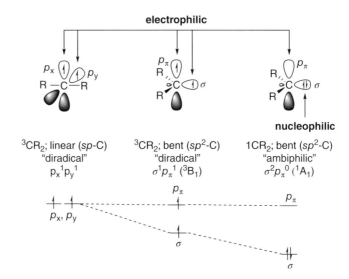

Figure 1.1
Relationship between the carbene bond angle, the nature of the frontier orbitals and singlet–triplet separation.

is intuitive to assign an *electrophilic* or *diradical* character to the carbene carbon (Figure 1.1). Singlet carbenes, on the other hand, contain a fully occupied σ orbital (closed-shell, nucleophilic) and an empty p_π orbital (electrophilic). The presence of both electrophilic and nucleophilic sites makes singlet carbenes formally *ambiphilic*.

$$M = 2\,S + 1,\, where\; S = \Sigma m_s = \tfrac{1}{2} \times n\left(unpaired\; electrons\right)$$

Equation 1.1
Definition of spin multiplicity for the determination of singlet and triplet state.

Whether a bent carbene adopts the singlet or triplet ground state is determined by the relative energies of the σ and p_π orbitals, which in turn is influenced by the direct substituents R at the carbene carbon. A large energy gap of at least 2 eV (~193 kJ/mol) between the σ orbital and the p_π orbital is required to stabilize a singlet ground state, whereas an energy difference of less than 1.5 eV (~145 kJ/mol) leads to a triplet ground state [6].

The relative energies of σ and p_π orbitals can also be influenced by the steric and electronic effects of the substituents on the carbene carbon atom. For instance, electron-withdrawing substituents (–I effect) inductively stabilize the σ orbital by enriching its *s* character and leave the p_π orbital essentially unchanged, thereby increasing the energy gap between the σ and p_π orbitals. Thus the singlet state is favored. On the other hand, electron donating groups (+I effect) decrease the energy gap between σ and p_π orbitals, which stabilizes the triplet state.

Besides inductive effects, which govern the ground-state spin multiplicity in carbenes, mesomeric effects of the R-substituents also play a crucial role by influencing the degree of bending in singlet carbenes. If the carbene carbon is attached to at least one π-accepting group Z (–M effect), for example, Z = COR, CN, CF$_3$, BR$_2$, or SiR$_3$, a linear or quasi-linear geometry is predicted. In this case, the initial degeneracy of the p_x and p_y is broken through π interactions with the Z substituents, therefore allowing for an unusual linear singlet state.

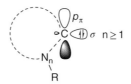

Figure 1.2
Minimum structural requirement for an N-heterocyclic carbene.

On the other hand, π-electron donating X substituents (+*M* effect), for example, X = N, O, P, S, and halogens, adjacent to the carbene center increase the energy of the p_π orbital of the carbene carbon atom. Since the σ orbital remains unchanged, the σ–p_π gap is increased, and hence a bent singlet state is favored.

N-heterocyclic carbenes (NHCs) are carbenes incorporated into heterocyclic rings that must contain at least one nitrogen atom. This minimum structural requirement for an NHC is highlighted in Figure 1.2. In addition to nitrogen, other heteroatoms may also be part of the ring system. Generally, they are bent singlet carbenes. Since they are a centerpiece of this book, the properties and electronic structures of NHCs will be discussed in more detail in Chapter 2.

1.2 Historical Overview of Carbenes, N-Heterocyclic Carbenes, and Their Complexes

The following section gives a brief overview on the milestones in the historical development of N-heterocyclic carbenes and their early use as ligands in organometallic chemistry. It is not the intention to provide a detailed account of short-lived carbene and carbenoid chemistry in general, which goes beyond the scope of this textbook, the focus of which is classical N-heterocyclic carbenes. The pioneering work of Jack Hine, William Doering, Philip Skell, Gerhard Closs, and later Robert A. Moss, Wolfgang Kirmse, Hideo Tomioka, and others on such highly reactive species are recounted elsewhere in detail [7], and will not be further mentioned here. Further, this section is divided into two parts. The first deals with the quest for free carbenes, while the second provides a brief summary dealing with the stabilization of carbenes, particularly of NHCs, by transition metals in an historical context. The milestones of these two separate quests are summarized in two timelines depicted in Figure 1.3 and Figure 1.4.

1.2.1 The Quest for Free Stable Carbenes

The quest for stable carbenes [8, 9], and in particular that for methylene (H_2C:) as the simplest possible representative began back in 1835 [10], when the French chemist Jean-Baptiste Dumas (1800–1884) tried to synthesize methylene by dehydration of methanol (CH_3OH) using phosphorus pentoxide or sulfuric acid. In those days, it was known that carbon can form compounds with oxygen and hydrogen in different oxidation states. In particular, these include carbon monoxide ($C^{II}O$), carbon dioxide ($C^{IV}O_2$) and methane ($C^{-IV}H_4$). Thus, the search for a carbon-hydrogen compound in which carbon adopts the missing intermediate –II oxidation state seemed a reasonable and doable task. In analogy

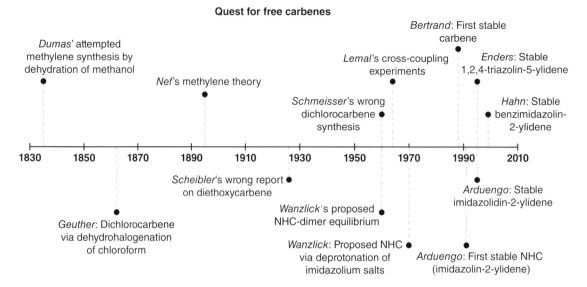

Figure 1.3
Milestones in the quest for free carbenes.

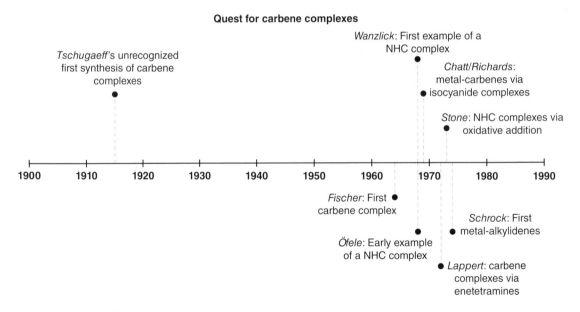

Figure 1.4
Milestones in the quest for carbene complexes.

to the aforementioned gaseous carbon compounds, Dumas' reaction indeed afforded a gaseous compound. Nowadays, we know that he had prepared dimethyl ether (CH_3OCH_3), which indeed had formed by dehydration, but under condensation of two methanol molecules (Scheme 1.1) instead of from a single molecule.

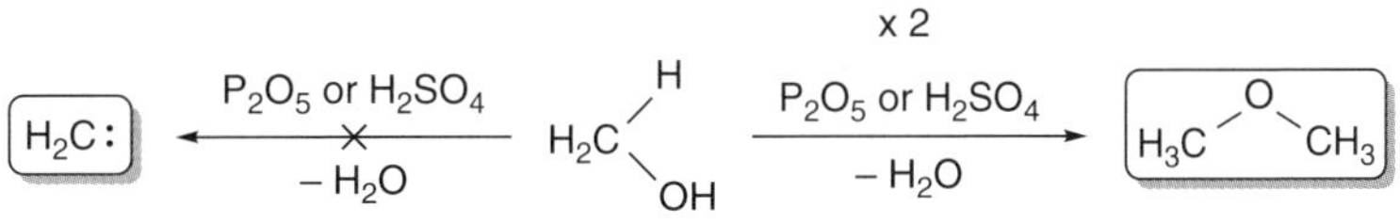

Scheme 1.1
Dumas' attempt to prepare methylene afforded dimethyl ether.

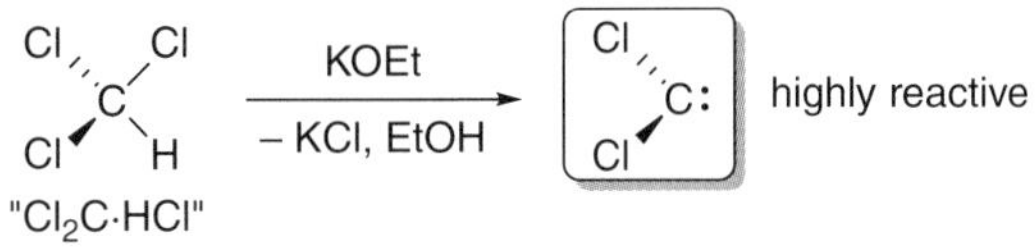

Scheme 1.2
Dehydrohalogenation of chloroform to give dichlorocarbene.

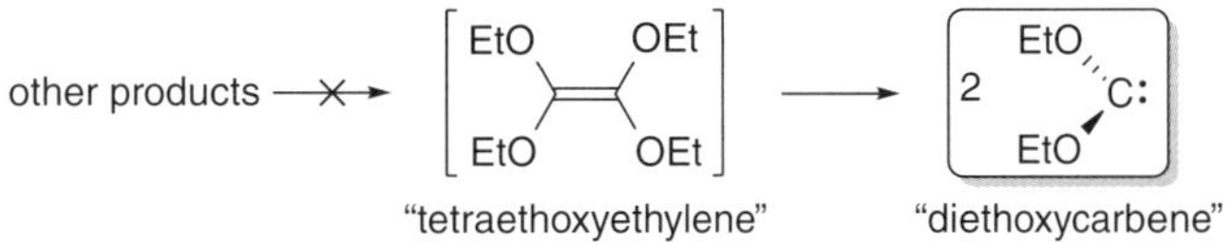

Scheme 1.3
Wrongly proposed formation of diethoxycarbene.

Pursuing a similar strategy, Anton Geuther (1833–1889) treated chloroform with potassium ethoxide solution in ethanol. He considered chloroform ($CHCl_3 =$ "$Cl_2C \cdot HCl$") as a simple HCl adduct of dichlorocarbene (Cl_2C:), and in 1862, he proposed the formation of the latter according to Scheme 1.2 [11]. This dehydrohalogenation reaction indeed yielded the dichlorocarbene, which under the given reaction conditions was, however, too reactive to be isolated and quickly decomposed to other products.

In 1895 and 60 years after Dumas report, Johann Ulric Nef (1862–1915) announced that he would tackle the preparation of methylene and its nitrogen free derivatives based on his previous work on derivatives of hydrogen cyanide [12]. He subsequently proposed a theory, in which methylene can be regarded as a suitable building block in organic chemistry and predicted its large scale synthesis [13]. Nevertheless, methylene remained elusive.

The next proposed preparation of a carbene that attracted attention was reported in 1926 by Helmut Scheibler (1882–1966) [14]. According to this report, the reaction of sodium ethoxide with certain esters would—after a complicated and wrong reaction sequence—eventually give tetraethoxyethylene, which further decomposes to two molecules of diethoxycarbene (Scheme 1.3). Interestingly, Scheibler already proposed the formation of a carbene dimer, which dissociates into free monomeric carbenes. This proposal was heavily criticized by the chemical community at that time and finally revoked by later studies.

Another preparation of the highly reactive dichlorocarbene that attracted the attention of the scientific community was reported in 1960 by Martin Schmeisser (1912–1981), who suggested that treatment of tetrachloromethane with activated carbon at very high temperatures would give dichlorocarbene among a range of other products (Scheme 1.4). The claimed "carbene" could according to the authors be condensed as a yellow solid in a cold trap immersed in liquefied air. Furthermore, subsequent reactions such as phosgene formation in air and insertions into double bonds were used as indications of the successful

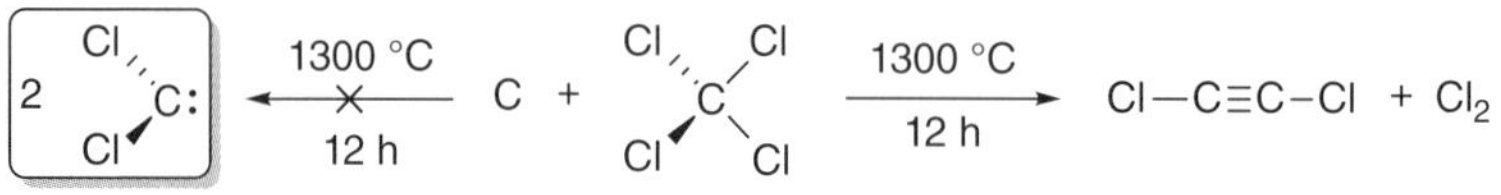

Scheme 1.4
Proposed dichlorocarbene synthesis by comproportionation of carbon with CCl_4.

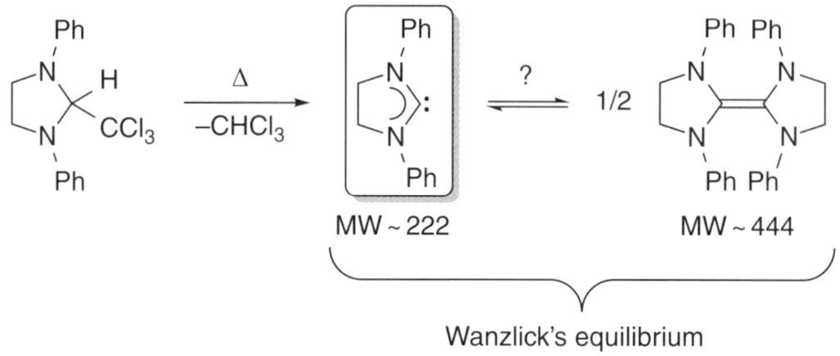

Wanzlick's equilibrium

Scheme 1.5
Wanzlick's proposed synthesis of an NHC and its monomer–dimer equilibrium.

dichlorocarbene formation. Further studies by his own team revealed that the activated carbon did not react with CCl_4, but catalyzed its decomposition. The yellow product isolated was later correctly identified as a mixture of dichloroacetylene and chlorine.

These numerous failed attempts to isolate free carbenes led the chemical community to think of carbenes only as fleeting intermediates that are too reactive to be isolated [7]. Thus, any further report on their isolation was skeptically regarded. It was during such a time when Hans-Werner Wanzlick (1917–1988), a former student of H. Scheibler, reported in 1960 that thermolysis of 1,3-diphenyl-2-trichloromethylimidazolidine would lead to an α-elimination of chloroform and formation of a carbene, which could be isolated as a colorless crystalline material [15]. The molecular weight determined for this material was 300 Da, which is in between the theoretical value for the free carbene and its dimer, and therefore Wanzlick assumed a monomer–dimer equilibrium (Scheme 1.5).

For Wanzlick's search for a free carbene, it was unfortunate that he was working on the fully saturated imidazolidine-based system. Nowadays, we know that the dimeric enetetramine form is more favorable for such systems. Although the elimination of chloroform was accepted, Wanzlick's proposed equilibrium, and therefore the existence of free carbenes in the reaction mixture, was heavily debated.

In 1964, David M. Lemal published results of cross-coupling experiments that provided evidence against the existence of Wanzlick's equilibrium and consequently ended Wanzlick's claim for the isolation of a free carbene [16]. Using NMR spectroscopy, Lemal and coworkers heated a mixture of two tetraaminoethylenes with very similar N-substituents in xylene under reflux for 2 h, and subsequently subjected the mixture to oxidation using $AgNO_3$ (Scheme 1.6). No oxidized derivatives of cross metathesis products were found, and instead, only symmetrical 2,2′-bis(imidazolinium) nitrates as oxidation products derived from the two starting materials were identified. Similar cross-coupling experiments and analysis by gas chromatography by Winberg and coworkers published in 1965 [17], confirmed Lemal's view that an equilibrium does not exist in the absence of any electrophiles.

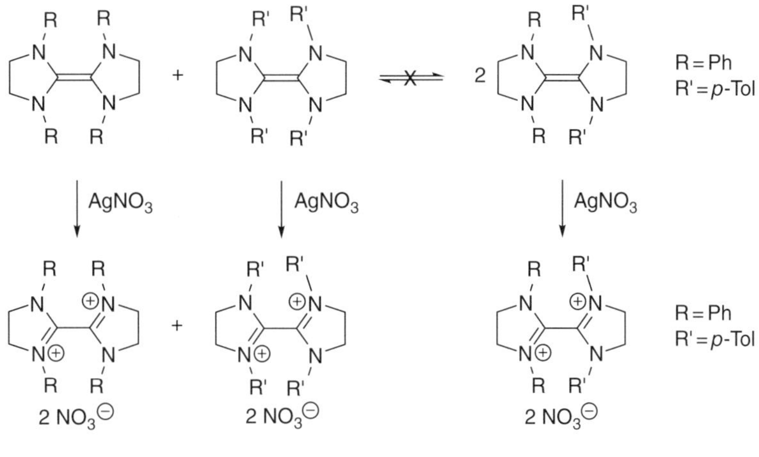

Scheme 1.6
Cross-coupling experiments conducted by D. M. Lemal and coworkers.

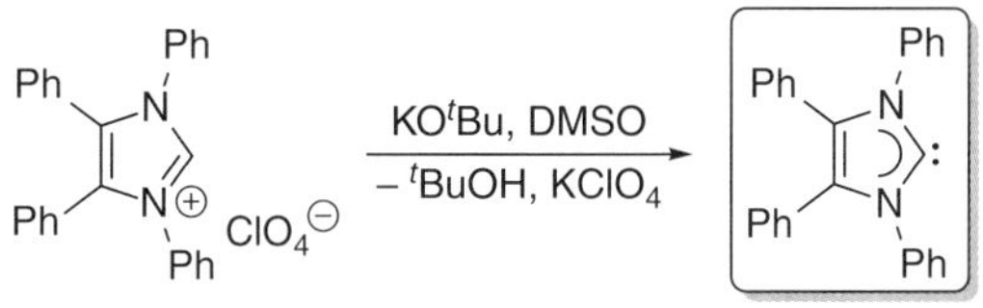

Scheme 1.7
Deprotonation of tetraphenylimidazolium salt to give a carbene.

In 1970, Wanzlick again proposed the formation of a free carbene that could be obtained by deprotonation of an imidazolium salt (Scheme 1.7) [18]. Although dimerization of this species was not observed, he unfortunately did not attempt to isolate the compound, which we nowadays know is indeed an N-heterocyclic carbene. Instead, subsequent reactions with transition metals and other electrophiles were studied with *in situ* generated carbenes.

Due to the many failures in isolating stable carbenes, this field of research remained relatively dormant during the next decade, until finally Guy Bertrand and his team in 1988 reported the isolation of the first free carbene that was stabilized by heavier main group elements. Thermolysis or photolysis of (trimethylsilyl)[bis(diisopropylamino)phosphino] diazomethane liberated dinitrogen and a new compound was obtained (Scheme 1.8), which showed properties and behavior similar to phosphaacetylenes, but peculiarly also that of carbenes [19]. The compound was reported to be stable for weeks at room temperature under an inert atmosphere. The ambiguity over whether the isolated compound was a carbene or phaspha-alkyne was finally addressed by a subsequent paper of the same group, where they provided further and convincing evidence for carbene-like reactivities of the species in question, such as insertion into double bonds and formation of oxirans with aldehydes [20]. These findings established [bis(diisopropylamino)phosphino]trimethylsilylcarbene as the first stable nucleophilic carbene that has been isolated and fully characterized in substance.

Bertrand's work demonstrated that carbenes can be isolated in pure form. Nevertheless, handling of such carbenes is difficult and requires very elaborate synthetic skills. So, it

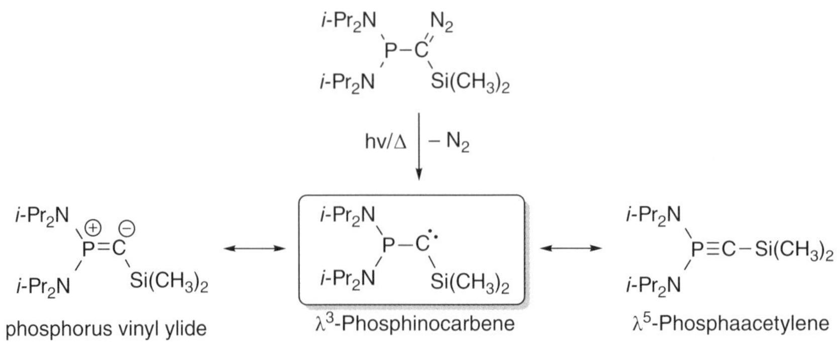

Scheme 1.8
Bertrand's [bis(diisopropylamino)phosphino]trimethylsilylcarbene and its relationship to the respective phospha-alkyne and ylide.

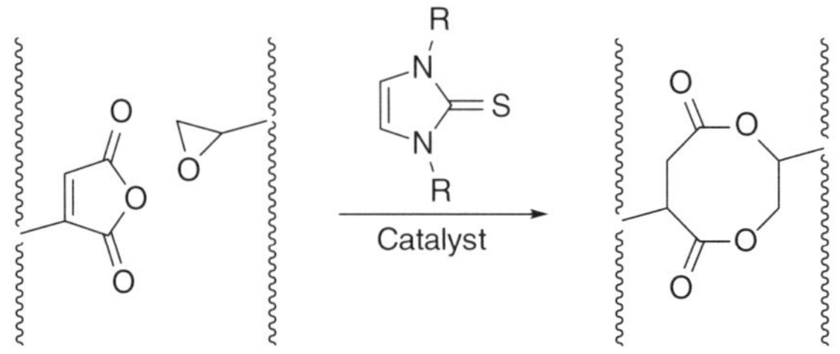

Scheme 1.9
Crosslinking of anhydride- and epoxide-functionalized low molecular weight polymers.

appears that the chemical community was just waiting for the next discovery, which would propel carbene chemistry from a niche area to mainstream research.

Around the same time as Bertrand's report on his first stable carbene, industrial chemists at DuPont had identified imidazolin-2-thiones as suitable organocatalysts for crosslinking of acid anhydride- and epoxide-functionalized low molecular weight polymers to higher molecular weight polymers for the production of watersoluble coatings for automotive use (Scheme 1.9).

Since imidazolin-2-thiones were not commercially available and older methodologies were too costly for preparation on an industrial scale, a newer cost efficient approach had to be developed. For this purpose, the old Wanzlick-type chemistry was re-investigated. The multi-component condensation reaction of one equivalent of gyoxal, two equivalents of primary amine and one equivalent of formaldehyde provided convenient access to imidazolium chloride salts after addition of hydrochloric acid. Deprotonation of such an imidazolium salt afforded an N-heterocyclic carbene *in situ*, which was trapped by addition of elemental sulfur to give the desired imidazolin-2-thiones (Scheme 1.10).

The reaction sequence was tested in the laboratory under an inert atmosphere to prevent decomposition of the *in situ* generated carbene prior to oxidation with sulfur. On an industrial scale using a 2000 L reactor, however, precautions to exclude moisture and air were impossible. Nevertheless, the product yields remained surprisingly high and comparable to those of lab scale experiments. It was therefore concluded that the supposedly reactive carbene intermediate must be relatively stable to survive such conditions.

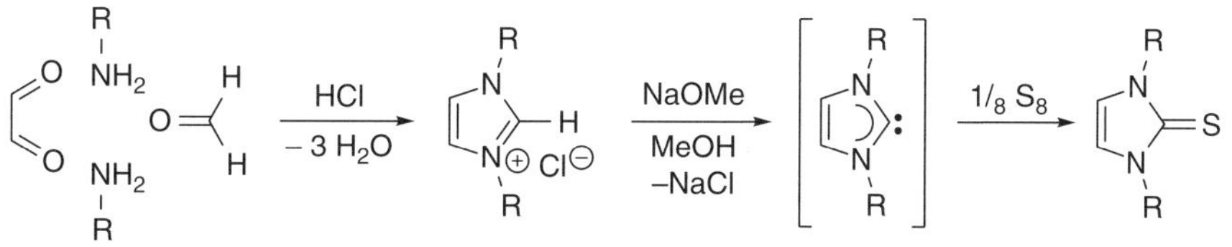

Scheme 1.10
Preparation of imidazoline-2-thiones as crosslinking organocatalysts.

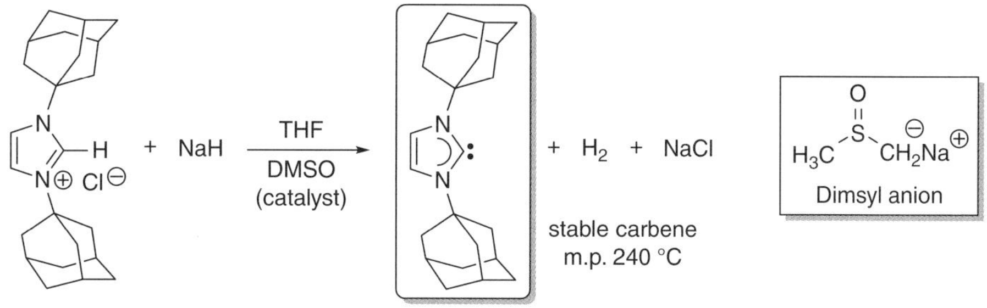

stable carbene
m.p. 240 °C

Dimsyl anion

Scheme 1.11
Preparation of the stable 1,3-di-1-adamentylimidazolin-2-ylidene (IAd) as the first representative of free NHCs.

Subsequently, Anthony J. Arduengo III and his team made the first attempts to isolate free imidazolin-2-ylidenes bearing sterically bulky N-adamantyl wing tip groups, and reported their findings in 1991. The respective imidazolium chloride was treated with sodium hydride (NaH) in tetrahydrofurane (THF) with catalytic amounts of dimethyl sulfoxide (DMSO), which generated the dimsyl anion as an intermediate base (Scheme 1.11). Dihydrogen gas (H_2) and sodium chloride (NaCl) formed as easily removable byproducts. Concentration of the THF filtrate afforded large colorless single crystals, which were subjected to single crystal X-ray diffraction analysis to evaluate the solid state molecular structure of the isolated compound. The results obtained confirmed that deprotonation of 1,3-di-1-adamentylimidazolium chloride had yielded the first free N-heterocyclic carbene, which is indefinitely stable if kept under an inert atmosphere. Intriguingly, it is also thermally very stable and melts without decomposition at 240 °C [21].

Following Arduengo's successful and seminal isolation of the first free NHC, there have been numerous other reports detailing various methods through which free NHCs can be isolated [22].

In 1995, the groups of Dieter Enders and J. Henrique Teles together reported the isolation and solid state molecular structure of the first 1,2,4-triazolin-5-ylidene. This carbene was prepared by the addition of sodium methoxide to 1,3,4-triphenyl-1,2,4-triazolium perchlorate in methanol, which afforded the neutral 5-methoxy-1,3,4-triphenyl-4,5-dihydro-1*H*-1,2,4-triazole. Upon heating to 80 °C under low pressure, the latter endothermically decomposes under α-elimination of methanol to form the stable 1,3,4-triphenyl-1,2,4-triazolin-5-ylidene (Scheme 1.12) [23]. Apparently, an additional heteroatom in the heterocyclic ring does not negatively affect the stability of the carbenes. Moreover, this carbene should become the first commercially available NHC. Notably, these researchers successfully applied, in principle, the same α-elimination approach that Wanzlick previously attempted in 1960.

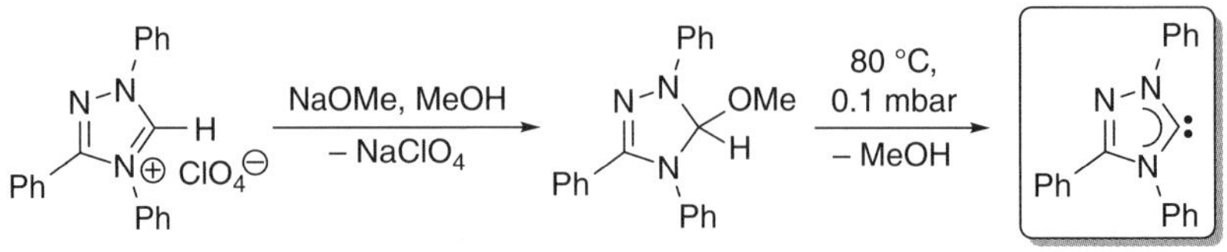

Scheme 1.12
Enders α-elimination approach to the first stable 1,2,4-triazolin-5-ylidene.

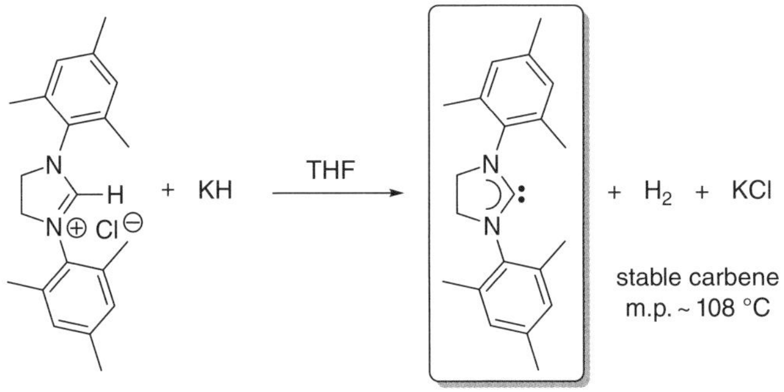

Scheme 1.13
Arduengo's isolation of the first imidazolidin-2-ylidene.

In the same year, Arduengo and coworkers also demonstrated that saturated imidazolidin-2-ylidenes can be isolated as monomeric species when sufficiently bulky N-substituents are applied to provide kinetic stability against dimerization to enetetramines.

Thus, by deprotonation of 1,3-dimesitylimidazolinium chloride with potassium hydride (KH) in dry THF, the researchers obtained the free and monomeric 1,3-dimesitylimidazolidin-2-ylidene (SIMes) as the first example of a saturated NHC (Scheme 1.13) [24]. This carbene was again structurally characterized by single crystal diffraction and showed a melting point of ~ 108 °C.

Finally, in 1999, F. Ekkehardt Hahn and coworkers reported the isolation and solid state structure of the first benzimidazolin-2-ylidene, completing the series of free NHCs for the four types of classical NHCs. The authors started with N,N′-dineopentyl-substituted *ortho*-phenylenediamines, which upon reaction with thiophosgene and triethylamine (NEt$_3$, to trap *in situ* generated hydrochloric acid) gave N,N′-dineopentylbenzimidazolin-2-thione. Reductive desulfurization using sodium/potassium alloy in toluene yielded the first free benzimidazolin-2-ylidene, which was also structurally characterized by single crystal X-ray diffraction (Scheme 1.14) [25]. Similar to saturated NHCs, the choice of sufficiently bulky wing tip groups is crucial to avoid dimerization to enetetramines.

It was now clear that stable NHCs of various backbones can be isolated in pure form. In retrospect, these later discoveries revealed that Wanzlick's hypothesis on stable NHCs was indeed correct, although he never isolated any by himself. Since the isolation of the first free NHCs 40 years after Wanzlick's proposal, the chemistry of NHCs has grown exponentially. Nowadays, free NHCs and many of their precursors are commercially available. They have become state-of-the-art organocatalysts and are routine ligands in organometallic chemistry and transition metal mediated catalysis. Applications of NHCs in other areas are beginning to surface, promising an even brighter future for these unique species.

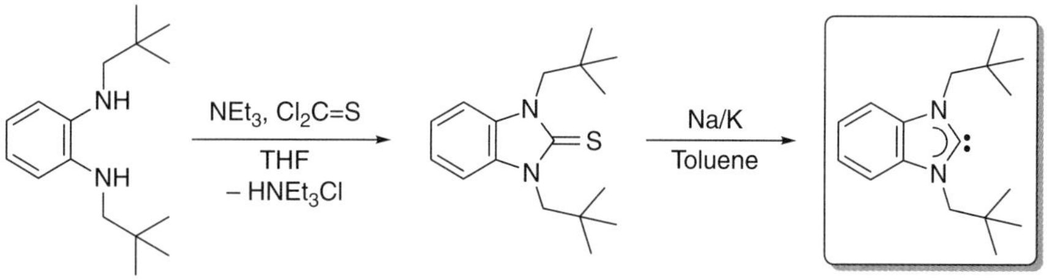

Scheme 1.14
Hahn's approach for the isolation of the first benzimidazolin-2-ylidene.

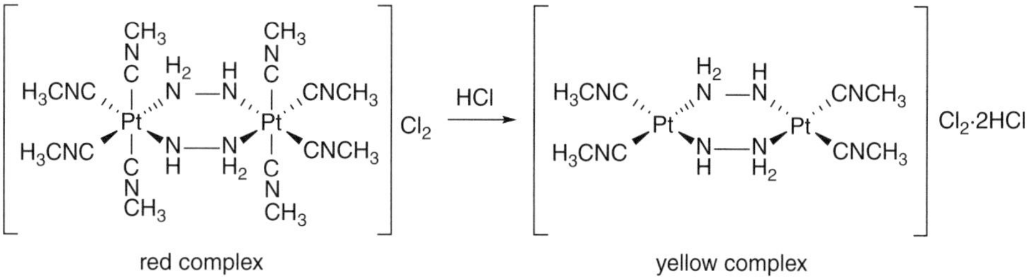

Scheme 1.15
Tschugajeff's wrongly proposed structures for his two colored complexes.

1.2.2 The Quest for Carbene Complexes

Compared to the quest for free carbenes, the historical events related to their stabilization by transition metals, and therefore attempts to isolate carbene complexes, were generally less debated. From today's point of view, the most obvious route to generate carbene complexes, and generally for all complexes, would be the reaction of a ligand with a chosen transition metal salt. However, such an approach for carbene complexes is historically insignificant, for the first free carbenes were only isolated in 1988 and 1991, respectively. Therefore, most successful pathways fall into the categories of metal-template-directed synthesis or *in situ* generation of free carbene in the presence of metal ions as carbene traps.

The first carbene complex was probably—but also unknowingly—prepared by Lew Alexandrowitsch Tschugajeff [26] (1873–1922) and coworkers in 1915 by the treatment of tetrakis(methylisocyanide)platinum(II) with hydrazine, which afforded a red crystalline complex. Protonation of this red complex with hydrochloric acid led to the release of methylisocyanide and the formation of yellow crystals. For both compounds, he wrongly proposed dimeric Pt^{II} species with bridging ligands, which were formally derived from the deprotonation of hydrazine (Scheme 1.15) [27].

Only a reinvestigation of Tschugajeff's complexes in 1970 by John E. Enemark and coworkers, that also included X-ray diffraction studies, revealed that these complexes were most likely the first carbene complexes to be synthesized [28]. The red and cationic acyclic monocarbene Pt(II) complex was formed by initial nucleophilic attack of hydrazine on coordinated carbon donors of two isocyanide ligands with concurrent proton shift. Protonation of the iminic nitrogen atom of the red complex with excess hydrochloric acid and isocyanide-chlorido ligand substitution led to reversible formation of the yellow and neutral dichlorido-dicarbene Pt^{II} complex (Scheme 1.16).

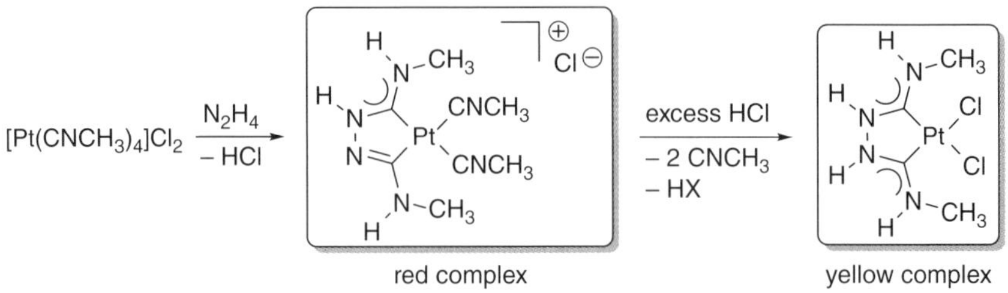

Scheme 1.16
Template-assisted approach to Tschugaeff's Pt^{II} carbene complexes.

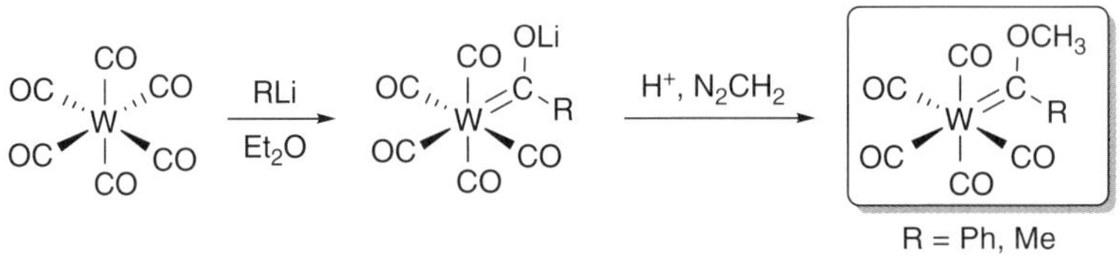

Scheme 1.17
Fischer's first carbene complex synthesis.

Notably, the attack of protic nucleophiles, such as alcohols or primary or secondary amines, on coordinated isocyanide ligands, has become a standard template-assisted approach to metal complexes, yielding acyclic and cyclic carbenes including NHCs.

However, the first person who knowingly prepared a metal-carbene complex was the German chemist and Nobel laureate Ernst Otto Fischer (1918–2007), working in Munich [29]. In 1964, he reported the attack of organolithium reagents on a coordinated carbonyl ligand in hexacarbonyltungsten(0). The resulting carbene complex anion could be precipitated as an ammonium salt or protonated with acids. Further reaction of the protonated complex with diazomethane gave stable complexes bearing either a methoxymethylcarbene or methoxyphenylcarbene ligand (Scheme 1.17) [30]. Such types of carbene complexes was named after their discoverer as *Fischer carbene complexes*. They typically contain metal centers in low to middle oxidation states, and the carbene donor is adjacent to at least one heteroatom-substituent. Furthermore, the co-ligands are usually good π-acceptors, and octahedral representatives generally fulfill the 18-electron rule. In terms of reactivity, Fischer carbene complexes typically undergo nucleophilic attack at the electrophilic carbene carbon.

In 1968 and four years after the publication of the first carbene complex by E. O. Fischer, the groups of Wanzlick in Berlin and Öfele in Munich almost simultaneously reported their work on the isolation of the first NHC complexes.

Wanzlick reacted 1,3-diphenylimidazolium perchlorate salt with mercury(II) acetate. Formally, the basic acetato ligands deprotonate the imidazolium salts to generate free carbenes, which immediately coordinate to the mercury center forming the cationic, linear 14-electron bis(1,3-diphenylimidazolin-2-ylidene)mercury(II) perchlorate complex (Scheme 1.18) [31].

In the same year, Karl Öfele reported a similar strategy for NHC complexes. In his case, *in situ* deprotonation occurred by heating the 1,3-dimethylimidazolium-pentacarbonylhydridochromate(0) complex salt, which under oxidative liberation of dihydrogen afforded the neutral pentacarbonyl(1,3-dimethylimidazolin-2-ylidene)chromium(0) complex (Scheme 1.19) [32].

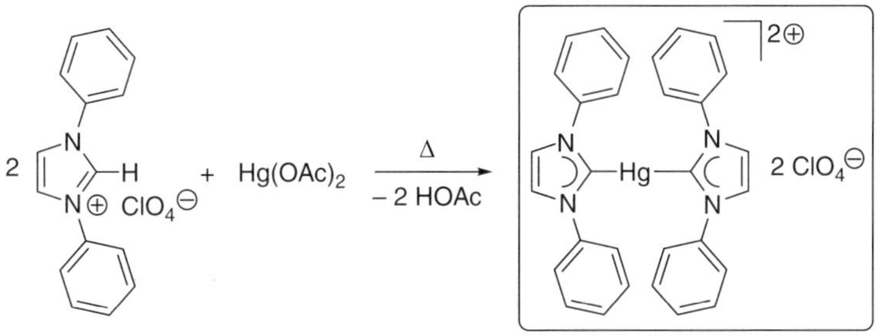

Scheme 1.18
Wanzlick's route to the first NHC complex.

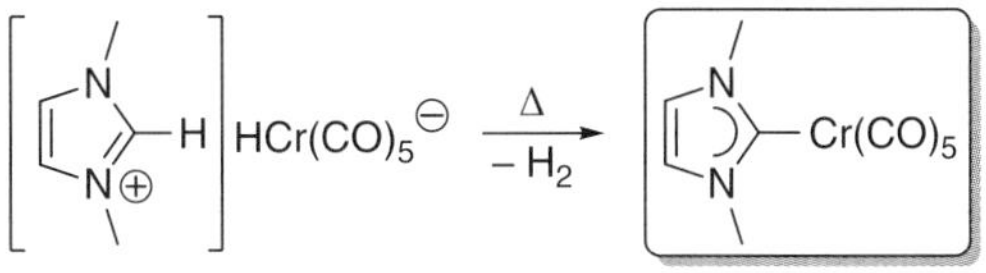

Scheme 1.19
Öfele's early NHC complex synthesis.

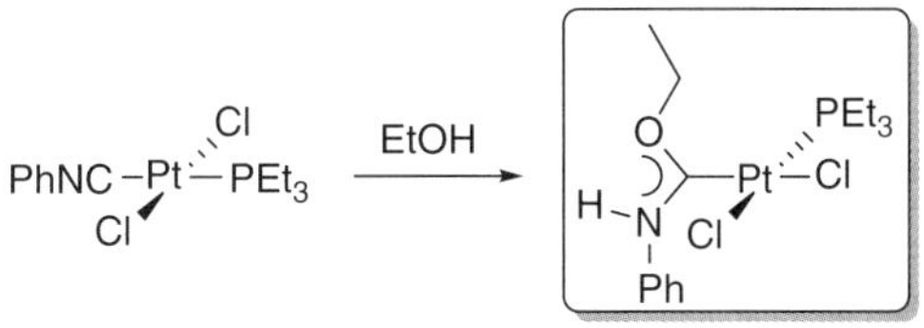

Scheme 1.20
Chatt and Richard's preparation of a carbene complex by nucleophilic attack on a coordinated isocyanide.

In 1969, more than 50 years after Tschugaeff's report on the reaction of protic nucleophiles with isocyanide complexes, Joseph Chatt (1914–1994) and Raymond L. Richards communicated a convenient route to carbene complexes by essentially the same approach. They reacted the complex *trans*-[PtCl$_2$(CNPh)(PEt$_3$)] with ethanol and obtained a white, sparingly soluble complex that turned out to be the mixed carbene/Phosphine complex *cis*-[PtCl$_2${C(OEt)NHPh}(PEt$_3$)] (Scheme 1.20) [33].

In 1971, Michael F. Lappert's (1928–2014) team demonstrated that carbene complexes can also be obtained by the reaction of carbene dimers with transition metal complexes. Since the existence of free carbenes was unknown at that time, this approach was the closest to an obvious reaction of the free ligand with a metal salt. The dimer of 1,3-diphenylimidazolidin-2-ylidene was chosen to react with a dimeric platinum(II)-phosphine complex in a 1:1 ratio to give two equivalents of the respective mixed phosphine-carbene Pt^{II} complex under chlorido-bridge-cleavage (Scheme 1.21) [34]. Later it was found that the use of bridged complex-dimers was not a necessity for the cleavage of electron-rich entetramines [35].

Shortly after in 1973, F. Gordon A. Stone (1925–2011) reported an elegant alternative route to carbene complexes via oxidative addition of simple 2-halo-azoles, for example,

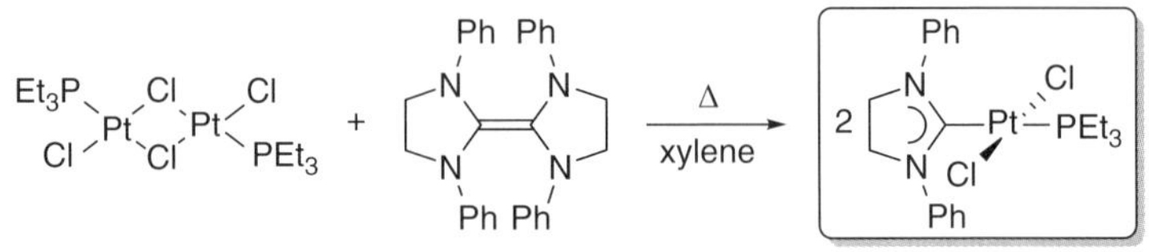

Scheme 1.21
Lappert's access to NHC complexes via cleavage of electron-rich entretramines with transition metal complexes.

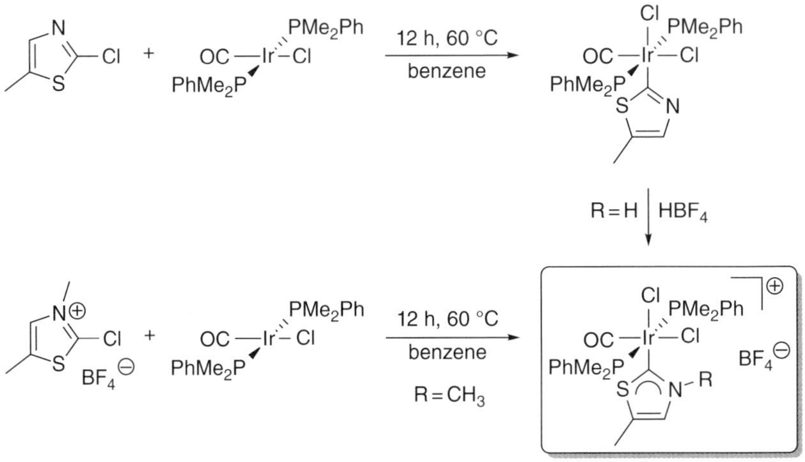

Scheme 1.22
Stone's pathway to NHC complexes via oxidative addition.

2-chloro-4-methylthiazole, 2-chlorobenzoxazole and 2-chlorobenzothiazole, to electron-rich, low-valent metal precursors under carbon-halogen bond activation. Neutral *C*-bound complexes were isolated as primary reaction products, which upon N-protonation with perchloric or tetrafluoroboric acid gave rise to cationic NHC complexes (Scheme 1.22, upper pathway) [36].

N-alkylation of the neutral *C*-bound complexes was unsuccessful, but complexes of N-alkylated NHC ligands could be directly obtained by oxidative addition using the N-alkylated azolium salts (Scheme 1.22, lower pathway) [37].

Up to this point in time, all carbene complexes reported contained singlet carbene ligands that bear a heteroatom adjacent to the carbene carbon, and can thus be broadly classified as Fischer carbene complexes. Complexes of carbenes with only hydrogen or alkyl substituents were unknown until Nobel laureate Richard R. Schrock reported the first example in 1974 [38]. Complexes of this type are referred to as *Schrock carbene (or alkylidene) complexes* and are distinctively different from Fischer carbene complexes.

The first preparation involved the reaction of dichlorido-trineopentyltantalum(V) with two equivalents of neopentyllithium in an attempt to investigate sterically bulky peralkyl complexes (Scheme 1.23). Instead, the first Schrock alkylidene complex was obtained as a very soluble orange crystalline material via α-hydrogen abstraction of a coordinated neopentyl ligand.

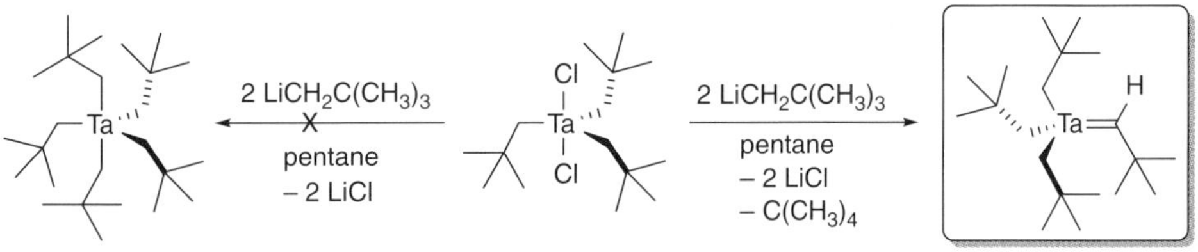

Scheme 1.23
The first synthesis of a Schrock alkylidene complex.

In contrast to Fischer carbene complexes, Schrock alkylidenes preferably coordinate to early transition metals in high oxidation states. Co-ligands in such species are typically good σ- or π-donors. The more nucleophilic carbene carbon in their complexes, which are often not 18-electron species, is more susceptible to electrophilic attack.

The wealth of synthetic routes to various carbene complexes discovered by these pioneers greatly contributed to the fundamental understanding of their organometallic chemistry. The discussion of bonding properties, applications, structure and activity relationship of typical Fischer carbenes and Schrock alkylidenes are beyond the scope of this work and can be found elsewhere [39].

Bonding analysis, ligand properties and many other (unusual) interactions observed in NHC complexes will be addressed in subsequent chapters.

References

[1] G. P. Moss, P. A. S. Smith, D. Tavernier, *Pure Appl. Chem.*, **1995**, *67*, 1307.
[2] Although valence bond theory has been criticized, the terms related to hybridization are still very useful in describing simple geometries.
[3] In simple terms, Hund's rule states that every orbital in a subshell is singly occupied with one electron before any one orbital is doubly occupied. Moreover, all electrons in singly occupied orbitals have the same spin.
[4] Reviews: (a) F. E. Hahn, M. C. Jahnke, *Angew. Chem. Int. Ed.* **2008**, *47*, 3122; (b) D. Bourissou, O. Guerret, F. P. Gabbaï, G. Bertrand, *Chem. Rev.* **2000**, *100*, 39.
[5] G. B. Schuster, *Adv. Phys. Org. Chem.* **1986**, *22*, 311.
[6] R. Hoffmann, G. D. Zeiss, G. W. Van Dine, *J. Am. Chem. Soc.* **1968**, *90*, 1485.
[7] (a) *Carbene Chemistry*, (Ed.: G. Bertrand), Marcel Dekker, New York, **2002**; (b) *Carben(oide), Carbine* (Ed.: M. Regitz), Houben-Weyl, Vol. *E19b*, Thieme, Stuttgart, **1989**; (c) R. A. Moss, M. Jones, Jr., *Carbenes*, Vol. *I*, Wiley, New York, **1973**, Vol. II **1975**; (d) W. Kirmse, *Carbene Chemistry*, 2nd ed., Academic Press, New York, **1971**; (e) *Advances in Carbene Chemistry*, (Ed. U. Brinker), Vol. *I*, JAI Press, Greenwich, (CT), **1994**, Vol. II, JAI Press, Stamford (CT), **1998**, Vol. III, Elsevier, Amsterdam, **2001**. (f) H. Tomioka, *Acc. Chem. Res.* **1997**, *30*, 315.
[8] A. J. Arduengo III, R. Krafczyk, *Chem. Unserer Zeit* **1998**, *32*, 6.
[9] A. J Arduengo III, *Acc. Chem Res.* **1999**, *32*, 913.
[10] J. B. Dumas, E. Péligot, *Ann. Chim. Phys.* **1835**, *58*, 5.
[11] A. Geuther, *Ann. Chem. Pharm.* **1862**, *123*, 121.
[12] J. U. Nef, *Justus Liebigs Ann. Chem.* **1895**, *287*, 265.
[13] J. U. Nef, *Justus Liebigs Ann. Chem.* **1897**, *289*, 202.
[14] H. Scheibler, *Ber. Dtsch. Chem. Ges.* **1926**, *59*, 1022.
[15] H.-W. Wanzlick, E. Schikora, *Angew. Chem.* **1960**, *72*, 494.
[16] D. M. Lemal, R. A. Lovald, K. I. Kawano, *J. Am. Chem. Soc.* **1964**, *86*, 2518.
[17] H. E. Winberg, J. E. Carnahan, D. D. Coffman, M. Brown, *J. Am. Chem. Soc.* **1965**, *87*, 2055.
[18] H.-J. Schönherr, H.-W. Wanzlick, *Liebigs Ann. Chem.* **1970**, *731*, 176.

[19] A. Igau, H. Grutzmacher, A. Baceiredo, G. Bertrand, *J. Am. Chem. Soc.* **1988**, *110*, 6463.

[20] A. Igau, A. Baceiredo, G. Trinquier, G. Bertrand, *Angew. Chem. Int. Ed. Engl.* **1989**, *28*, 621.

[21] A. J. Arduengo, III, R. L. Harlow, M. Kline, *J. Am. Soc. Chem.* **1991**, *113*, 361.

[22] (a) Herrmann, W. A.; Köcher, C. *Angew. Chem. Int. Ed.* **1997**, *36*, 2162. (b) Douthwaite, R. E.; Haussinger, D.; Green, M. L. H.; Silcock, P. J.; Gomers, P. T.; Martins, A. M.; Danopoulos, A. A. *Organometallics* **1999**, *18*, 4584. (c) Herrmann, W. A.; Köcher, C.; Goossen, L. J.; Artus, G. R. J. *Chem. Eur. J.* **1996**, *2*, 1627. (d) Kuhn, N.; Kratz, T. *Synthesis* **1993**, 561. (e) Kjellin, G.; Sandström, J. *Acta Chem. Scand.* **1969**, *23*, 2879. (f) Hahn, F. E.; Wittenbecher, L.; Noese, R.; Bläser, D. *Chem. Eur. J.* **1999**, *5*, 1931. (g) Scholl, M.; Ding, S.; Lee, C. W.; Grubbs, R. H. *Org. Lett.* **1999**, *1*, 953.

[23] D. Enders, K. Breuer, G. Raabe, J. Runsink, J. H. Teles, J.-P. Melder, K. Ebel, S. Brode, *Angew. Chem. Int. Ed. Engl.* **1995**, *34*, 1021.

[24] A. J. Arduengo, III, J. R. Goerlich, W. Marshall, *J. Am. Chem. Soc.* **1995**, *117*, 11027.

[25] F. E. Hahn, L. Wittenbecher, R. Boese, D. Bläser, *Chem. Eur. J.* **1999**, *5*, 1931.

[26] This Russian name is also often found spelled as "Tschugajew" or "Chugaev".

[27] (a) L. Tschugajeff, M. Skanawy-Grigorjewa, *J. Russ. Chem. Soc.* **1915**, *47*, 776; (b) L. Tschugajeff, M. Skanawy-Grigorjewa, A. Posnjak, *Z. Anorg. Allg. Chem.* **1925**, *148*, 37.

[28] (a) A. Burke, A. L. Balch, J. H. Enemark, *J. Am. Chem. Soc.* **1970**, *92*, 2555; (b) W. M. Butler, J. H. Enemark, *Inorg. Chem.* **1971**, *10*, 2416; (c) W. M. Butler, J. H. Enemark, J. Parks, A. L. Balch, *Inorg. Chem.* **1973**, *12*, 451.

[29] The Nobel Prize in Chemistry 1973 was awarded jointly to Ernst Otto Fischer and Geoffrey Wilkinson *"for their pioneering work, performed independently, on the chemistry of the orga-nometallic, so called sandwich compounds"* (refer to: http://www.nobelprize.org/nobel_prizes/chemistry/laureates/1973/(accessed October 7, 2016))

[30] E. O. Fischer, A. Maasböl, *Angew. Chem.* **1964**, *76*, 645; *Angew. Chem. Int. Ed. Engl.* **1964**, 580.

[31] H.-W. Wanzlick, H. Schönherr, *Angew. Chem.* **1968**, *80*, 154; *Angew. Chem. Int. Ed. Engl.* **1968**, *7*, 141.

[32] K. Öfele, *J. Organomet. Chem.* **1968**, *12*, P42.

[33] E. M. Badley, J. Chatt, R. L. Richards, G. A. Sim, *Chem. Commun.*, **1969**, 1322.

[34] D. J. Cardin, B. Çetinkaya, M. F. Lappert, Lj. Manojlovic-Muir, K. W. Muir, *Chem. Commun.*, **1971**, 400.

[35] D. J. Cardin, B. Çetinkaya, M. F. Lappert, *Chem. Rev.* **1972**, *72*, 545.

[36] P. J. Fraser, W. R. Roper, F. G. A. Stone, *J. Organomet. Chem.* **1973**, *50*, C54.

[37] P. J. Fraser, W. R. Roper, F. G. A. Stone, *J. Chem. Soc., Dalton Trans.* **1974**, 102.

[38] R. R. Schrock, *J. Am. Chem. Soc.* **1974**, *96*, 6796.

[39] For example: (a) K.H. Doetz, H. Fischer, P. Hofmann, F.R. Kreißl, U. Schubert, K. Weiß: *Transition Metal Carbene Complexes*, Verlag Chemie, Weinheim, **1983**; (b) *Advances in Metal Carbene Chemistry* (Ed.: U. Schubert), Kluwer, Dordrecht, **1989**; (c) *Metal Carbenes in Organic Synthesis* (Ed.: K.H. Doetz), Springer, Heidelberg, **2004**.

2 General Properties of Classical NHCs and Their Complexes

N-heterocyclic carbenes (NHCs) [1] are, as the name suggests, neutral, cyclic organic compounds with at least one nitrogen atom within the ring and one divalent carbene atom bearing a lone pair of electrons. They can therefore be regarded as a special subclass of the more general Fischer carbenes. However, conventional Fischer carbenes are considered relatively weak σ–donors and their strong metal-carbene bond is a result of enhanced π–backdonation from the low-valent metal center. Due to this fact, Lewis structures of Fischer carbene complexes are commonly drawn with an $M = C_{carbene}$ double bond. On the other hand, NHCs are very strong electron donors with high nucleophilicity and have therefore often been compared to trialkylphosphines. This view has also been substantiated by experimental and theoretical studies. However, it must be stressed that they are much more than simply phosphine substitutes. Their donating abilities surpass those of phosphines, and while many transition metal phosphine complexes suffer from metal-phosphine bond degradation, metal-NHC bonds are known to be much stronger. Many NHC complexes are thus stable toward air and moisture, which facilitates their applications. The implications of stronger metal-NHC bonds can be seen in catalytic applications. In contrast to metal-phosphine mediated transformations, where addition of excess phosphine ligands is often required, well-defined metal-NHC complexes can be employed usually without the need for ligand excess. While the handling of most transition metal-NHC complexes does not require the exclusion of air and moisture, it must be highlighted that the large majority of free NHCs are not air stable. Due to their general high basicity, many can react with moisture (H_2O) in a typical Brønsted acid-base reaction to yield back their respective salt precursors. Direct reactions of NHCs with oxygen (O_2) are rare, but they can form zwitterionic adducts with carbon dioxide (CO_2).

The unique properties of NHCs, those of their complexes, and the ease of their preparation are the reasons for the general popularity of NHCs in current synthetic chemistry. Since the seminal isolation of the very first stable NHC by Arduengo III, NHC chemistry has experienced exponential growth (Figure 2.1).

The most common NHCs are five-membered cyclic diaminocarbenes, in which the carbene carbon is flanked by two nitrogen atoms. As depicted in Chart 2.1, cyclic diaminocarbenes can be further classified into four different classical types dependent on their backbone, which can be either saturated (**I**), unsaturated (**II**), benzannulated (**III**), or containing a third nitrogen atom (**IV**).

Chart 2.1 also depicts the two most common alternative ways (**A** and **B**) to represent NHCs with neutral Lewis structures. For simplicity reasons, mesomeric resonance forms

The Organometallic Chemistry of N-heterocyclic Carbenes, First Edition. Han Vinh Huynh.
© 2017 John Wiley & Sons Ltd. Published 2017 by John Wiley & Sons Ltd.

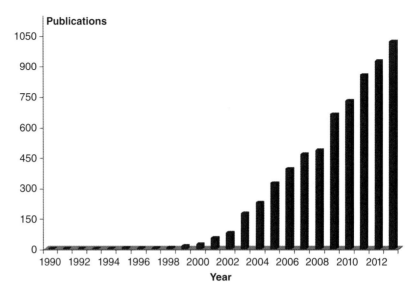

Figure 2.1
Exponential growth of publications on N-heterocyclic carbenes according to a search with
SciFinder® [2].

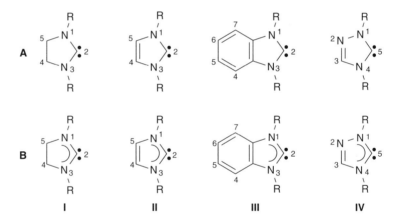

Chart 2.1
Conventional Lewis structures and numbering scheme for the four types of classical N-heterocyclic
carbenes.

with charge separation for both free NHCs and their complexes are not taken into account,
although some of these may reflect the real bonding picture more appropriately. Throughout
this textbook, alternative **B** will generally be used to draw Lewis structures of NHCs
to emphasize a certain degree of delocalization due to the stabilizing *"push-pull effect"*
(see Section 2.1), which also leads to N–C bonds that are shorter than a single bond, but
still longer than a typical N=C double bond. Alternative **A** will only be used in Section 2.3,
which deals with the nomenclature of NHCs. There the emphasis is on the IUPAC definition
of a carbene, which only bears two substituents at the carbene center, and the "half circle"
is omitted to avoid confusion with a real double bond.

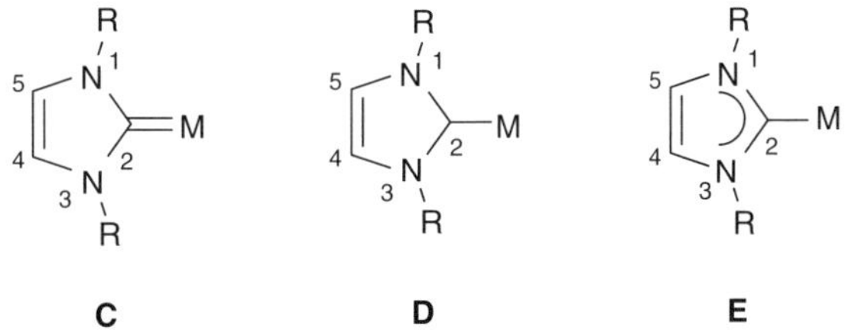

Chart 2.2
Conventional Lewis structures for imidazolin-2-ylidene metal complexes.

As depicted in Chart 2.2, metal-NHC bonds are also drawn in different ways in the literature (C, D, and E). In version **C**, the carbene carbon fulfills the octet rule. However, this representation would imply enhanced π–backdonation to the NHC, which for most complexes in not the case. Furthermore, X-ray structural data suggests that the M–C$_{carbene}$ bonds are generally closer to a single bond, and that the N–C bonds are shorter than single bonds. On the other hand, the M–C$_{carbene}$ "single bond character" is reflected in version **D**, although the N–C bonds are still treated as single bonds. Moreover, the carbene carbon could be easily confused with a sp^3–hybridized carbon center bearing an additional hydrogen atom, which is not the case.

Form **D** appears to be the most suitable among the three conventional versions of metal-NHC Lewis structures and will be used throughout this book. The "half-circle" correctly reflects a certain degree of N–C–N electron delocalization (*push-pull effect*), which results in slightly shorter N–C bonds. Also, the carbene center is sp^2–hybridized, and the metal-carbene bond is drawn as a single bond.

2.1 Stabilization in NHCs (*Push-Pull Effect*)

Due to the presence of two π-electron donating nitrogen atoms adjacent to the carbene carbon, NHCs are typical singlet carbenes, where the two electrons are located in the σ orbital, while the p_π orbital perpendicular to the carbene plane is formally vacant. Among various types of carbenes, NHCs are unusually stable and can be stored for extended periods of time under an inert atmosphere. Their unusual stability is attributed to the nitrogen atoms, which exhibit both a *negative inductive effect* (–*I* effect) and a *positive mesomeric effect* (+*M* effect) on the carbene carbon. The –*I* effect is due to a higher electronegativity of the nitrogen atom compared to that of carbon and leads to a withdrawal (i.e. "*pull*") of electron density at the carbene center, stabilizing the σ orbital as a result. The +*M* effect, on the other hand, is due to the presence of a lone pair at the nitrogen atom, which allows delocalization and π-donation (i.e. "*push*") into the formally vacant p_π-orbital of the carbene carbon enhancing its electron density (Figure 2.2). Consequently this increases the relative energy of the p_π orbital. These two effects, in combination described as a "*push-pull effect*," increase the energy gap between the two frontier orbitals, and therefore stabilize the singlet ground state.

In classical NHCs, the carbene center is adjacent to two nitrogen atoms, and both contribute equally to this combined "*push-pull effect*." This results in several resonance structures, which also enhances the stability of NHCs. The electron delocalization across the NCN moiety is typical for NHCs and leads to two essentially equal N–C$_{carbene}$ bonds.

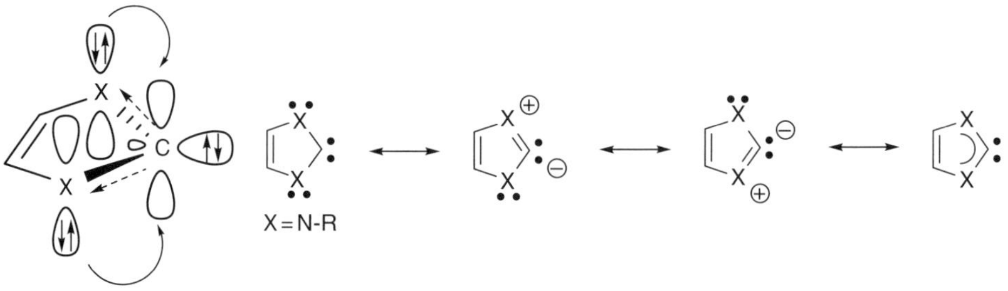

Figure 2.2
Combined "Push-pull effect" in imidazole-derived NHCs due to $-I$ effect (dotted arrow) and $+M$ effect (bowed arrow).

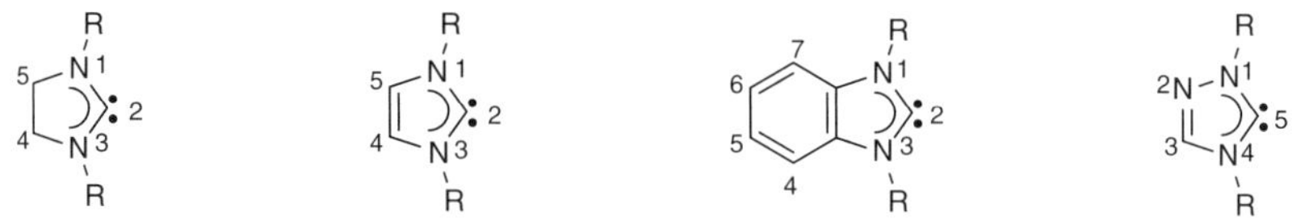

imidazolidin-2-ylidene	imidazolin-2-ylidene	benzimidazolin-2-ylidene	triazolin-5-ylidene

	I	II	III	IV
Topology	sat., non-arom.	unsat., arom.	benzannulated, arom.	unsat., arom.
$\delta(\mathbf{C2})$	~238–245 ppm	~211–221 ppm	~223–232 ppm	~210–215 ppm
$\angle$**N–C–N**	~105°	~101°	~103°	~100°
Reactivity	dimers favoured	monomers favoured	monomers and dimers depending on R-groups	monomers favoured

Figure 2.3
Comparison among the common four types of classical NHCs.

In order to emphasize this delocalization, all Lewis structures of classical NHCs in this textbook will be drawn with a half circle involving the NCN moiety.

2.2 Backbone Differences and Their Implications

The four most common types of classical NHCs that we are concerned with differ in their backbones, which understandably has an influence on their individual donating abilities in accordance with different inductive effects (see Section 2.5). In addition there are also characteristic topological, structural and spectroscopic features that set them apart. A comparison of some key features of NHCs is summarized in Figure 2.3. In terms of topology, imidazolidin-2-ylidene (**I**) with a saturated backbone stands out. Among the four typical NHCs, it is the only one that cannot exhibit Hückel aromaticity, since delocalization is limited to the NCN fragment. The unsaturated or benzannulated backbones of the remaining three types (**II–IV**) allow for conjugation, and at least formally, their five-membered and 6π electron containing carbene rings can be considered to have aromatic character. The aromatic character of the five-membered is ring is understandably smaller for NHCs of type **III** compared to the other two, since it already has a fully delocalized 6π aryl system that may not extend to the NCN moiety. In terms of increasing aromaticity, the NHCs are therefore ranked in the order **I** < **III** < **IV** ∼ **II**. Nevertheless and maybe surprisingly, studies have also indicated that—in all three cases—the π-electron delocalization across the NCN moiety is not fully extended to the backbone [3].

The most important spectroscopic signature for NHCs is the ^{13}C NMR chemical shifts of the carbene carbon atoms. Generally, these are found in the downfield region >200 ppm. The most deshielded carbene resonances are observed for saturated imidazolidin-2-ylidenes (**I**, ∼238–245 ppm) followed by benzannulated benzimidazolin-2-ylidenes (**III**, ∼223–232 ppm) and unsaturated imidazolin-2-ylidenes (**II**, ∼211–221 ppm). The carbene signals for triazolin-5-ylidenes (**IV**, ∼210–215 ppm) are found in the same range as the latter, which suggests that replacing one CH with an isolobal N fragment does not influence the ^{13}C$_{carbene}$ signature to a great extent. Notably, this sequence of increasing upfield shift for the ^{13}C$_{carbene}$ signal is the same as that observed for aromaticity.

It is also interesting to note that benzannulated benzimidazolin-2-ylidenes (**III**) represent a mid-point between unsaturated imidazolin-2-ylidene (**II**) and the saturated imidazolidin-2-ylidene (**I**). For example, this is evidenced by their ^{13}C NMR spectroscopic chemical shift range for the carbene carbon, which falls in between those for imidazolin-2-ylidenes and imidazolidin-2-ylidenes. The same intermediate position was also observed for the degree of aromaticity of the five-membered ring and the range of their N–C–N bond angles [4].

With respect to dimerization, saturated imidazolidin-2-ylidenes are relatively unstable and tend to form dimers following a Lewis acid (i.e. proton or alkali-metal ions) catalyzed mechanism, for example (SIMe)$_2$ and (SIPh)$_2$ (Chart 2.3) [4,5]. Monomeric free carbenes can only be obtained under strict exclusion of any Lewis acid traces or with very bulky N-substituents, e.g. SIPr and SI^{t}Bu [6]. Unsaturated imidazolin-2-ylidenes and triazolin-5-ylidenes, on the other hand, are more stable and are generally isolated as monomers even with small N-methyl substituents (e.g. IMe, Chart 2.4) [7]. Notably, the triazolin-5-ylidene (tria) is the first free NHC that was commercialized. To enforce dimerization to tetrazafulvalenes, two imidazolin-2-ylidene fragments must be doubly-bridged using a short propylene (C3) spacer [8]. Extension of the two bridges by only one methylene unit leads to the formation of stable and doubly C4-bridged dicarbenes.

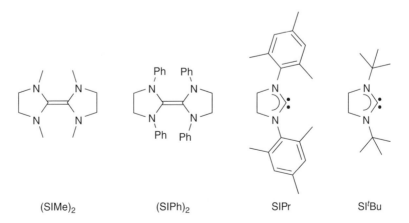

| (SIMe)$_2$ | (SIPh)$_2$ | SIPr | SI^{t}Bu |

Chart 2.3
"Dimeric" entetramines and monomeric imidazolidin-2-ylidenes.

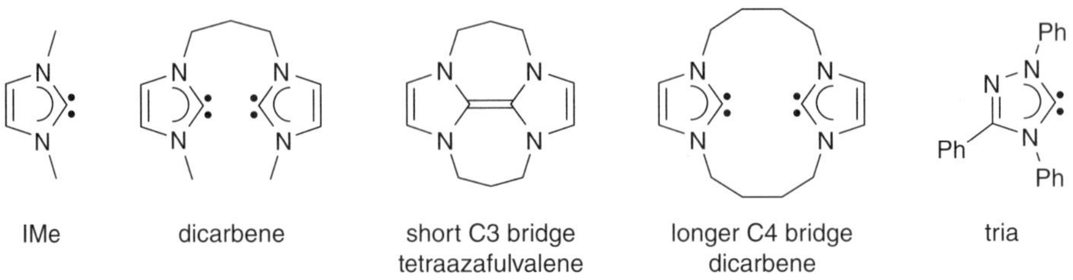

Chart 2.4
Free mono-, dicarbenes and enforced dimers of imidazolin-2-ylidenes.

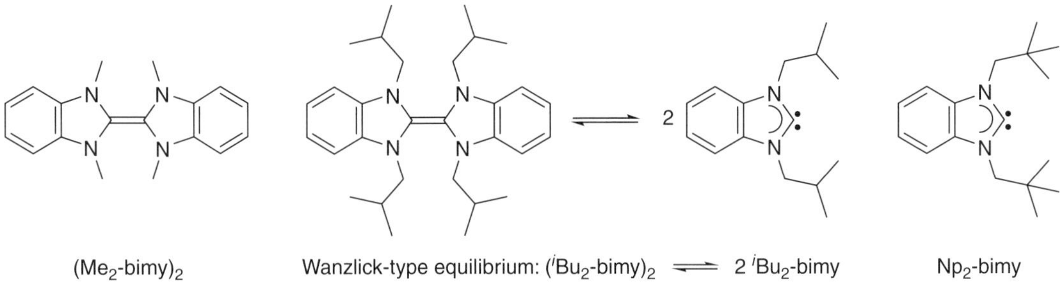

Chart 2.5
Mono- and dimeric benzimidazolin-2-ylidenes and a proposed "Wanzlick-type" equilibrium.

Benzimidazolin-2-ylidenes are in this respect again intermediate, since both monomers and dimers can be observed depending on the N-substituents. With N-methyl substituents, tetraazafulvalenes, such as $(Me_2$-bimy$)_2$, are observed (Chart 2.5). The requirement for steric demand as a prerequisite for monomeric carbenes appears to be less strict as compared to that in saturated imidazolidin-2-ylidenes. For example, the monomeric benzimidazolin-2-ylidene Np_2-bimy is obtained with bulky but yet flexible N-neopentyl groups at the nitrogen, although the substituents should possess sufficient rotational freedom across the N-CH$_2$ bond to allow for dimerization to occur. Removal of only one methyl group leads to N-isobutyl substituted N-heterocycles. Interestingly, both monomeric $^iBu_2$-bimy and dimeric $(^iBu_2$-bimy$)_2$ are observed, and it has been proposed that a "Wanzlick-type" equilibrium exists between these two species.

Benzimidazolin-2-ylidenes therefore have the topology of unsaturated imidazolin-2-ylidenes, but resemble saturated imidazolidin-2-ylidenes in their reactivities. The combination of features unique to type **I** and **II** NHCs make them highly fascinating carbenes to study.

2.3 Dimerization of Carbenes

The dimerization process of singlet and triplet carbenes must be different due to their different electronic situation. Triplet carbenes like methylene dimerize via overlap of two singly occupied sp^2 hybrid orbitals (σ^1), which leads to a C–C σ–bond. Overlap of the two singly occupied p_π-orbitals (p_π^{-1}) completes the process and leads to the formation of the π–bond giving overall a classical planar alkene (Figure 2.4). In the case for methylene a planar ethene molecule is obtained upon dimerization.

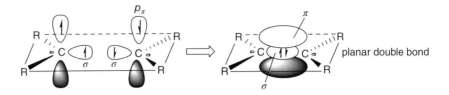

Figure 2.4
Dimerization of triplet carbenes to give olefins.

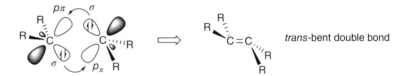

Figure 2.5
Dimerization of singlet carbenes leading to a *trans-bent* bond.

The dimerization of ground state singlet carbenes, on the other hand, is less straightforward. One possibility is the interaction of the lone pair in the σ-orbital of one carbene with the formally vacant p_π-orbital of the second carbene. The dimer thus formed cannot be planar in such a case, and a so-called *trans-bent* doubly bonded species is obtained (Figure 2.5). Heavier element analogs of ethene, such as Si_2H_4, Ge_2H_4, and Sn_2H_4, are examples of *trans-bent* species [9]. On the other hand, dimers of carbenes exhibiting *trans-bent* bonds are less common.

According to the Carter–Goddard–Malrieu–Trinquier (CGMT) Model [10] the strength of the C=C double bond (E_{TBE}) in formal dimers of singlet carbenes, such as NHCs **I–IV**, can be estimated. This is done by considering the bond strength of a regular double bond (Figure 2.4), usually that of ethylene ($720 \pm 8\,kJmol^{-1}$) [11], minus the sum of the singlet–triplet energy gap ($\Sigma\,\Delta E_{S \to T}$) for both carbene parts (Equation 2.1).

$$E_{TBE} = 720\,kJmol^{-1} - \Sigma\Delta E_{S \to T}$$

Equation 2.1

The calculated singlet–triplet energy gap for 1,3-dimethylimidazolin-2-ylidene (IMe) of type **II** is approximately $355\,kJmol^{-1}$. The resulting bond energy for **II**=**II** dimer would merely be $10\,kJmol^{-1}$ (Table 2.1), which can be used to explain why type **II** NHCs generally do not dimerize easily. Furthermore, dimerization of two unsaturated carbenes would also lead to formal loss of aromaticity of two 6π systems, making it a thermodynamically unfavorable process.

The singlet–triplet gap of $311\,kJmol^{-1}$ calculated for NHCs of type **I** is smaller due to the saturation of the N-heterocyclic ring. Application of Equation 2.1 gives a reasonable bond energy of $98\,kJmol^{-1}$ for entetramines **I**=**I**, which have often been isolated, but are difficult to dissociate into carbenes in a unimolecular way. Since type **I** NHCs are not aromatic, no sacrificial process with respect to de-aromatization is required, and stable dimers are easily obtained.

Annulated NHCs of type **III** have an intermediate singlet–triplet energy gap of about $331\,kJmol^{-1}$. They also take an intermediate position in terms of aromatic character.

Table 2.1 Calculated singlet–triplet energies of common NHCs and estimated bond energies for their respective olefins.

Type	NHC	$\Delta E_{S \to T}$ [kJmol^{-1}] [12]	E_{TBE} (NHC=NHC) [kJmol^{-1}]
I	SIMe	311	98
II	IMe	355	10
III	Me$_2$-bimy	331	58
IV	TMe	355	10

Thus, the formation and dissociation of dibenzotetraazafulvalenes **III=III** are primarily determined by the steric bulk of the N-substituents. Finally, for 1,2,4-triazolin-5-ylidenes of type **IV**, the same values as those for type **II** are obtained, highlighting once again that substitution of CH with an isolobal N fragment in NHCs has only an overall small effect on the reactivity. In accordance with these values, dimers of the type **IV=IV** are not observed.

Although thermodynamically stable dimers are observed for certain types of less stable diaminocarbenes, a classic mechanism of their formation leading to *trans bent* species as mentioned in a previous section has been debated. For example, it has been noted by Roger W. Alder and coworkers that the energy barriers for dimerization of diaminocarbenes are generally very high [13]. Consequently, very few would dimerize by approach of two free carbenes. Instead, an electrophile-catalyzed process (Lewis acid such as a proton or metal ion) is much more plausible. Such a sequence may start with protonation of a diaminocarbene, which would lead to a formamidinium ion. The latter is attacked by a second diaminocarbene under C–C bond formation, and a protonated carbene dimer is obtained. In a final step, another basic carbene removes the proton from the protonated dimer, which re-generates the formamidinium salt with release of the dimer as the desired product (Scheme 2.1).

In line with this concept, a protonated dimer could be isolated from deprotonation of 1,3-diethylimidazolidinium hexafluorophosphate with only one half-equivalent of lithium bis(trimethylsilyl)amide (LiHMDS) (Scheme 2.2) [13]. This compound can be regarded as an intermediate in the formation of carbene dimers.

Another experiment that points to a proton-catalyzed dimerization process came from the group of F. E. Hahn [14]. They studied the slow dimerization of unsymmetrically

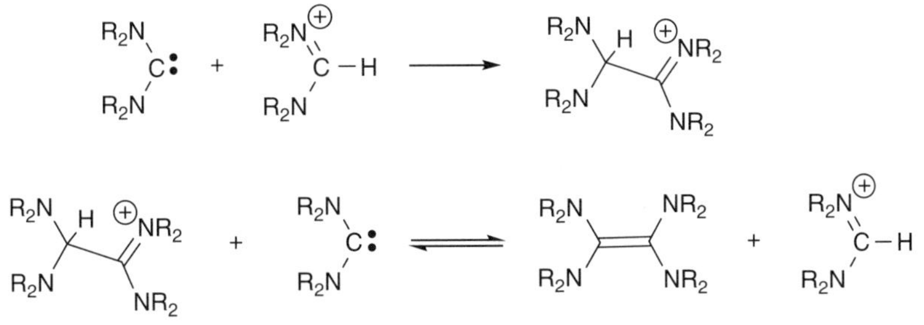

Scheme 2.1
Formamidinium-promoted dimerization of diaminocarbenes.

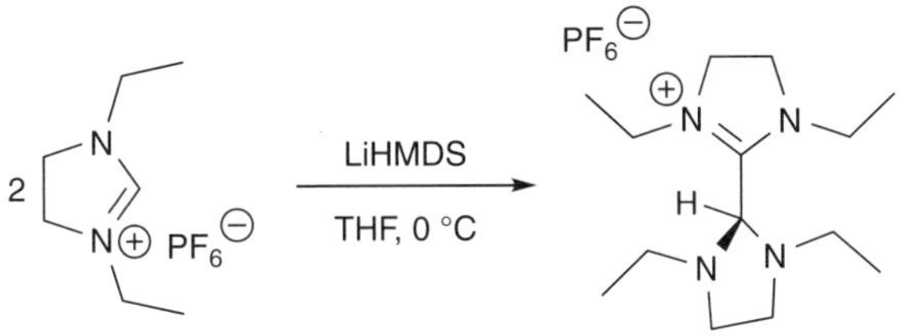

Scheme 2.2
Formation of a protonated dimer as an intermediate.

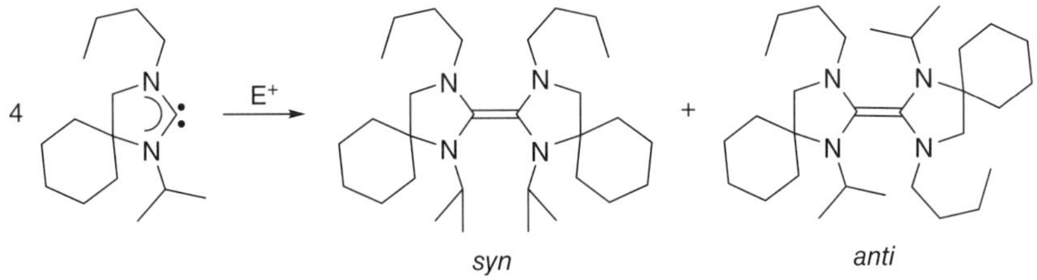

Scheme 2.3
Slow dimerization of unsymmetrically substituted imidazolidin-2-ylidenes.

substituted imidazolidin-2-ylidenes by solution NMR spectroscopy in [D$_8$]THF, which occurred only after several weeks (Scheme 2.3). The ^{13}C NMR signal for the free carbene is observed at 236.0 ppm, which is distinctively different from the equivalent signals of the *syn*- and *anti*-dimers at 130.0 ppm and 128.6 ppm, respectively. However, if the same carbene was stored under rigorous exclusion of moisture and over a sodium/potassium alloy under argon, dimerization could be completely suppressed even after several weeks.

The dissociation of dimers into monomeric carbenes is simply the reversed sequence of this reaction, and may also operate in an electrophile-catalyzed fashion.

2.4 Nomenclature of N-Heterocyclic Carbenes

Although the nomenclature of N-heterocyclic carbenes has been addressed in an early seminal review article [1a], it is evident from the current literature that there is still a lot of confusion about the correct naming. Most of the later articles simply list the author's preferred choice of names without providing detailed insights on how those are derived. As with the nomenclature for coordination compounds, many researchers avoid the IUPAC recommendations for convenience, which may result in misunderstanding and—in the worst case—even fuels misconceptions. In this section, we are addressing this issue in detail with the aim of providing a clearer picture of the naming principles of NHCs.

In systematic name formation according to *IUPAC* [15], the term "carbene" is the name of the parent hydride H_2C: (i.e. methylene). Substitution of the hydrogen atoms by other single heteroatoms X give rise to simple derivatives of the formula X_2C: for which the name "carbene" can be retained. Thus the species Cl_2C: is named dichlorocarbene. However, if there is more than one carbon in the carbene molecule the suffix –*n–ylidene* (n = number) has to be used. Here, the term *ylidene* simply refers to the carbene carbon, and the number *n* indicates its position within the molecule, for example prop-2-en-1-ylidene (n = 1) is the name for the $H_2C=CHCH$: molecule.

N-heterocyclic carbenes are therefore also ylidenes. The stem of the name reflects the nature of the heterocycle that contains the carbene carbon. It should be noted at this point that N-heterocyclic carbenes obviously belong to the general class of heterocycles, for which a systematic IUPAC nomenclature (i.e. the Hantzsch-Widman system) has been established. Here we only focus on a simplified nomenclature of common five-membered rings with at least one nitrogen atom, for which the degree of unsaturation (or the number of double bonds) is indicated by the suffixes –*ole* (fully unsaturated with two double bonds), –*oline* (unsaturated with one double bond) and –*olidine* (saturated without any double bond). Alternatively, and in addition to the use of suffixes, prefixes such as *dihydro–* or *tetrahydro–* can be attached to the name of the parent unsaturated heterocycle (e.g. az*ole*) to indicate the degree of saturation. The application of these rules on common N-heterocycles, from which most classical NHCs are derived, is demonstrated in Table 2.2.

In order to apply the Hantzsch-Widman system to the naming of NHCs, the number of double bonds in the respective N-heterocycle needs to be identified. Table 2.3 depicts the accepted Lewis structures (alternative **A**) for the four most common types of NHCs **I–IV**, which all contain the carbene carbon adjacent to two nitrogen atoms. The fully saturated NHC of type **I** clearly stands out, as it does not contain any double bond, while the five-membered NHCs **II–IV** all contain formally one double bond. The lone pairs at each nitrogen atom are not drawn. Accordingly, type **I** NHCs should be called imidaz*olidin*-2-ylidenes (or *tetrahydro*imidazol-2-ylidene), while types **II–IV** are correctly referred to as imidaz*olin*-2-ylidenes (or *2,3-dihydro*imidazol-2-ylidene, **II**), benzimidaz*olin*-2-ylidenes (or *2,3-dihydro*benzimidazol-2-ylidene, **III**), and 1,2,4-triaz*olin*-5-ylidenes (or *4,5-dihydro*-1,2,4-triazol-5-ylidene, **IV**), respectively.

It is worth noting that the correct nomenclature for NHCs has already been applied in the original NHC publications by Wanzlick, Öfele, and other pioneers in this field. However, it is also true that nomenclature in chemistry, as with all languages, is subject to language drifts. At this point, it must be highlighted again, that this correct naming of NHCs is less often followed in the current literature, leading to confusion, since multiple names are used for identical species by different authors. In many publications nowadays, the most common NHCs of type **II** resulting from the deprotonation of imidazolium salts, which in turn are derived from imidazole, are trivially called "imidazol-2-ylidene." According to the Hantzsch-Widman system, this name would wrongly imply an N-heterocyclic carbene with *two* double bonds, although formally only *one* is present. A five-membered ring NHC with two double bonds is simply not possible. The same is often applied to NHCs derived from benzimidazole (**III**), triazole (**IV**) and imidazoline (**I**), which are simply referred to as "benzimidazol-, triazol-, and imidazolin-ylidenes."

Table 2.2   Common N-heterocycles and their nomenclature according to the Hantzsch-Widman system.

Heterocycle	No. of double bonds	Name 1	Name 2
	2	1*H*-imidaz*ole*	–
	1	2-imidaz*oline*	4,5-dihydro-1*H*-imidazole
	1	4-imidaz*oline*	2,3-dihydro-1*H*-imidazole
	0	imidaz*olidine*	tetrahydroimidazole
	2	1*H*-1,2,4-triazole	–
	1	1,2,4-triazoline	4,5-dihydro-1*H*-triazole
	0	1,2,4-triazolidine	tetrahydro-1,2,4-triazole
	2	1*H*-benzimidazole	–
	1	benzimidazoline	2,3-Dihydro-1*H*-benzoimidazole

Table 2.3 Four types of classical N-heterocyclic carbenes and their nomenclature according to the Hantzsch-Widman system.

Type	NHC	No. of double bonds	Name 1	Name 2
I	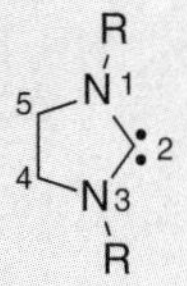	0	Imidazolidin-2-ylidene	*tetrahydro*imidazol-2-ylidene
II	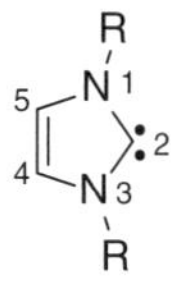	1	Imidazolin-2-ylidene	*2,3-dihydro*imidazol-2-ylidene
III	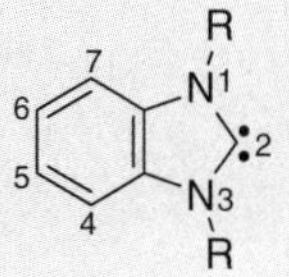	1	Benzimidazolin-2-ylidene	*2,3-dihydro*benz-imidazol-2-ylidene
IV	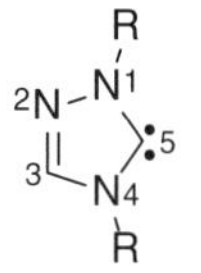	1	1,2,4-Triazolin-5-ylidene	*4,5-dihydro*-1,2,4-triazol-5-ylidene

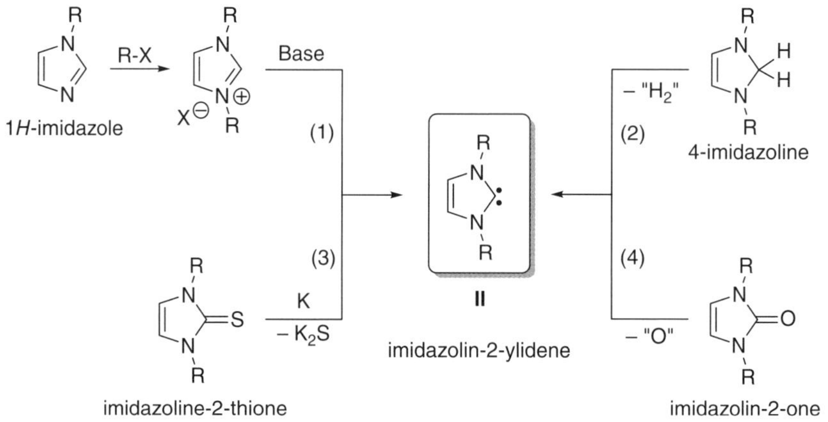

Scheme 2.4
Different formal pathways to imidazoln-2-ylidenes.

As mentioned above, the origin of this "trivial" naming is certainly related to the most common synthetic pathway to NHCs. For example, the preparation of type **II** NHCs usually starts with the *N*-alkylation of mono-substituted imidazoles with an electrophile R–X, which affords 1,3-disubstituted imidazolium salts as direct NHC precursors. Subsequent deprotonation at C2 with a suitable base affords the free NHC **II** as depicted in Scheme 2.4, route (1). Although this route starting from an imidazole is ubiquitous, it is at least formally conceivable that the same free NHC can also be obtained by direct elimination of dihydrogen (H_2) from 4-imida*zoline* following pathway (2). In this process, the number of double bonds in the heterocycles remains unchanged. This is also the case for the experimentally viable pathway (3), which involves reduction of imidazoline-2-thiones with elemental potassium under elimination of metal sulfides. In analogy, a deoxygenation of imidazolin-2-ones is formally conceivable as well. Pathways (2)–(4) clearly emphasize the relationship of NHC **II** to other derivatives of imidazolines and highlight once again that "imidazolin-2-ylidene" is the correct name for NHC **II**.

The other classical NHCs **I**, **III**, and **IV** can be analyzed in a similar way and their relationships to some other derivatives are summarized in Scheme 2.5. Furthermore, application of this nomenclature can also be extended to non-classical NHCs derived from other heterocycles, such as pyrazoles, 1,2,3-triazoles, tetrazoles, etc. Throughout this textbook, the systematic nomenclature according to the Hantzsch-Widman system (Table 2) will be used for all NHCs, their derivatives, and their complexes.

2.5 Electronic Properties of NHCs and Different Electronic Parameters

N-heterocyclic carbenes are known as strongly donating ligands. In addition, their N-substituents can be varied easily, which allows tuning of their stereoelectronic properties. The detailed knowledge of the electron donating abilities and nucleophilicities of NHCs and their steric bulk has obvious implications on their applications in organo- and transition metal-mediated catalysis. Therefore, the study of their stereoelectronic properties has attracted considerable attention.

2.5.1 Tolman's Electronic Parameter (TEP) and Related Carbonyl Based Systems

The Tolman electronic parameter or simply TEP is the best-known experimental method for the quantification of ligand donor abilities. It is named after Chadwick A. Tolman, who developed the methodology in 1970 for the evaluation of numerous tertiary phosphines. The TEP is the value for the carbonyl A_1 IR stretching frequency measured for tetrahedral Ni^0 complexes of the type $[Ni(CO)_3PR_3]$ [16], which can be prepared by reaction of free phosphines with tetracarbonylnickel(0), $[Ni(CO)_4]$, under substitution of one carbonyl ligand (Scheme 2.6). The basis for this method is the special property of carbonyl ligands to act as very good π-acceptors, and the unique range of carbonyl IR stretching frequencies (~1800–2200 cm^{-1}), which can be regarded as independent from other vibrations in the molecule. A strongly donating phosphine ligand would increase the electron density in the complex, and therefore enhance π-backdonation to the carbonyl ligands. This in turn weakens the CO bond, and thus a smaller CO wavenumber indicates a stronger donating phosphine ligand.

K. Öfele and Wolfgang A. Herrmann were the first to prepare an NHC analog of Tolman's complexes $[Ni(CO)_3(IMe)]$ by reacting the free dimethylimidazolin-2-ylidene ligand (IMe) with $[Ni(CO)_4]$, although they did not systematically study the TEPs of NHCs [17]. Based on the A_1 carbonyl stretch at 2055 cm^{-1} measured in hexane, one can conclude that IMe is stronger or at least as strong as the strongest phosphine ligand in Tolman's study, that is, P^tBu_3 ($A_1 = 2056$ cm^{-1}, CH_2Cl_2). Steven P. Nolan and coworkers extended this study to the most common NHC ligands and obtained the TEP values from CH_2Cl_2 and hexane shown in Chart 2.6 [18]. Unfortunately, for the two most bulky NHCs in this series I^{t}Bu and IAd,

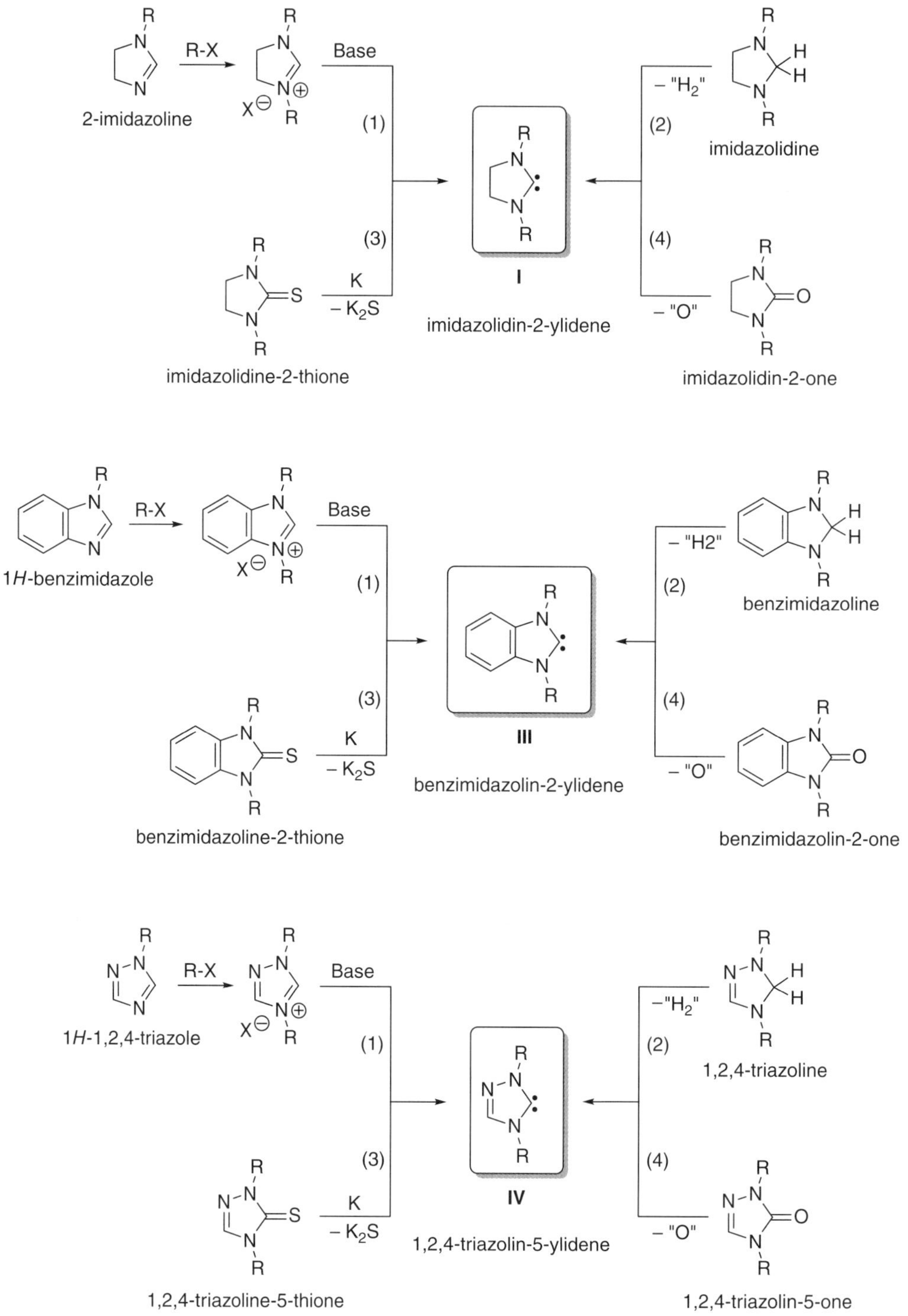

Scheme 2.5
Different formal pathways to azolinylidenes **I**, **III**, and **IV**.

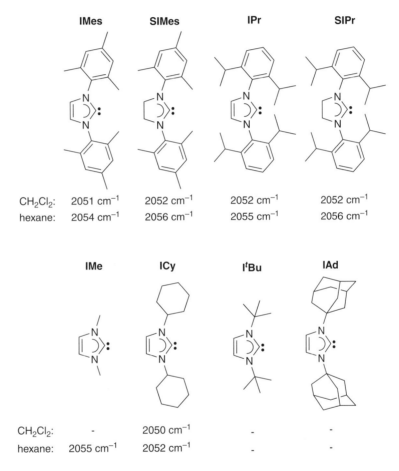

Scheme 2.6
Preparation of tricarbonyl complexes of Ni⁰ for TEP determination.

Chart 2.6
Experimental TEP values for some common NHC ligands.

the required tetrahedral [Ni(CO)₃(NHC)] complexes could not be obtained. In these two cases, elimination of two carbonyl ligands took place, and three-coordinated [Ni(CO)₂(NHC)] complexes were isolated instead that are not suitable for TEP determination.

Comparison of the TEP values leads to the conclusion that NHCs are generally stronger donors than even the most basic trialkylphosphines. A critical comparison among the NHCs, however, did not reveal any clear trend. Although, ICy shows the smallest carbonyl wavenumber, as expected, it is apparent that TEP cannot properly resolve substituent effects in NHCs. These results also show that only values measured in the same solvent can be compared. More striking is the fact that the TEP value for IMes/IPr is smaller than

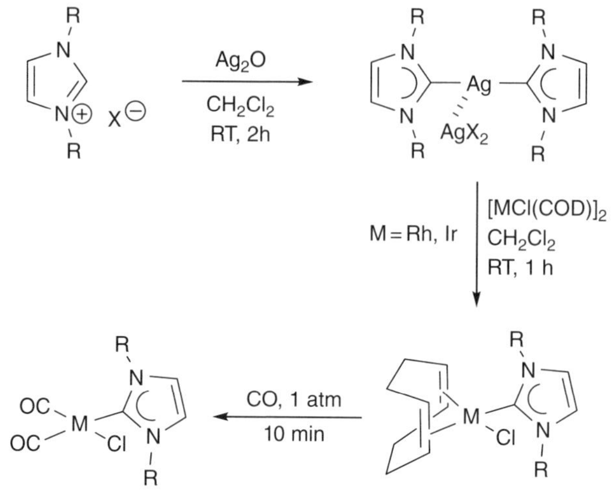

Scheme 2.7
Preparation of Rh[I] and Ir[I] NHC complexes with carbonyl ligands.

that for SIMes/SIPr, which would imply that unsaturated imidazolin-2-ylidenes are stronger donors compared to their saturated analogs. Intuitively, this statement is difficult to comprehend, since sp^2 carbon atoms are more electronegative than saturated sp^3 carbon atoms. Based on the more positive inductive effects of their saturated backbones, SIMes/SIPr should therefore be slightly superior donors compared to their unsaturated analogs.

Since the preparation of diagnostic [Ni(CO)$_3$(NHC)] complexes requires the handling of highly toxic and gaseous [Ni(CO)$_4$] as well as air- and moisture sensitive free carbenes, there was demand for an alternative and less toxic metal system.

In 2003, Robert H. Crabtree and coworkers reported the facile preparation of Rh[I] and Ir[I] NHC complexes via transmetallation from silver-carbene complexes as an extension of Ivan J. B. Lin's approach [19]. The carbene transfer occurs with concurrent bridge-cleavage reaction of chlorido-bridged [MCl(COD)]$_2$ dimers (M = Ir or Rh), which affords the mononuclear [MCl(COD)(NHC)] complexes (Scheme 2.7) [20]. The desired carbonyl derivatives *cis*-[MCl(CO)$_2$(NHC)] are then easily synthesized by passing CO gas through solutions of the COD complexes at one atmosphere.

By analogy to TEP, the carbonyl ligands in complexes of the type *cis*-[MCl(CO)$_2$(NHC)] are sensitive to the electron density at the metal, which in turn is influenced by the NHC co-ligands. Therefore, measurement of the carbonyl stretch would in principle allow for the donor strength determination of the NHC ligand.

Moreover, with values known for structurally analogous phosphine compounds of the type *cis*-[IrCl(CO)$_2$(PR$_3$)] a correlation to TEP was made possible. However, it must be noted that this correlation was made using values that were obtained in different solvents (Table 2.4), which as we have noted earlier do have a significant influence on the stretching frequencies. The carbonyl stretches for most Ir-phosphine complexes were measured in chloroform, while those of NHC ligands and TEPs were obtained using dichloromethane solutions. With exclusion of the outlying phosphites, a first reasonably good linear regression equation ($R^2 = 0.91$) was obtained (Equation 2.2) that allows interconversion of TEP and values for the *cis*-[IrCl(CO)$_2$(NHC)] system (Figure 2.6).

Table 2.4 Data used for correlation of Ir-system with TEP.

L	$\tilde{v}_{av}(CO)[cm^{-1}]$ for $[IrCl(CO)_2 (L)]$	TEP [cm^{-1}]	Solvent
PCy$_3$	2028	2056.4	CH$_2$Cl$_2$
P^{i}Pr$_3$	2031.5	2059.2	CH$_2$Cl$_2$
PEt$_3$	2037.5	2061.7	CHCl$_3$
P(p-CH$_3$C$_6$H$_4$)$_3$	2039	2066.7	CHCl$_3$
PMe$_2$Ph	2041.5	2065.3	CHCl$_3$
PPh$_3$	2043.5	2068.9	CHCl$_3$
P(OBu)$_3$	2044	2077	CS$_2$
PMePh$_2$	2044	2067	CHCl$_3$
P(OPh)$_3$	2049	2085.3	CHCl$_3$

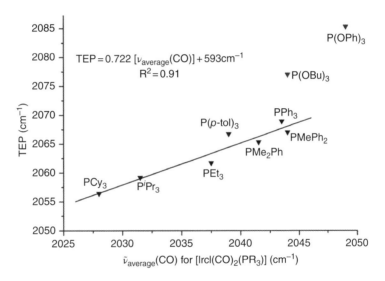

Figure 2.6
Correlation of the average $\tilde{v}(CO)$ values for $[IrCl(CO)_2L]$ complexes with TEP with exclusion of outlying values for P(OBu)$_3$, P(OPh)$_3$ complexes.

$$TEP = 0.722\left[\tilde{v}_{av/Ir}\left(CO\right)\right] + 593\,cm^{-1}$$

Equation 2.2
For conversion of Ir^I into Ni^0 based data.

Nolan and coworkers extended and significantly improved this Ir^I/TEP correlation ($R^2 = 0.97$, Equation 2.3) by including more *cis*-$[IrCl(CO)_2(NHC)]$ (NHC = SIPr, SIMes, IPr, IMes, ICy) complexes with NHCs, for which experimental TEPs have previously been determined [21]. Despite this improvement, a criticism remains that the same phosphine data set measured in chloroform was used in this correlation.

$$TEP = 0.8475\left[\tilde{v}_{av/Ir}\left(CO\right)\right] + 336.2\,cm^{-1}$$

Equation 2.3
For conversion of Ir^I into Ni^0 based data.

Since in the meantime, a large amount of data for *cis*-[RhCl(CO)$_2$(NHC)] was also available, Wolf and Plenio used values determined under precisely the same conditions to correlate the Rh^I to the Ir^I system and vice versa. With a regression coefficient of $R^2 = 0.97$, values of the two metal systems could be interconverted using their initial Equation 2.4 and Equation 2.5 [22].

$$\tilde{v}_{av/Ir}\left(CO\right)=0.8695\left[\tilde{v}_{av/Rh}\left(CO\right)\right]+250.7\,cm^{-1}$$

Equation 2.4
For conversion of Rh^I into Ir^I based data.

$$\tilde{v}_{av/Rh}\left(CO\right)=1.116\left[\tilde{v}_{av/Ir}\left(CO\right)\right]-139.7\,cm^{-1}$$

Equation 2.5
For conversion of Ir^I into Rh^I based data.

These two initial regression equations, established in their original work, were later expanded into Equation 2.6 and Equation 2.7 ($R^2 = 0.98$) by adding more data points from literature. Furthermore, using the same data, a new regression equation to convert Rh^I values to TEP was introduced as well (Equation 2.8) [23].

$$\tilde{v}_{av/Ir}\left(CO\right)=0.9441\left[\tilde{v}_{av/Rh}\left(CO\right)\right]+98.9\,cm^{-1}$$

Equation 2.6
For conversion of Rh^I into Ir^I based data.

$$\tilde{v}_{av/Rh}\left(CO\right)=1.035\left[\tilde{v}_{av/Ir}\left(CO\right)\right]-56.9\,cm^{-1}$$

Equation 2.7
For conversion of Ir^I into Rh^I based data.

$$TEP=0.8001\left[\tilde{v}_{av/Rh}\left(CO\right)\right]+420\,cm^{-1}$$

Equation 2.8
For conversion of Rh^I into Ni^0 based data.

Currently, the Ir^I and Rh^I systems are the most widely used methods to determine the electronic properties of NHCs. Almost every new NHC ligand is routinely evaluated using these methodologies. However, care must be taken to ensure a valid comparison. Only when the measurements are carried out under the exact same conditions, can a comparison be made. This is particularly worth mentioning since IR data are measured in as KBr pellets [24,25], as a film, or often in solutions. For the last case, the use of the same solvent is crucial. Moreover, while most scientists are using the average of the two observed CO stretches for comparison, others rely on only one of the two stretches [26]. Furthermore, systems with different halido ligands, for example, chlorido, bromide, and iodido, are not comparable.

A critical evaluation of the application of these methodologies is presented in the following paragraphs. The averaged carbonyl stretching frequencies obtained in dichloromethane solutions for representatives of the four common NHC types **I–IV** using *cis*-[RhI(CO)$_2$(NHC)] complexes is given in Table 2.5 [27]. In order to avoid any steric

Table 2.5 Comparison of carbonyl stretches among the four common NHC ligands I–IV.

$\tilde{v}_{\mathrm{av/Rh}}(\mathbf{CO})[\mathbf{cm^{-1}}]$	2036	2037	2038	2042
cis-[**RhI(CO)**$_2$(**NHC**)] [27]				

Table 2.6 Comparison of the $\tilde{v}_{\mathrm{av}}(\mathrm{CO})$ values obtained using Ni^0, Ir^I, and Rh^I systems for the four most popular NHC ligands.

Entry		IMes	SIMes	IPr	SIPr
1	$\tilde{v}_{\mathrm{av/Ni}}(\mathrm{CO})[\mathrm{cm^{-1}}]$ [Ni(CO)$_3$(NHC)] [18]	2050.7	2051.5	2051.5	2052.2
2	$\tilde{v}_{\mathrm{av/Ir}}(\mathrm{CO})[\mathrm{cm^{-1}}]$ *cis*-[IrCl(CO)$_2$(NHC)] [21]	2023.1, 2023 [28]	2024.6, 2024.5 [28]	2023.9	2024.9
3	$\tilde{v}_{\mathrm{av/Rh}}(\mathrm{CO})[\mathrm{cm^{-1}}]$ *cis*-[RhCl(CO)$_2$(NHC)] [22]	2038.5	2040.5	2037.5 [29]	NA

influences all ligands carry the small N-methyl substituent. Even with a very optimistic resolution for solution IR data (e.g. standard deviation of $2\,\mathrm{cm^{-1}}$), it can be seen that the differences in the averaged carbonyl stretching frequencies among **I–III** are not significant. Within a deviation of 3σ the values obtained for all four NHC complexes are with $\Delta\tilde{v}_{\mathrm{av}} = 6\,\mathrm{cm^{-1}}$ —essentially equal. It appears that influences due to variations in the backbone, which must exist, cannot be accurately evaluated using the Rh^I carbonyl system.

The literature data obtained for the donor strength comparison of the four most popular NHC ligands IMes, SIMes, IPr and SIPr using all three carbonyl based systems is summarized in Table 2.6. The first notable observation is that the relevant IR data is reported to the tenth decimal place. Generally, this is not common practice. As mentioned above, the resolution of IR solution data is not within this limit, which can also be seen in the broadness of IR bands in the spectra.

In all three metal systems, there is no significant difference among the ligands. If one were to disregard the importance of proper data analysis and standard deviation, the interpretation of the IR data obtained purely based on backdonation to CO ligands can give some very surprising results that are difficult to explain. For example, unsaturated IMes always gives rise to smaller values than saturated SIMes, and the same observation applies to IPr and SIPr. Is it therefore conclusive that IMes or IPr are indeed stronger donors compared to their saturated counterparts SIMes and SIPr, despite the fact that sp^3 carbon atoms in the backbone are less electronegative than sp^2-hybridized analogs? If we compare IMes and IPr that solely differ in the N-aryl substituents, contradicting results are obtained with different metal systems. Based on the Ni^0 and Ir^I systems, IPr could be considered a slightly weaker or equally donating ligand as IMes. However, for the Rh^I system, the inverse was observed, and IPr appears to be the stronger donor by $3 \, cm^{-1}$. One must note that these data were reported by different groups, but the same type of complex was used, and the measurements were made in the same solvent.

Overall, it can be concluded that comparison of NHC ligands on a finer level, that is with small differences, can often not be done with carbonyl based systems due to the limited resolution of IR spectroscopy in general. One has to be aware that other more complicated interactions may be at play in such electron-rich carbonyl complexes. For example, non-negligible back donation of electron density to the NHC may also occur. The results obtained must be critically discussed for every ligand before any conclusion regarding its donor strength can be made. The carbonyl based methodologies are however still very useful and can provide good results when sufficiently different ligands are compared. A general guideline on what changes would contribute to a sufficient difference cannot be given, and this is left to the experimenter who should make the decision based on chemical intuition, experience, and knowledge of significance in data analysis.

2.5.2   Lever's Electronic Parameter (LEP)

While the Tolman electronic parameter is dominant in organometallic chemistry, electronic influences of classical Werner-type ligands (mainly N, O, S, halido, and pseudo-halido donors) are usually tabulated in the form of the Lever electronic parameter (LEP), having been established using redox potentials of $Ru^{II/III}$ metal complexes. With the intention of providing an easy means to predict redox potentials of inorganic complexes, Alfred Beverley Philip Lever introduced a general formula (Equation 2.9) in 1990, which was based on the assumption that ligand contributions to the overall redox potential are additive [30]. Summation of experimental ligand electrochemical parameters E_L with consideration of values specific to the spin state (S_m) and redox couple of the respective metal (I_m) gives the predicted redox potential of the complex. In most cases, this simple formula gave realistic results that are close to the experimentally determined redox potentials.

$$E_{redox} = S_m \left(\sum E_L (L) \right) + I_m$$

Equation 2.9
Lever's general formula to calculate redox potentials of complexes.

The E_L values of individual ligands reflect their relative capacity to stabilize a metal in a certain oxidation state. Ligands with smaller E_L values can therefore stabilize the Ru^{III} state in the $Ru^{II/III}$ couple better than those with larger E_L values. In other words, their sum would give rise to a lower redox potential, which in turn is related to a more facile oxidation process. Although E_L values are not a direct measure of the donating ability of a ligand, they are nevertheless related in a certain way [31]. The drawback is that only complexes that exhibit reversible or quasi-reversible redox chemistry can be considered if one would like to establish an E_L value of a new ligand. Moreover, the presence of non-innocent ligands, which interfere in the metal-redox processes, must be avoided. Consequently, the choice of suitable complex probes is reduced. The requirement for electrochemical setups, which are also relatively less common, has further contributed to the fact that the electronic properties of only a small number of NHCs have been evaluated in this manner.

By assuming insignificant wing tip substituent effects, Martin Albrecht and coworkers determined the first Lever electrochemical parameter for unsaturated imidazolin-2-ylidenes using piano-stool carbene complexes for the $Fe^{II/III}$ redox couple. Surprisingly, the LEP found for imidazolin-2-ylidenes of $E_L = 0.29$ is similar to that of pyridine ($E_L = 0.25$, Table 2.7), which according to the authors points to a considerable amount of π-backbonding from electron-rich Fe^{II} centers to the NHC ligand [32].

As already highlighted, the availability of LEP values for NHC ligands is limited. However, the net donating abilities of NHCs can easily be compared by directly ranking the electrochemical redox potential of analogous NHC complexes. In relation to LEP, a stronger donating NHC should give rise to a smaller redox potential in complexes that otherwise bear the same metal and co-ligands among all comparators.

Attempts have been made to determine the redox potentials of Ir^I and Rh^I coordination compounds. Complexes of the type cis-[IrCl(CO)$_2$(NHC)] show an irreversible electrochemistry, probably due to loss of CO during the redox process. However, their cis-[IrCl(COD)(NHC)] complex precursors and the respective Rh^I counterparts turn out to be suitable redox probes [22,28].

Using these complexes, Plenio and coworkers determined remote substituent effects on the donating ability of N,N′-diaryl substituted imidazolin- and imidazolidin-2-ylidenes, and representative data obtained from cis-[RhCl(COD)(NHC)] complexes are summarized in Table 2.8 [22]. In addition to substituent effects, electrochemical studies could also

Table 2.7 Lever's electrochemical parameter for selected common ligands.

	Cl^-	$CF_3CO_2^-$	NCS^-	H_2O	py	R (NHC)	CH_3CN	PPh_3	$P(OMe)_3$	CO
E_L	−0.24	−0.15	−0.06	0.04	0.25	0.29	0.34	0.39	0.42	0.99

Table 2.8 Redox potentials $E_{1/2}$ [V] and peak separation $E_a - E_c$ [mV] of the cis-[RhCl(COD)(NHC)] complexes (referenced vs. Fc/Fc$^+$ $E_{1/2}$ = 0.46 V, 0.1 M TBAPF$_6$ in CH$_2$Cl$_2$).

	Imidazolin-2-ylidenes		Imidazolidin-2-ylidenes	
R	$E_{1/2}$ [V]	$E_a - E_c$ [mV]	$E_{1/2}$ [V]	$E_a - E_c$ [mV]
NEt$_2$	0.718	76	0.651	78
OC$_{12}$H$_{25}$	0.836	82	0.785	92
Me	0.833	76	0.791	76
H	0.855	82	0.817	78
Br	0.926	74	0.903	90
SOtol	0.923	62	0.961	84

corroborate that saturated imidazolidin-2-ylidenes are—as one would expect—indeed stronger net donors than unsaturated imidazolin-2-ylidenes. This also highlights that the electrochemical approach is in some cases superior to carbonyl based methodologies in terms of sensitivity and accuracy.

2.5.3   Huynh's Electronic Parameter (HEP)

A very important spectroscopic method for characterization in general NHC chemistry is ^{13}C NMR spectroscopy. Similar to the carbonyl stretch in IR spectroscopy, the ^{13}C NMR signal of the carbene carbon is diagnostic and found in a very downfield region, where it is usually not obstructed by resonances of other groups. The carbene carbon of free NHCs resonates at >200 ppm and a significant upfield shift of this signal is observed upon binding to a transition metal. Lappert has already noted in an early work that the ^{13}C$_{carbene}$ NMR signal is sensitive to the co-ligand in *trans* position. The deshielding effect of this *trans* co-ligand on the carbene carbon increases with its *trans* influence, which in turn is related to its donating ability [33]. Herrmann later reported that carbene signals generally experience an increasing upfield shift with increasing Lewis acidity of the bound complex fragment [34]. In agreement with this notion, M. V. Baker and coworkers prepared a series of gold(I) NHC complexes [AuX(I^{t}Bu)], for which they found a good correlation of the carbene chemical shift with the net donor ability of the anionic co-ligand (Chart 2.7) [35]. Based on the increasing downfield chemical shift of the I^{t}Bu carbene carbon, the anionic ligands were ranked in the following order of increasing donor strength: ONO_2^- < OAc$^-$ < NCO$^-$ < N$_3^-$ < Cl$^-$ < Br$^-$ < SCN$^-$ < NCS$^-$ < SeCN$^-$ < I$^-$ < CN$^-$ < CH$_3^-$. Similarly, K.-R. Pörschke observed such dependency in palladium allyl complexes of the type [Pd(η^3-allyl)X(NHC)] [36] (Chart 2.7). Comparison of the ^{13}C$_{carbene}$ signals recorded for these complexes led to the increasing donor strength/nucleophilicity series: PF$_6^-$ < CH$_2$Cl$_2$ ≈ BF$_4^-$ < OTf ≈ THF < H$_2$O < Cl$^-$, CH$_3^-$.

The same sensitivity of the carbene signal on the electronic situation at the complex fragment was also noted by Huynh and coworkers, which led them to explore this concept for the donor strength determination of NHCs and various other ligands [37]. For this purpose, the dimeric Pd complex [PdBr$_2$(^{i}Pr$_2$-bimy)]$_2$ was subjected to a simple bridge-cleavage reaction with a range of neutral ligands L, which straightforwardly afforded the complex probes of the type *trans*-[PdBr$_2$(^{i}Pr$_2$-bimy)L] in good yields (Scheme 2.8). In such complexes, the electronic properties of the ligand L in question influence the ^{13}C$_{carbene}$ NMR signal of the ^{i}Pr$_2$-bimy reporter ligand. A stronger donating ligand L gives rise to a more downfield ^{13}C NMR resonance of the ^{i}Pr$_2$-bimy carbene carbon. For the donor strength

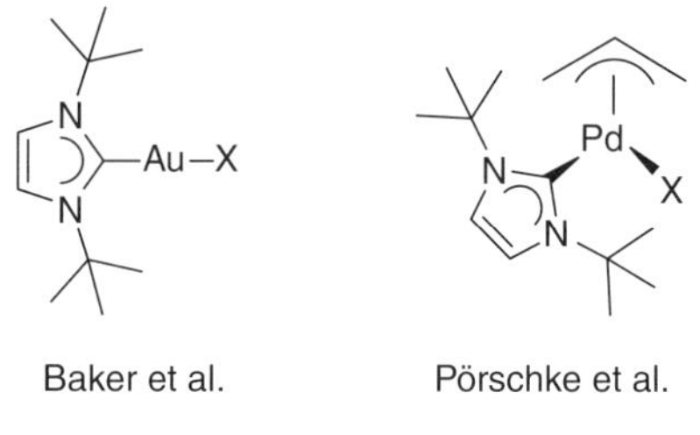

Baker et al.   Pörschke et al.

Chart 2.7
NHC complexes studied by Baker and Pörschke.

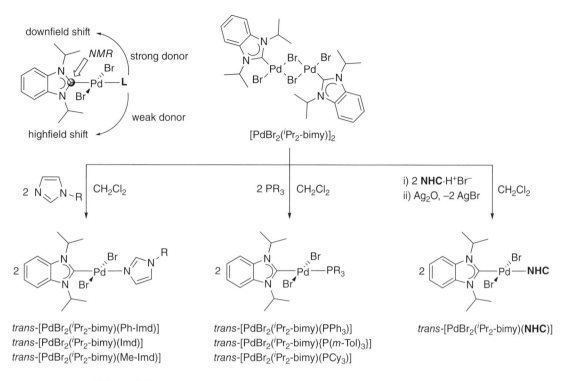

Scheme 2.8

Preparation of Pd^{II} complexes for the donor strength determination by ^{13}C NMR spectroscopy.

determination of NHCs, new hetero-bis(NHC) complexes of the general formula *trans*-[PdBr$_2$(^{i}Pr$_2$-bimy)(NHC)] bearing two different NHCs are required. Thus, a simple one-pot procedure was developed, following which a mixture of the respective NHC precursor salts and the Pd^{II} dimer was treated with silver oxide. Azolium bromides were chosen as carbene precursors to avoid halide scrambling.

In order to solely evaluate the backbone effects derived from different heterocycles and to minimize any additional steric or electronical interferences of the N-substituents, four identically *N,N′*-dibenzyl-substituted NHCs of types **I–IV** were chosen for the preparation of the respective hetero-bis(NHC) complex probes. To further prevent solvent effects and to facilitate reasonably fast measurements, the ^{13}C NMR spectra of all complexes were determined using concentrated solutions in CDCl$_3$ and referenced to the solvent signal set at 77.7 ppm.

Correct assignment of the two expected carbene signals occurring in these hetero-bis(NHC) complexes to the respective two different NHCs is facilitated by heteronuclear multiple-bond correlation (HMBC) NMR spectroscopy (Figure 2.7), in which cross peaks correlating the carbene carbon with protons in the respective N-substituent or ligand backbone allow for an unambiguous assignment of the correct carbene donors.

For difficult cases or complexes with low solubility in CDCl$_3$, the feasibility of ^{13}C labeling the carbene donor in the ^{i}Pr$_2$-bimy ligand was also explored. For this purpose, H^{13}COOH can be condensed with 1,2-phenylenediamine to give ^{13}C-2 labeled benzimidazole. Dialkylation of the latter with an excess of isopropyl bromide affords 1,3-diisopropyl-^{13}C-2-benzimidazolium bromide (Scheme 2.9) as a precursor to the ^{13}C$_{carbene}$ labeled dimeric Pd^{II} complex [PdBr$_2$(^{i}Pr$_2$-^{13}C-2-bimy)]$_2$.

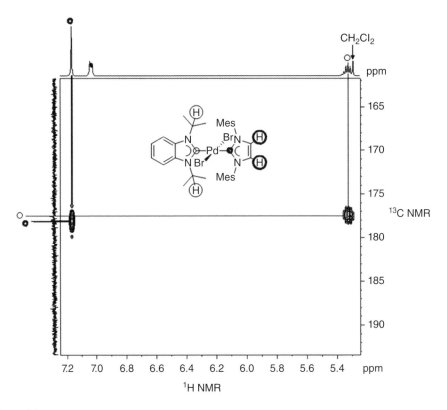

Figure 2.7
HMBC NMR spectrum of *trans*-[PdBr$_2$(^{i}Pr$_2$-bimy)(IMes)].

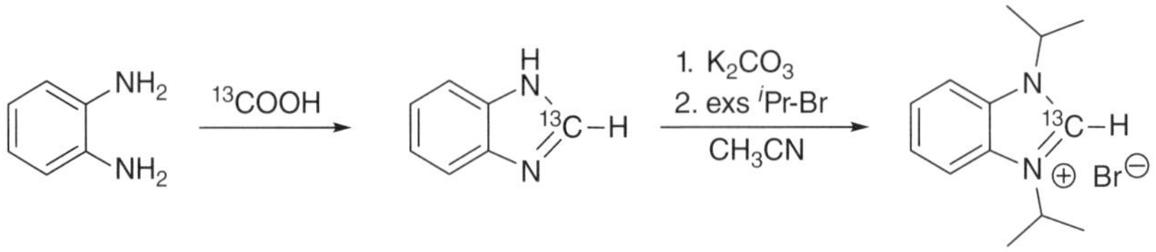

Scheme 2.9
Preparation of ^{13}C-2 labeled 1,3-diisopropylbenzimidazolium bromide.

   A comparison of the ^{i}Pr$_2$-bimy carbene signals in respective hetero-bis(NHC) complexes indeed shows decreasing donor ability for the NHC co-ligands in the order **I** > **II** > **III** > **IV** (Figure 2.8). This correlation is remarkable considering the simplicity of this system, and the resulting trend is reasonable when the inductive effects of the varying backbones adjacent to the NCN moiety are considered. Clearly, the greater +*I* effect of two sp^3 carbon atoms in saturated **I** over two sp^2 carbon atoms in unsaturated **II** translates into a stronger carbene donor of the former. Benzannulation in **III** further weakens the +*I* effect, and intro-duction of an additional electronegative nitrogen atom in **IV** even results in a –*I* effect giving rise to the weakest donor in this series. Notably, carbonyl based systems have not been useful for differentiating the donor abilities of saturated imidazolidin-2-ylidenes versus the unsaturated imidazolin-2-ylidenes [18,21,22].

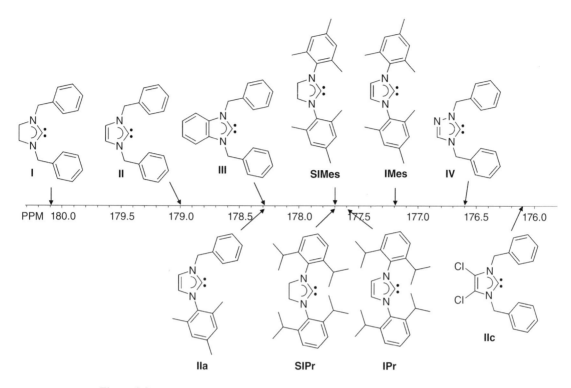

Figure 2.8
Comparison of donor abilities of some common NHCs on the ^{13}C NMR scale.

To further test the sensitivity of this method for the determination of substituent effects in saturated and unsaturated NHCs, hetero-bis(NHC) complexes with the ^{i}Pr$_2$-bimy spectator and one of the four currently most popular carbenes **IMes**, **SIMes**, **IPr**, and **SIPr** bearing aromatic N-substituents were studied. In addition, a mixed 1-mesityl-3-benzyl substituted NHC **IIa** was included to close the gap between **II** and **IMes**.

A comparison of the ^{13}C$_{\text{carbene}}$ signals for the constant ^{i}Pr$_2$-bimy ligand indeed revealed a measurable influence of the N-substituents on the donor ability of NHC ligands in Pd^{II} complexes (Figure 2.8). The stepwise substitution of benzyl with mesityl groups going from **II** (179.0 ppm) to **IIa** (178.3 ppm) and finally to **IMes** (177.2 ppm) leads to a stepwise highfield shift corroborating decreasing donor ability of the carbenes in the order **II** > **IIa** > **IMes**. Again this observation is reasonable and can be explained by the greater +*I* effect of alkyl versus aryl substituents. It is also interesting to note that carbenes **IIa** and **III** supposedly have the same donor strength. Thus, the change of one substituent from benzyl to mesityl in imidazolin-2-ylidenes has a comparable effect as a benzannulation. For the more popular carbenes the order of decreasing donor strength has been established as **SIPr** ≈ **SIMes** > **IPr** > **IMes**. Notably, the change from saturated to unsaturated NHCs bearing N-aryl groups has smaller impact on the electronic properties than for those with N-alkyl substituents. A likely reason is the presence of two opposing effects in SIMes/SIPr (i.e. the donating effect of the backbone and the withdrawing effect from the aryl groups). Overall, the ^{13}C NMR data confirm that saturated NHCs are in general stronger donors than their unsaturated analogs, which is also in line with the calculated relative basicity of related

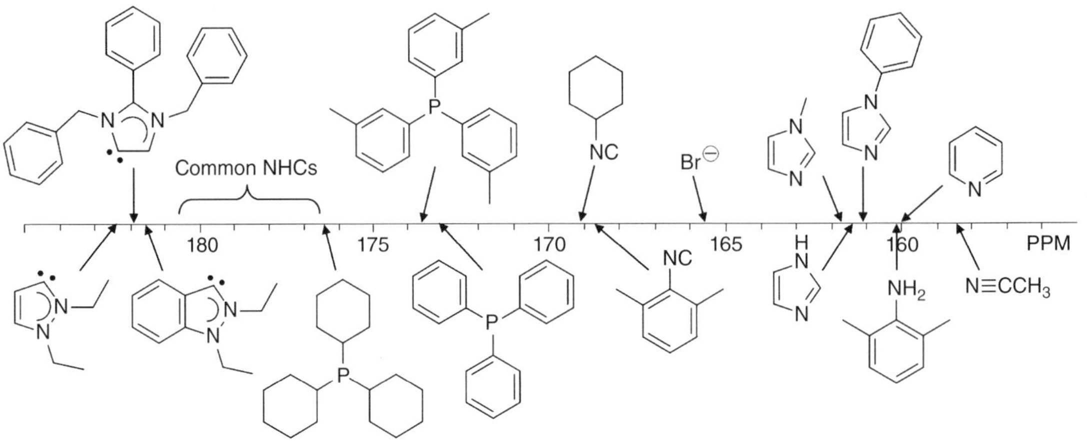

Figure 2.9
Donor abilities of selected Werner-type and organometallic ligands on a unified ^{13}C NMR spectroscopic scale.

NHCs [38]. Modification of the donating ability of imidazolin-2-ylidenes can also be carried out by introducing electron-withdrawing groups in C4/5 positions. This is demonstrated with 1,3-dibenzyl-4,5-dichloroimidazolin-2-ylidene **IIb**, which induces an even more upfield chemical shift of the ^{i}Pr$_2$-bimy probe compared to triazolin-5-ylidene **IV**, and therefore is an even weaker donor.

In addition to NHCs ligands, a range of other organometallic and classical Werner-type ligands was evaluated for their donating abilities. Huynh's electronic parameter therefore allows donor strength comparison of both organometallic and Werner-type ligands on a unified ^{13}C NMR spectroscopic scale (Figure 2.9).

2.6 Steric Properties of NHCs

The second important parameter of a given ligand is the steric bulk that it imposes on the metal complex fragment upon binding. The steric characteristics of ligands have major implications on the reactivity of their respective complexes, often overruling their electronic influences. A number of steric parameters have been developed based on experiments or calculations [39]. The most often applied steric parameter in organometallic chemistry was again developed by Chadwick Tolman in 1977 primarily for phosphine ligands [40]. The so-called *Tolman cone angle* for various phosphines was originally determined from simple 3-D space filling models, where the "metal-phosphine" distance was adjusted to 2.28Å (Figure 2.10). From Figure 2.10, it can also been seen that the R-substituents of phosphines generally point away from the metal center. The cone angle θ (theta) indicates the approximate amount of "space" that the respective ligand consumes about the metal center. The bigger R-groups of a bulkier phosphine would be more spread out giving rise to a larger cone and a bigger θ value. However, the use of the angle θ is limited to ligands, which bind to the metal in cone shape manner. Examples for such ligands include triarylarsines, the cyclopentadienyl (cp) ligand and its derivatives.

Table 2.9 summarizes the Tolman cone angles for a few selected common phosphine ligands. Notably, T. E. Müller and D. P. Mingos determined the crystallographic cone

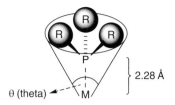

Figure 2.10
Tolman cone angle θ for symmetrical phosphines.

Table 2.9   Values for average crystallographic cone angles and Tolman cone angles θ.

Ligand	PPh_3	$PMePh_2$	PMe_2Ph	PMe_3	PEt_3	PCy_3
Tolman cone angle (°)	145 ± 2	136 ± 2	122 ± 2	118 ± 2	132 ± 2	170 ± 2
Crystallographic cone angle (°)	148.2 (4.9)	134.5 (5.0)	119.9 (5.4)	111.1 (2.4)	137.3 (5.0)	160.1 (5.1)

angles of phosphines using solid state molecular structures of more than 4000 complexes and concluded that these are quite similar to values determined by Tolman's simple method (Table 2.9) [41]. Their study also demonstrated that Tolman cone angles are not absolute values and do vary dependent on the number of the ligands (substituent effects, etc.) and the nature of other co-ligands present in the complex. They also vary dependent on the metal centers. Nevertheless, averaged values are useful for the estimation of the steric bulk of a given ligand.

Contrary to phosphines, NHCs do not coordinate to metal centers in a cone shape manner. In the first instance, the two wing tip N-substituents would always point in the direction of the metal-NHC vector due to the sp^2 hybridization of the nitrogen atoms. Variations in these substituents should therefore have a greater impact on the steric crowding about the first coordination sphere. Application of the Tolman cone angle is thus not feasible for NHCs.

In order to quantify the steric demand of NHC ligands, Cavallo and Nolan have introduced a parameter, which they termed "percentage of buried volume" ($\%V_{Bur}$) [42]. This parameter measures the relative volume that a given ligands occupies in the first coordination sphere of a metal center (Figure 2.11). The greater the $\%V_{Bur}$, the bulkier is the NHC ligand. Since its introduction, $\%V_{Bur}$ has undergone several refinement steps, which essentially optimized the set of atom radii (Bondi radii scaled by 1.17) and the radius defining the first coordination sphere (3.5 Å) used for the calculations. The resulting algorithm has been incorporated into a user-friendly and freely available interface on the internet [43]. Using Crystallographic Information (CIF) or Chem3D Cartesian files, any user can determine the $\%V_{Bur}$ of any NHC.

Table 2.10 reports $\%V_{Bur}$ values of some selected common NHC ligands that were obtained using this interface. It is obvious that the sterics of an NHC can be tuned by choice of the wing tip substituent. Small groups such as simple linear alkyl chains will lead to less bulky NHCs, while 3-dimensionally spaced out substituents such as tertiary butyl or adamantyl impose the most steric bulk. In this respect, aryl groups are intermediate, since the aromatic ring is twisted almost perpendicularly out of the carbene ring

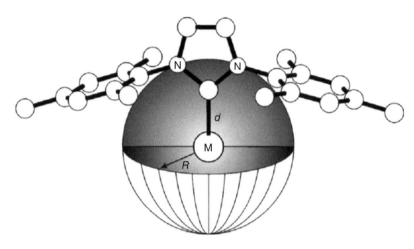

Figure 2.11
Graphical representation of the sphere used to calculate $\%V_{Bur}$. Poater [43]. Reprodcued with permission from John Wiley & Sons.

Table 2.10 $\%V_{Bur}$ values for selected common NHC ligands. Bondi radii scaled by 1.17 for atoms; radius of the sphere $R = 3.5$Å [43].

R	(imidazolidine)	(imidazole)	(benzimidazole)
CH₃	25.4	24.9	25.1
ᵗBu	36.2	35.5	38.9
(adamantyl)	36.6	36.1	40.8
(phenyl)	31.6	30.5	30.2
(mesityl)	32.7	31.6	31.2
(diisopropylphenyl)	35.7	33.6	31.9

plane. Differences among different types of common NHCs are, although not negligible, predictably less severe. By comparison of $\%V_{\text{Bur}}$ values, the conclusion can be drawn that for any given set of identical wing tip R-substituents, unsaturated imidazolin-2-ylidenes are generally less bulky than their saturated imidazolidin-2-ylidene counterparts. This is understandable, since the NCN angle is larger for the latter, which in turn brings the N-substituents closer to the metal center. For benzannulated NHCs, on the other hand, no clear trend can be observed. Here additional repulsive interactions between the R-group and the aryl backbone can significantly increase the steric bulk at the metal center. Modification at the backbone therefore offers the opportunity to fine-tune the steric bulk of an NHC ligand.

2.7 Structural Diversity of NHC Ligands and Their Complexes

The popularity of NHCs is not solely due to their strong σ donating abilities, but is undoubtedly also linked to the ease of their preparation and modification. We have already seen that modifications in the backbone and wing tip groups can measurably alter the stereoelectronic properties of NHCs. In addition, solubilities of their complexes can be influenced as well. The choice of N-substituents is almost unlimited, which leaves a lot of room for the imagination and creativity of the experimentalist. Almost every kind of substituent can be attached to the nitrogen atoms, and large libraries of NHCs are already known, demonstrating their structural variety. Therefore, only a small selection of examples can be mentioned in the following sections to highlight their structural diversity.

2.7.1 Donor-Functionalized NHCs

In addition to simple N-alkyl or N-aryl substituents that give rise to simple monodentate NHC ligands, one can also easily attach groups that contain heteroatoms. Substituents containing heteroatoms capable of coordination can give rise to ditopic or donor-functionalized NHCs [44]. Depending on the nucleophilicity of the donor functions, complexes with dangling tethers can result or κ^2C,D chelate-complexes can be obtained, where D denotes the nature of the bound heteroatom. In addition to the variation of the donor heteroatoms, modifications of the tether length, the NHC backbone, and the second wing tip group can be carried out (Chart 2.8). Early works in this area were dominated by NHCs functionalized with nitrogen and oxygen donors as a marriage of organometallic with harder Werner-type ligands [45,46]. Pyridine [46,47], amine [45], imine [48] and amido [49] groups have been studied extensively among the nitrogen functions, while ethers [45], esters [46,50], and alkoxido [51] groups dominate among the oxygen donors. The combination of phosphine and NHC donors results in organometallic ligands with two strongly donating and soft functions [45,52], which upon coordination can give electron-rich κ^2C,P complexes. The incorporation of soft sulfur functions, on the other hand, has been less studied [53]. This is particularly surprising given the great versatility that sulfur donors demonstrate in classical coordination chemistry. The reason for this could be the unpleasant odor that is associated with sulfur chemistry, and the sensitivity of many sulfur compounds to oxidation that has delayed the development of preparative methods. Thioether [54] and thiolato [55] were the first sulfur donor functions to be studied in this respect.

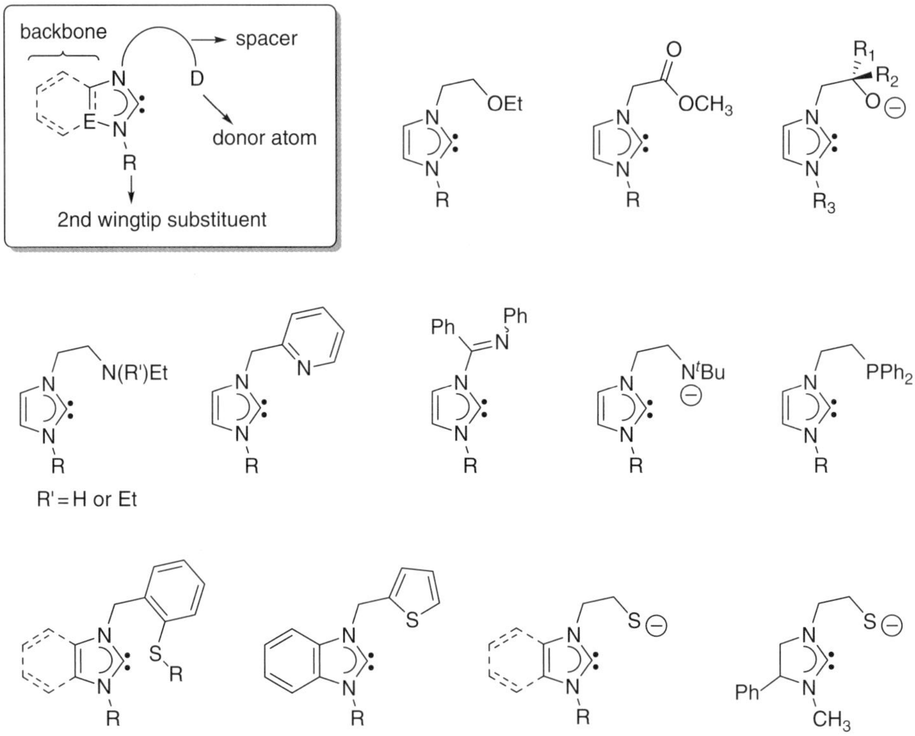

Chart 2.8
Selected examples for N, O, P, and S-functionalized NHC ligands.

2.7.2 Multidentate NHCs

Linking two or more NHCs at the nitrogen atoms is straightforward and gives rise to di- or polycarbenes, which can increase complex stability due to the chelate effect [56]. The spacers in such ligands and thus the size of the resulting metallacycle can be modified, which can also affect complex stability. Further variations in the two NHC backbones and the two additional wing tip groups can be used to fine-tune the electronic and steric properties of the diNHC ligand (Chart 2.9).

Simple alkyl bridged dicarbenes were among the earliest examples due to their easy preparation, and they have been used to prepare diNHC metal chelates [57]. Even complexes of hetero-diNHCs, in which two different types of NHC are linked, have been prepared [58]. It has been observed that the length of the alkyl bridge will not affect the bite angle to a great extent. Instead, the dihedral angle between the NHC and the coordination plane in square planar complexes is easily influenced, whereby a longer and more flexible alkyl bridge is better suitable for the preferred perpendicular orientation (i.e. a dihedral angle of ~90°) of the NHC plane with respect to the square coordination plane [58,59]. Furthermore, it was found that complexes of methylene (C1) and propylene (C3) bridged diNHC ligands could be prepared in better yields as compared to ethylene (C2) linked diNHCs. This observation is attributed to the unfavorable geometry of the ethylene spacer for a *cis*-chelating coordination mode [58,60].

Connecting two NHC moieties via a rigid aromatic spacer can lead to chelating, bridging, or pincer-type diNHC ligands. For example, the choice of a 1,2-phenylene bridge still

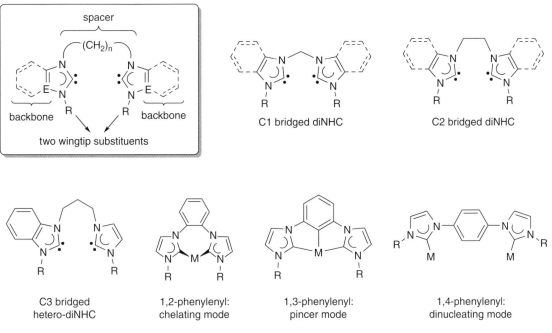

Chart 2.9
Selected examples for diNHC and hetero-diNHC ligands.

allows for a *cis*-chelating $\kappa^2 C^2$ coordination [61]. A diNHC with 1,4-phenylene spacer [62], on the other hand, is incapable of chelation and gives dinuclear complexes. An intermediate position is taken by 1,3-phenylene [63] linked diNHCs, which can afford dinuclear or CCC-pincer complexes after aryl C-H activation (Chart 2.9).

2.7.3 Pincer-Type NHC Ligands

When a donor-functionalized spacer (D) is chosen to bridge two NHCs moieties, unique linear tridentate CDC-pincer-type ligands can be obtained (Chart 2.10) [55]. Early examples were combinations of two NHCs with a central donor unit, such as 2,6-lutidinyl [64] or 1,3-xylenyl [64a,65], which result in CNC or anionic CCC-pincer ligands that are believed to increase the robustness of the complex. The rigidity of the pincer ligand can be further increased by changing the central donor unit to 2,6-pyridyl [66] and 1,3-phenylene [63], respectively, which gives essentially planar pincer-type ligands. On the other hand, more flexible pincer ligands can be prepared by choosing aliphatic amines (CNC) [67], ethers (COC) [68] or thioether (CSC) [69] bridges.

Donor-functionalization (D) of both N-substituents, on the other hand, gives DCD pincer-type ligands that contain a central NHC unit and two additional donors. Complexes of DCD pincer-type ligands bearing common donor functions, such as phosphines [70], amines [71], pyridyl [72], alkoxido [73] and thiolato [55] have been prepared.

NHC ligands with a pincer topology offer many possibilities for variations, and it is easily imaginable that these are substantially increased by extension to unsymmetrical pincers. Indeed this is topic of current research.

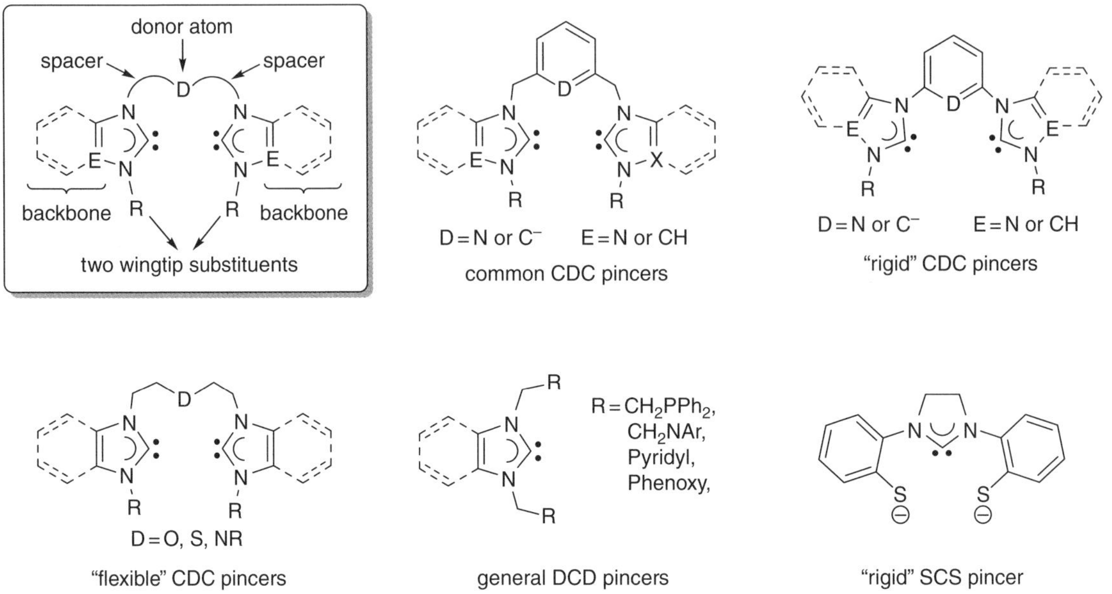

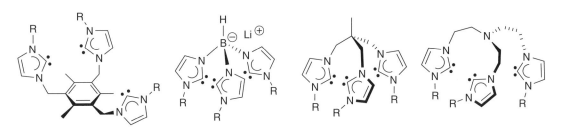

Chart 2.10
Selected examples of pincer-type NHC ligands.

Chart 2.11
Tripodal triimidazolin-2-ylidene ligands.

2.7.4 Tripodal and Macrocyclic Ligands

As mentioned earlier in this section, linking three donors linearly can give pincer-type ligands. On the other hand, when three donors are connected in a branched fashion, tripodal ligands are obtained that are tailor-made for a facial coordination mode. Surprisingly, the first tripodal NHC ligand had already been reported in 1994 [74]. In this ligand three imidazolin-2-ylidenes are connected via methylene groups to a central aromatic unit in a 1,3,5-substitution pattern. Another early example is the direct isomer of the well-studied tris(pyrazolyl)borate ligand, which therefore carries a negative charge [75]. Variations in the central linker led to the development of more flexible tripodal triimidazolin-2-ylidenes [76], which can be tri- or tetradentate in the case when the central linker carries an additional donor atom (Chart 2.11). Compared to the coordination chemistry of pincer-type NHC ligands, that of their tripodal counterparts is relatively less investigated, although complexes of unique properties have been obtained.

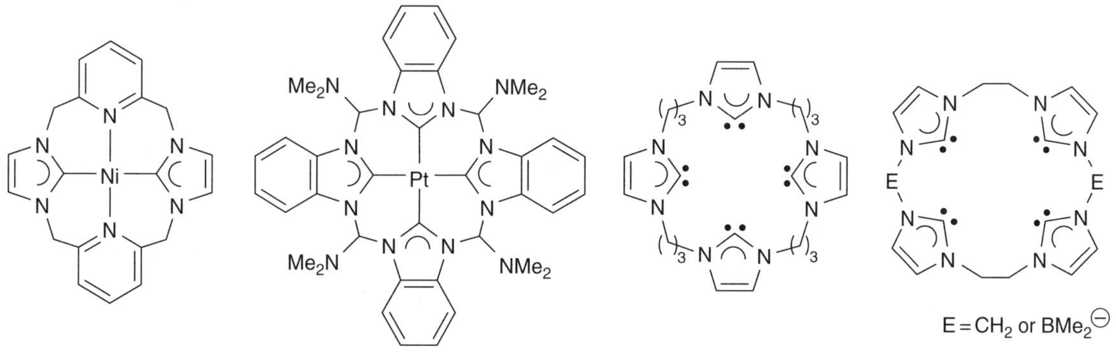

Chart 2.12
Macrocyclic NHC complexes and macrocyclic tetracarbene ligands.

Another less investigated, but unique class of NHC ligands are macrocyclic tetracarbene ligands. Although a number of cyclic azolium salts are known, only a handful of these are suitable for binding a single central metal atom in a macrocyclic manner. The first example of a complex bearing a macrocyclic NHC ligand was reported by M. Baker and coworkers [77]. This C_2N_2-macrocyle contains two alternatingly linked NHC and pyridine moieties. F. E. Hahn and coworkers later reported the template-directed preparation of the first complex with a C4-macrocycle containing purely NHC donors [78]. Since then, a few tetraimidazolium salts with different spacers have been reported that were used to prepare complexes of macrocyclic tetracarbenes of various transition metals (Chart 2.12) [79].

The study of macrocyclic NHC complexes, although still in its infancy, has already demonstrated a broad scope in terms of structures and reactivities, which warrants future endeavors.

References

[1] For general reviews: (a) W. A. Herrmann, C. Köcher, *Angew. Chem. Int. Ed. Engl.* **1997**, *36*, 2162; (b) D. Bourissou, O. Guerret, F. P. Gabbaï, G. Bertrand, *Chem. Rev.* **2000**, *100*, 39; (c) F. E. Hahn, M. C. Jahnke, *Angew. Chem. Int. Ed.* **2008**, *47*, 3122.

[2] https://scifinder.cas.org/(accessed October 7 2016)

[3] C. Heinemann, T. Müller, Y. Apeloig, H. Schwarz, *J. Am. Chem. Soc.* **1996**, *118*, 2023.

[4] F. E. Hahn, L. Wittenbecher, D. Le Van, R. Fröhlich, *Angew. Chem. Int. Ed.* **2000**, *39*, 541.

[5] M. K. Denk, A. Thadani, K. Hatano, A. J. Lough, *Angew. Chem. Int. Ed. Engl.* **1997**, *36*, 2607.

[6] A. J. Arduengo III, J. R. Goerlich, W. J. Marshall, *J. Am. Chem. Soc.* **1995**, *117*, 11027.

[7] A. J. Arduengo III, H. V. Rasika Dias, R. L. Harlow, M. Kline, *M. J. Am. Chem. Soc.* **1992**, *114*, 5530.

[8] T. A. Taton, P. Chen, *Angew. Chem. Int. Ed. Engl.* **1996**, *35*, 1011.

[9] J. P. Malrieu, G. Trinquier, *J. Am. Chem. Soc.* **1989**, *111*, 5916.

[10] M. Driess, H. Grützmacher, *Angew. Chem. Int. Ed.* **1996**, *35*, 828.

[11] S. I. Miller, *J. Chem. Educ.* **1978**, *55*, 778

[12] Values for **II** and **IV** are found in J. C. Bernhammer, G. Frison, H. V. Huynh, *Chem. Eur. J.* **2013**, *19*, 12892. Values for **I** and **III** were calculated at the same B3LYP/aug-cc-pVTZ level according to the reference for accurate comparison.

[13] For a review see: R. W. Alder, M. E. Blake, L. Chaker, J. N. Harvey, F. Paolini, J. Schütz, *Angew. Chem. Int. Ed.* **2004**, *43*, 5896.

[14] (a) F. E. Hahn, M. Paas, D. Le Van, T. Lügger, *Angew. Chem. Int. Ed.* **2003**, *42*, 5243. (b) F. E. Hahn, M. Paas, D. Le Van, R. Fröhlich, *Chem. Eur. J.* **2005**, *11*, 5080.

[15] *IUPAC. Compendium of Chemical Terminology*, 2nd ed. (the "Gold Book"). Compiled by A. D. McNaught and A. Wilkinson. Blackwell Scientific Publications, Oxford (**1997**). XML on-line corrected version: http://goldbook.iupac.org (accessed October 7, 2016) (2006–) created by M. Nic, J. Jirat, B. Kosata; updates compiled by A. Jenkins. ISBN 0-9678550-9-8. doi:10.1351/goldbook.

[16] (a) C. A. Tolman, *J. Am. Chem. Soc.* **1970**, *92*, 2953; (b) C. A. Tolman, *Chem. Rev.* **1977**, *77*, 313.

[17] K. Öfele, W. A. Herrmann, D. Mihalios, M. Elison, E. Herdtweck, W. Scherer, J. Mink, *J. Organomet. Chem.* **1993**, *459*, 177.

[18] (a) R. Dorta, E. D. Stevens, C. D. Hoff, S. P. Nolan, *J. Am. Chem. Soc.* **2003**, *125*, 10490, (b) R. Dorta, E. D. Stevens, N. M. Scott, C. Costabile, L. Cavallo, C. D. Hoff, S. P. Nolan, *J. Am. Chem. Soc.* **2005**, *127*, 2485.

[19] H. M. J. Wang and I. J. B. Lin, *Organometallics* **1998**, *17*, 972.

[20] A. R. Chianese, X. Li, M. C. Janzen, J. W. Faller, R. H. Crabtree, *Organometallics* **2003**, *22*, 1663.

[21] R. A. Kelly III, H. Clavier, S. Giudice, N. M. Scott, E. D. Stevens, J. Bordner, I. Samardjiev, C. D. Hoff, L. Cavallo, S. P. Nolan, *Organometallics* **2008**, *27*, 202.

[22] S. Wolf, H. Plenio, *J. Organomet. Chem.* **2009**, *694*, 1487.

[23] T. Dröge, F. Glorius, *Angew. Chem. Int. Ed.* **2010**, *49*, 6940.

[24] A measurement of the [RhCl(CO)$_2$(SIMes)] complex as a KBr pellet gives the reported $\tilde{v}_{av/Rh}$(CO) value of 2038.5 cm^{-1},[23] as compared to a value of 2040.5 cm^{-1} obtained for the same complex in CH$_2$Cl$_2$ solutions.[20]

[25] K. Denk, P. Sirsch, W. A. Herrmann, *J. Organomet. Chem.* **2002**, *649*, 219.

[26] A. Fürstner, M. Alcarazo, H. Krause, C. W. Lehmann, *J. Am. Chem. Soc.* **2007**, *129*, 12676.

[27] W. A. Herrmann, J. Schütz, G. D. Frey, E. Herdtweck, *Organometallics* **2006**, *25*, 2437.

[28] S. Leuthäußer, D. Schwarz, H. Plenio, *Chem. Eur. J.* **2007**, *13*, 7195.

[29] T. Sato, Y. Hirose, D. Yoshioka, S. Oi, *Organometallics* **2012**, *31*, 6995.

[30] A. B. P. Lever, *Inorg. Chem.* **1990**, *29*, 1271.

[31] S. S. Fielder, M. C. Osborne, A. B. P. Lever, W. J. Pietro, *J. Am. Chem. Soc.* **1995**, *117*, 6990.

[32] L. Mercs, G. Labat, A. Neels, A. Ehlers, M. Albrecht, *Organometallics* **2006**, *25*, 5648.

[33] D. J. Cardin, B. Cetinkaya, E. Cetinkaya, M. F. Lappert, E. W. Randall, E. Rosenberg, *J. Chem. Soc., Dalton Trans.*, **1973**, 1982.

[34] W. A. Herrmann, O. Runte, G. Artus, *J. Organomet. Chem.* **1995**, *501*, C1.

[35] M. V. Baker, P. J. Barnard, S. K. Brayshaw, J. L. Hickey, B. W. Skelton, A. H. White, *Dalton Trans.* **2005**, 37.

[36] E. S. Chernyshova, R. Goddard, K.-R. Pörschke, *Organometallics* **2007**, *26*, 3236.

[37] H. V. Huynh, Y. Han, R. Jothibasu, J. A. Yang, *Organometallics* **2009**, *28*, 5395.

[38] Magill, A. M.; Cavell, K. J.; Yates, B. *J. Am. Chem. Soc.* **2004**, *126*, 8717.

[39] Some selected steric parameters include: Winstein-Holness parameter (A-value), Interference values, Charton/Taft values and Tolman cone angles. For a comparison of these parameters and further references see: K. C. Harper, E. N. Bess, M. S. Sigman, *Nature Chem.* **2012**, *4*, 366.

[40] C. A. Tolman, *Chem. Rev.* **1977**, *77*, 313.

[41] T. E. Müller, D. T. P. Mingos, *Trans. Met. Chem.* **1995**, *20*, 533.

[42] A. C. Hillier, W. J. Sommer, B. S. Yong, J. L. Petersen, L. Cavallo, S. P. Nolan, *Organometallics* **2003**, *22*, 4322.

[43] A. Poater, B. Cosenza, A. Correa, S. Giudice, F. Ragone, V. Scarano, L. Cavallo, *Eur. J. Inorg. Chem.* **2009**, 1759.

[44] (a) O. Kühl, *Chem. Soc. Rev.* **2007**, *36*, 592; (b) A. T. Normand, K. J. Cavell, *Eur. J. Inorg. Chem.* **2008**, 2781; (c) O. Kühl, *Functionalised N-Heterocyclic Carbene Complexes*, John Wiley & Sons, New York, **2010**.

[45] W. A. Herrmann, C. Köcher, L. J. Gooßen, G. R. J. Artus, *Chem. Eur. J.* **1996**, *2*, 1627.

[46] D. S. McGuinness, K. J. Cavell, *Organometallics* **2000**, *19*, 741.

[47] A. A. D. Tulloch, Andreas A. Danopoulos, R. P. Tooze, S. M. Cafferkey, S. Kleinhenz, M. B. Hursthouse, *Chem. Commun.* **2000**, 1247.

[48] K. S. Coleman, S. Dastgir, G. Barnett, M. J. P. Alvite, A. R. Cowley, M. L. H. Green, *J. Organomet. Chem.* **2005**, *690*, 5591.

[49] S. A. Mungur, S. T. Liddle, C. Wilson, M. J. Sarsfield, P. L. Arnold, *Chem. Commun.* **2004**, 2738.

[50] H. V. Huynh, J. Wu, *J. Organomet. Chem.* **2009**, *694*, 323.

[51] P. L. Arnold, M. Rodden, K. M. Davis, A. C. Scarisbrick, A. J. Blake, C. Wilson, *Chem. Commun.* 2004, **1612**.

[52] C. Yang, H. M. Lee, S. P. Nolan, *Org. Lett.* **2001**, *3*, 1511.

[53] (a) M. Bierenstiel, E. D. Cross, *Coord. Chem. Rev.* **2011**, *255*, 574; (b) D. Yuan, H. V. Huynh, *Molecules* **2012**, *17*, 2491.

[54] (a) H. Seo, H. Park, B. Y. Kim, J. H. Lee, S. U. Son, Y. K. Chung, *Organometallics* **2003**, *22*, 618; (b) H. V. Huynh, C. H. Yeo, G. K. Tan, *Chem. Commun.* **2006**, 3833; (c) A. Ros, D. Monge, M. Alcarazo, E. Alvarez, J. M. Lassaletta, R. Fernandez, *Organometallics* **2006**, *25*, 6039.

[55] D. Sellmann, W. Prechtel, F. Knoch, M. Moll, *Organometallics* **1992**, *11*, 2346. (b) J. A. Cabeza, I. del Rio, M. G. Sanchez-Vega, M. Suarez, *Organometallics* **2006**, *25*, 1831. (c) D. Yuan, H. V. Huynh, *Organometallics* **2010**, *29*, 6020.

[56] M. Poyatos, J. A. Mata, E. Peris, *Chem. Rev.* **2009**, *109*, 3677.

[57] (a) W.A. Hermann, M. Elison, J. Fischer, C. Köcher, G. R. J. Artus, *Angew. Chem.* **1995**, *107*, 2602; (b) R. E. Douthwaite, D. Haüssinger, M. L. H. Green, P. J. Silcock, P. T. Gomes, A. M. Martins, A. A. Danopoulos, *Organometallics* **1999**, *18*, 4584; (c) F. E. Hahn, M. Foth, *J. Organomet. Chem.* **1999**, *585*, 241.

[58] H. V. Huynh, R. Jothibasu, *J. Organomet. Chem.* **2011**, *696*, 3369.

[59] F. E. Hahn, T. von Fehren, T. Lügger, *Inorg. Chim. Acta* **2005**, *358*, 4137.

[60] H. V. Huynh, R. Jothibasu, *Eur. J. Inorg. Chem.* **2009**, **1926**.

[61] M. Albrecht, R. H. Crabtree, J. Mata, E. Peris, *Chem. Commun.* **2002**, 32.

[62] L. Mercs, A. Neels, H. Stoeckli-Evans, M. Albrecht, *Dalton Trans.* **2009**, 7168.

[63] G. T. S. Andavan, E. B. Bauer, C. S. Letko, T. K. Hollis, F. S. Tham, *J. Organomet. Chem.* **2005**, *690*, 5938.

[64] (a) S. Gründemann, M. Albrecht, J. A. Loch, J. W. Faller, R. H. Crabtree, *Organometallics* **2001**, *20*, 5485; (b) F. E. Hahn, M. C. Jahnke, V. Gomez-Benitez, D. Morales-Morales, T. Pape, *Organometallics* **2005**, *24*, 6458; (c) H. V. Huynh, C.-S. Lee, *Dalton Trans.* **2013**, *42*, 6803.

[65] F. E. Hahn, M. C. Jahnke, T. Pape, *Organometallics* **2007**, *26*, 150.

[66] (a) J. C. C. Chen, I. J. B. Lin, *J. Chem. Soc. Dalton Trans.* **2000**, 839; (b) E. Peris, J. A. Loch, J. Mata, R. H. Crabtree, *Chem. Commun.* **2001**, 201; (c) T. Tu, J. Malineni, K. H. Dötz, *Adv. Synth. Catal.* **2008**, *350*, 1791; (d) X. W. Li, F. Chen, W. F. Xu, Y. Z. Li, X. T. Chen, Z. L. Xue, *Inorg. Chem. Commun.* **2011**, *14*, 1673.

[67] R. E. Douthwaite, J. Houghton, B. M. Kariuki, *Chem. Commun.* **2004**, 698.

[68] D. J. Nielsen, K. J. Cavell, B. W. Skelton, A. H. White, *Organometallics* **2006**, *25*, 4850.

[69] (a) H. V. Huynh, D. Yuan, Y. Han, *Dalton Trans.* **2009**, 7262; (b) D. Yuan, H. Tang, L. Xiao, H. V. Huynh, *Dalton Trans.* **2011**, *40*, 8788.

[70] H. M. Lee, J. Y. Zeng, C.-H. Hu, M.-T. Lee, *Inorg. Chem.* **2004**, *43*, 6822.

[71] L. P. Spencer, S. Winston, M. D. Fryzuk, *Organometallics* **2004**, *23*, 3372.

[72] V. J. Catalano, M. A. Malwitz, A. O. Etogo, *Inorg. Chem.* **2004**, *43*, 5714.

[73] H. Aihara, T. Matsuo, H. Kawaguchi, *Chem. Commun.* **2003**, 2204.

[74] H. V. Rasika Dias, W. Jin, *Tetrahedron Lett.* **1994**, *35*, 1365.

[75] (a) U. Kernbach, M. Ramm, P. Luger, W. P. Fehlhammer, *Angew. Chem. Int. Ed. Engl.* **1996**, *35*, 310; (b) R. Fränkel, C. Birg, U. Kernbach, T. Habereder, H. Nöth, W. P. Fehlhammer, *Angew. Chem. Int. Ed.* **2001**, *40*, 1907.

[76] X. Hu, I. Castro-Rodriguez, K. Meyer, *Organometallics* **2003**, *22*, 3016; (b) X. Hu, I. Castro-Rodriguez, K. Meyer, *J. Am. Chem. Soc.* **2003**, *125*, 12237.

[77] M. V. Baker, B. W. Skelton, A. H. White, C. C. Williams, *Organometallics* **2002**, *21*, 2674.

[78] F. E. Hahn, V. Langenhahn, T. Lügger, T. Pape, D. Le Van, *Angew. Chem. Int. Ed.* **2005**, *44*, 3759.

[79] (a) N. J. Findlay, S. R. Park, F. Schoenebeck, E. Cahard, S. Zhou, L. E. A. Berlouis, M. D. Spicer, T. Tuttle, J. A. Murphy, *J. Am. Chem. Soc.* **2010**, *132*, 15462; (b) H. M. Bass, S. A. Cramer, A. S. McCullough, K. J. Bernstein, C. R. Murdock, D. M. Jenkins, *Organometallics* **2013**, *32*, 2160.

3 Synthetic Aspects

3.1 General Routes to Azolium Salts as NHC Precursors

Most N-heterocyclic carbenes are generated from azolium salts as NHC precursors. There are numerous ways by which azolium salts can be prepared [1], and their variations seem endless. The detailed descriptions of all these are outside the scope of this textbook. Therefore, only the most common routes to ligand precursors of the classical NHCs belonging to types **I–IV** will be summarized in the following sections. Variations in the parent backbone of these will also not be considered. It must also be noted that the reaction conditions described here are very general and should be used as general guidelines. As with most chemical compounds, optimization of reaction conditions is often required for the best reaction outcome of each individual azolium salt.

3.1.1 N-Alkylation of Neutral Azoles

One of the easiest and most straightforward ways to prepare azolium salts is the stepwise N-alkylation of the parent azoles, which are often low cost starting materials. The parent azoles for the four classical NHCs of types **I–IV** of concern in this textbook are depicted in Table 3.1 along with their atom numbering scheme and nomenclature. In order to obtain 1,3-dialkylazolium compounds, alkylation at both nitrogen atoms of the heterocycle is required. Scheme 3.1 depicts a general pathway, and the important steps of the dialkylation process. The first N-alkylation often requires the addition of a base (e.g. NaOH, K_2CO_3), which deprotonates the N1–H proton giving an azolate as a stronger nucleophile. This step is usually carried out in an aprotic, polar solvent (e.g. acetonitrile or tetrahydrofurane) to assist in the dissolution of the base and the azole without interfering with the deprotonation process. Upon removal of the proton, the negative charge undergoes conjugation with the initial "N3=C2 double bond," and the two nitrogen atoms become equivalent because of this delocalization. Subsequent nucleophilic substitution with a suitable electrophile R–X (e.g. alkyl halide) furnishes the neutral 1-(R)-substituted azole. The leaving group X^- usually forms a salt MX with the metal cation of the base. This base-assisted reaction is usually facile and can be carried out at room temperature. By convention the newly substituted nitrogen will be atom number one of the new heterocycle. The N3 nitrogen still has one electron lone pair available for a second nucleophilic substitution with the same or a different electrophile R′–X. However, due to the reduced N-nucleophilicity of the neutral heterocycle compared to the azolate, heating is usually required to affect quaternization in the second step. A solvent of choice should be a less polar one with high boiling point (e.g. toluene), which can still dissolve the neutral 1-alkylazole as well as the electrophile, but not the

The Organometallic Chemistry of N-heterocyclic Carbenes, First Edition. Han Vinh Huynh.
© 2017 John Wiley & Sons Ltd. Published 2017 by John Wiley & Sons Ltd.

Table 3.1 Classical N-heterocyclic carbenes and their parent azoles.

Type	NHC	Parent azole	Nomenclature of the azole
I			2-imidazoline or 4,5-dihydro-1*H*-imidazole
II			1*H*-imidazole
III			1*H*-benzimidazole
IV			1*H*-1,2,4-triazole

desired ionic final product. Upon reaction, a dialkylazolium cation is obtained, with its positive charge balanced by the negative charge of the leaving group X$^-$ of the electrophile. In an ideal case, this organic salt precipitates from the reaction, and isolation and purification is greatly simplified.

The dialkylation with the same electrophile can be conveniently carried out in a one-pot reaction. An excess of R–X (>2 equivalents) is usually heated with the parent azole and a suitable base in an aprotic and polar solvent. The resulting organic 1,3-dialkylazolium salt can often be separated from the inorganic byproducts by filtration due to its relatively better solubility in polar organic solvents, and its isolation can be achieved by removal of all volatiles from the filtrate.

When the reaction is carried out in two steps with different electrophiles R–X and R′–X unsymmetrically 1,3-disubstituted azolium salts are obtained, and the priority of the R or R′ group determines the correct numbering scheme used for the nomenclature. In the given example (Scheme 3.1), R has been given the higher priority over R′, and therefore an 1-(R),3-(R′)-azolium salt is obtained. Selected examples of azolium salts and their naming are depicted in Chart 3.1. The alternative and also common "N,N′-nomenclature" is given in brackets.

For unsymmetrical azoles, such as triazoles, the numbering of atoms is already determined by the nature of the heterocycle. In such cases, deprotonation leads to an anion, in which the nitrogen atoms are not equivalent, and unselective alkylation leads to a mixture of isomers. For example, the base-assisted first alkylation of 1*H*-1,2,4-triazole will therefore yield isomeric mixtures of 1-(R)-1,2,4-triazole and 4-(R)-1,2,4-triazole (Scheme 3.2). Among these two products, the 1-alkyl isomer is statistically expected to be the major isomer, since attack at both adjacent nitrogen atoms N1 and N2 will lead to the same

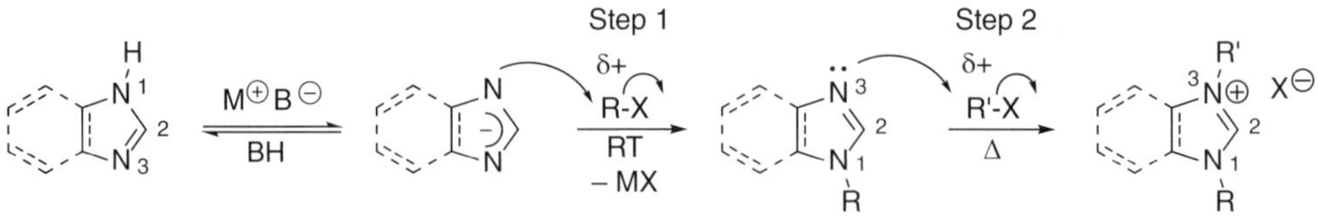
Step 1
Step 2
M⊕ B⊖
BH
δ+
R-X
RT
– MX
R'-X
Δ
X⊖

Scheme 3.1
Synthesis of dialkylazolium salts by stepwise N-alkylation of azoles.

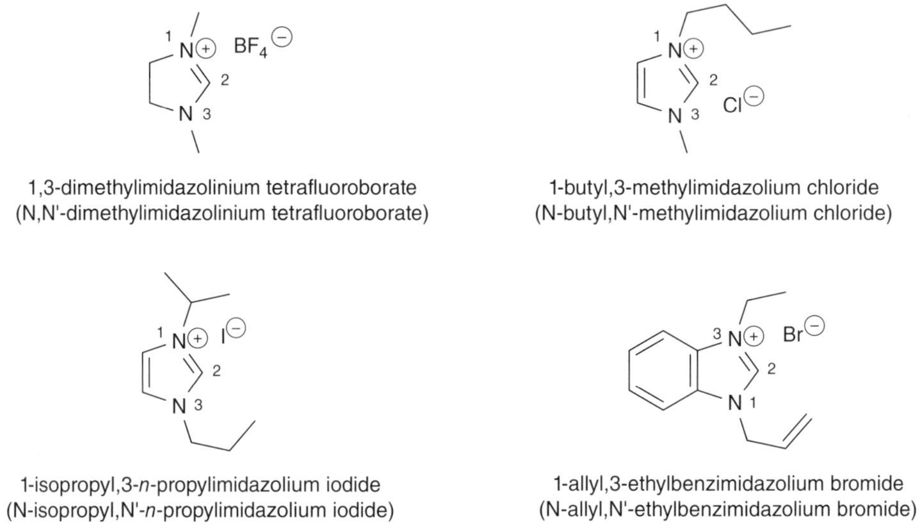
1,3-dimethylimidazolinium tetrafluoroborate
(N,N'-dimethylimidazolinium tetrafluoroborate)
1-butyl,3-methylimidazolium chloride
(N-butyl,N'-methylimidazolium chloride)
1-isopropyl,3-n-propylimidazolium iodide
(N-isopropyl,N'-n-propylimidazolium iodide)
1-allyl,3-ethylbenzimidazolium bromide
(N-allyl,N'-ethylbenzimidazolium bromide)

Chart 3.1
Selected azolium salts and their naming.

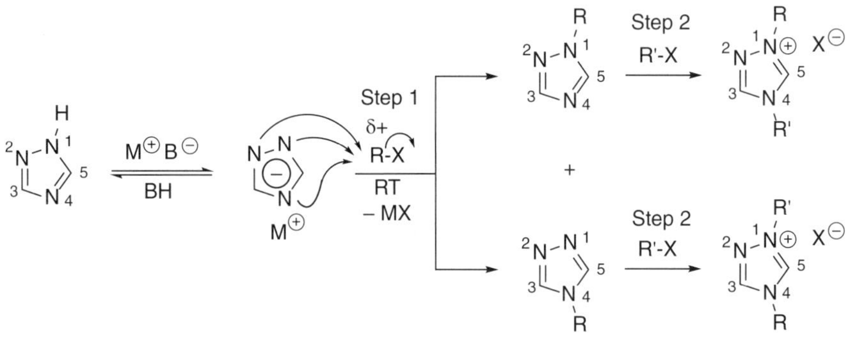
M⊕ B⊖
BH
Step 1
δ+
R-X
RT
– MX
Step 2
R'-X
X⊖

Scheme 3.2
Isomeric mixtures in the N-alkylation of 1*H*-1,2,4-triazole.

product, as opposed to the single attack of N4 leading to the 4-isomer. The isomeric mixture is not of much concern, if the same electrophile is used for the second alkylation step. However, their separation at this stage is crucial if another electrophile is to be used for the second alkylation step.

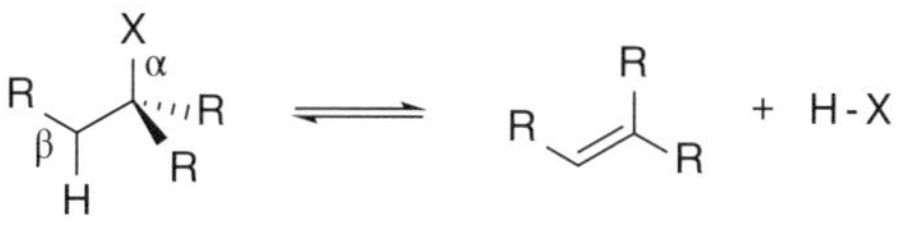

Scheme 3.3
Elimination side reactions occurring with secondary/tertiary alkyl halides.

Primary alkyl halides R–X (X=Cl, Br, I) are the most commonly used electrophiles for N-alkylation in NHC chemistry giving N-alkyl heterocycles in good yield. Their reactivity increases in the order Cl<Br<I in line with the leaving group ability of the anions. Stronger electrophiles such as alkyl tosylates, mesylates, triflates, dialkyl sulfates and trialkyloxonium (Meerwein) salts are also often used for N-alkylation in more demanding cases. However, inert conditions are required for their usage in order to prevent hydrolysis.

Reactions involving the use of secondary and tertiary alkyl halides are less straightforward due to the stepwise increased tendency of these electrophiles to undergo competing base-promoted elimination reactions if a β-hydrogen is present (3°>2°>1° alkyl halides). The elimination reaction leads to the formation of an olefin and an acid HX, which is subsequently consumed by the reaction with any base present in the mixture driving the reaction to the product side, thus preventing N-alkylation of the azole (Scheme 3.3). In this respect, it is important to note that the azoles are nitrogen bases in their own right, which can promote such elimination reactions.

In addition, the increased steric bulk that comes with secondary and tertiary alkyl halides needs to be factored in. It is easily comprehensible that this will decrease the N-alkylation reaction rate. In order to obtain reasonable yields of azolium salts with secondary/tertiary alkyl substituents a prolonged reaction time, a weaker base, and a relatively less polar solvent are often a necessity. Aryl halides do not react in typical substitution reactions known for alkyl halides, and arylation of azoles require metal catalysis (see Section 3.1.3, Scheme 3.18).

In most cases, the formation of an azolium salt is accompanied by a pronounced and characteristic downfield shift of the C2–H proton in the ^{1}H NMR spectrum (δ>9 ppm) in comparison to the equivalent signal in the neutral azole precursor, which indicates an increased acidity of this C2–H proton in the cationic and aromatic heterocycle. The results of this deshielding effect accompanying the second alkylation is shown in Figure 3.1, which compares the ^{1}H NMR spectrum of the cationic 1,3-diisopropylbenzimidazolium bromide with that of its neutral precursor of 1-isopropylbenzimidazole. Upon formation of the salt, the C2–H proton experiences a significant downfield shift from 7.98 ppm initially to 11.24 ppm. In comparison, changes of the all other protons are less noteworthy. In most cases, this proton is also the most acidic site in the azolium salt, and carbene generation occurs straightforwardly by its removal by a basic reagent. Therefore, the disappearance of this characteristic downfield signal is usually diagnostic for NHC or NHC complex formation.

3.1.2 Multicomponent Condensation Reactions

Unsaturated imidazolin-2-ylidenes are by far the most popular among all NHCs, and imidazolium salts are their common source. In addition to N,N′-dialkylation of imidazoles, imidazolium salts can also be obtained by a multicomponent reaction involving a 1,2-dicarbonyl compound (e.g. glyoxal or butane-2,3-dione), primary amines and paraformaldehyde, followed by an acidic workup in two steps overall (Scheme 3.4) [2][3]. This methodology is particularly useful, since it easily allows the introduction of aromatic wing tip groups,

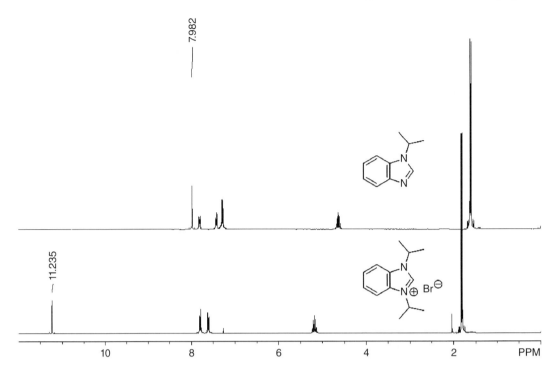

Figure 3.1
Stacked plot of ¹H NMR spectra of 1-isopropylbenzimidazole (top) and 1,3-diisopropylbenzimida-
zolium bromide (bottom) recorded in CDCl$_3$ at room temperature.

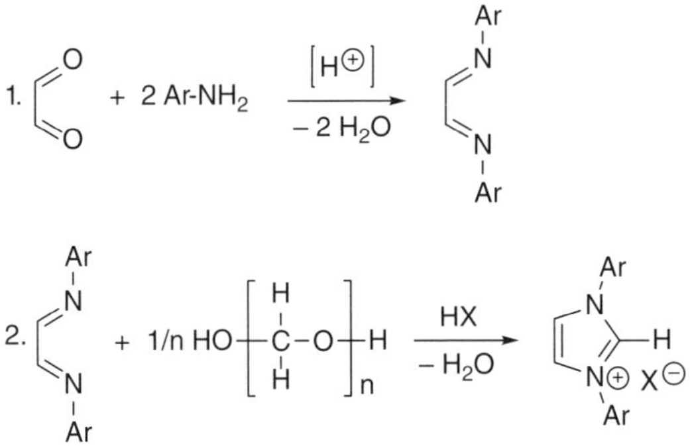

Scheme 3.4
Two-step synthetic sequence for the preparation of 1,3-diarylimidazolium salts.

which is difficult to achieve via nucleophilic substitution of azoles. Most commonly,
glyoxal is condensed with two equivalents of anilines to yield 1,2-diimines (diazabutadienes)
under liberation of two molecules of water in the first step. This reaction is routinely carried
out in alcoholic solvents, where the diimines usually precipitate, and the rate can be further
accelerated by addition of formic or acetic acid. Since the imine formation is generally reversible,
attempts to prepare unsymmetrical diimines only result in complicated mixtures, therefore
limiting the versatility of this approach. Nevertheless, the resulting symmetrical diimines

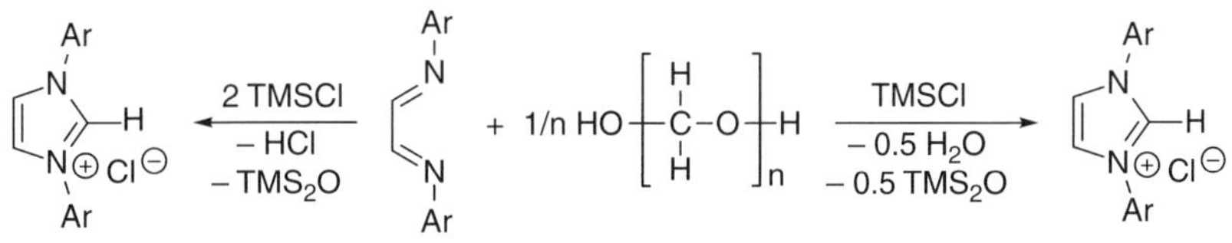

Scheme 3.5
Use of TMSCl as a cost saving alternative chloride source.

are already very useful chelating nitrogen donors in their own right, but further condensation with formaldehyde as a source for the C_1 precarbenic unit under acidic conditions finally affords the 1,3-diarylimidazolium salt as a precipitate in a second step. Instead of formaldehyde solutions, the polymeric and solid form paraformaldehyde is conveniently used, which liberates the formaldehyde *in situ* and is easier to handle. Although the second step formally eliminates water as well, it was found that additional water in the mixture is detrimental to the reaction outcome, which is in accordance with Le Chatelier's principle. Therefore, the use of relatively dry solvents is a necessity for better yields.

The choice of the acid HX is important since it determines the counter-anion X^- in the resulting imidazolium salt. In most cases, a 4 M HCl solution in dry dioxane is used, and imidazolium chlorides are obtained as the reaction products. These can subsequently be subjected to straightforward anion metathesis if other counter-anions are desired.

Instead of the relatively expensive 4 M HCl solution in dioxane, the much cheaper chlorotrimethylsilane (TMSCl) can be used as an alternative source for the chloride anion (Scheme 3.5) [4]. By controlling the stoichiometry of the TMSCl, the acidity of the mixture and thus the amounts of undesired decomposition or byproducts can be controlled. When more than one equivalent of TMSCl is added, the mixture will become HCl acidic, while the addition of only one equivalent assures neutral reaction conditions and reduces undesired side reactions. In addition, the formation of strong silicon–oxygen bonds in the TMS$_2$O byproduct provides an additional driving force, and the imidazolium salts can be obtained in better yields.

Notably, the precursors to the most popular unsaturated NHCs with aromatic wing tip groups IMes and IPr are prepared via one of these routes using glyoxal and 2,4,6-trimethylaniline or 2,6-diisopropylaniline, respectively. A proposed mechanism for the imidazolium salt formation via 1,5-diploar 6π electrocyclization is depicted in Scheme 3.6.

The preparation of imidazolium salts bearing secondary or tertiary alkyl substituents is also possible by using the respective aliphatic amines. The precursor to the bulky 1,3-di-*tert*-butylimidazolin-2-ylidene (I^{t}Bu) ligand is prepared via this route. Here, a one-pot, two-step procedure, where the addition of reactants follows a different sequence, was found to be beneficial for a good yield. Accordingly, two equivalents of *tert*-butylamine are reacted with paraformaldehyde first, supposedly forming the respective aminal *in situ*, which upon subsequent acidification and further condensation with glyoxal affords the desired 1,3-di-*tert*-butylimidazolium salt (Scheme 3.7) [5]. In this second general approach, the precarbenic unit is installed first, followed by its capping with a suitable backbone.

A slight variation of the above procedure can also be used to prepare mono-N-substituted imidazoles, which can then be quaternized to the respective salts by using electrophiles (see Section 3.1.1) [6]. For this purpose, glyoxal is reacted with only one equivalent of primary amine to form an iminoacetaldehyde in the first step. Subsequent condensation with an ammonium source (e.g. ammonium chloride) and cyclization with formaldehyde gives the

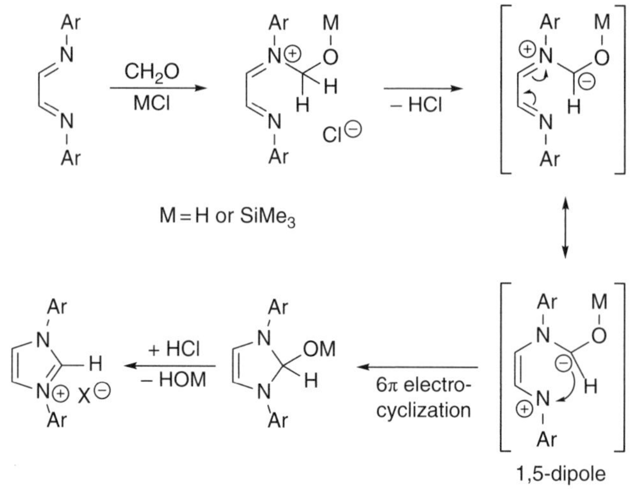

Scheme 3.6
Proposed mechanism for the condensation of diazabutadienes with formaldehyde via 1,5-diploar 6π electrocyclization.

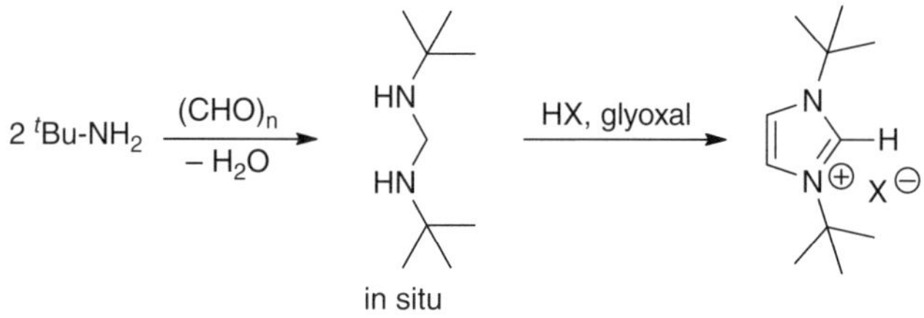

Scheme 3.7
Preparation of a 1,3-di-*tert*-butylimidazolium salt.

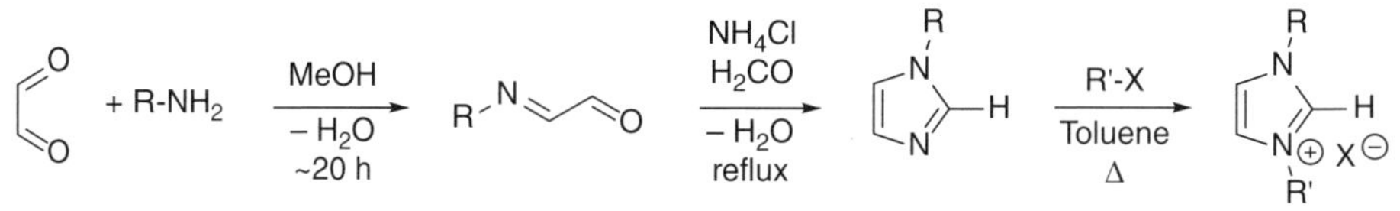

Scheme 3.8
Preparation of 1-aryl- or 1-alkylimidazoles and subsequent quaternization.

mono-N-substituted imidazole (Scheme 3.8). The latter can react with various electrophiles to give an unsymmetrically substituted imidazolium salt. Both aliphatic and aromatic amines have been employed, although a one-pot procedure can also be applied for aliphatic amines [7].

3.1.3   Cyclization of Diamines

Secondary diamines can be easily cyclized by reaction with trialkyl orthoformate, most commonly HC(OEt)$_3$ and HC(OMe)$_3$, under acidic conditions to give the azolium salts. In the case of N,N′-disubstituted ethylenediamines such an approach will lead to imidazolinium salts with a saturated C4/C5 backbone as precursors for imidazolidin-2-ylidenes, while the analogous reaction of ortho-phenylenediamines would lead to benzimidazolium salts as precursors for benzimidazolin-2-ylidenes. The making of the suitably substituted diamines is therefore a key step in this approach, and the following sections give an overview on some general preparative pathways.

Reaction of Amines with Electrophiles

The most straightforward way to prepare substituted diamines is by either (i) reacting the parent primary ethylenediamine with two equivalents of alkyl halides or (ii) reacting 1,2-dihaloethanes with two equivalents of primary amines. However, over-alkylation to tertiary amines can easily occur in both ways, if the correct stoichiometry and other reaction conditions, for example, solvent, addition of base, and so on, are not carefully controlled. To illustrate this, we can consider the general reaction of a primary amine (1° amine) with an alkyl halide R–X (electrophile). In this reaction (reaction a, Scheme 3.9), the 1° amine reacts as a nucleophile and attacks the α-carbon atom of the alkyl halide under elimination of one equivalent of the acid H–X. The secondary amine product (2° amine) is usually more basic (reaction with protons) and nucleophilic (here reaction with other carbon-electrophiles) than the 1° amine used as the starting material, and can either react as a base with the acid to the 2° ammonium salt (reaction b, Scheme 3.9) or as a competing nucleophile with excess alkyl halide to the tertiary amine (3° amine, reaction c, Scheme 3.9), which also can be protonated to the respective 3° ammonium salt. The formation of the 2° ammonium salt is preferred when no additional base is added and deactivates further N-alkylation. Poly-alkylation can also be lessened when any local excess of the alkyl halide is avoided. With this in mind and a careful choice of reaction conditions, this strategy can be exploited to prepare ethylenediammonium salts for ring-closure with trialkyl orthoformate.

Accordingly, two equivalents of amino alcohols react with one equivalent of 1,2-dibromoethane in the absence of solvent and under pressure and elevated temperatures to give the ethylenediammonium salts that can be isolated as solids from the reaction mixture [8]. The free substituted diamine can be obtained after additional treatment with NaOH solution, which can be subjected to a final ring-closure step with triethyl orthoformate and a proton source to yield chiral imidazolinium salts with alcohol functions (Scheme 3.10) [9].

In a similar way, 1,2-dibromoethane reacts with mesityl-, adamantyl- or 1-naphthylamine in a 1:2 stoichiometric ratio to the respective diammonium salt, which upon direct treatment with trimethyl orthoformate affords the respective imidazolinium salts as precursors to the carbenes SIMes, IAd, and INap (Scheme 3.11) [10].

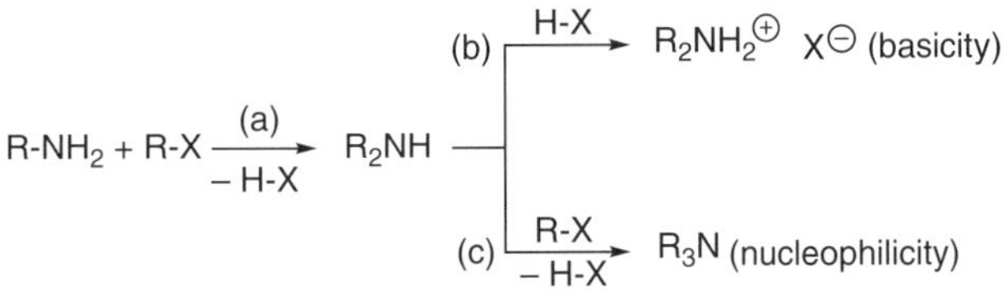

$$R\text{-}NH_2 + R\text{-}X \xrightarrow[- H\text{-}X]{(a)} R_2NH$$

(b) $\xrightarrow{H\text{-}X}$ $R_2NH_2^{\oplus}$ $X^{\ominus}$ (basicity)

(c) $\xrightarrow[- H\text{-}X]{R\text{-}X}$ R_3N (nucleophilicity)

Scheme 3.9
Alkylation of amines using an alkyl halide.

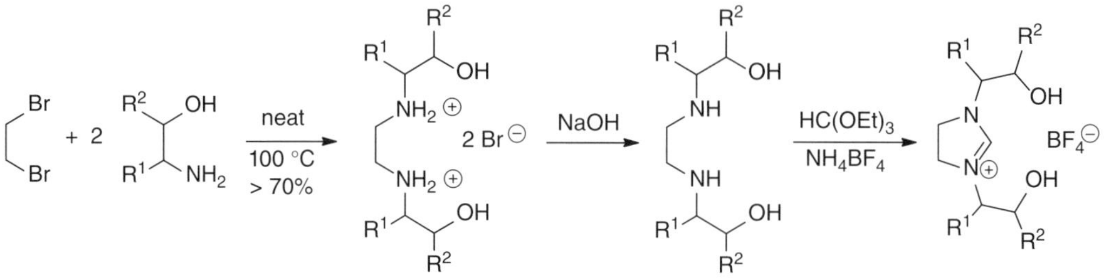

Scheme 3.10
Preparation of imidazolinium salts with chiral N-substituents from 1,2-dibromoethane and aminoalcohols.

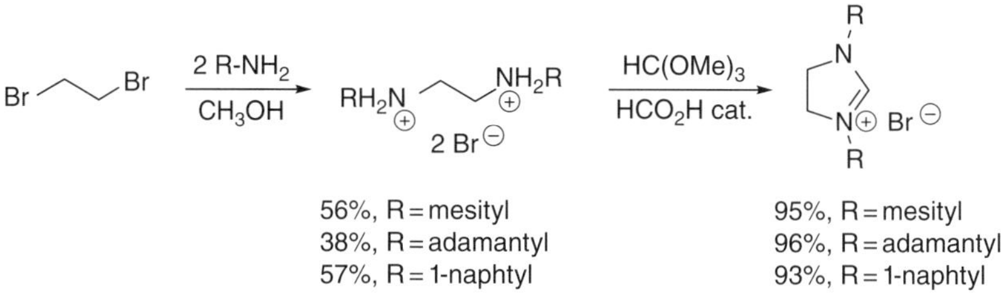

56%, R = mesityl
38%, R = adamantyl
57%, R = 1-naphtyl

95%, R = mesityl
96%, R = adamantyl
93%, R = 1-naphtyl

Scheme 3.11
Preparation of symmetrical imidazolinium salts via primary amines and 1,2-dibromoethane.

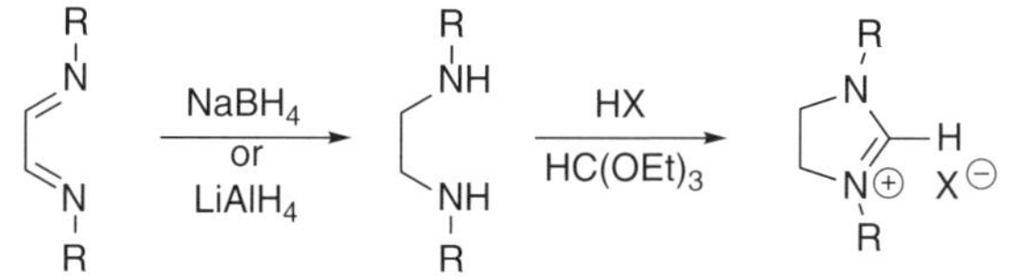

Scheme 3.12
Diamines via reduction of diazabutadienes and subsequent cyclization.

Reductive Amination

Another way to access suitable diamines for the cyclization process is by reduction of diazabutadienes, that in turn are easily prepared by condensation of two equivalents primary amine with glyoxal. For this purpose, the primary diimine condensation products have to be reduced to the respective diamines (Scheme 3.12). Often this can be simply achieved using standard reducing agents such as $NaBH_4$ or $LiAlH_4$. In the final step, the substituted ethylenediamines undergo ring-closure with trialkyl orthoformate and a proton source to the respective carbene precursors. This pathway is particular useful for the introduction of aromatic N-substituents. Using this 3-step condensation-reduction-cyclization sequence, Arduengo and coworkers prepared the nowadays very popular SIMes and SIPr N-heterocyclic carbene ligands, and this route is still the most popular one for the synthesis of imidazolinium salts [11].

Alternatively, suitable diamines can also be prepared via reductive amination of ethylenediamine as a low cost starting material. The reaction of ethylenediamine with aldehydes or ketones affords the respective ethylenediimines, which upon analogous reduction leads to

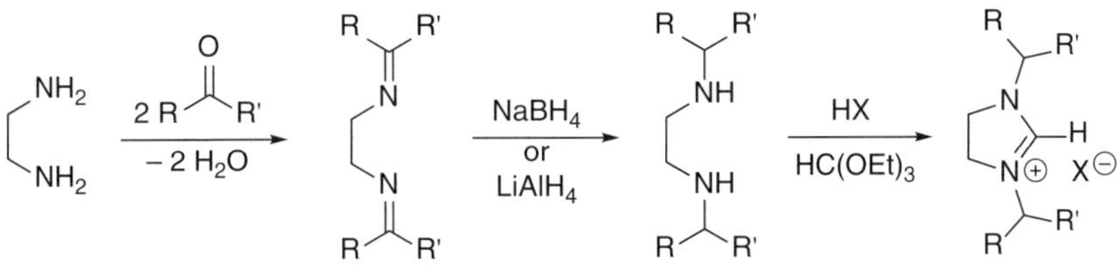

Scheme 3.13
Reductive amination of ethylendiamine and subsequent ring-closure.

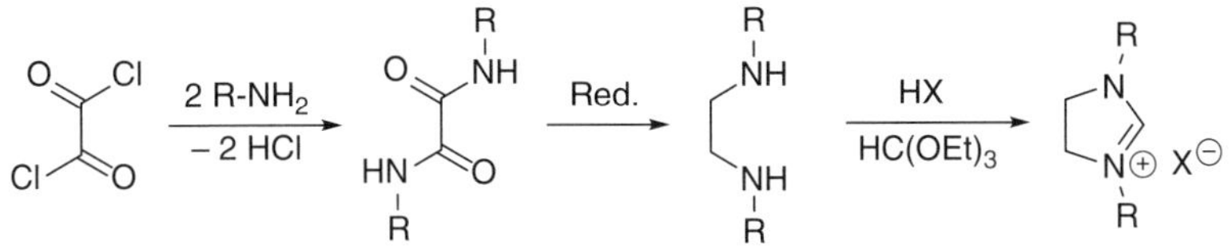

Scheme 3.14
General pathway to imidazolinium salts from oxalyl chloride.

disubstituted ethylenediamines for subsequent ring-closure reactions (Scheme 3.13) [12]. In contrast to the "diazabutatiene" pathway described above, this approach only allows for the installation of aliphatic wing tip groups.

Employing aldehydes in the first condensation step will lead to salts with primary N-alkyl substituents, while ketones give rise to secondary wing tip groups with secondary α-carbon atoms adjacent to the nitrogen atom. Through the use of unsymmetrical ketones (R = R′), chiral centers can be installed at these α-positions. However, no selectivity can be achieved by using standard reducing agents, and complicated mixtures will result, which would have to be resolved by other means. Furthermore, unsymmetrical salts cannot be obtained due to the reversibility of the condensation process.

Amide Reduction

A third general method to prepare diamines as precursors to imidazolinium salts involves the reduction of diamides. In analogy to the reductive amination process described in the previous section, a dicarbonyl compound or a primary diamine can be used as starting material. In general, the reaction of inexpensive oxalyl chloride with two equivalents of primary amines and a sacrificial base, for example NEt_3, yields the N,N′-dialkyl- or N,N′-diaryloxalamide in a straightforward manner. The purpose of the sacrificial base is to absorb the hydrochloric acid byproduct, which could otherwise deactivate the primary amine leading to lower yields. Particularly for expensive amine substrates, this should be factored in. On the other hand, an excess of a low cost amine can be conveniently used, which serves both as reactant and sacrificial base. Most diamides are air and moisture stable, which facilitates purification and storage. Diamide reduction can then be accomplished with strongly reducing agents like $LiAlH_4$ or BH_3, and cyclization is commonly achieved via routine treatment with trialkyl orthoformate (Scheme 3.14).

In contrast to the C=N double bond in imines, the C–N bonds in amides are stable and not easily hydrolyzed. This offers the opportunity for the preparation of unsymmetrical carbene precursors by stepwise introduction of two different primary amines. The preparation

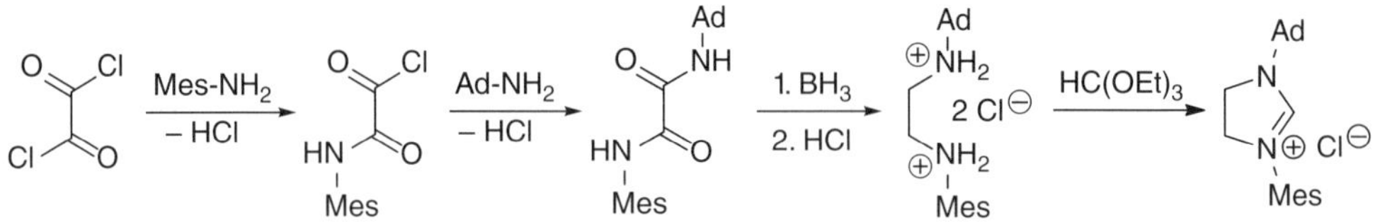

Scheme 3.15
Synthesis of 1-adamantyl-3-mesitylimidazolinium chloride.

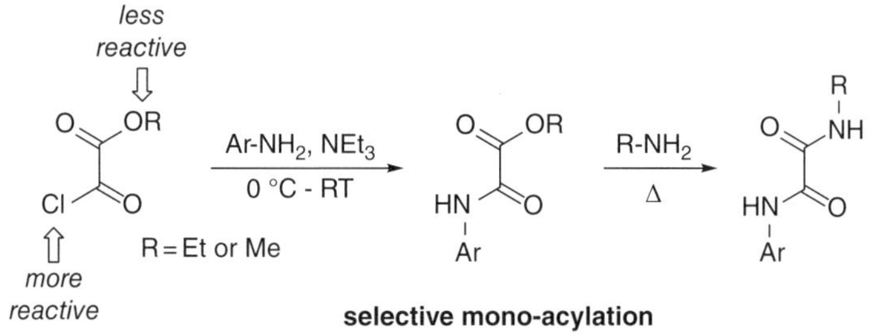

Scheme 3.16
Alkyl chlorooxoacetate as a useful starting material for mono-acylated intermediates.

of 1-adamantyl-3-mesitylimidazolinium chloride as a representative example is shown in Scheme 3.15 [13]. In the first step, oxalyl chloride is reacted with only one equivalent of mesityl amine, which yields the monoamide. A local excess of mesityl amine, which would lead to formation of a symmetrical diamide can be prevented by slow addition of the amine to the dichloride. The desired mono-acylation product is then transformed into the unsymmetrical diamide by addition of the adamantyl amine, which is then easily reduced, acidified, and cyclized by routine methods.

In order to avoid diacylation in the key step, the stoichiometry and the addition of the first amine has to be carefully controlled. Therefore, this methodology has been further improved by the use of ethyl or methyl chlorooxoacetate as the dicarbonyl component. These compounds contain two distinct different reactive sites offering a greater selectivity for the mono-acylated intermediates (Scheme 3.16) [14].

Here, the carboxylic acid chloride moiety is much more reactive than the ester-moiety and will selectively react with the primary amine to the mixed mono-amide/ester derivative, which can be isolated and purified (Scheme 3.16). Reaction at the ester-moiety to give the unsymmetrical diamide for further reduction requires harsher conditions and can be achieved by reaction with the second amine under elevated temperatures. Using this improved methodology, the yields of unsymmetrically substituted imidazolinium salts can be increased.

By "reversing" the carbonyl and amine components, the origin of the backbone and N-substituents can be reversed. Any primary 1,2-diamine could be used as the source for the backbone instead of oxalyl chloride, and the N-substituents of the salt are introduced via the acid chlorides. However, this approach does not allow for the introduction of aryl or secondary and tertiary alkyl groups as N-substituents. On the other hand, salts with a greater diversity in their heterocyclic backbones can be prepared, which is a distinct advantage over the "oxalyl chloride route." The use of 1,2-phenylenediamine

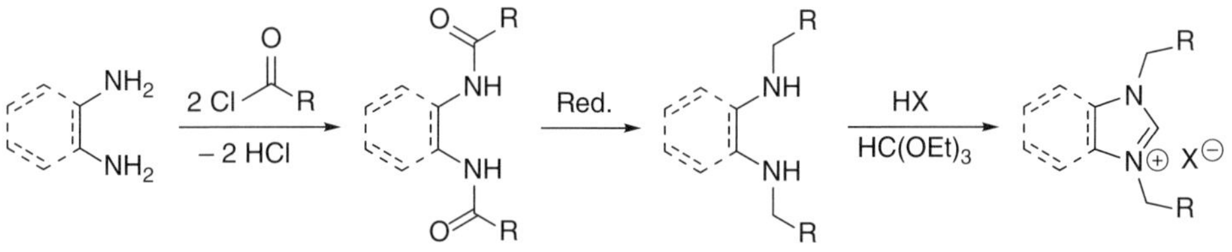

Scheme 3.17
General pathway to imidazolinium and benzimidazolium salts via amidation of 1,2-diamines.

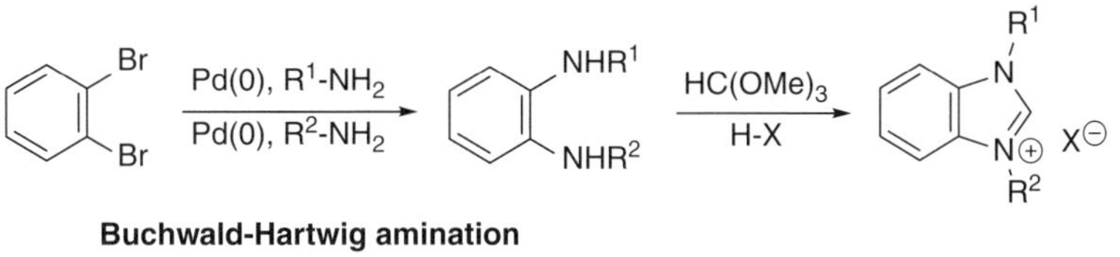

Buchwald-Hartwig amination

Scheme 3.18
Benzimidazolium salts via Buchwald-Hartwig amination and cyclization.

for example allows for the preparation of benzimidazolium salts via a 3-step sequence involving (i) reaction with two equivalents of acid chlorides to the 1,2-phenylenediamides; (ii) their reduction to the substituted 1,2-di-R-phenylenediamines; and finally (iii) the cyclization of the latter under acidic conditions to the respective 1,3-di-R-azolium salt (Scheme 3.17).

Metal Catalyzed Aminations

While the simple reaction of a 1° amine with an alkyl halide as the electrophile easily furnishes the respective 2° amine, non-activated aryl halides do not react to the analogous aromatic compounds. The efficient preparation of 2° (or 3°) aromatic amines from a 1° (or 2°) amine and an aryl halide usually requires the help of transition metal catalysts and is nowadays broadly known as the Buchwald-Hartwig amination [15]. By employing this reaction, substituted 2° ortho-phenylenediamines can be easily prepared as precursors to benzimidazolium salts. In a typical case, C–N coupling is achieved by heating 1,2-dibromobenzene, a primary amine and a strong base (e.g. NaO^{t}Bu) in toluene with a catalytic amount of [Pd$_2$dba$_3$] as the palladium(0) source and BINAP as the supporting diphosphine ligand (Scheme 3.18) [16].

The resulting symmetrical (R^1=R^2) or unsymmetrical (R^1≠R^2) 2° ortho-phenylenediamines can undergo straightforward cyclization to the benzimidazolium salts under routine conditions, when R-substituents are aliphatic. For aromatic R groups, cyclization requires much harsher conditions due to the low nucleophilicity at the amine-nitrogen atoms. Even so, some aromatic groups like mesityl do not cyclized, and benzannulated analogs of the popular IMes or IPr carbenes are not accessible via this route [17].

Eventually, 1,3-dimesitylbenzimidazolium tetrafluoroborate could be prepared in a low yield by a rather unconventional route via the intermediacy of dihydrophenazine formed upon oxidation of N,N′-dimesityl-1,2-benzenediamine with sodium periodate on wet silica gel, followed by electrocyclic ring-closure (Scheme 3.19) [18].

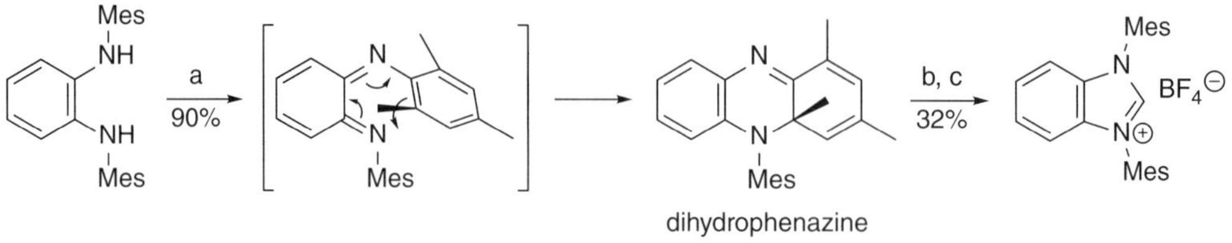

Scheme 3.19
Synthesis of 1,3-dimesitylbenzimidazolium tetrafluoroborate.

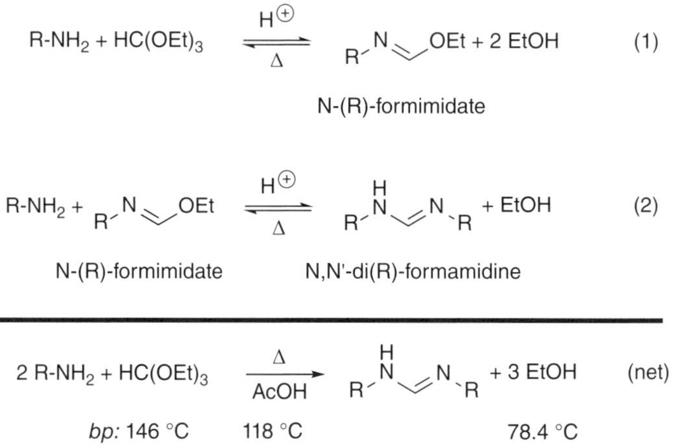

Scheme 3.20
Preparation of symmetrically disubstituted formamidines.

3.1.4 Cyclization of Formamidines

As detailed in the previous subsection, trialkyl orthoformates are often used as a cyclization agent in conjunction with an acid in the final step of the preparation of imidazolinium or benzimidazolium salts. In a different approach, they can also be directly used to react with two equivalents of primary amine to yield formamidines. The latter are also useful precursors to azolium salts as they can be easily cyclized by reaction with a bis(electrophile), such as an dihalide.

In general, the condensation reaction of two equivalents of primary amines with one molecule of triethyl orthoformate occurs readily in the presence of acetic acid as the catalyst and liberates three molecules of ethanol. This overall reaction can be split into two individual steps that are in dynamic equilibrium (Scheme 3.20). In the first step (1), the amine reacts acid-catalyzed with the orthoester to form the formimidate, which can be isolated when the orthoformate is used in excess with respect to the amine. In the presence of additional amine, the formimidate is rapidly converted to the stable formamidine following step (2) [19]. The preparation of symmetrical aromatic formamidines can be straightforwardly achieved using only catalytic amounts of acid in a one-pot procedure. Alkyl-substituted formamidines, on the other hand, are more difficult to prepare, and a stoichiometric amount of acid has been reported to be essential for good yields. Apparently, a full equivalent of acid would lead to protonation of the aliphatic formamidines to the formamidinium acetates, thus driving the reaction to the right [20].

Furthermore, removal of the alcohol by distillation shifts the equilibrium of the reaction to the product side, and the yield can be increased according to Le Chatelier's principle. The difference in boiling points between ethanol and the other components in this mixture is an additional advantage.

The synthesis of unsymmetrically substituted formamidines is also possible. The introduction of the first amine is easy and follows step (1) as described above. The formimidate is isolated and then further reacted with the second amine to yield the unsymmetrical formamidine [21]. However, it is crucial that step (2) is carried out in the absence of any acid to prevent reversibility, which would otherwise result in scrambling of the two N-substituents and formation of undesired symmetrical byproducts (Scheme 3.21).

Cyclization of the substituted formamidines can be achieved by reaction with an bis(electrophile), which typically provides the backbone of the carbene precursor. Therefore, this methodology can potentially lead to a large variety of cyclic formamidinium salts extending NHC chemistry beyond the classical four types **I–IV** of our focus.

To obtain the salt precursor to the popular saturated SIMes ligand, N,N'-dimesitylformamidine is simply subjected to prolonged heating in neat dichloroethane (DCE) according to method A in Scheme 3.22. The 1,3-dimesitylimidazolium chloride is obtained in 49% yield based on the formamidine. The reason for the moderate yield is that half of the formamidine is required to react as a sacrificial base to form the simple formamidinium chloride as the byproduct. Nevertheless, the latter can be easily re-isolated and recycled for subsequent reactions. In the presence of diisopropylethylamine $\{NEt(^iPr)_2\}$ as an additional base, the reaction proceeds smoothly giving near-quantitative yields of the imidazolinium chloride (method B, Scheme 3.22) [22].

By replacing the dihalo alkane with α-halo ketones, precursors to unsaturated NHCs with a wide range of 4,5-substitution patterns can be prepared by an alkylation-condensation pathway [23]. This one-pot, two-step synthetic route commences with a base-assisted "clipping" of the α-halo ketone onto the formamidine to give the hydroxyimidazolinium chloride.

61%

Ar = 2,4,6-trimethylphenyl; 92%
Ar = 2,6-diisopropylphenyl; 91%

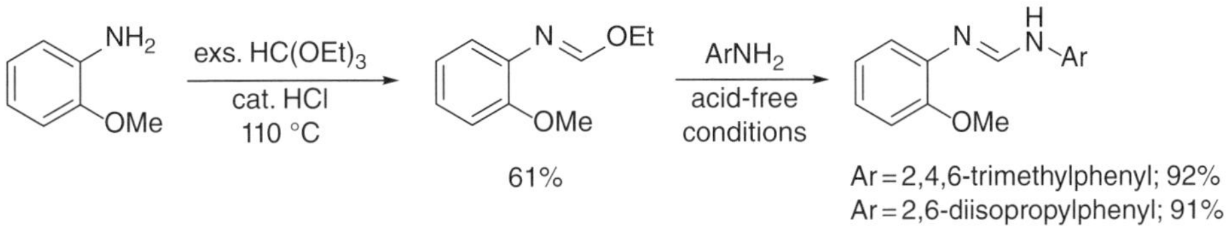

Scheme 3.21
Two-step preparation of unsymmetrical aromatic formamidines.

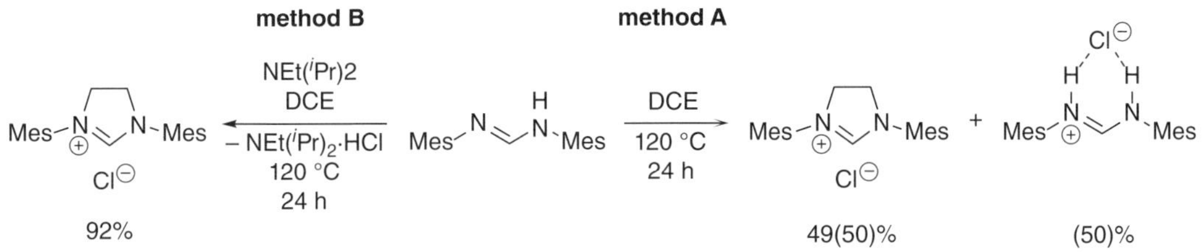

method B

method A

92%

49(50)%

(50)%

Scheme 3.22
Preparation of 1,3-dimesitylimidazolinium chloride by bis(electrophilic) attack of the respective dimesitylformamidine.

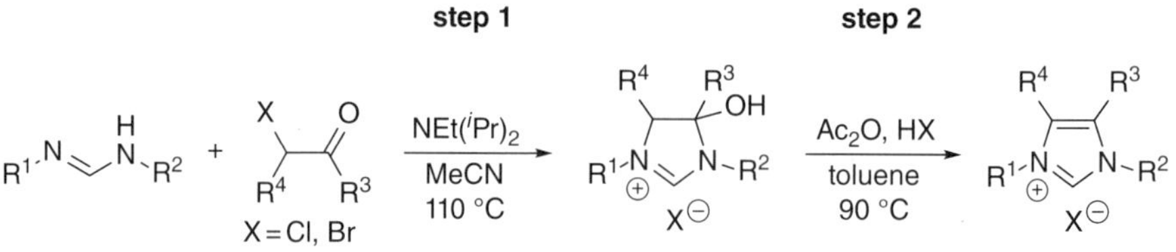

Scheme 3.23
Preparation of 4,5-substituted imidazolium salts from formamidines and α-halo ketones.

In accordance to the reaction involving dihalo alkanes, diisopropylethylamine was found to be the most suitable base for this reaction (Scheme 3.23).

In the second step, the hydroxyimidazolinium chloride is dehydrated to the desired imidazolium chloride with the aid of acetic anhydride (Ac$_2$O) and a mineral acid (H–X). The use of the anhydride provides driving force for the dehydration by liberating acetic acid upon reaction with the eliminated water molecule making the reaction irreversible.

3.2   General Routes to Free NHCs

Compared to other carbenes, N-heterocyclic carbenes have an enhanced stabilization through the "*push-pull effect*" (see Chapter 2.1) from the adjacent nitrogen atoms. Nevertheless, most of them are still air sensitive species, and their preparation and handling usually requires an inert nitrogen or argon atmosphere. Furthermore, dry and aprotic solvents are a necessity in order to prevent protonation of the very basic NHCs, which is the most common decomposition pathway of NHCs. The reaction of a free NHC (**A**) with any potential proton source HB is a typical reversible Brønsted acid-base reaction, which gives rise to an organic salt (**B**), where the conjugate base B$^-$ ends up as the counter-anion balancing the charge of the azolium cation. Therefore, the acidity of HB (or inversely the basicity of B$^-$) is a determining factor for the likeliness of such reactions. In addition, direct insertion of the carbene into the B–H bond can occur, which yields a neutral azoline or azolidine (**C**) with a saturated sp^3 hybridized C2 carbon atom (Scheme 3.24) [24]. Both species **B** and **C** can undergo further reactions. For example, benzimidazolium or imidazolinium salts (**B**) can catalyze the dimerization of the respective free NHCs to entetramines (see Chapter 2.3).

The nucleophilicity and basicity of B$^-$ is also important, depending on which further reactions of the resulting azolium cation are feasible. In the case of water (B=OH), the stability of the azolium hydroxide is affected by the amount of water molecules present [25]. Excess water molecules reduce the basicity of the hydroxide anion by hydrogen bonding and thus indirectly stabilize the azolium cation. When only a trace of water is present, the hydroxide anion lacks hydrogen bonding making it a much stronger base, and reversible deprotonation can occur, which returns the free NHC. Subsequently, the direct insertion of this NHC into an O–H bond of a water molecule gives rise to a tertiary alcohol of type **C**. The latter usually collapses into thermodynamically favorable ring-opened formamides (Scheme 3.25).

This hydrolysis reaction is relatively slow for unsaturated imidazolin-2-ylidenes, which has been attributed to their potential 6π-aromaticity. Saturated imidazolidin-2-ylidenes lack this stabilization and hydrolyze rapidly to the respective aminoethylformamides (Scheme 3.26). The reactivity of benzimidazolin-2-ylidenes with respect to hydrolysis is

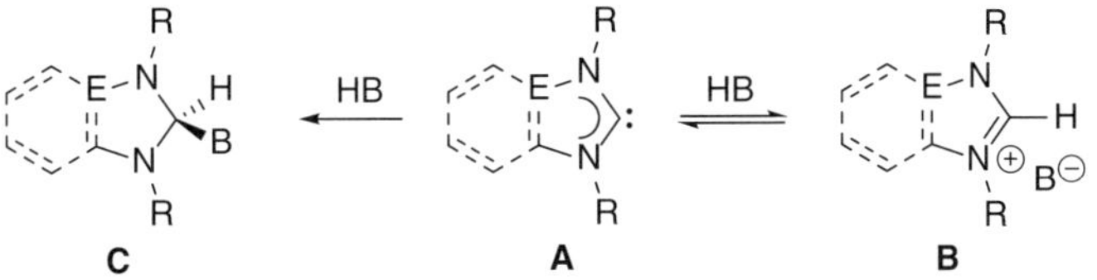

Scheme 3.24
Major decomposition pathways of free NHCs.

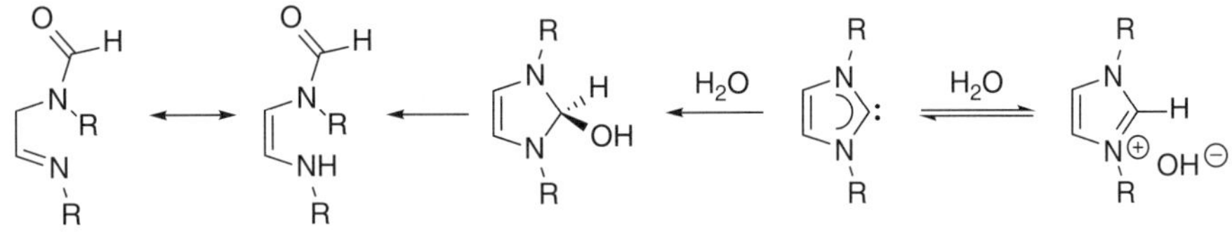

Scheme 3.25
Hydrolysis of imidazolin-2-ylidenes.

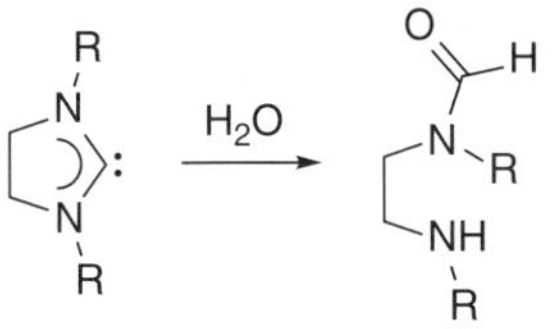

Scheme 3.26
Hydrolysis of imidazolidin-2-ylidenes.

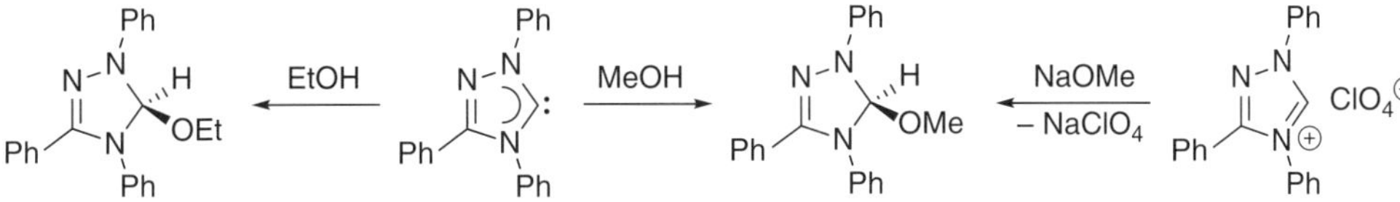

Scheme 3.27
Formation of 5-alkoxy-1,2,4-triazolines by insertion of a 1,2,4-triazolin-5-ylidene into alcohols or addition of an alkoxide to a triazolium salt.

expected to be intermediate to that of their unsaturated and saturated congeners, while 1,2,4-triazolin-5-ylidenes should hydrolyze similarly to imidazolin-2-ylidenes (see Chapter 2.2).

Free NHCs can also directly insert into an O–H bond of alcohols (B = OR) or an N–H bond of amines (B = HNR or NR$_2$) [26]. For example, 1,3,4-triphenyl-1,2,4-triazolin-5-ylidene reacts with methanol or ethanol under formation of the respective 5-alkoxytriazolines (Scheme 3.27). Notably, this insertion into methanol was found to be reversible, and the free carbene can be generated by heating the triazoline under reduced pressure to 80 °C [26]. A two-step reaction pathway to triazolines is also feasible that involves protonation of the NHC followed by nucleophilic addition of an alkoxide to the resulting azolium cation.

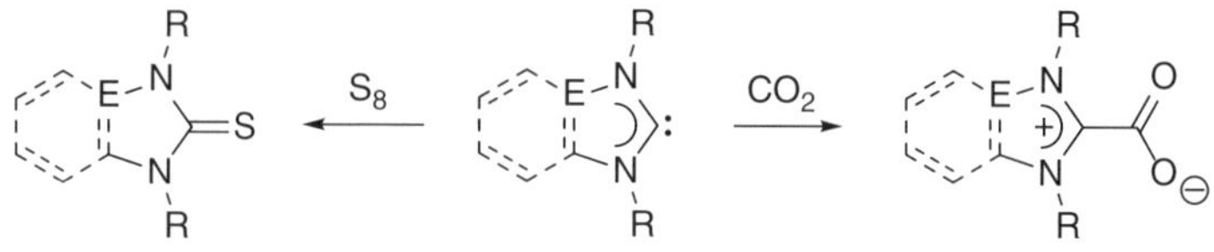

Scheme 3.28
Reaction of NHCs with carbon dioxide or sulfur under formation of zwitterionic adducts or thiones.

While NHCs react rapidly with proton sources, they usually do not get oxidized by aerial oxygen. Unsaturated imidazolin-2-ylidenes and saturated imidazolidin-2-ylidenes have been reported to be inert to oxygen and other oxidants such as CuO, Cu_2O, or HgO even after prolonged heating [24]. This observation is surprising, since NHCs or their dimers are known to undergo oxidation by elemental sulfur (S_8) to form thiones (Scheme 3.28) [26,27,28]. In contrast, 1,3,4-triphenyl-1,2,4-triazolin-5-ylidene was observed to react with >30% oxygen to give the respective urea derivative [26].

Free NHCs can also act as Lewis basees and attack electrophiles such as CO_2 to form, for example azolium carboxylates as zwitterionic adducts (Scheme 3.28) [29]. In some cases, imidazolium-2-carboxylates have been used as a convenient NHC source for further reactions, although there is no direct evidence for the occurrence of free NHCs in these reactions [30]. Indeed, the electrophilic attack of a free nucleophilic NHC on a carbonyl group of aldehyde or ketone derivatives forms the basis and initial step of most organocatalytic applications involving NHCs [31]. Since the concentration of CO_2 in the air is relatively low, this pathway is insignificant for the decomposition of NHCs in air. It can thus be concluded that the air sensitivity of NHCs is neither due to aerial oxygen, nor to aerial CO_2, but primarily due to the humidity in air.

3.2.1 NHCs via Deprotonation of Azolium Salts

The most common method to generate free NHCs is by deprotonation of their azolium salt precursors in aprotic solvents using a strong base. The negative inductive *"pull effect"* exerted by the two adjacent nitrogen atoms makes the C2-H proton in the common azolium precursors (C5-H for 1,2,4-triazolium salts) the most acidic and reactive site of the heterocycle. Electronically, deprotonation therefore occurs favorably at this position, which gives rise to *normal* NHCs. The driving force for this typical Brønsted acid-base reaction is the formation of the respective weaker conjugate acid-base pair. Since free NHCs are themselves strong bases, even stronger bases are required for this reaction. Despite their general low solubility in organic solvents, the use of inorganic bases is particularly useful, since it leads to the energetically favorable formation of ionic solids as byproducts. The lattice energy and the precipitation of the salt byproduct from the reaction mixture drive the reaction to the product side. The use of metal hydride bases can even make the reaction irreversible by the liberation of volatile hydrogen gas, which is removed from the reaction system. A selection of suitable bases and solvents that are commonly used is given in Table 3.2.

Arduengo and coworkers prepared the first free NHC by using a combination of NaH and catalytic amounts of DMSO in THF [32]. Here, the actual base for the deprotonation of the imidazolium salt was proposed to be the dimsyl anion, which is obtained *in situ* by the action of NaH on DMSO (Scheme 3.29). The addition of catalytic amounts of DMSO also helps to in the dissolution of the otherwise insoluble NaH. The hydrogen gas formed

Table 3.2 Selected bases and solvents used for the deprotonation of azolium salts.

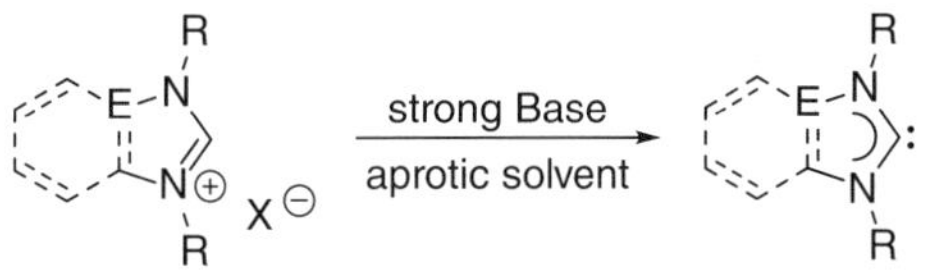

Bases	Solvents
NaH or KH	benzene
NaO^{t}Bu or KO^{t}Bu	toluene
M[N(SiMe$_3$)$_2$] (M = Li, Na, or K)	Tetrahydrofurane (THF)
BuLi	diethyl ether
Li[N(^{i}Pr)$_2$]	Hydrocarbons
	Dimethylsulfoxide (DMSO)

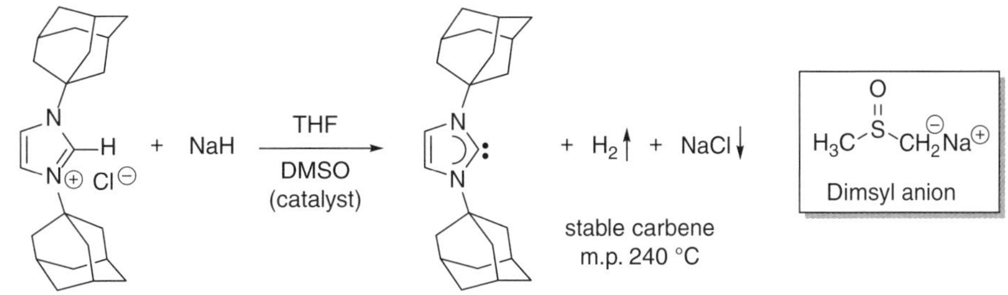

Scheme 3.29
Preparation of the stable 1,3-di-1-adamentylimidazolin-2-ylidene (IAd) as the first representative of free NHCs.

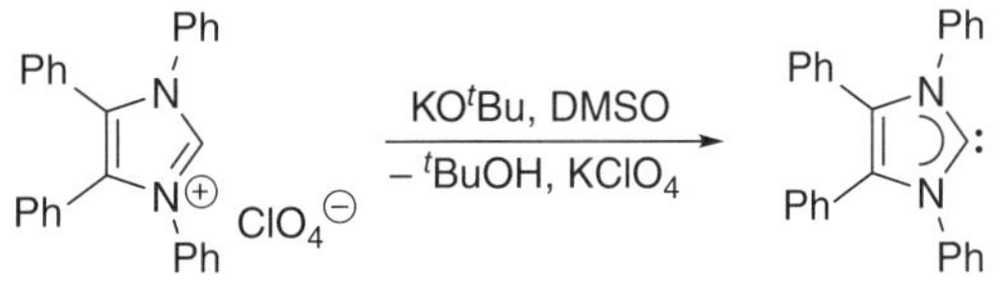

Scheme 3.30
Deprotonation of tetraphenylimidazolium salt to give a carbene.

is automatically removed from the reaction mixture, and the insoluble NaCl byproduct can be easily filtered off. Subsequent removal of the volatile solvent from the filtrate under vacuum afforded the free stable NHC.

Even earlier, Wanzlick and coworkers employed KO^{t}Bu as a strong and bulky base to deprotonate a tetraphenylimidazolium salt in dry DMSO (Scheme 3.30) giving the NHC, *tert*-butanol, and potassium perchlorate [28]. The authors did not attempt to isolate the free NHC, but used it *in situ* for subsequent reactions. The use of the polar and aprotic solvent DMSO facilitates dissolution of the NHC precursor salt for a homogenous reaction without interfering with the deprotonation process.

However, DMSO has also a very high boiling point of 189 °C, and vacuum distillation is usually required for solvent removal and isolation of the free NHC. Nowadays, KO^{t}Bu

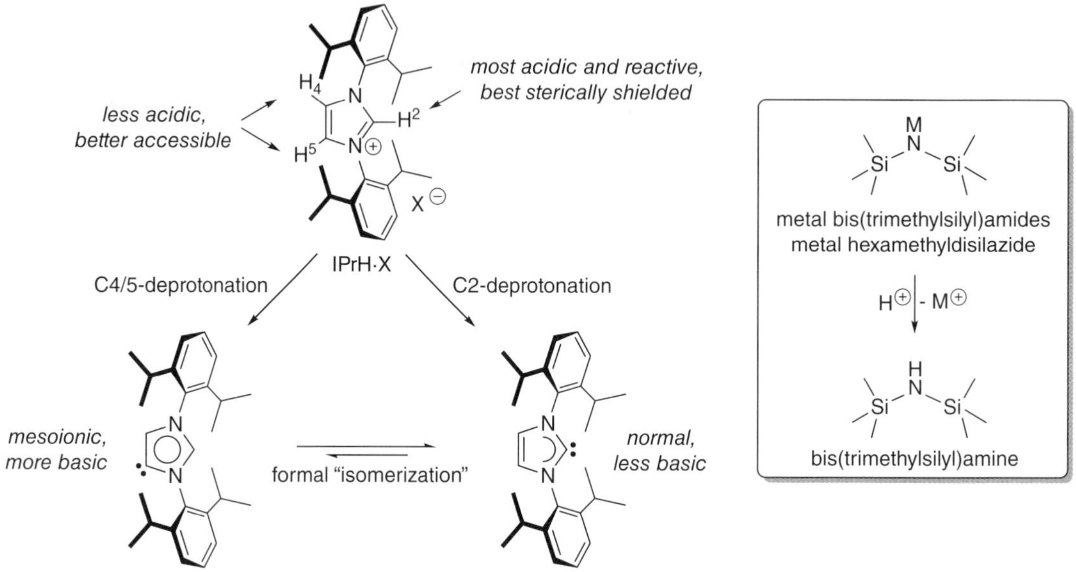

Scheme 3.31
Competing reactive sites for deprotonation in IPrH·X and Lewis structure of M[N(SiMe$_3$)$_2$] or MHMDS.

in THF is a more preferably used combination for the generation of free NHCs [33]. In addition to having a lower boiling point of 66 °C, THF is also comparatively less polar, which facilitates the precipitation of salt byproducts as an additional driving force for the reaction.

In more difficult cases, sterically demanding bases such as metal bis(trimethylsilyl) amides (also known as metal hexamethyldisilazides, i.e. M[N(SiMe$_3$)$_2$] or MHMDS; M = Li, Na, or K) have often been successfully used to make NHCs. The two trimethylsilyl groups also impart better solubility of these bases in organic solvents, which increases reaction rates. Intuitively, the use of a bulky base can lead to competitive proton abstraction of imidazolium salts at the sterically less shielded C4 (i.e. H^4) or C5 (i.e. H^5) positions leading to the formation of so-called *mesoionic* (or *abnormal*) NHCs. However, the proton at C2 (i.e. H^2) is the most acidic and significantly more reactive, and its removal yields more stable *normal* NHCs. Furthermore, *normal* NHCs are less basic compared to their *mesoionic* counterparts and are thus more favorable in formal isomerization reactions.

The reaction products resulting from deprotonation at competing reactive sites in an imidazolium salt and their relation to each other is exemplified in Scheme 3.31 using the IPrH·X salt precursor. Maybe more important is that M[N(SiMe$_3$)$_2$] reacts to form a sterically demanding secondary amine, bis(trimethylsilyl)amine. It has been noted before that free NHCs can react with alcohols or amines by insertion into O–H or N–H bonds, respectively. Such insertion reactions are understandably less likely to occur with bulky amines, such as HN(SiMe$_3$)$_2$, which may arguably be the main reason for the successful application of M[N(SiMe$_3$)$_2$] bases in the making of free NHCs. For example, quantitative NHC insertion into the N–H bond of morpholine has been reported by Enders and coworkers [26], while no example of NHC insertion into bis(trimethylsilyl)amine was observed yet (Scheme 3.32).

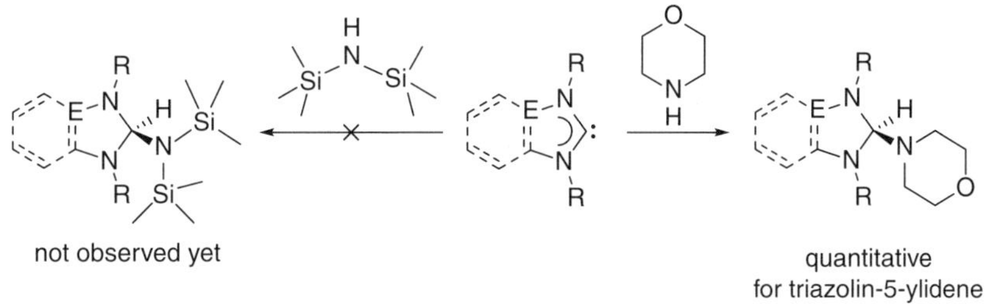

not observed yet

quantitative
for triazolin-5-ylidene

Scheme 3.32
Sterically controlled insertion of NHCs into N–H bonds of amines.

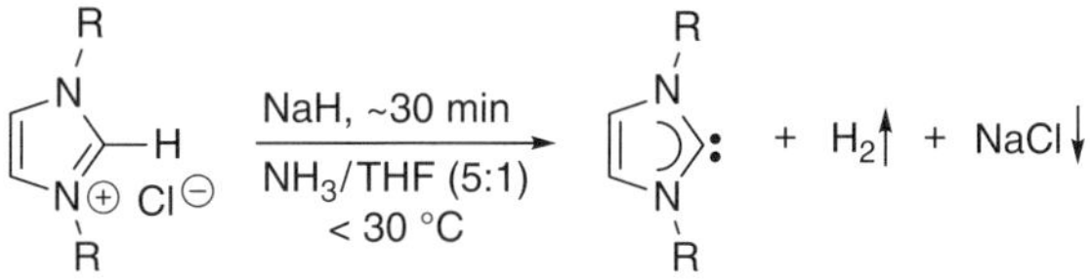

Scheme 3.33
Rapid deprotonation of imidazolium salts by NaH in NH$_3$/THF solvent mixture.

Another combination that has been successfully used to prepare NHCs from azolium salts involves the use of MH or MNH$_2$ (M = Na or K) in a liquid ammonia/THF (5:1) solvent mixture at < 30 °C [34]. The use of liquid ammonia ensures homogeneity of the reaction, since both metal hydrides/amides and azolium salts are soluble in this media. The reaction rate is therefore much faster, and completion is achieved within minutes (Scheme 3.33). Due to the mild reaction condition, selective C2 deprotonation is achieved and undesired side reactions are avoided. Using this method, chiral and functionalized NHCs could be prepared in high yields. After reaction, the liquid ammonia is simply evaporated by slowly warming to room temperature, which leaves a THF suspension containing the inorganic salt as side product and the soluble NHC. Isolation of the latter can be achieved by filtration and subsequent removal of the solvent from the filtrate.

Although this method is fast and mild, it is rarely used nowadays, probably due to the difficult handling of the pungent and hazardous liquid ammonia.

3.2.2 NHCs via Reduction of Thiones

NHCs readily react with elemental sulfur to yield the respective heterocyclic thiones. Inversely, free NHCs are also assessable by reduction of the latter using suitably strong reductants. Kuhn and Kratz first reported that reaction of tetraalkylimidazolin-2-thiones with potassium in boiling THF affords the respective tetraalkylimidazolin-2-ylidene in good yield [35]. The thiones are easily obtained by condensation of thioureas with 3-hydroxy-2-butanone in boiling 1-hexanol (Scheme 3.34). The temperature at which THF boils is sufficient to melt the potassium, which in turn is required to ensure reasonable reaction rates (~4 h). Formation of the insoluble byproduct potassium sulfide drives the reaction forward, and a solution of the NHC in THF can be easily obtained by simple filtration.

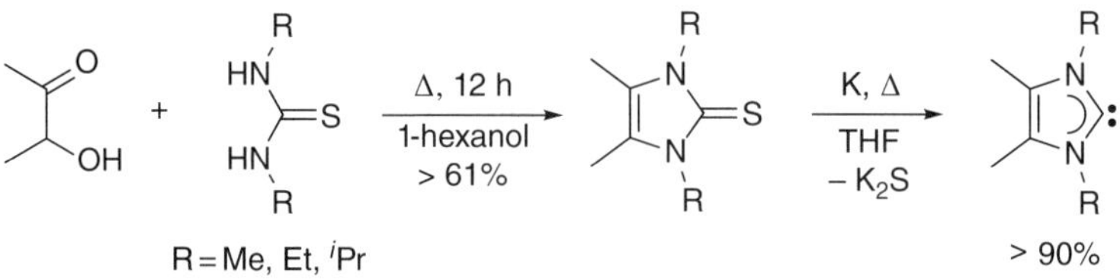

Scheme 3.34
Preparation of imidazolin-2-ylidenes by reduction of their thiones.

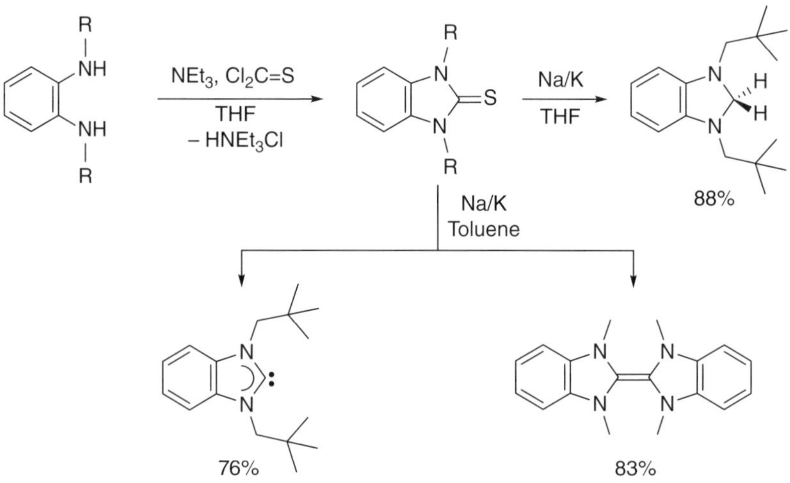

Scheme 3.35
Room temperature reduction of benzimidazolin-2-thiones using Na/K alloy.

The necessity for elevated temperatures is, however, a limiting factor, and only thermally stable NHCs can be prepared using this methodology.

Hahn and coworkers modified this procedure by employing a sodium-potassium mixture in toluene. The mixing of the two reactive alkali metals leads to a lowering of the melting point, making the alloy a liquid at room temperature, which allows for reduction of thiones at reduced temperatures. Consequently, thermally less stable NHCs can be obtained, although prolonged reaction times of ~20 days are required under these conditions. In addition, the choice of the correct solvent is crucial. For example, reduction of benzimidazolin-2-thiones with Na/K alloy in THF leads to "over-reduction" to benzimidazolines, while the use of more inert toluene successfully affords the desired free benzimidazolin-2-ylidene or its dimer depending on the N-substituent (Scheme 3.35) [36].

The major disadvantage, however, is the very long reaction time of ~20 days. Lappert and coworkers further improved the thione reduction methodology by employing potassium-graphite (C_8K) as a much more reactive reducing agent. Using C_8K in THF, the reduction of 1,3-dineopentylbenzimidazolin-2-thione is complete within one day [37], and NHC isolation involves filtration and removal of the solvent from the filtrate (Scheme 3.36).

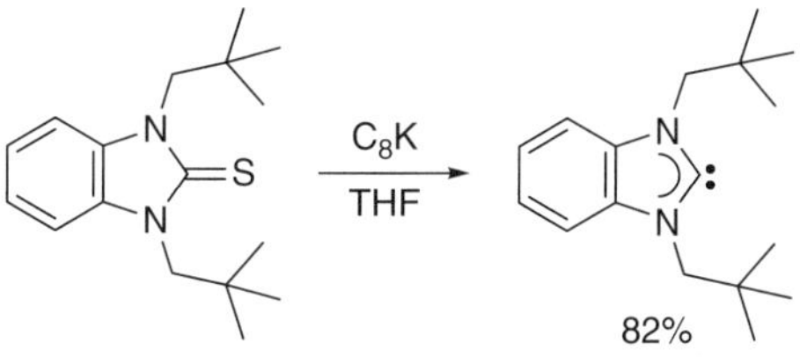

Scheme 3.36
Room temperature reduction of a benzimidazolin-2-thione using C_8K.

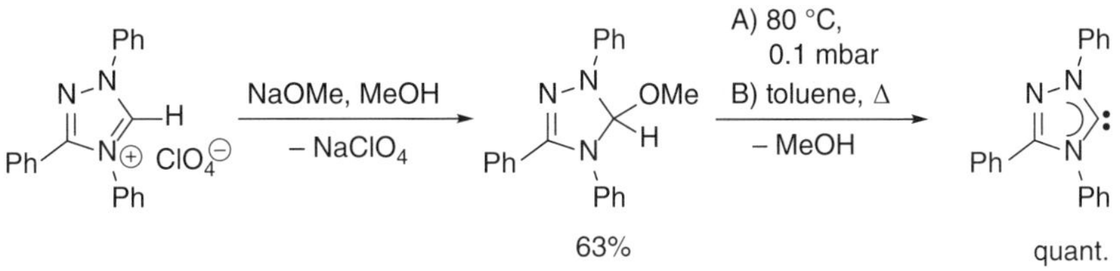

Scheme 3.37
Preparation of a triazolin-5-ylidene by MeOH elimination from methoxytriazoline.

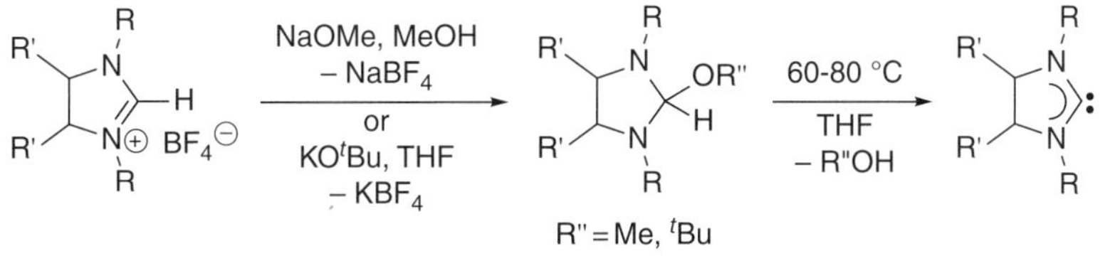

Scheme 3.38
Preparation of imidazolidin-2-ylidenes by alcohol elimination from 2-alkoxyimidazolidines.

3.2.3 NHCs via α-Elimination of Small Molecules

Attempts to prepare carbenes from α-elimination of small molecules had already been carried out at the very early stages of carbene chemistry (see Chapter 1). However, real evidence for NHC formation by this method was only reported in 1995 by Enders and Teles. Nucleophilic addition of methoxide at the C5 position of a triazolium perchlorate salt gave the respective 5-methoxytriazoline. Heating of the latter to 80 °C under reduced pressure (A) or in dry toluene under reflux (B) liberates methanol and generates the free triazolin-5-ylidene in quantitative yield (Scheme 3.37) [26]. Although this reaction is reversible, it is entropically favored as it produces two molecules from one. The activation energy is provided thermally, and removal of the volatile methanol *in vacuo* further shifts the reaction to the product side. In general, the formation of volatile and highly stable α-elimination byproducts is energetically favorable and highly advantageous for the isolation of free NHCs, since such byproducts can also be removed with ease.

The extension of this alcohol elimination method for the *in situ* preparation of saturated imidazolidin-2-ylidenes was reported by Grubbs and coworkers. The free NHCs, generated from heating of the respective NHC-alcohol adducts in THF (Scheme 3.38), were not isolated, but immediately trapped by irreversible formation of stable Ru complexes [38].

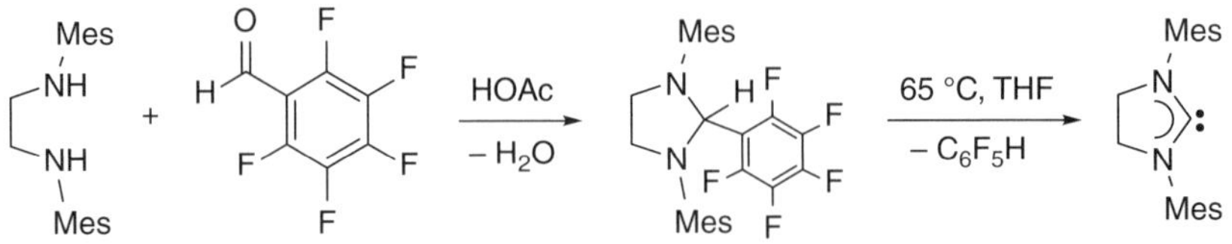

Scheme 3.39
Preparation of SIMes by α-elimination of pentafluorobenzene.

Waymouth and Hedrick modified the procedure and employed 2-aryl or -alkylimidazolidines as precursors [39]. These can be easily obtained by reactions of N,N'-diarylethylenediamines with aromatic aldehydes. In particular, condensation of dimesitylethylenediamine with pentafluorobenzaldehyde affords 1,3-dimesityl-2-pentafluorophenyl-imidazolidine, which was found to be the best and air stable precursor (Scheme 3.39). Heating of this neutral compound to only 65 °C in THF affords *in situ* 1,3-dimesitylimidazolidin-2-ylidene (SIMes) under liberation of stable pentafluorobenzene. The SIMes solution in THF can be directly used for further reactions, such as organocatalytic lactide polymerization or coordination to transition metals.

3.3   General Synthetic Routes to NHC Complexes

Most free NHCs are air-sensitive species. However, they can be stabilized by formation of complexes or adducts, where the NHC generally acts as a neutral two-electron donor. The first NHC complexes had already been knowingly and independently prepared as early as 1968 by Wanzlick and Öfele [40,41]. Nowadays, NHC complexes cover all the transition metals as well as a large number of main group elements. The discovery that many NHC complexes are useful catalysts for various organic transformations contributed much to their popularity, and was a major driving force for the development and advancement of their coordination chemistry [42]. A small selection of reactions that have been catalyzed by NHC complexes is given in Table 3.3.

In addition to applications in catalysis, many NHC coordination compounds show a broad structural diversity. Versatile structures of NHC complexes bearing monodentate NHCs, bidentate chelating NHCs or donor-functionalized NHCs, tripodal NHCs or pincer-type NHCs, etc, have been reported, and current research is also concerned with their use as ligands in metallo-based supramolecular chemistry [43,44]. Moreover, some NHC complexes possess antibiotic, antitumor, and antimicrobial properties making potential biological applications feasible [45]. The exploration of NHC complexes in material science is also just beginning to unfold.

The organometallic compounds of NHCs have often been compared to those of phosphines, since both ligands are strong electron donors. However, it is commonly accepted that metal-NHC bonds are generally stronger than metal-phosphine bonds, which is a factor that has significantly contributed to the success of NHC ligands. The bond dissociation energies (BDE) of selected NHCs and phosphines in their Ni^0 an Ru^{II} complexes are compared in Table 3.4 [46,47]. It can be seen that the BDE values for NHCs are in most cases significantly greater than those of their phosphine analogs. In addition, the BDEs of NHCs are also greater than those of CO and pyridine ligands [48].

Thus, well-defined NHC complexes are often employed as catalysts. On the other hand, a large excess of phosphine ligand is often required to counter oxidative metal-phosphine bond degradation in catalytic applications involving phosphine complexes.

Table 3.3 A selection of reactions catalyzed by NHC complexes.

Heck and Suzuki couplings	Kumada coupling
Aryl amination	Stille coupling
Amide α-Arylation	C-H activation
Hydrosilylation	Hydrogenation, hydroformylation
Olefin metathesis	Furan synthesis and alkyne coupling
Metathesis cross-coupling	Olefin cyclopropanation
Sonogashira coupling	Arylation and alkenylation of aldehydes
Ethylene/carbon monoxide copolymerization	Reduction of aryl halides
Asymmetric catalysis	Atom-transfer radical polymerization

Table 3.4 BDE values (in kcal mol^{-1}) for CO and L (NHC or phosphine) determined by calorimetric studies and calculations.

Entry	Ligand	BDE of CO in $[Ni(CO)_3L]^a$	BDE of L in $[Ni(CO)_2L]^a$	BDE of L in $[Cp*RuCl(L)]^b$
1	IMes	28.3	46.5	15.6
2	ICy	27.0	46.3	21.2
3	SIMes	26.8	47.2	16.8
4	IPr	26.7	45.4	11.1
5	SIPr	25.6	46.1	12.1
6	I^{t}Bu	13.3	44.3	–
7	IAd	7.6	46.5	6.8
8	PPh$_3$	30.4	30.0	–
9	P^{t}Bu$_3$	27.4	34.3	–
10	PCy$_3$	–	–	10.5

^{a}Calculated data from Ref [46].
^{b}Experimental data from Ref [47].

NHC complex formation involves donation of the carbene lone pair of electrons to a Lewis acid, which substantially reduces the probability for the n → π* transition (see Chapter 2.2). The latter corresponds to a mixing of the carbene lone pair σ into the formally empty p_π and is the main contributor to the paramagnetic shielding term, which is the reason for the very downfield shift of the carbene signal in free NHCs. The unavailability of the lone pair σ for such mixing reduces the paramagnetic contribution to the shielding term, and the carbene carbon becomes significantly shielded upon metal coordination. As a result, upfield shifts for the carbene signal to <190 ppm is observed in the complexes, which can be used as a sensitive probe to determine successful complexation of NHCs.

X-ray crystallographic studies revealed that the metal-NHC bond is close to an elongated single bond indicative of a comparatively limited amount of π-backdonation from the metal center to the carbene atom. For this reason, Lewis structures of metal-NHC complexes are more accurately drawn with a single bond. Nevertheless, it must be noted that the π-accepting properties of NHCs are not negligible [49], especially when complexes of low-valent metal centers are considered.

Along the way, a number of preparative methods toward NHC complexes have been established, and the most commonly used ones are summarized in Scheme 3.40. These

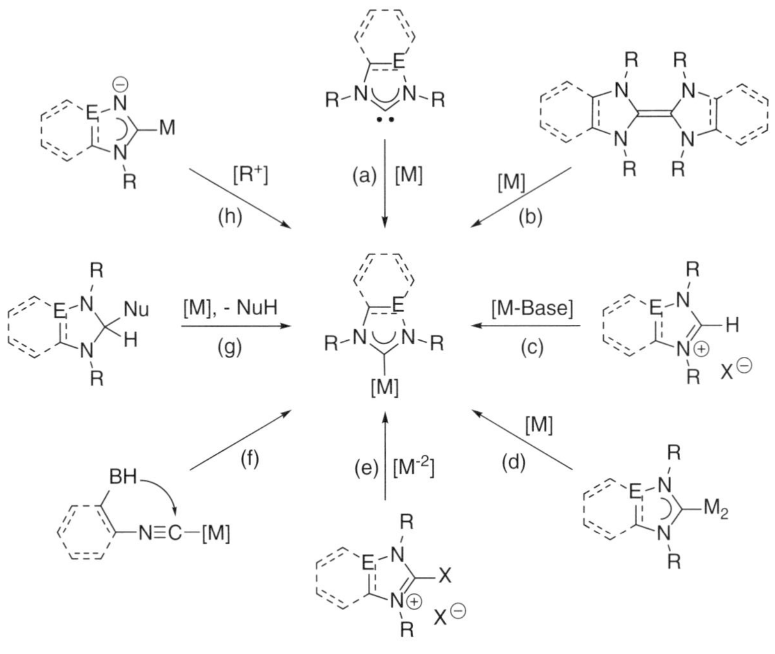

$X^{\ominus}$ = counter anion; B = NH, O, S; X = CH$_3$, H, halogen

Scheme 3.40
Major synthetic routes to NHC complexes.

include (a) coordination of free NHCs; (b) cleavage of electron-rich entetramines by transition metals; (c) *in situ* deprotonation of azolium salts in the presence of transition metals; (d) metal carbene transfer routes (Ag, Mo, Cu, etc.); (e) oxidative addition of an azolium salt to a low-valent metal complex; (f) metal-template synthesis using isocyanide complexes as precursors; (g) NHC complexes by small molecule elimination; and (h) NHC complexes by protonation/alkylation of azolyl complexes. The individual methods are briefly described in the following paragraphs.

3.3.1 Coordination of Free NHCs

The most straightforward access to NHC complexes is the reaction of the free ligand with suitable metal precursors, which usually occurs under displacement of other ligands that are already present in the coordination sphere of the metal precursor (Scheme 3.41) [50].

In order to apply this method, the NHCs to be introduced must be stable to a certain extent. Therefore, this method is less suitable for complexes of benzimidazolin-2-ylidenes or imidazolidin-2-ylidenes with smaller N-substituents, for these NHCs often dimerize rapidly. Due to the stronger metal-NHC bond, a large number of two-electron donating ligands can be displaced. These include neutral, weakly coordinating ligands and solvent molecules, such as amines, ethers, thioethers, nitriles, and sulfoxides (e.g. THF, DMSO, DMF, acetonitrile, tetrahydrofurane, dimethylsulfide, tetrahydrothiophene, and pyridine). But also anionic halido ligands (Cl, Br, I) and even strong donors, such as carbon monoxide (see Chapter 2.2) and phosphine ligands, can be displaced by NHCs.

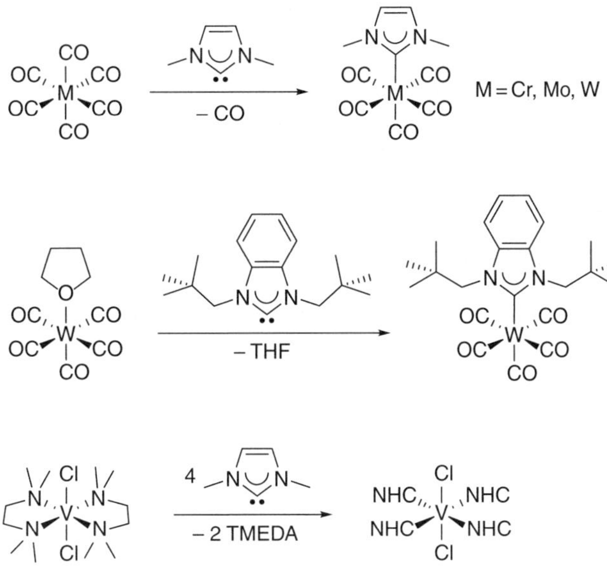

Scheme 3.41
Metal-NHC complexes by ligand displacement reactions.

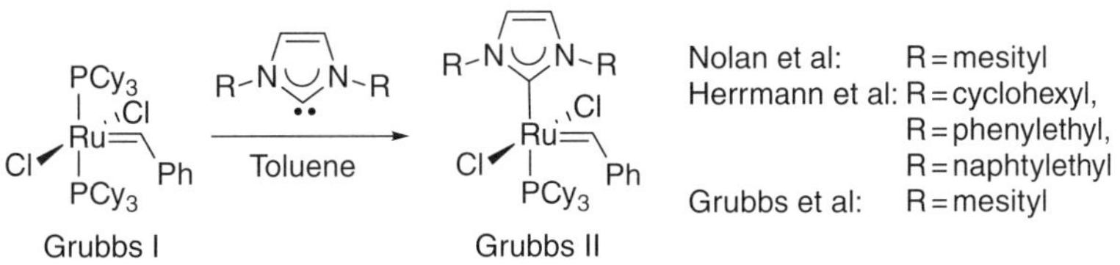

Scheme 3.42
Preparation of early versions of Grubbs II catalysts containing imidazolin-2-ylidenes.

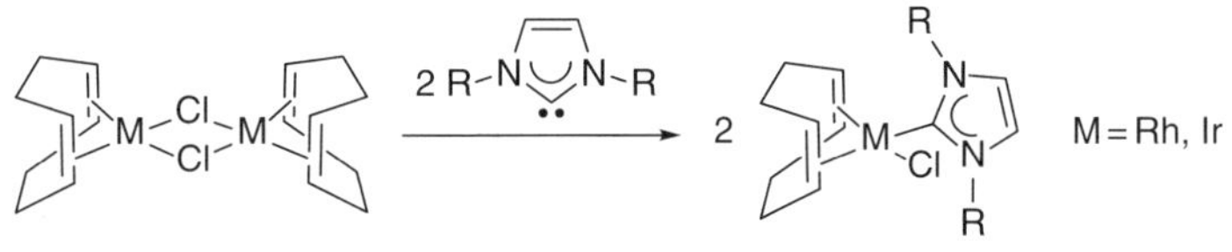

Scheme 3.43
Rh and Ir NHC complexes by bridge-cleavage reactions.

One prominent example is an early attempt to prepare NHC containing Ru-catalyst of the Grubbs type for olefin metathesis reactions. In the same year, the groups of Nolan [51], Herrmann [52] and Grubbs [53] independently published their results on the reaction of imidazolin-2-ylidenes with the first generation Grubbs catalyst *trans*-[RuCl$_2$(CHPh)(PCy$_3$)$_2$]. This reaction proceeded under substitution of one tricyclohexylphosphine ligand with an imidazolin-2-ylidene giving early versions of the second generation Grubbs catalysts *trans*-[RuCl$_2$(CHPh)(NHC)(PCy$_3$)] of enhanced stability, which contain mixed NHC/Phosphine donor sets (Scheme 3.42).

Ligand displacement is not required when NHCs react with coordinately unsaturated dimeric complexes. Here, an enthalpically favorable bridge-cleavage reaction occurs, whereby coordinatively saturated metal centers are obtained (Scheme 3.43).

3.3.2 Cleavage of Electron-Rich Entetramines by Transition Metals

A number of saturated and benzannulated NHCs rapidly dimerize to the respective entetramines, and are thus not available as free ligands. Nevertheless, Lappert and coworkers demonstrated that these electron-rich entetramines can be directly used to prepare a large number of metal-NHC complexes (Scheme 3.44) [54].In fact, NHC complexes of many transition metals were first prepared using this methodology. The best results are obtained when this olefin cleavage reaction is carried out with coordinatively unsaturated and electrophilic metal complexes such as $[PtCl_2(PEt_3)]_2$ [55]. The workup particularly for such bridge splitting reactions is straightforward as no other products are formed. The entetramines can also displace neutral (e.g. CO) or anionic (halido) ligands from the coordination sphere of the metal. For example, simple metal carbonyls can be heated with electron-rich olefins to give the respective complexes [56]. It has been proposed that the displacement reaction is initiated by a rate-determining N-coordination step of the entetramine to the metal.

This method not only broadens the scope of complexes to saturated imidazolidin-2-ylidenes, but is also a very useful tool for the access of benzimidazolin-2-ylidene complexes. For example, Hahn and coworkers demonstrated that room temperature reaction of PdI_2 with a bridged dibenzotetraazafulvalene in THF affords the di(NHC) complex via insertion of the PdI_2 moiety (Scheme 3.45) [57].

3.3.3 *In Situ* Deprotonation of Azolium Salts in the Presence of Transition Metals

By far the most commonly used method for the preparation of NHC complexes involves the *in situ* deprotonation of azolium salts by a suitable inorganic or organic base in the presence of a metal complex or metal salt. The major advantage of this method is that handling

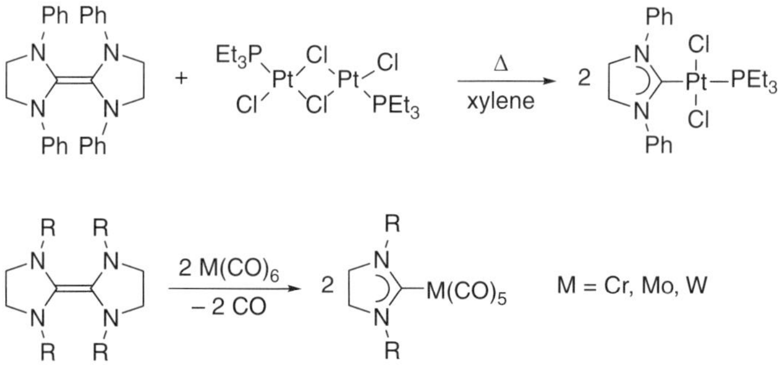

Scheme 3.44
Cleavage and ligand displacement reactions involving entetramines.

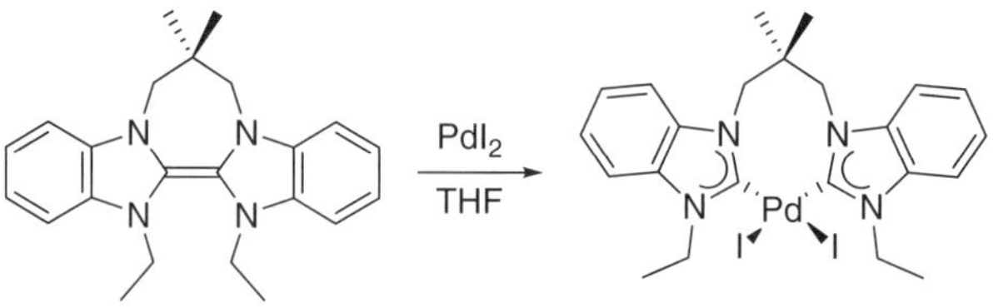

Scheme 3.45
Direct insertion of PdI_2 into a dibenzotetraazafulvalene.

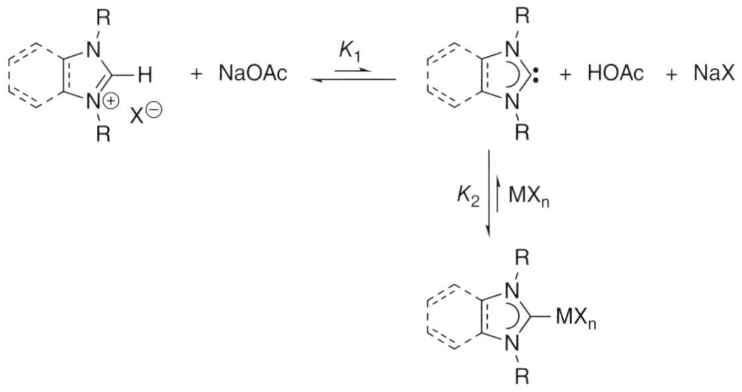

Scheme 3.46
Dynamic acid-base (K_1) and complexation (K_2) equilibria.

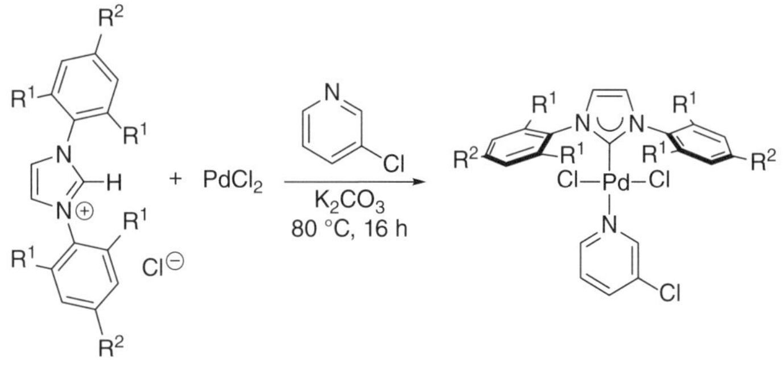

Scheme 3.47
One-pot preparation of PEPPSI catalysts.

of air/moisture sensitive free carbenes and other reactants can often be avoided. For example, the basicity requirement in this method is substantially relaxed, which is in stark contrast to the generation and isolation of free NHCs, where very strong and often air sensitive bases are a necessity. Here, simple and relatively weaker bases, such as acetates, carbonates, and amines can already be used. These bases by themselves (e.g. NaOAc) are generally not strong enough to provide significant amounts of free NHCs, and the equilibrium constant K_1 in Scheme 3.46 is understandably very small. The equilibrium for such a reaction lies almost entirely on the side of the educts, that is, on the side of the azolium salt.

In the presence of a metal Lewis acid, however, even minute amounts of free NHC generated can be immediately trapped by thermodynamically favorable NHC complex formation, which not only prevents NHC decomposition, but more importantly drives product formation. The complex formation constant K_2 is expected to be large, which means that the equilibrium for the complex formation lies on the side of the NHC-metal complex.

One more recent example is the one-pot preparation of the so-called PEPPSI [58] catalysts developed in the group of Mike Organ [59]. In this procedure, a mixture of the imidazolium salt, $PdCl_2$ and K_2CO_3 is heated in neat 3-chloropyridine. The K_2CO_3 base deprotonates the imidazolium salt to generate the free carbene *in situ*, which can directly coordinate to the Pd^{II} center (Scheme 3.47). 3-chloropyridine acts as a solvent, but notably,

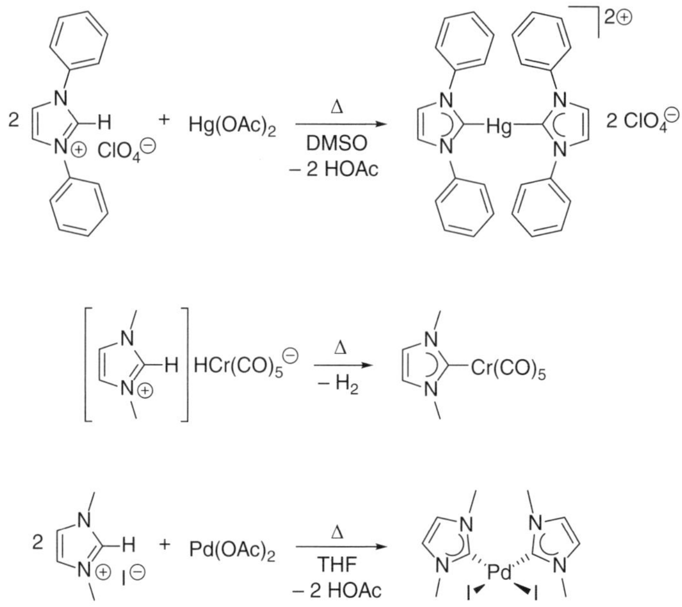

Scheme 3.48
Selected examples of *in situ* deprotonation using basic metal precursors.

it is also a source of the co-ligand, which completes the square planar coordination sphere. In such reactions, it is important that the azolium salt and the metal precursor contain the same halides to prevent product mixtures due to scrambling.

In an elegant variation, metal precursors bearing basic ligands such as Pd(OAc)$_2$ or [Ir(μ-OR)(COD)]$_2$ (COD = cyclooctadiene) can be directly used without addition of any external base. In fact, the first NHC complexes were obtained by the groups of Wanzlick [40] and Öfele [41] in 1968 using this methodology long before the first free NHC was isolated (Scheme 3.48). Wanzlick and Schönherr heated two equivalents of 1,3-diphenylimidazolium perchlorate with mercury(II) acetate in DMSO. The two acetato ligands deprotonate the imidazolium salts, and the two equivalents of the carbene can directly coordinate to Hg^{II} giving rise to a linear and dicationic d^{10} bis(NHC) metal complex, the charge of which is balanced by two perchlorate anions.

Öfele heated a complex salt consisting of a 1,3-dimethylimidazolium cation and a pentacarbonylhydridochromate(0) anion under vacuum. Here, the basic hydrido ligand deprotonates the imidazolium cation, which liberates dihydrogen gas, and the 1,3-dimethylimidazolin-2-ylidene thus generated directly coordinates to the Cr^0 center giving a neutral complex.

The third example was reported by Herrmann and coworkers [60]. Heating of two equivalents of 1,3-dimethylimidazolium iodide with Pd(OAc)$_2$ in boiling THF affords a Pd^{II} bis(NHC) complex. Similar to Wanzlick's example, the acetato ligands act as the base to generate the NHC ligands. In addition, the iodide anions of the imidazolium salt are also good ligands, and binding of these to the Pd^{II} center affords a neutral and square planar complex. Although, the authors only isolated the *cis* isomer, the formation of the *trans* isomer cannot be ruled out. A similar reaction involving 1,3-dimethylbenzimidazolium iodide indeed gave a mixture of *cis* and *trans* isomers, which could be separated due to their different solubilities [61].

3.3.4 Carbene Transfer Routes

Certain preformed metal-NHC complexes obtained by either one of the other routes can transfer the NHC to another metal (transmetallation). The most prominent variation employs Ag-NHC complexes as carbene transfer agents and was first reported by Lin and coworkers in 1998 [62]. The Ag-carbene complexes are generated by *in situ* deprotonation of azolium salts using silver(I) oxide (Ag_2O) as a basic metal precursor, but other silver compounds such as carbonates (Ag_2CO_3) or acetates (AgOAc) can also be used. In order to maintain the correct stoichiometry, two equivalents of carbene precursor are reacted with one equivalent of Ag_2O at ambient temperature. This reaction can be regarded as a 2-step-1-pot process as depicted in Scheme 3.49. The dibasic oxide anion can deprotonate a total of two acidic C2-hydrogen atoms of the azolium salts generating two NHC ligands and thereby liberating one molecule of water into the mixture. The resulting Ag-NHC species are thus relatively moisture stable.

Due to the relatively labile nature of the silver-carbene bond, these complexes can serve as mild carbene transfer reagents to many other metals, such as palladium and gold (Scheme 3.50), and thus afford a wide range of metal complexes. This method has become a standard procedure in the synthesis of NHC complexes nowadays, due to its mild reaction conditions, which normally give rise to less decomposition and fewer byproducts.

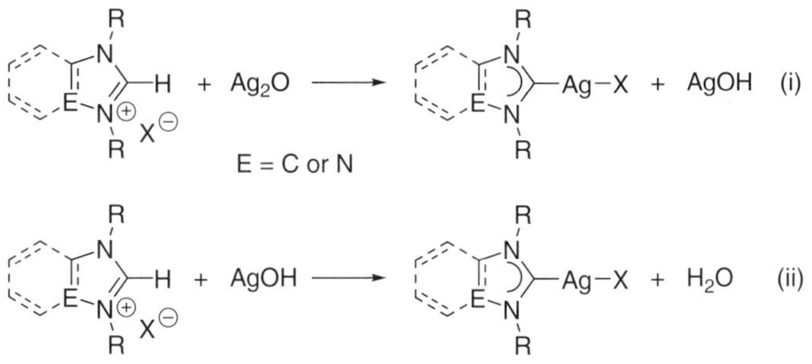

Scheme 3.49
Reaction of azolium salts with Ag_2O under elimination of water.

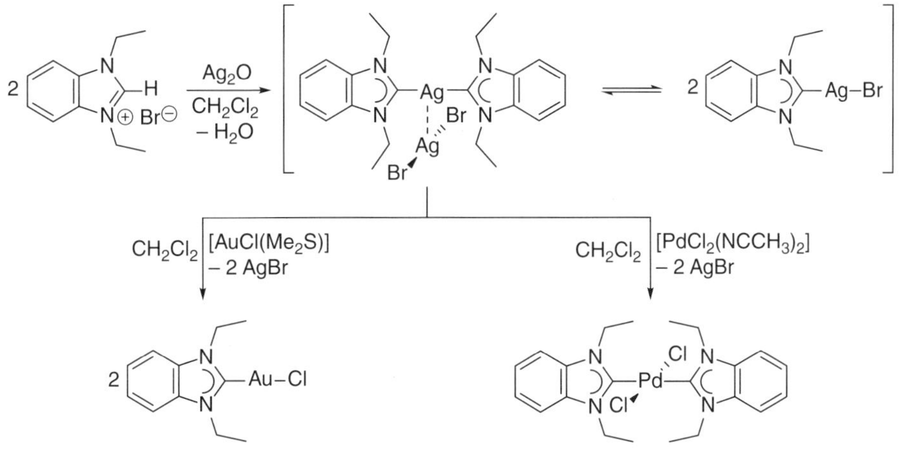

Scheme 3.50
Transfer of the NHC from Ag[I] to Au[I] and Pd[II].

Two early examples of such transfers are depicted in Scheme 3.50. The reaction of 1,3-diethylbenzimidazolium bromide with Ag_2O affords either two neutral molecules of a bromido-mono(NHC) Ag^I complex or a salt consisting of a bis(NHC) Ag^I complex cation, which is balanced by a dibromido-argentate(I) anion via argentophilic interactions. These two forms may be in equilibrium with each other, and in the present case single crystals of the dinuclear silver species were obtained, and its molecular structure was determined by X-ray diffraction [62]. In solution this dinuclear species dissociates into two neutral monocarbene complexes as indicated by NMR spectroscopy. Another type of Ag^I-NHC complex that is often encountered contains two three-coordinate Ag^I-centers that are bridged by two halido ligands, that is, $[(NHC)Ag(\mu-X)_2Ag(NHC)]$. Notably, the structural diversity of Ag-NHC complexes is determined by the solvent of crystallization, and species of a more complicated nature and a higher nuclearity have also been obtained in the meantime [63].

Treatment of the Ag-NHC mixture with chlorido-dimethylsulfidegold(I) or bis(acetonitrile) dichloridopalladium(II) successfully transfers the NHC ligand to the Au^I or Pd^{II} center, respectively, affording chlorido-(NHC)gold(I) and dichlorido-bis(NHC)palladium(II) complexes. The driving force for this reaction is the precipitation of insoluble Ag-halides, the stability of which increases in the order AgCl<AgBr<AgI. The resulting metal complexes are therefore likely to retain the lighter halido ligand [64]. To ensure a successful carbene transfer, some source of halide should be present in the reaction mixture.

When 1,3-diethylbenzimidazolium hexafluorophosphate is used instead, the bis(NHC) Ag^I complex is formed as the sole expected product. Carbene transfer from this species to chlorido-dimethylsulfidegold(I) leads directly to the formation of bis(NHC) Au^I complex with formation of AgCl (Scheme 3.51).

In addition to the mild reaction conditions, the Ag-carbene transfer method also offers the advantage that it can usually be carried out without precautions to exclude air and moisture. This facilitates the making of NHC complexes tremendously and contributed much to the popularity of this method.

Carbene transfers from metal-NHC complexes other than silver are also known. For example, Liu and coworkers reported the transfer of saturated imidazolidin-2-ylidene ligands from their pentacarbonyltungsten(0) complexes to Pd^{II}, Pt^{II}, Rh^I and Au^I (Scheme 3.52) [65]. The NHC transfer agents were prepared by cyclization of the respective metal isocyanide complexes via nucleophilic attack on the bound isocyanide carbon atom. Although the mechanism of these reactions is not known in detail, the authors have reported that the transfer occurs with liberation of carbon monoxide and decomposition of the carbene transfer agent to some unknown tungsten species. In addition, the N-substituents on the NHC influence the rate of the reaction as well as the structure of the products.

Although less investigated, Cu–NHC complexes have also been shown to act as NHC transfer reagents. In 2009, Albrecht *et al.* provided the first example on Cu–NHC transmetallation,

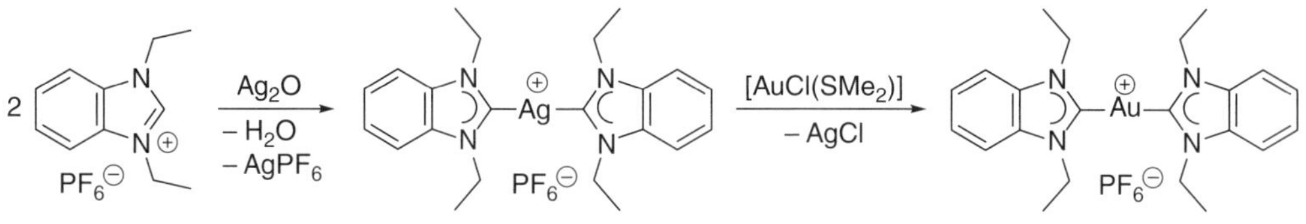

Scheme 3.51
Formation of bis(NHC) Au(I) complex by Ag-carbene transfer.

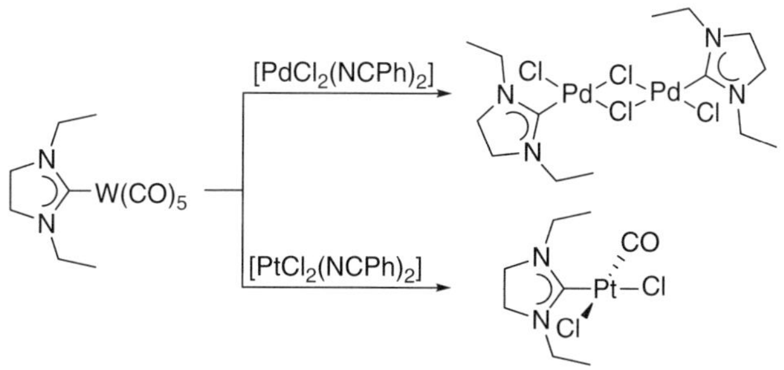

Scheme 3.52
Carbene transfer from W⁰ to Pdᴵᴵ and Ptᴵᴵ.

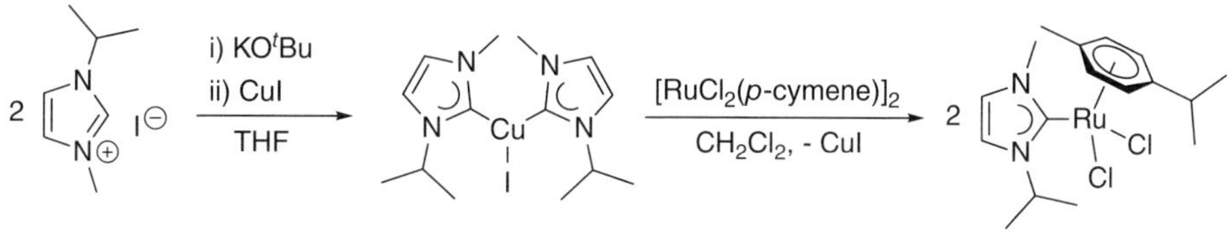

Scheme 3.53
NHC transfer from copper to ruthenium.

whereby the carbene ligand was transferred from copper to ruthenium under enthalpically favorable bridge-cleavage of the dimeric ruthenium complex precursor (Scheme 3.53) [66]. The three-coordinate Cu-NHC transfer agent was synthesized by *in situ* deprotonation of imidazolium salts by KO^{t}Bu in the presence of copper(I) iodide in THF.

The preparation of suitable Cu-NHC complexes has since been improved by employing copper(I) oxide (Cu_2O) in analogy to the use of Ag_2O described earlier. Subsequently, the carbene transfer from copper has also been exploited for the preparation of NHC complexes of gold, palladium, and nickel [67]. Interestingly and in a very specific case, the carbene transfer from copper to silver has also been observed [68], although data from computational calculations suggest that the metal-NHC dissociation energy of coinage metals decreases in the order Au>Cu>Ag [69]. The result of an "inverse" transfer emphasizes the fact that bond energy considerations cannot be regarded as the sole factor that determines a reaction outcome. Indeed, most reactions are in a dynamic equilibrium. In this case, the precipitation of the Ag-NHC complex drives the "inverse transfer" at the expense of metal-NHC bond energy. As with Ag-NHC complexes, the relatively labile nature of Cu-NHC bonds is the reason for NHC migration, and similarly, the precipitation of insoluble cuprous halide generated during carbene transfer provides additional driving force for the transmetallation reaction [66,67a].

3.3.5 Oxidative Addition of an Azolium Salt to a Low-Valent Metal Complex

This methodology was initially reported by Stone as early as in 1973 [70] and subsequently re-investigated in detail by Cavell and coworkers [71]. It involves the direct oxidative addition of C–X (X = alkyl, H, halogen) bonds in neutral azole compounds to a low oxidation-state metal precursor such as [Pd(PPh$_3$)$_4$] and [Ni(COD)$_2$] as the initial step. Furthermore, it has been reported that the activation barrier for oxidative addition decreases in the order C–R>C–H>C–halogen. As observed for aryl halide substrates for palladium catalyzed C-C coupling reactions, the reactivity among the halide substituents increases in the order Cl<Br<I. It must be noted, however, this this sequence is not necessarily applicable in reactions involving nickel, where the activation of C–Cl bonds is often more favorable. The hard-soft-acid-base concept can be used to explain this observation. Upon oxidative addition across a C–Cl bond with nickel, a new Ni–Cl bond is formed, which is favorable due to hard-hard interactions.

By employing strongly electron donating ligands, such as phosphines, the electron-richness of the already low-valent metal center can be further increased, which enhances the rate of oxidative addition. As illustrated in Scheme 3.54, the oxidative addition of the neutral 2-chloro-4-methylthiazole molecule to a square planar Vaska-type Ir^I complex initially affords an octahedral Ir^{III} complex bearing a new formally anionic thiazolato ligand as well as an additional chlorido ligand. Subsequent protonation using a strong acid leads to the formation of cationic Ir^{III} carbene complex with an unusual NH-substituted thiazolin-2-ylidene ligand. Modification of this 2-step procedure by employing N-alkyl azolium salts in the oxidative addition directly leads the carbene complexes in a single step (Scheme 3.55). Nevertheless, interesting NH-substituted carbene complexes are not accessible via this modification, which highlights the complementary nature of the two routes.

The oxidative addition of C2-halo azolium salts was also extended to zero-valent and electron-rich tetrakis(triphenylphosphine)complexes of group 10 transition metals yielding cationic mixed NHC/phosphine complexes.

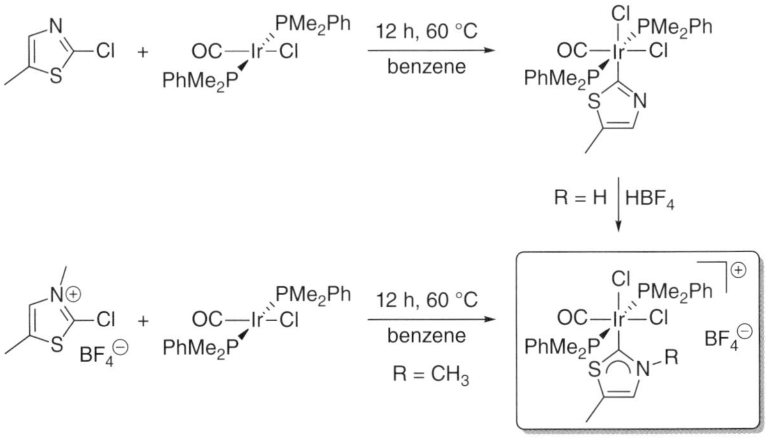

Scheme 3.54
Oxidative addition of a chlorothiazole to a Vaska-type complex and subsequent protonation.

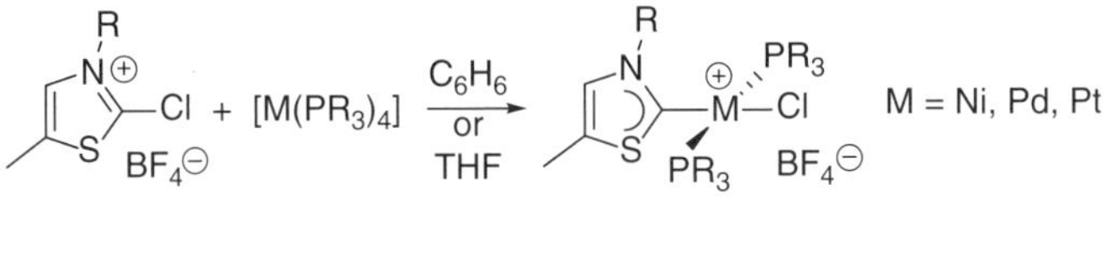

Scheme 3.55
Oxidative addition of 2-chlorothiazolium salts to zero-valent group 10 metals.

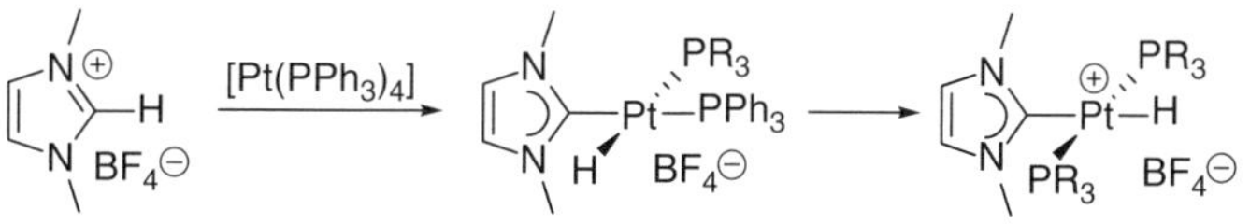

Scheme 3.56
Oxidative addition of 1,3-dimethylimidazolium tetrafluoroborate to Pt^0 and subsequent *cis-trans* isomerization.

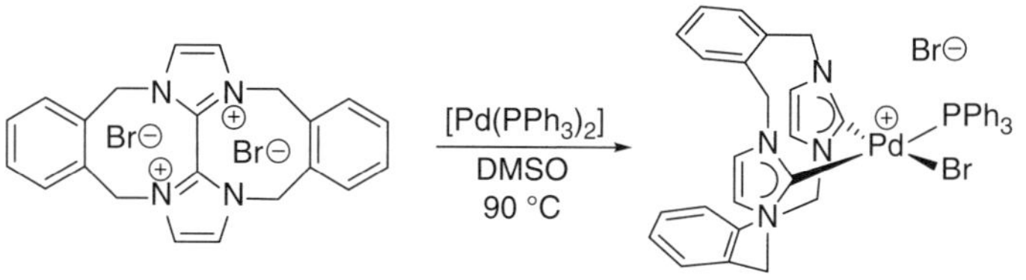

Scheme 3.57
Oxidative addition of a strained "C2–C2 linked, doubly-bridged" dimidazolium salt with Pd^0.

When halide free azolium salts are employed, oxidative addition across a C–H bond occurs and hydrido/carbene complexes can be obtained. In the example depicted in Scheme 3.56 involving the [Pt(PPh$_3$)$_4$] precursor, a *cis*-configured complex is first formed upon oxidative addition. When a solution of this complex is left to stand, isomerization to the more favorable *trans* complex is observed [71].

Due to the high energy required for the activation of C–alkyl bonds, access to NHC complexes via oxidative addition of C2-alkyl azolium salts is uncommon. Nonetheless, Baker and coworkers could activate a special "C2–C2 linked, doubly-bridged" diimidazolium salt with Pd^0 in DMSO at elevated temperatures (Scheme 3.57) [72]. This rare example of a concerted C–C activation of a dimidazolium salt directly led to a cationic dicarbene Pd^{II} complex in a single step. The driving force for this reaction is believed to be the relief of ring strain in the precursor salt that occurs upon palladation and dicarbene formation.

3.3.6 Metal-Template Synthesis Using Isocyanide Complexes as Precursors

Metal isocyanide complexes can be used as templates for the preparation of carbene complexes. The first example for this methodology was unknowingly discovered by Tschugajeff and Skanawy-Grigorjewa as early as 1915 (Scheme 3.58) [73].

The chemistry has later been extensively developed by the groups of Chatt [74], Beck [75], Fehlhammer [76] and Michelin [77]. The formation of the carbene ligand occurs via nucleophilic attack of a proton base HX (X = OR, SR, RNH) at the coordinated C-donor of the isocyanide ligand (Scheme 3.59) effectively resulting in a 1,2-addition across a formal C≡N triple bond. The intermolecular nucleophilic addition leads to acyclic carbenes (route a), whereas NHC complexes can be obtained by intramolecular cyclization of suitably functionalized isocyanide ligands (route b). Different ways of preparing functionalized isocyanide complexes have been developed and are summarized in Scheme 3.60. The nitrogen atom of isocyanic acid or isocyanide complexes is still sufficiently reactive to nucleophilically ring-open strained epoxides or aziridines to give O- or N-functionalized isocyanide complexes. In an alternative pathway, ethylenediamine or ethanolamine can

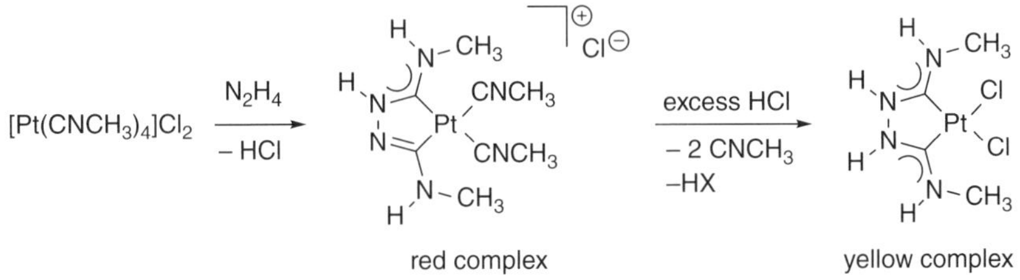

Scheme 3.58
Template-assisted approach to Tschugaeff's Pt^{II} carbene complexes.

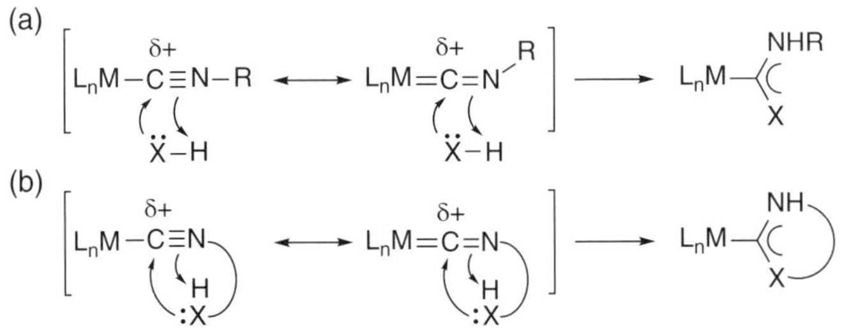

Scheme 3.59
Acyclic and cyclic aminocarbenes by nucleophilic attack on coordinated isocyanides.

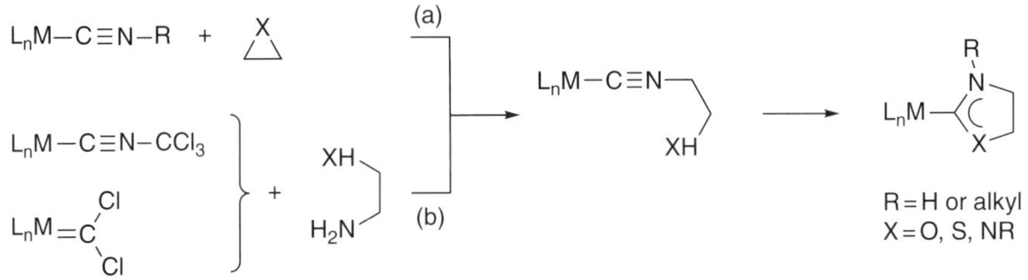

Scheme 3.60
Early pathways to functionalized isocyanide complexes as intermediates to NHC complexes.

react with trichloromethyl isocyanide [78] or dichlorocarbene [79] complexes giving similar products. While the reaction involving the dichlorocarbene ligand releases HCl, the fate of the "NCCl$_3$" leaving group in the trichloromethyl isocyanide ligand is less straightforward. It has been feasibly proposed that "NCCl$_3$" is eliminated as a carbonic acid derivative HN=C(Nu)$_2$ (e.g. guanidine with diamines) upon reaction with the nucleophiles.

Liu and coworkers reported another interesting approach to such complexes, which involves the reaction of an amino-phosphinimine with readily available carbonyl complexes [80]. It is likely that the isocyanide ligand is formed via initial attack of the carbonyl carbon by the lone pair of the phosphinimine nitrogen. This gives rise to a Wittig-type dipolar intermediate that collapses via a transient four-membered state into the phosphine oxide and the isocyanide complex. The amino-functionalized isocyanide complex cyclizes and forms a rare NH,NH-imidazolidin-2-ylidene complex. After N-alkylation, which can occur via N-H deprotonation using NaH and subsequent nucleophilic substitution with

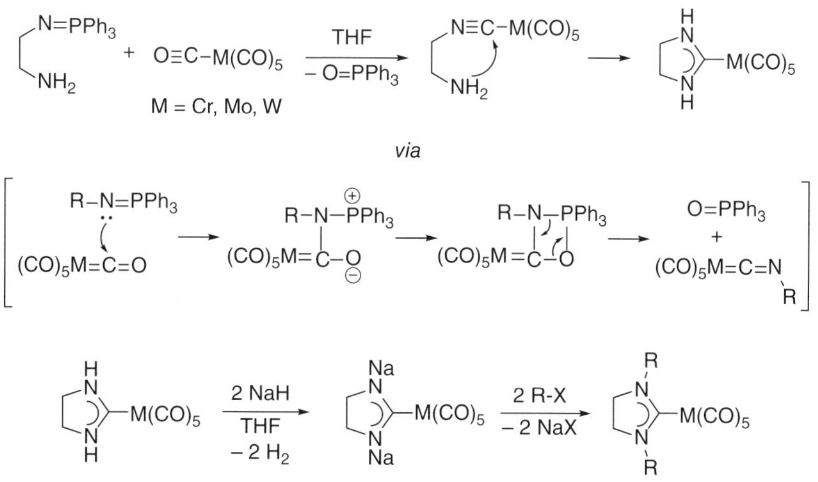

Scheme 3.61
Functionalized isocyanide complexes via reaction of amino-phosphinimine with metal carbonyls.

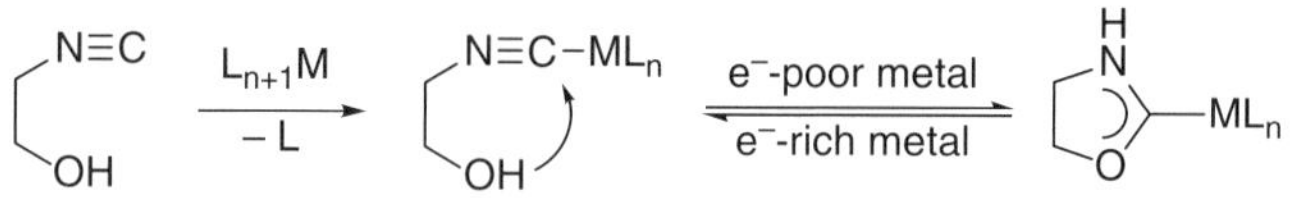

Scheme 3.62
Intramolecular cyclization of a hydroxoethyl isocyanide at metal to an oxazolidin-2-ylidene complex.

alkyl halides, these NHC/carbonyl complexes have also been used as carbene transfer agents to other transition metals (Scheme 3.61) [65].

The use of pre-functionalized isocyanide ligands is more straightforward. For example, 2-hydroxyethyl isocyanide already contains the nucleophilic hydroxy group. Its coordination to a metal complex fragment can occur via ligand displacement. The isocyanide ligand is activated by an electron-deficient metal center, which leads to spontaneous cyclization via intramolecular nucleophilic attack of the OH-function, and a new oxazolidin-2-ylidene complex is obtained (Scheme 3.62) [81].

When the metal center is electron rich, enhanced π-backdonation occurs, which deactivates the isocyanide carbene for nucleophilic attack. In such a scenario, no cyclization and carbene formation is observed. Instead, complexes with stable functionalized isocyanide ligands are obtained.

While 2-hydroxyethyl isocyanide is stable as a free species, its benzannulated derivatives 2-hydroxyphenyl isocyanides are unstable and always undergo ring-closure to benzoxazoles even in the absence of any metal sources. However, lithiation of the latter by *n*BuLi leads to ring-opening, and the resulting phenyl isocyanide can be *O*-protected by tris(trimethyl) silylchloride to give the siloxyphenyl isocyanide as a masked hydroxyphenyl isocyanide [82]. Coordination of the siloxy derivative to electrophilic metal centers and subsequent *O*-deprotection via O–SiMe$_3$ cleavage, for example by a methanolic KF solution, leads to instantaneous cyclization, if the metal center is electron poor (Scheme 3.63). Carbene formation will not occur if the metal center is electron rich and the amount of π-backbonding is substantial.

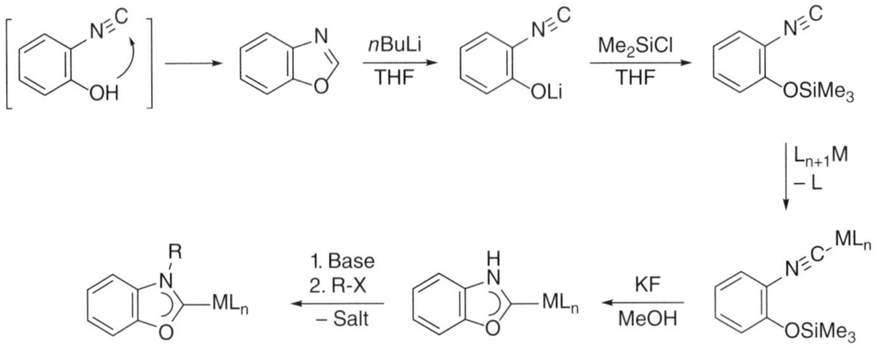

Scheme 3.63
Preparation of 2-(trimethylsiloxy)phenyl isocyanide and its use in NHC complex formation.

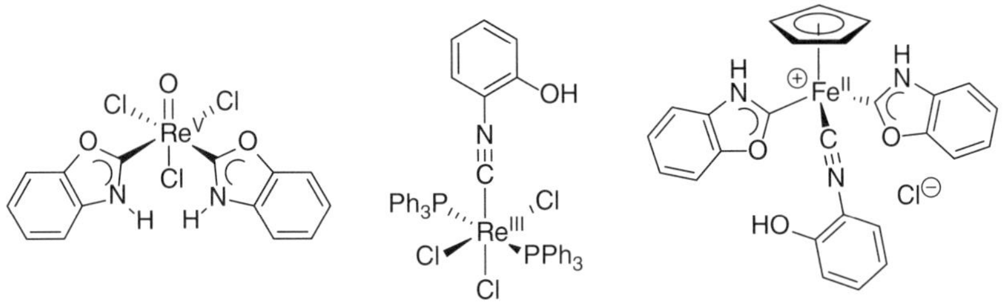

Chart 3.2
Complexes with "open" or "cyclized" 2-hydroxyphenyl isocyanide ligands.

Complexes that nicely demonstrate this behavior are depicted in Chart 3.2. In an electron poor Re^V metal center, intramolecular cyclization is facile, which leads to the formation of two NH,O-carbene ligands. A Re^{III} metal center is more electron rich, particularly when the electron density is increased by two strongly donating phosphine ligands, and therefore no cyclization is observed [83]. The hydroxyphenyl isocyanide ligand, that otherwise does not exist in free form, is stabilized via metal coordination. The third example is a Fe(II) complex that has formed from the initial coordination of three isocyanide ligands. The stepwise cyclization with concurrent formation of new strong donor carbene ligands makes the metal center more and more electron rich. After two NHC ligands have been formed, the metal center is sufficiently electron-rich, and the resulting enhanced d→π* backbonding to the third isocyanide ligand prevents its cyclization [84].

The deprotection of these silylethers is facile with fluoride reagents, since Si–F bonds are about 30 kcal mol^{-1} stronger than Si–O bonds. The formation of stronger Si–F bonds therefore drives the deprotection reaction. In analogy to the reaction sequence depicted in Scheme 3.61, the resulting NH-carbene complexes can be deprotonated with a base, and reaction with electrophiles lead to N-alkylated benzoxazolin-2-ylidene complexes. Using this methodology, a large number of transition metal complexes have been obtained [85].

In order to access complexes bearing classical cyclic five-membered diaminocarbenes, β-amino-functionalized isocyanides are required. However, neither the aliphatic nor the benzannulated isocyanides of this type are known to be stable [86]. On the other hand, both

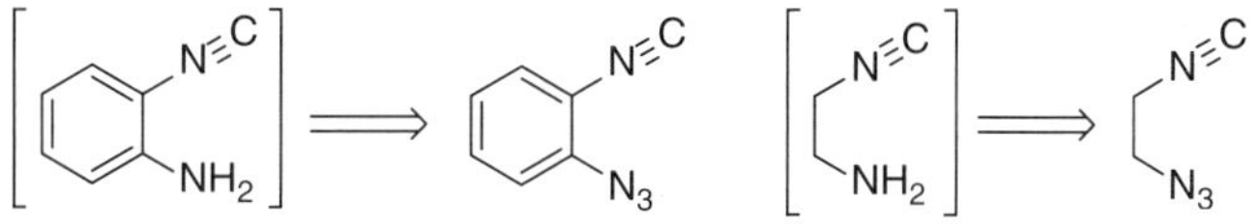

Scheme 3.64
Elusive β-amino-functionalized isocyanides and their β-azido-functionalized synthetic equivalents.

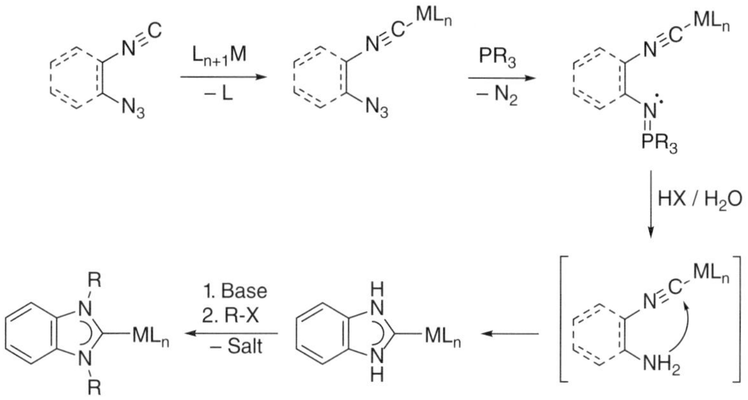

Scheme 3.65
General template-directed cyclization of β-azido-functionalized isocyanides.

aromatic and aliphatic β-azido-functionalized isocyanides are synthetically accessible via a short reaction sequence, and have been used as synthetic equivalents for these synthons (Scheme 3.64).

These can be straightforwardly coordinated to metal centers. The generation of the β-amino-function is realized by a Staudinger [87] reaction with phosphines, which yields iminophosphine derivatives. Hydrolysis of the latter with an aqueous mineral acid (HX) liberates the primary amino group. Subsequent nucleophilic attack at the isocyanide carbon donor and proton shift completes the cyclization giving a NH,NH-carbene complex, that can be easily N-alkylated by treatment with a base and alkyl halide (Scheme 3.65) [88].

The cyclization of functionalized isocyanides represents a useful alternative to other methodologies. Using this approach, rare NH-carbenes are accessible that cannot be obtained by routes involving deprotonation of azolium salts. Although very versatile, this method is still comparatively less popular, which can be traced back to the often more tedious preparation of suitable functionalized isocyanides that require sufficiently skilled experimentalists. In particular, organic azides, that are potentially explosive, have to be handled with caution.

3.3.7   NHC Complexes by Small Molecule Elimination

Previously, we have seen that free carbenes can be generated by thermally induced α-elimination of small molecules from suitable precursors. Using this pathway, Enders and coworkers prepared the very first 1,2,4-triazolin-5-ylidene. An extension of this method to the preparation of NHC complexes is easily achievable, when the reaction is carried out in the presence of transition metals. One prominent example is the preparation of second generation Grubbs catalysts that contain saturated imidazolidin-2-ylidene ligands. Heating a benzene solution of Grubbs I catalyst and the 2-alkoxy-1,3-dimesitylimidazolidine liberates the free saturated carbene, which directly coordinates to the Ru^{II} center via displacement of one tricyclohexylphosphine ligand (Scheme 3.66) [38]. The reaction is enthalpically driven due to the formation of the stronger Ru–NHC bond compared to the initial Ru–PCy_3 bond. Moreover, the liberation of one molecule of phosphine and alcohol each also contributes favorably to the entropy term. The resulting Grubbs II system with a saturated NHC outperforms those with unsaturated imidazolin-2-ylidenes, which has been attributed to the stronger donating abilities of saturated NHCs.

Waymouth and coworkers reported a similar strategy, which could be applied to prepare palladium(II) NHC complexes. Instead of alcohol adducts of carbenes, they employed 1,3-dimesityl-2-pentafluorophenyl-imidazolidine as a NHC precursor. Heating a solution of the imidazolidine and η^3-allylchloridopalladium(II) dimer in toluene afforded a Pd^{II}-SIMes complex under bridge-cleavage condition and with release of pentafluorobenzene [39]. The reaction is driven by the formation of the stable benzene derivative and the saturation of the coordination sites at palladium (Scheme 3.67).

As described in Chapter 3.1, free NHCs can react with excess carbon dioxide to give zwitterionic compounds. Reversely, these azolium-2-carboxylates can formally release carbon dioxide upon heating and transfer the nucleophilic NHC to metal centers. Crabtree and coworkers exploited this observation and developed this route to prepare transition metal complexes by simply heating imidazolium-2-carboxylates in the presence of suitable metal precursors [89]. One way to prepare these NHC transfer agents is by reaction of *in situ* generated NHC with carbon dioxide under high pressure (20–50 atm) [29b,90]. More convenient is the preparation by heating N-alkylimidazoles in neat dimethylcarbonate, which leads to N-methylation and C2-carboxylation in one pot (Scheme 3.68) [91].

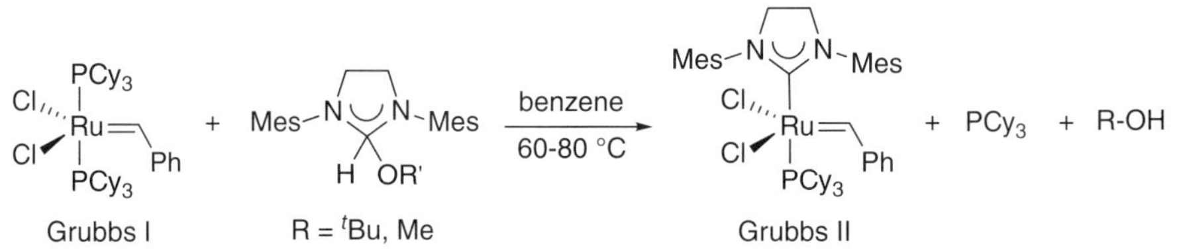

Scheme 3.66
Preparation of a Grubbs II catalyst by α-elimination of alcohol.

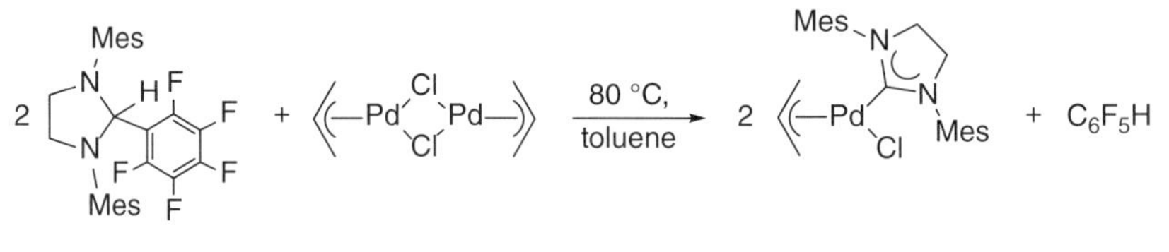

Scheme 3.67
Preparation of a Pd-SIMes complex by α-elimination of pentafluorobenzene.

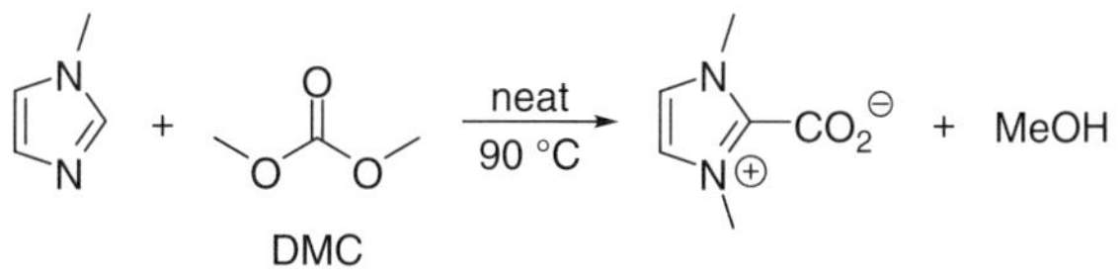

Scheme 3.68
Preparation of 1,3-dimethylimidazolium-2-carboxylate using DMC.

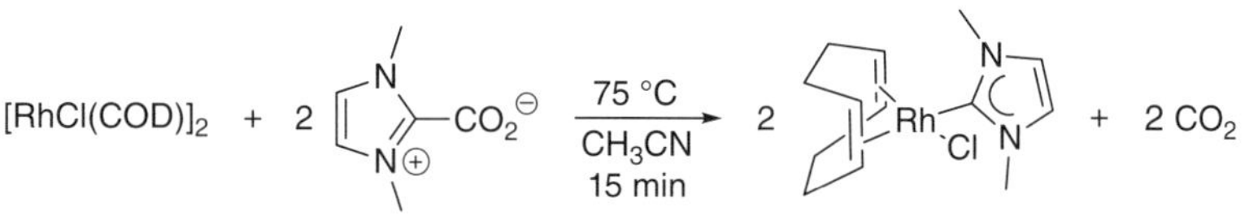

Scheme 3.69
NHC transfer from imidazolium-2-carboxylate to Rh(I).

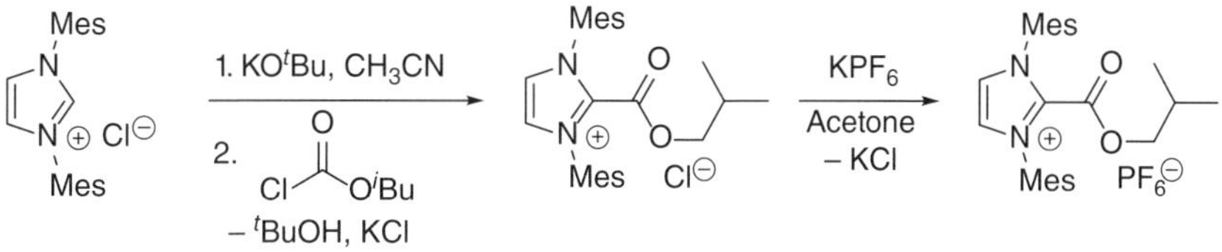

Scheme 3.70
Preparation of an imidazolium-2-carboxylate ester.

The resulting zwitterions are air and moisture stable, and precautions to exclude air or moisture are usually not required in the NHC-metallation process, making this method a convenient alternative. The carbene can be transferred to a range of metal centers including Rh^I, Ir^I, Ru^{II}, Pt^{II}, and Pd^{II}.

One example is depicted in Scheme 3.69. Heating two equivalents of 1,3-dimethylimida-zolium-2-carboxylate with μ-chlorido(1,5-cyclooctadiene)rhodium(I) dimer leads to the formation of two molecules of chlorido(1,5-cyclooctadiene)(1,3-dimethylimidazolin-2-ylidene)rhodium(I) complex in high yields and with release of two carbon dioxide molecules in only 15 min. The bridge-cleavage reaction and the formation of two new Rh-NHC bonds contribute favorably to the reaction enthalpy, while the release of gaseous carbon dioxide increases the reaction entropy. When the reaction is carried out at room temperature, completion is observed after 40 min, which is in agreement with an expected decreased reaction rate at lower temperature. The reaction can be carried out in wet solvents, and no imidazolium salts have been observed, which excludes the formation of free NHC as intermediates. Instead, a detailed computational study suggests that NHC complex formation is preceded by O-coordination of the carboxylato group. Subsequent carbene formation and decarboxylation occur intramolecularly at the metal center.

A drawback of the use of imidazolium-2-carboxylates is their limitation to carbene ligands with N-methyl substituents. However, the scope of this methodology could be broadened by the use of the respective imidazolium-2-carboxylate esters, which are also stable carbene sources and readily available by deprotonation of imidazolium salts and reaction with isobutyl chloroformate (Scheme 3.70).

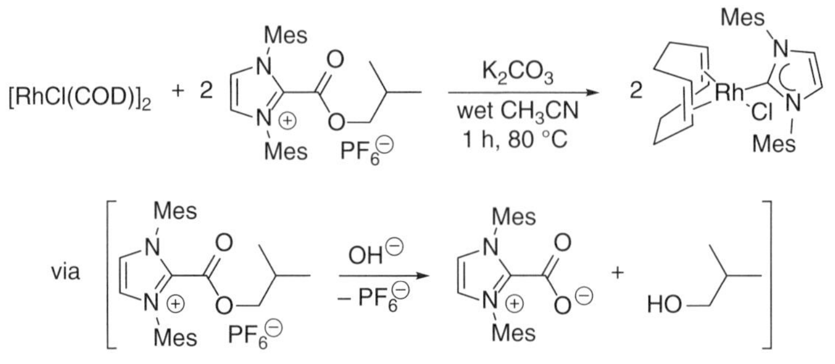

Scheme 3.71
Synthesis of a Rh^I-NHC complex using an imidazolium-2-carboylate ester.

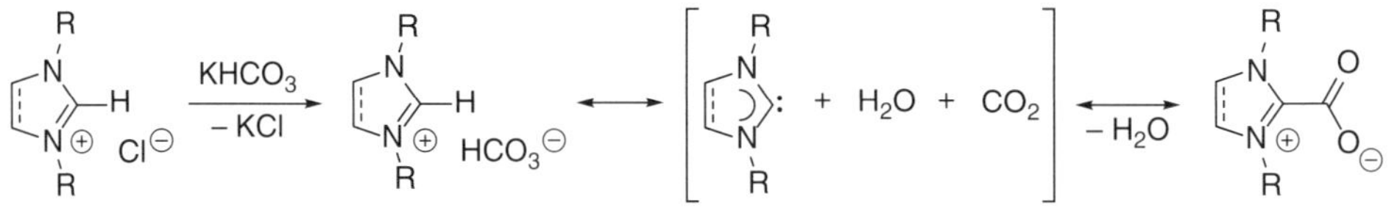

Scheme 3.72
Preparation of azolium bicarbonates and their relationship to azolium-2-carbonates.

However, the NHC transfer from the respective esters to transition metals requires harsher reaction conditions and longer reaction times. Furthermore, potassium carbonate and wet solvents facilitate the reaction, which suggests that *in situ* hydrolysis of the ester to the respective 2-carboxylate takes place. The latter is the actual NHC transfer agent (Scheme 3.71). The need for hydrolysis also explains the harsher reactions conditions and longer reaction time that are required for a good yield of the metal complex.

A further variation reported by Taton and Vignolle involves the use of azolium bicarbonate salts as a source for NHCs. These air and moisture stable salts are easily synthesized by anion metathesis of azolium halides with potassium bicarbonate ($KHCO_3$). In dry DMSO-d_6 solution, they were observed to be in a dynamic equilibrium with the respective azolium-2-carboxylates. Based on a computational study, it was proposed that the NHC-CO_2 adduct formation is initiated by *in situ* deprotonation of the acidic azolium C2–H group by the basic bicarbonate counter-anion, which gives the free NHC and carbonic acid as intermediates. The latter collapses into one molecule of water and carbon dioxide each, which in turn reacts with the intermediate free NHC to form the carboxylate derivative (Scheme 3.72) [92].

Although attempts to isolate any free NHCs were to no avail, the application of azolium bicarbonates as a convenient NHC source for organometallic chemistry was demonstrated by the successful preparation of palladium and gold complexes. Simple stirring of the azolium bicarbonate salt with either 0.5 equiv. of allyl(chlorido)palladium(II) dimer or chlorido(dimethylsulfide)gold(I) at room temperature or with moderate heating to 50 °C afforded the allyl(chlorido)(NHC)palladium(II) or the chlorido(NHC)gold(I) complexes in high yields (Scheme 3.73). In addition, this NHC source could also be used for organocatalytic applications.

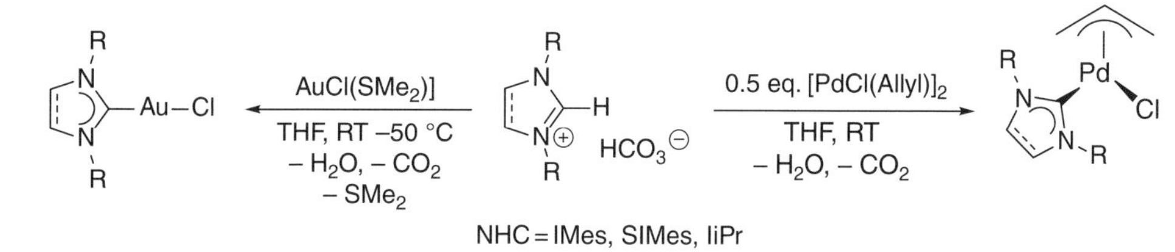

Scheme 3.73
Palladium(II) and gold(I) complexes from azolium bicarbonates.

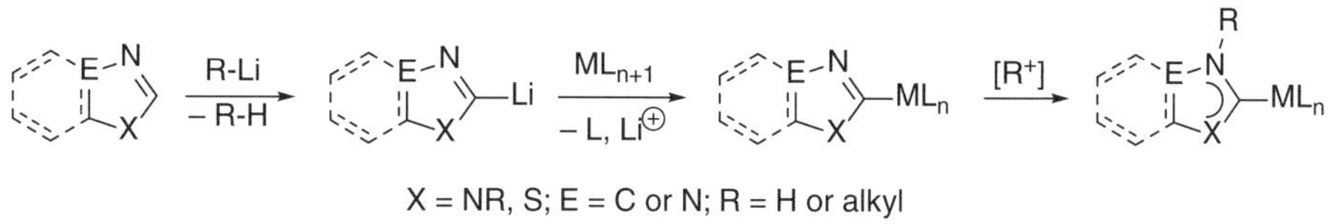

X = NR, S; E = C or N; R = H or alkyl

Scheme 3.74
Lithiation of azoles and their use in the preparation of NHC complexes.

M = Cr, Mo, W

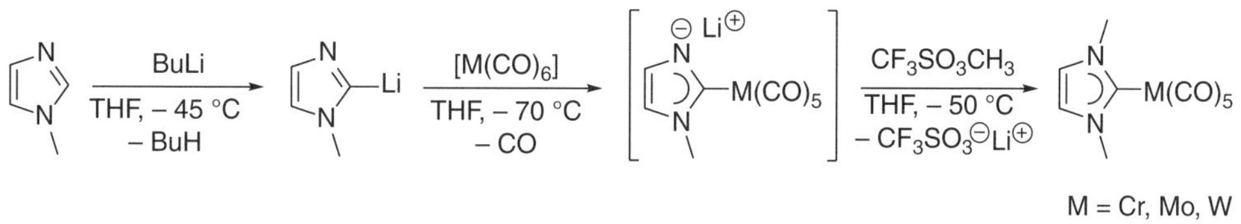

Scheme 3.75
Preparation of group metal-carbenes by alkylation of azolyl complexes.

3.3.8 NHC Complexes by Protonation/Alkylation of Azolyl Complexes

Another method for the preparation of NHC complexes that has received relatively less attention makes use of lithiated azoles. These are easily available by deprotonation of neutral azoles or N-alkylazoles with alkyl lithium reagents. The deprotonation occurs selectively at C2 (or C5 in 1,2,4-triazoles), which is the most acidic proton of the heterocycle (Scheme 3.74).

The lithiated azoles react readily with transition metal precursors to form azolato complexes, where the newly introduced azolato ligand carries a negative charge. This transmetallation process usually occurs under displacement of another neutral or anionic ligand from the initial metal source. The displacement of halido ligands is particularly favorable due to the elimination of Li-halide salts and the lattice energy contribution associated with their formation. In the final step, the complex undergoes protonation or alkylation at the nucleophilic N-atom of the azolato ligand, which furnishes either the neutral NH or N-alkyl-substituted NHC ligand. Strong electrophiles and acids with weakly coordinating anions are usually required for this last step.

Raubenheimer and coworkers developed and contributed extensively to this type of chemistry, and numerous metal carbene complexes were prepared by this route [93]. One example depicted in Scheme 3.75 starts with the C2-lithiation of 1-methylimidazole with *n*-butyl lithium at low temperature. Subsequent reaction with group 6 metal carbonyls proceeds with displacement of one carbonyl ligand and yields anionic complexes. These azolyl complexes are then directly N-alkylated to the respective NHC complexes using methyl trifluoromethansulfonate as a strong electrophile.

It is interesting to note that the azolyl complexes were drawn as "anionic carbene" complexes in the original publication, where the negative charge is located at the nitrogen atom rather than at the carbon donor. The question as to whether the heterocyclic ligand in such complexes should be regarded as an anionic carbene (neutral donor, resonance form **A**) or as a classical azolato ligand (anionic donor, resonance form **B**) is still under debate and topic of a more recent study (Scheme 3.76) [94].

In addition to the reaction of lithiated azoles with metal sources, such azolyl complexes can also be obtained by oxidative addition of neutral 2-haloazoles to low-valent metal precursors (Scheme 3.54).

Alternatively, tetrazolyl derivates can be obtained by a metal template-directed approach, which does not require stringent exclusion of moisture and air (Scheme 3.77). For example,

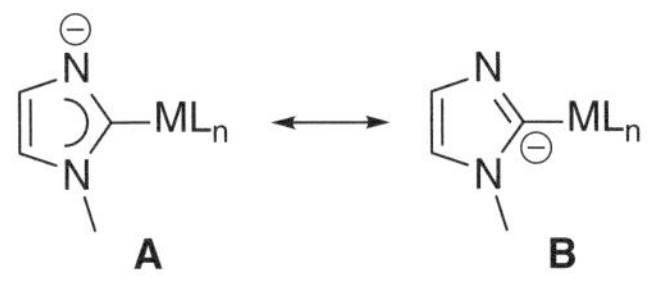

Scheme 3.76
Anionic NHC (A) and azolato (B) resonance structures of azolyl complexes.

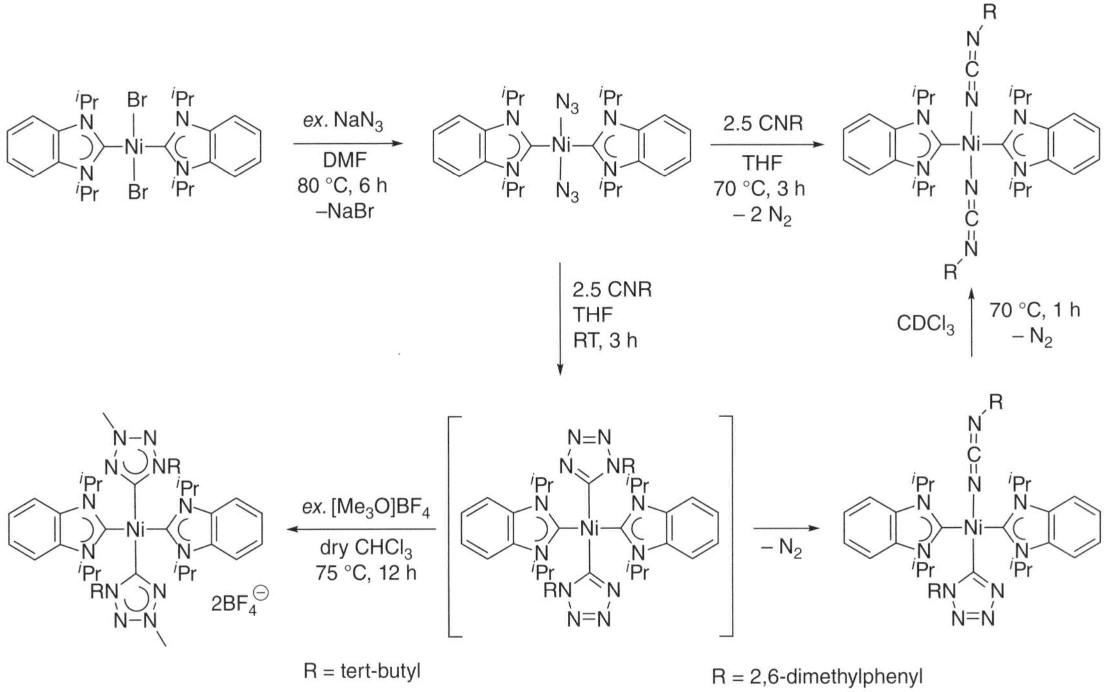

R = tert-butyl

R = 2,6-dimethylphenyl

Scheme 3.77
Template-directed approach to tetrazolyl and tetrazolin-5-ylidene complexes.

azido complexes can be obtained by ligand substitution reactions of bromido complexes with sodium azide. Reaction of the resulting azido complexes with isocyanide leads to a Huisgen-type dipolar cycloaddition reaction, which yields the tetrazolyl complexes. Depending on the reaction conditions and the nature of the isocyanide ligand, dinitrogen elimination from the tetrazolyl complexes can be affected affording carbodiimido complexes. On the other hand, N-alkylation of the anionic tetrazole-derived ligand gives stable tetrazolin-5-ylidene complexes [95].

References

[1] L. Benhamou, E. Chardon, G. Lavigne, S. Bellemin-Laponnaz, V. César, *Chem. Rev.* **2011**, *111*, 2705.

[2] (a) A. J. Arduengo, Preparation of 1,3-Disubstituted Imidazolium Salts. U.S. Patent 5,077,414, December 31, 1991.(b) S. P. Nolan, Synthesis of 1,3-Disubstituted Imidazolium Salts. U.S. Patent 7,109,348, September 19, 2006.

[3] (a) J. Cooke, O. C. Lightbody, *J. Chem. Educ.* **2011**, *88*, 88. (b) X. Bantreil, S. P. Nolan, *Nature Protocols* **2011**, 6, 69.

[4] L. Hintermann, *Beilstein J. Org. Chem.* **2007**, *3*, No. 22.

[5] (a) A. J. Arduengo III, U.S. Patent 5,077,414, 1991. (b) W. A. Herrmann, V. P. W. Böhm, C. W. K. Gstöttmayr, M. Grosche, C.-P. Reisinger, T. Weskamp, *J. Organomet. Chem.* 2001, *617–618*, 616.

[6] (a) A. J. Arduengo III, F. P. Gentry Jr., P. K. Taverkere, H. E. Simmons III, US Patent 6,177,575, January 2001. (b) J. Liu, J. Chen, J. Zhao, Y. Zhao, L. Liang, H. Zhang, *Synthesis* **2003**, 2661.

[7] A. A Gridnev, I. M. Mihaltseva, *Synth. Commun.* **1994**, *24*, 1547.

[8] V. Santes, E. Gomez, R. Santiallan, N. Farfan, Synthesis **2001**, 235.

[9] V. Jurcík, M. Gilani, R. Wilhelm, *Eur. J. Org. Chem.* **2006**, 5103.

[10] S. P. Roche, M.-L. Teyssot, A. Gautier, *Tetrahedron Lett.* **2010**, *51*, 1265.

[11] A. J. Arduengo III, R. Krafczyk, R. Schmutzler, H. A. Craig, J. R. Goerlich, W. J. Marshall, M. Unverzagt, *Tetrahedron* **1999**, *55*, 14523.

[12] H. V. Huynh, Y. Han, R. Jothibasu, J. A. Yang, *Organometallics* **2009**, *28*, 5395.

[13] M. B. Dinger, P. Nieczypor, J. C. Mol, *Organometallics* **2003**, *22*, 5291.

[14] (a) A. W Waltman, R. H. Grubbs, *Organometallics* **2004**, *23*, 3105. (b) G. C. Vougioukalakis, R. H. Grubbs, *Organometallics* **2007**, *26*, 2469.

[15] For recent selected reviews see: (a) D. S. Surry, S. L. Buchwald, *Chem. Sci.* **2011**, *2*, 27. (b) J. F. Hartwig, *Acc. Chem. Res.* **2008**, *41*, 1534.

[16] (a) F. M. Rivas, U. Riaz, A. Giessert, J. A. Smulik, S. T. Diver, *Org. Lett.* **2001**, *3*, 2673. (b) F. M. Rivas, U. Riaz, S. T. Diver, *Tetrahedron: Asymmetry* **2000**, *11*, 1703.

[17] (a) C. J. O'Brien, E. A. B. Kantchev, G. A. Chass, N. Hadei, A. C. Hopkinson, M. G. Organ, D. H. Setiadi, T.-H. Tang, D.-C. Fang, *Tetrahedron* **2005**, *61*, 9723; (b) D. M. Khramov, A. J. Boydston, C. W. Bielawski, *Org. Lett.* **2006**, *8*, 1831; (c) J. A. V. Er, A. G. Tennyson, J. W. Kamplain, V. M. Lynch, C. W. Bielawski, *Eur. J. Inorg. Chem.* **2009**, 1729.

[18] Y. Borguet, G. Zaragoza, A. Demonceau, L. Delaude, *Adv. Synth. Catal.* **2012**, *354*, 1356.

[19] R. M. Roberts, R. H. DeWolfe, *J. Am. Soc. Chem.* **1954**, *76*, 2411.

[20] E. C. Taylor, W. A. Ehrhart, *J. Org. Chem.* **1963**, *28*, 1108.

[21] R. H. DeWolfe, *J. Org. Chem.* **1962**, *27*, 490.

[22] K. M. Kuhn, R. H. Grubbs, *Org. Lett.* **2008**, *10*, 2075.

[23] K. Hirano, S. Urban, C. Wang, F. Glorius, *Org. Lett.* **2009**, *11*, 1019.

[24] M. K. Denk, J. M. Rodezno, S. Gupta, A. J. Lough, *J. Organomet. Chem.* **2001**, *617–618*, 242.

[25] O. Hollóczki, P. Terleczky, D. Szieberth, G. Mourgas, D. Gudat, L. Nyulászi, *J. Am. Chem. Soc.* **2011**, *133*, 780.

[26] D. Enders, K. Breuer, G. Raabe, J. Runsink, J. H. Teles, J. P. Melder, K. Ebel, S. Brode, *Angew. Chem. Int. Ed. Engl.* **1995**, *34*, 1021.

[27] H.-W. Wanzlick, B. König, *Chem. Ber.* **1964**, *97*, 3513.

[28] H.-J. Schönherr, H.-W. Wanzlick, *Liebigs Ann. Chem.* **1970**, *731*, 176.

[29] (a) N. Kuhn, M. Steimann and G. Weyers, *Z. Naturforsch., Teil B*, **1999**, *54*, 427; (b) H. A. Duong, T. N. Tekavec, A. M. Arif, J. Louie, *Chem. Commun.* **2004**, 112.

[30] (a) A. M. Voutchkova, L. N. Appelhans, A. R. Chianese, R. H. Crabtree, *J. Am. Chem. Soc.* **2005**, *127*, 17624; (b) M. Azouri, J. Andrieu, M. Picquet, P. Richard, B. Hanquet, I. Tkatchenko, *Eur. J. Inorg. Chem.* **2007**, 4877.

[31] D. Enders, O. Niemeier, A. Henseler, *Chem. Rev.* **2007**, *107*, 5606.

[32] A. J. Arduengo, III, R. L. Harlow, M. Kline, *J. Am. Chem. Soc.* **1991**, *113*, 361.

[33] A. J. Arduengo III, H. V. R. Dias, R. L. Harlow and M. Kline, *J. Am. Chem. Soc.* **1992**, *114*, 5530.

[34] W. A. Herrmann, C. Köcher, L. J. Goossen, G. R. J. Artus, *Chem. Eur. J.* **1996**, *2*, 1627.

[35] N. Kuhn, T. Kratz, Synthesis **1993**, 561.

[36] F. E. Hahn, L. Wittenbecher, R. Boese, D. Bläser, *Chem. Eur. J.* **1999**, *5*, 1931.

[37] B. Gehrhus, P. B. Hitchcock, M. F. Lappert, *J. Chem. Soc., Dalton Trans.* **2000**, 3094.

[38] M. Scholl, S. Ding, C. W. Lee, R. H. Grubbs, *Org. Lett.* **1999**, *1*, 953.

[39] G. W. Nyce, S. Cishony, R. M. Waymouth, J. L. Hedrick, *Chem. Eur. J.* **2004**, *10*, 4073.

[40] H.-W. Wanzlick, H. Schönherr, *Angew. Chem. Int. Ed. Engl.* **1968**, *7*, 141.

[41] K. Öfele, *J. Organomet. Chem.* **1968**, *12*, P42.

[42] W. A. Herrmann, *Angew. Chem. Int. Ed.* **2002**, *41*, 1290.

[43] (a) A. Rit, T. Pape, F. E. Hahn, *J. Am. Chem. Soc.* **2010**, *132*, 4572; (b) A. Rit, T. Pape, A. Hepp, F. E. Hahn, *Organometallics* **2011**, *30*, 334.

[44] (a) H. V. Huynh, W. Sim, C. F. Chin, *Dalton Trans.* **2011**, *40*, 11690. (b) Q. Teng, H. V. Huynh, *Chem. Commun.* **2015**, *51*, 1248.

[45] For recent general reviews of metal-NHC complexes as anticancer agents, see: (a) W. Liu, R. Gust, *Chem. Soc. Rev.* **2013**, *42*, 755; (b) L. Oehninger, R. Rubbiani, I. Ott, *Dalton Trans.* **2013**, *42*, 3269; (c) A. Gautier, F. Cisnetti, *Metallomics* **2012**, *4*, 23.

[46] R. Dorta, E. D. Stevens, N. M. Scott, C. Costabile, L. Cavallo, C. D. Hoff, S. P. Nolan, *J. Am. Chem. Soc.* **2005**, *127*, 2485.

[47] (a) A. C. Hillier, W. J. Sommer, B. S. Yong, J. L. Petersen, L. Cavallo, S. P. Nolan, *Organometallics* **2003**, *22*, 4322; (b) L. Jafarpour, S. P. Nolan, *J. Organomet. Chem.* **2001**, *617–618*, 17.

[48] L. Mercs, G. Labat, A. Neels, A. Ehlers, M. Albrecht, *Organometallics* **2006**, *25*, 5648.

[49] (a) S. Díez-González, S. P. Nolan, *Coord. Chem. Rev.* **2007**, *251*, 874; (b) U. Radius, F. M. Bickelhaupt, *Coord. Chem. Rev.* **2009**, *253*, 678; (c) H. Jacobsen, A. Correa, A. Poater, C. Costabile, L. Cavallo, *Coord. Chem. Rev.* **2009**, *253*, 687; (d) L. Benhamou, E. Chardon, G. Lavigne, S. Bellemin-Laponnaz, V. César, *Chem. Rev.* **2011**, *111*, 2705.

[50] (a) K. Öfele, W. A. Herrmann, D. Mihalios, E. Herdtweck, W. Scherer, J. Mink, *J. Organomet. Chem.* **1993**, *459*, 177; (b) W. A. Herrmann, K. Öfele, M. Elison, F. E. Kühn, P. W. Roesky, *J. Organomet. Chem.* **1994**, *480*, C7; (c) F. E. Hahn, L. Wittenbecher, R. Boese, D. Bläser, *Chem. Eur. J.* **1999**, *5*, 1931.

[51] J. Huang, E. D. Stevens, S. P. Nolan, J. L. Petersen, *J. Am. Chem. Soc.* **1999**, *121*, 2674.

[52] T. Weskamp, F. J. Kohl, W. Hieringer, D. Gleich, W. A. Herrmann, *Angew. Chem. Int. Ed.* **1999**, *38*, 2416.

[53] M. Scholl, T. M. Trnka, J. P. Morgan, R. H. Grubbs, *Tetrahedron Lett.* **1999**, *40*, 2247.

[54] (a) D. J. Cardin, B. Çetinkaya, M. F. Lappert, *Chem. Rev.* **1972**, *72*, 545; (b) M. F. Lappert, *J. Organomet. Chem.* **1988**, *358*, 185; (c) M. F. Lappert, *J. Organomet. Chem.* **2005**, *690*, 5467.

[55] D. J. Cardin, B. Çetinkaya, M. F. Lappert, Lj. Manojlović-Muir; K. W. Muir, *J. Chem. Soc., Chem. Commun.* **1971**, 400.

[56] P. B. Hitchcock, M. F. Lappert, P. L. Pye, *J. Chem. Soc., Dalton Trans.* **1977**, 2160.

[57] F. E. Hahn, T. von Fehren, T. Lügger, *Inorg. Chim. Acta* **2005**, *358*, 4137.

[58] PEPPSI is an abbreviation for Pyridine-Enhanced Precatalyst Preparation Stabilization and Initiation.

[59] C. J. O'Brien, E. A. B. Kantchev, C. Valente, N. Hadei, G. A. Chass, A. Lough, A. C. Hopkinson, M. G. Organ, *Chem. Eur. J.* **2006**, *12*, 4743.

[60] W. A. Herrmann, M. Elison, J. Fischer, C. Köcher, G. R. J. Artus, *Angew. Chem. Int. Ed. Engl.* **1995**, *34*, 2371.

[61] H. V. Huynh, J. H. H. Ho, T. C. Neo, L. L. Koh, *J. Organomet. Chem.* **2005**, *690*, 3854.

[62] H. M. J. Wang, I. J. B. Lin, *Organometallics* **1998**, *17*, 972.

[63] (a) J. C. Garrison, W. J. Youngs, *Chem. Rev.* **2005**, *105*, 3978; (b) J. C. Y. Lin, R. T. W. Huang, C. S. Lee, A. Bhattacharyya, W. S. Hwang, I. J. B. Lin, *Chem. Rev.* **2009**, *109*, 3561.

[64] The original publication [62] gives [AuBr(NHC)] instead of [AuCl(NHC)] as the product. This is less likely, since formation of AgBr is more favorable than that of AgCl.

[65] S.-T. Liu, T.-Y. Hsieh, G.-H. Lee, S.-M. Peng, *Organometallics* **1998**, *17*, 993.

[66] G. Venkatachalam, M. Heckenroth, A. Neels, M. Albrecht, *Helv. Chim. Acta* **2009**, *92*, 1034.

[67] (a) M. R. L. Furst, C. S. J. Cazin, *Chem. Commun.* **2010**, *46*, 6924; (b) A. C. Badaj, G. G. Lavoie, *Organometallics* **2012**, *31*, 1103; (c) V. Gierz, C. Maichle-Mössmer, D. Kunz, *Organometallics* **2012**, *31*, 739; (d) C. Chen, H. Qiu, W. Chen, *J. Organomet. Chem.* **2012**, *696*, 4166.

[68] X. Liu, R. Pattacini, P. Deglmann, P. Braunstein, *Organometallics* **2011**, *30*, 3302.

[69] C. Boehme, G. Frenking, *Organometallics* **1998**, *17*, 5801.

[70] (a) P. J. Fraser, W. R. Roper, F. G. A. Stone, *J. Organomet. Chem.* **1973**, *50*, C54; (b) P. J. Fraser, W. R. Roper, F. G. A. Stone, *J. Chem. Soc., Dalton Trans.* **1974**, 102.

[71] (a) D. S. McGuinness, K. J. Cavell, B. F. Yates, *Chem. Commun.* **2001**, 355; (b) D. S. McGuinness, K. J. Cavell, B. F. Yates, B. W. Skelton, A. H. White, *J. Am. Chem. Soc.* **2001**, *123*, 8317; (c) D. C. Graham, K. J. Cavell, B. F. Yates, *Dalton Trans.* **2007**, 4650.

[72] M. V. Baker, D. H. Brown, V. J. Hesler, B. W. Skelton, A. H. White, *Organometallics* **2007**, *26*, 250.

[73] (a) L. Tschugajeff, M. Skanawy-Grigorjewa, *J. Russ. Chem. Soc.* **1915**, *47*, 776; (b) L. Tschugajeff, M. Skanawy-Grigorjewa, A. Posnjak, *Z. Anorg. Allg. Chem.* **1925**, *148*, 37.

[74] E. M. Badley, J. Chatt, R. L. Richards, G. A. Sim, *Chem. Commun.* **1969**, 1322.

[75] W. Beck, W. Weigand, U. Nagel, M. Schaal, *Angew. Chem. Int. Ed. Engl.* **1984**, *23*, 377.

[76] Fehlhammer, W. P.; Fritz, M. Chem. Rev. **1993**, *93*, 1243.

[77] (a) R. Bertani, M. Mozzon, R. A. Michelin, *Inorg. Chem.* **1988**, *27*, 2809; (b) R. Bertani, M. Mozzon, R. A. Michelin, F. Benetollo, G. Bombieri, T. J. Castilho, A. J. L. Pombeiro, *Inorg. Chim. Acta* **1991**, *189*, 175.

[78] (a) G. Beck, W. P. Fehlhammer, *Angew. Chem. Int. Ed. Engl.* **1988**, *27*, 1344; (b) W. P. Fehlhammer, S. Ahn, G. Beck, *J. Organomet. Chem.* **1991**, *411*, 181; (c) W. P. Fehlhammer, G. Beck, *J. Organomet. Chem.* **1989**, *369*, 105.

[79] P. J. Brothers, W. R. Roper, *Chem. Rev.* **1988**, *88*, 1293.

[80] C.-Y. Liu, D.-Y. Chen, G.-H. Lee, S.-M. Peng, S.-T. Liu, *Organometallics* **1996**, *15*, 1055.

[81] (a) K. Bartel, W. P. Fehlhammer, *Angew. Chem. Int. Ed. Engl.* **1974**, *13*, 599; (b) U. Kernbach, W. P. Fehlhammer, *Inorg. Chim. Acta* **1995**, *235*, 299; (c) W. P. Fehlhammer, K. Bartel, B. Weinberger, U. Plaia, *Chem. Ber.* **1985**, *118*, 2220.

[82] P. Jutzi, U. Gilge, *J. Organomet. Chem.* **1983**, *246*, 159.

[83] F. E. Hahn, L. Imhof, *Organometallics* **1997**, *16*, 763.

[84] F. E. Hahn, M. Tamm, *J. Chem. Soc. Chem. Commun.* **1995**, 569.

[85] (a) M. Tamm, F. E. Hahn, *Coord. Chem. Rev.* **1999**, *182*, 175; (b) F. E. Hahn, V. Langenhahn, D. Le Van, M. Tamm, L. Wittenbecher, T. Lügger, *Heteroatom Chem.* **2002**, *13*, 540; (c) M. Basato, R. A. Michelin, M. Mozzon, P. Sgarbossa, A. Tassan, *J. Organomet. Chem.* **2005**, *690*, 5414.

[86] A. C. Dumke, T. Pape, J. Kösters, K.-O. Feldmann, C. Schulte to Brinke, F. E. Hahn, Organometallics **2013**, 32, 289.

[87] H. Staudinger, J. Meyer, Helv. Chim. Acta **1919**, 2, 635

[88] (a) F. E. Hahn, V. Langenhahn, N. Meier, T. Lügger, W. P. Fehlhammer, *Chem. Eur. J.* **2003**, *9*, 704; (b) F. E. Hahn, V. Langenhahn, T. Pape, *Chem. Commun.* **2005**, 5390.

[89] A. M. Voutchkova, M. Feliz, E. Clot, O. Eisenstein, R. H. Crabtree, *J. Am. Chem. Soc.* **2007**, *129*, 12834.

[90] (a) I. Tommasi, F. Sorrentino, *Tetrahedron Lett.* **2006**, *47*, 6453; (b) A. Tudose, A. Demonceau, L. Delaude, *J. Organomet. Chem.* **2006**, *691*, 5356.

[91] J. D. Holbrey, W. M. Reichert, I. Tkatchenko, E. Bouajila, O. Walter, I. Tommasi, R. D. Rogers, *Chem. Commun.* **2003**, 28.

[92] M. Fèvre, J. Pinaud, A. Leteneur, Y. Gnanou, J. Vignolle, D. Taton, K. Miqueu, J.-M. Sotiropoulos, *J. Am. Chem. Soc.* **2012**, *134*, 6776.

[93] (a) H. G. Raubenheimer, S. Cronje, *J. Organomet. Chem.* **2001**, *617–618*, 170; (b) H. G. Raubenheimer, F. Scott, S. Cronje, P. H. van Rooyen, K. Psotta, *J. Chem. Soc. Dalton Trans.* **1992**, 1009; (c) H. G. Raubenheimer, G. J. Kruger, A. van A. Lombard, L. Linford, J. C. Viljoen, *Organometallics* **1985**, *4*, 275; (d) H. G. Raubenheimer, Y. Stander, E. K. Marais, C. Thompson, G. J. Kruger, S. Cronje, M. Deetlefs, *J. Organomet. Chem.* **1999**, *590*, 158.

[94] T. Kösterke, J. Kösters, E.-U. Würthwein, C. Mück-Lichtenfeld, C. Schulte to Brinke, F. Lahoz, F. E. Hahn, *Chem. Eur. J.* **2012**, *18*, 14594.

[95] R. Jothibasu, H. V. Huynh, *Organometallics* **2009**, *28*, 2505.

4 Group 10 Metal(0)-NHC Complexes

N-heterocyclic carbene (NHC) complexes of group 10 transition metals, in particular those of palladium, are among the most popular and widely studied in the organometallic chemistry of NHCs. One of the contributing factors to their popularity is undoubtedly the fact that complexes of these metals are often highly active catalysts for a wide range of organic transformations. Therefore, the chemistry of these metals is relatively well understood, and a number of methodologies have been developed for their preparation. Commonly, most NHC complexes of this group can be found in the oxidation states of 0 or +II. Oxidation states of +I or +IV are known, but still quite rare and will thus not be covered explicitly.

Many typical transition metal catalyzed C–C coupling reactions commence with an oxidative addition to a zero-valent group 10 metal center, which is stabilized by an ancillary spectator ligand such as a tertiary phosphine. The strongly donating ligand also makes the metal center more electron rich, thereby facilitating oxidative addition. Since NHCs and phosphines share similarities in their electronic properties, much of the metal-NHC chemistry described here has been modeled after that of the older and more established phosphine congeners.

In this chapter, some common synthetic NHC chemistry related to group 10 metals in their zero-valent oxidation state will be discussed.

4.1 Nickel(0)-NHC Complexes

4.1.1 Reactions of Enetetramines and Free NHCs

One of the earliest reports about the preparation of nickel(0) NHC complexes had already appeared in 1977. In this approach, Lappert and coworkers made use of electron-rich entetramines as the source for saturated NHCs. They described how the reaction of such carbene dimers with tetracarbonylnickel(0) in non-polar solvents and at elevated temperature afforded different mixed carbonyl-NHC complexes of zero-valent nickel that can be controlled by stoichiometry (Scheme 4.1) [1].

Reacting the olefin with two equivalents of the nickel(0) precursor gave the tetrahedral and 18-electron tricarbonyl(imidazolidin-2-ylidene)nickel(0) complex $[Ni(CO)_3(NHC)]$ in a high yield of ~90% as white crystals. The replacement of the comparatively weaker donating carbonyl ligand with a saturated NHC as a very strong electron donor is favorable and at the same enhances backbonding from the metal center to the remaining three carbonyl ligands. This strengthens the respective metal-CO bonds, and as a result, solutions of these monocarbene nickel(0) complexes in hexane show enhanced thermostability even up to ~60 °C. Nevertheless, substitution of a second carbonyl ligand by another strong

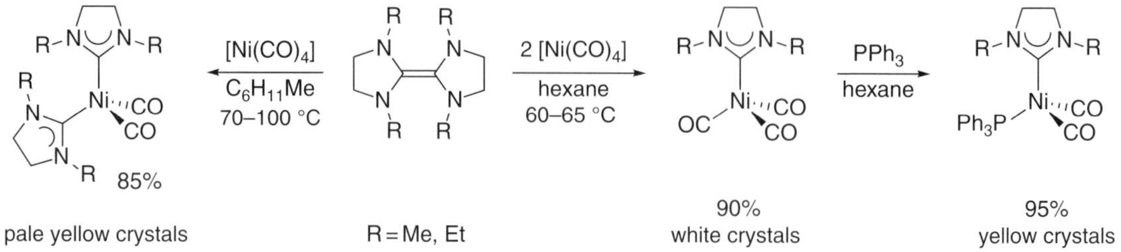

Scheme 4.1
Preparation of tetrahedral carbonyl containing nickel(0)-NHC complexes by cleavage of enetetramines.

Table 4.1   Comparison of the carbonyl IR stretching frequencies (cm^{-1}) in nickel(0)-NHC complexes.

	[Ni(CO)$_3$(NHC^R)]	[Ni(CO)$_2$(NHC^R)(PPh$_3$)]	[Ni(CO)$_2$(NHC^R)$_2$]
R = Me	2058, 1978[a]	1978, 1903[b]	1938, 1865[b]
R = Et	2055, 1976[a]	1972, 1908[b]	1935, 1859[a]

[a] measured in hexane;
[b] measured in methylcyclohexane;
Note: $\tilde{v}$[Ni(CO)$_4$]^2 = 2057 cm^{-1}; $\tilde{v}$[Ni(CO)$_2$(PPh$_3$)$_2$]^3 = 2007, 1952 cm^{-1}.

donor is facile. For example, addition of triphenylphosphine to a hexane solution of the tricarbonyl nickel(0) complex leads to immediate evolution of CO and formation of heteroleptic mixed NHC/phosphine complexes of the formula [Ni(CO)$_2$(NHC)(PPh$_3$)], which precipitate as yellow solids.

The preparation of bis(carbene)nickel(0) complexes of the type [Ni(CO)$_2$(NHC)$_2$] can be accomplished in a single step by prolonged heating of a 1:1 mixture of the enetetramine with tetracarbonylnickel(0) in methylcyclohexane, which liberates two molecules of carbon monoxide. The combination of two strong NHC donors and two strong π-acceptors on nickel(0) further increases the stabilization of the complex. Thus, the pale yellow bis(NHC) complex is even more thermally stable than its monocarbene counterpart, and does not decompose in solution up to 150 °C.

The IR spectroscopic data for the carbonyl stretches obtained for the three types of nickel(0)-NHC complexes are summarized in Table 4.1. Their comparison shows that the tricarbonyl-mono(NHC) complexes have the largest CO wavenumbers and are therefore the least electron-rich ones in this series. Substitution of a second carbonyl with a triphenylphosphine ligand leads to an increase in electron density resulting in stronger backdonation to the carbonyls as indicated by a decrease in their stretching frequencies. The bis(carbene) complexes contain the highest electron density as evidenced by the smallest wavenumbers, which supports the view that NHCs are clearly superior to phosphines in terms of electron donation to metal centers.

The same conclusions can be drawn from comparison of the ^{13}C NMR spectroscopic data for the carbene and the carbonyl carbon donors listed in Table 4.2. Thus substitution of the first CO ligand from the homoleptic complex precursor leads to a significant downfield shift of the carbonyl carbon from 191.6 to 199.1 ppm, while the resonance for the carbene donor is observed at 214.4 ppm. The bis(NHC) complex is expected to be more electron rich, which is substantiated by further downfield shifts of the resonances for both carbon donors to 205.4 and 223.4 ppm, respectively. Comparison of the carbonyl chemical

Table 4.2 Comparison of ^{13}C NMR spectroscopic data (ppm) for the carbene carbon and the carbonyl donor measured in C_6D_6.

	$^{13}C_{carbene}$	^{13}CO
[Ni(CO)$_4$][4]	–	191.6
[Ni(CO)$_3$(NHC^{Et})]	214.4	199.1
[Ni(CO)$_2$(NHC^{Et})$_2$]	223.4	205.4
[Ni(CO)$_2$(PPh$_3$)$_2$][4b]	–	199.4[a]

[a] measured in CDCl$_3$

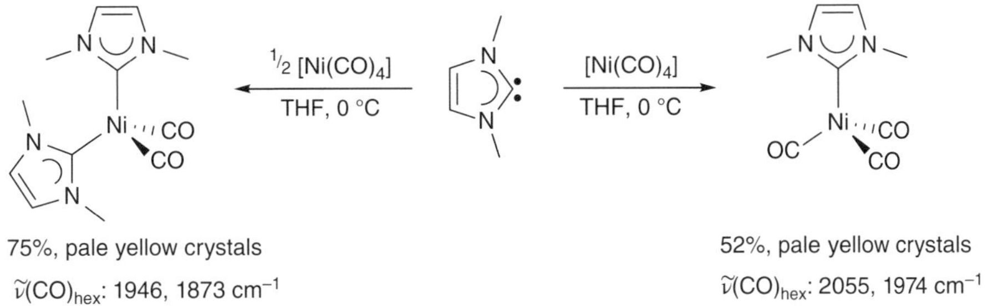

75%, pale yellow crystals

$\tilde{\nu}$(CO)$_{hex}$: 1946, 1873 cm^{-1}

52%, pale yellow crystals

$\tilde{\nu}$(CO)$_{hex}$: 2055, 1974 cm^{-1}

Scheme 4.2
Preparation of carbonyl containing nickel(0)-IMe complexes using the free NHC.

shift in the bis(NHC) with that observed for the bis(phosphine) nickel(0) complex at higher field (i.e. 199.4 ppm) leaves no doubt that the latter is less electron dense, providing further evidence that NHCs are stronger donors than phosphines.

In analogy to the reaction with enetetramines, tetrahedral Nickel(0) complexes of unsaturated 1,3-dimethylimidazolin-2-ylidenes can also be obtained by reacting the free NHCs with [Ni(CO)$_4$] as reported by Öfele in 1993 (Scheme 4.2) [5].

The use of one equivalent of the proligand IMe affords the mono-NHC-tricarbonyl compound [Ni(CO)$_3$(IMe)], whereas two equivalents of IMe displaces two carbonyl ligands giving the bis(NHC)-dicarbonyl complex [Ni(CO)$_2$(IMe)$_2$]. Both nickel(0)-NHC complexes were isolated as pale yellow crystalline materials. Notably, these reactions could be successfully carried out in tetrahydrofurane at 0 °C, which indicates that the direct ligand substitution requires less activation energy and is thus more facile than those involving the carbene dimers described above. This is intuitive as no activation energy for the cleavage of the C=C double bond is required in the latter case. The tetrahedral geometry of the bis(NHC) nickel(0) complex [Ni(CO)$_2$(IMe)$_2$] was also confirmed by an X-ray-diffraction study on single crystals obtained from a saturated solution in diethyl ether at −30 °C. The IR spectroscopic data for the carbonyl stretches reported for these two complexes confirm the expected greater electron density of the bis(NHC) complex. A more detailed IR spectroscopic comparison of the mono-NHC complex [Ni(CO)$_3$(IMe)] with Lappert's analog [Ni(CO)$_3$(SIMe)] bearing a saturated 1,3-dimethylimidazolidin-2-ylidene (SIMe) measured in the same solvent, however, is less conclusive as the latter unexpectedly shows larger wavenumbers for the equivalent stretches. This discrepancy may indicate some limitation in the use of IR spectroscopy for the accurate determination of NHC donor strengths on a finer level [6].

Using the same methodology reported by Öfele, the groups of Hermann [7] and in particular Nolan [8] later prepared a small library of analogous [Ni(CO)$_3$(NHC)] complexes

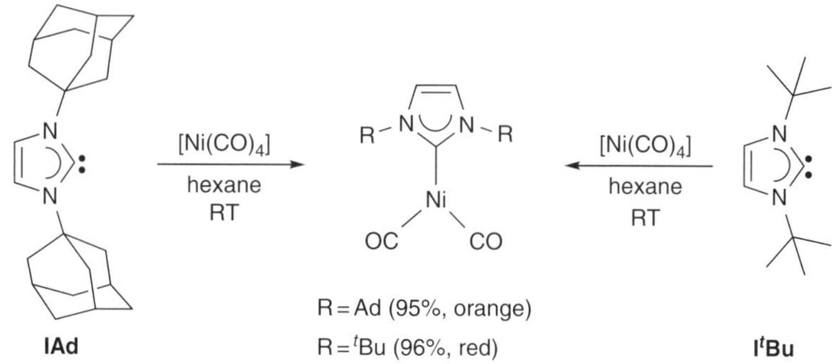

Scheme 4.3
Preparation of tricoordinated [Ni(CO)$_2$(NHC)] complexes with bulky NHCs.

primarily for the determination of TEP values for the respective NHCs. In the course of this study, Nolan and coworkers also noted that very bulky NHCs, such as IAd and I^{t}Bu, would replace two carbonyl ligands from [Ni(CO)$_4$] leading to rare tricoordinated, 16-electron nickel(0) complexes of the formula [Ni(CO)$_2$(NHC)], which were isolated as orange and red solids, respectively (Scheme 4.3).

Attempts to isolate the desired (but elusive) tricarbonyl complexes [Ni(CO)$_3$(IAd)] or [Ni(CO)$_3$(I^{t}Bu)] for TEP evaluation of these NHCs were futile. The different reactivity of these two NHCs compared to, for example, IMes, SIMes, IPr, SIPr, or ICy, was ascribed to their increased steric bulk in line with their larger values for %V_{Bur} (see Chapter 2). This notion is also supported by calculation of their carbonyl bond dissociation energy (BDE) in [Ni(CO)$_3$(NHC)] complexes. Carbonyl BDE values for complexes with IAd, I^{t}Bu, and also SI^{t}Bu (BDE $\approx$ 10 kcal mol^{-1}) were found to be much lower than those for the other NHCs (BDE > 25 kcal mol^{-1}) indicating that complexes containing the first group of NHCs are more likely to release a CO molecule. The similar values obtained for I^{t}Bu and SI^{t}Bu, which are sterically similar but electronically different, supports the view that steric pressure is the major contributor for the observed reactivity difference. Single crystal X-ray diffraction analysis (Figure 4.1) revealed that the Ni–C$_{carbene}$ distances range from 1.960–1.979 Å for the [Ni(CO)$_3$(NHC)] complexes, while an averaged value of 1.955 Å was noted for the two [Ni(CO)$_2$(NHC)] complexes.

Apart from the heteroleptic 18- and 16-electron nickel(0) complexes described so far, a number of homoleptic 14-electron homoleptic bis(carbene)nickel(0) complexes have also been prepared. The first representative among these is the nickel(0) complex [Ni(IMes)$_2$], that was isolated by Arduengo and coworkers in the mid-1990s in a relatively high yield of 83% (Scheme 4.4) [9]. The group drew inspiration from analogous homoleptic two-coordinate copper(I) and silver(I) NHC complexes with bulky substituents that had been synthesized by them previously. The complex was synthesized by the reaction of one equivalent of 18-electron bis(1,5-cyclooctadiene)nickel(0) [Ni(COD)$_2$] with two equivalents of free 1,3-dimesitylimidazolin-2-ylidene (IMes) in tetrahydrofuran (THF) at room temperature according to Scheme 4.4.

The two NHCs simply replace the weaker COD ligands, and the geometry at the nickel center changes from initially tetrahedral to eventually linear. Due to the air-sensitivity of IMes, [Ni(COD)$_2$] and the resulting nickel(0)-NHC complex, the reaction has to be carried out under a dry nitrogen atmosphere. Dark violet single crystals were grown from a saturated solution in hexane at −25 °C, and analysis of these by X-ray diffraction confirmed the

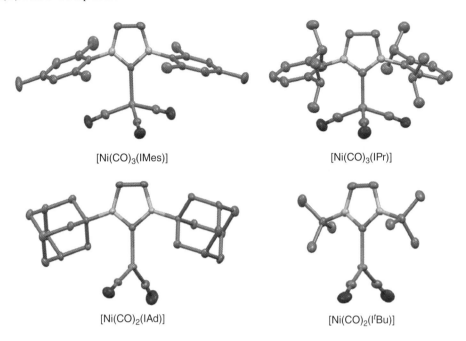

Figure 4.1
Representative molecular structures for tri- and dicarbonyl nickel(0)-NHC complexes.

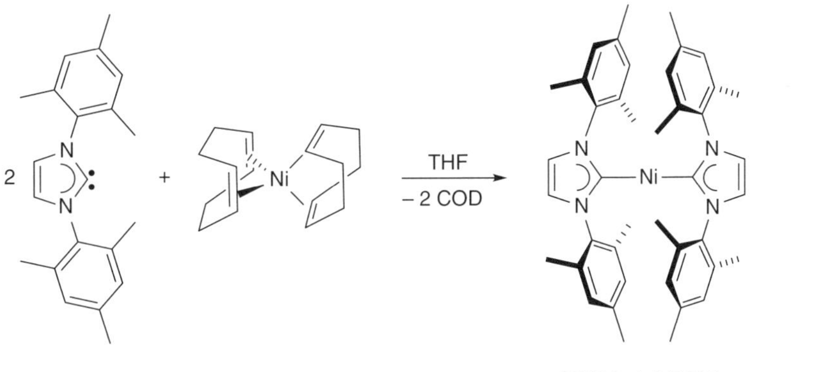

Scheme 4.4
Preparation of the homoleptic, linear nickel(0)-NHC complex [Ni(IMes)$_2$].

linear coordination geometry. It is interesting to note that the Ni–C$_{carbene}$ distances of 1.827(6) and 1.830(6)Å are significantly shorter than those reported for the 18-electron complexes [Ni(CO)$_3$(NHC)] discussed above. Based on this observation, the conclusion can be made that a certain amount of $d_\pi \rightarrow p_\pi$ backdonation from the electron-rich zerovalent metal to the carbene carbon is operative in this homoleptic complex.

The same methodology has also been straightforwardly applied for the preparation of [Ni(IPr)$_2$], which also contains two imidazolin-2-ylidenes with bulky and aromatic N-substituents [10]. The preparation of homoleptic bis(imidazolin-2-ylidene) nickel(0) complexes with aliphatic N-substituents is more challenging, and the first example has been prepared by Cloke and coworkers using an unusual metal vapor synthesis (MVS) approach [11]. The method involves evaporation of a metal pellet with an electron beam gun followed by co-condensation with the ligand vapor at −196 °C (Scheme 4.5). The ligand 1,3-di-*tert*-butylimidazolin-2-ylidene (I^{t}Bu) was chosen, due to the ease of its synthesis and its suitably high vapor pressure at low temperature, thus meeting the requirement for MVS [12]. Although an excess of ligand was introduced into the system, the yield of the brown microcrystalline product [Ni(I^{t}Bu)$_2$] amounts to only 10% (Figure 4.2, left).

Due to the extreme conditions and specialized equipment required, this methodology has limited common applicability. Nevertheless, it must be noted that later attempts to prepare [Ni(I^{t}Bu)$_2$] by conventional solution phase synthesis involving ligand substitution of [Ni(COD)$_2$] with I^{t}Bu were futile. Instead these reactions proceed with *tert*-butyl cleavage from the I^{t}Bu ligand to give different mono(NHC)-stabilized Ni complexes in different oxidation states [13]. Based on these observations one can make the conclusion that ligand substitution from [Ni(COD)$_2$] is only suitable for NHCs with aromatic wing tip groups.

The increased reactivity of zero-valent nickel complexes containing imidazolin-2-ylidenes with aliphatic N-substituents can be rationalized by their increased electron density compared to related NHC complexes with N-aryl substituents. Further support for this concept is provided by the reaction of two equivalents of 1,3-diisopropylimidazolin-2-ylidene (I^{i}Pr) with [Ni(COD)$_2$] reported by the group of Radius [14]. The two isopropyl

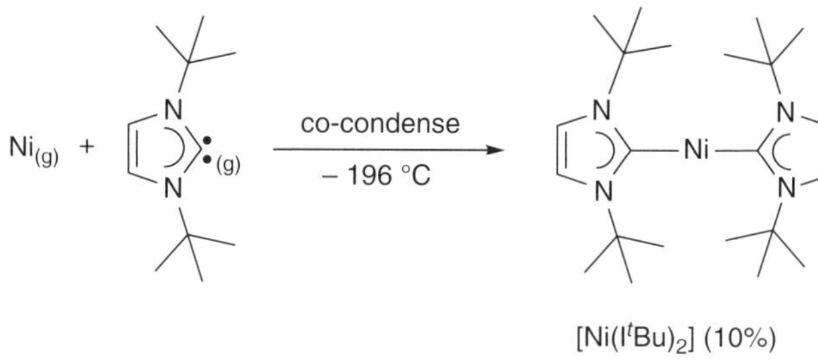

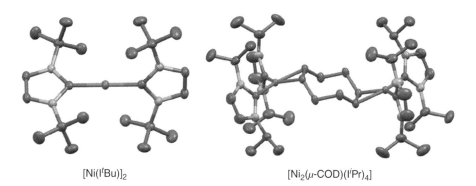

[Ni(I^{t}Bu)$_2$] (10%)

Scheme 4.5
Metal vapor synthesis of [Ni(I^{t}Bu)$_2$].

[Ni(I^{t}Bu)]$_2$ [Ni$_2$(μ-COD)(I^{i}Pr)$_4$]

Figure 4.2
Molecular structures of [Ni(I^{t}Bu)$_2$] representing a mononuclear bis(NHC) nickel(0) complex (left) and a COD-bridged tetra(NHC) dinickel(0) complex (right).

N-substituents in I^{i}Pr make the ligand less sterically demanding and also less electron rich compared to I^{t}Bu. Nevertheless, it is still much more electron rich compared to the aryl-substituted counterparts IMes or IPr. The reaction of I^{i}Pr with [Ni(COD)$_2$] gives a quite different outcome, with incomplete COD replacement. As a major product, an air sensitive COD-bridged, dinuclear nickel(0) complex was isolated (Scheme 4.6).

Each nickel(0) center in this complex is ligated by two I^{i}Pr and one olefin unit of the now bridging 1,5-cyclooctadiene (Figure 4.2, right). In this arrangement, the excess electron density at each nickel(0) center can be efficiently backdonated into the π* molecular orbitals of the diolefin, which is a better π-acceptor compared to the I^{i}Pr ligands. Overall, this enhances Ni–COD bonding, which also stabilizes the complex by reducing otherwise vacant coordination sites. The greater coordination number in this complex also leads to elongated Ni–C$_{carbene}$ bond distances of 1.906(3) and 1.904(3) Å compared to those in the linear and homoleptic complexes [Ni(IMes)$_2$] and [Ni(I^{t}Bu)$_2$] (Figure 4.2, left).

Nickel(0)-NHC complexes of carbenes derived from benzimidazole are very rare. The only example of a bis(benzimidazolin-2-ylidene)nickel(0) complex was reported by the group of Lappert in 1999 [15]. They followed Arduengo's procedure and reacted two equivalents of the free 1,3-dineopentylbenzimidazolin-2-ylidene with [Ni(COD)$_2$] and isolated the violet target compound [Ni(Np$_2$-bimy)$_2$] in 82% yield (Scheme 4.7). The signal for the carbene carbon was detected at 207.6 ppm in C$_6$D$_6$.

Surprisingly, well-defined nickel(0)-complexes bearing NHCs derived from 1,2,4-triazoles are unknown thus far [16]. Nevertheless, they have been generated *in situ* by mixing [Ni(COD)$_2$] with two equivalents of Ender's carbene (1,3,4-triphenyl-1,2,4-triazolin-5-ylidene) and studied as catalyst for the dehydrogenation of ammonia–borane [17].

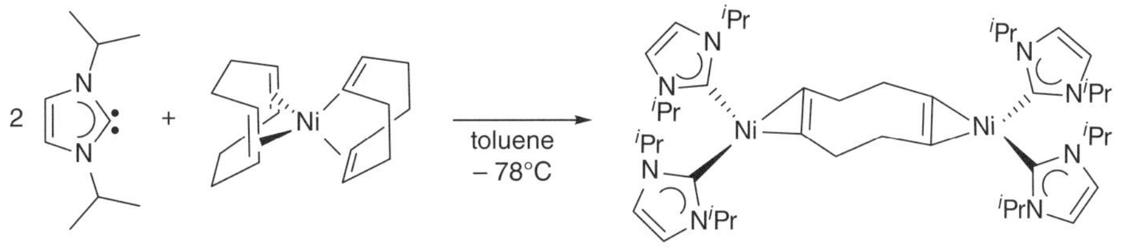

Scheme 4.6
Preparation of a COD-bridged dinuclear nickel(0)-NHC complex.

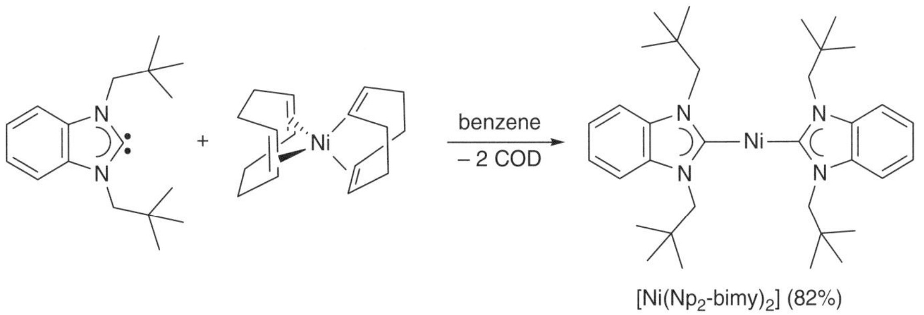

[Ni(Np$_2$-bimy)$_2$] (82%)

Scheme 4.7
Preparation of bis(1,3-dineopentylbenzimidazolin-2-ylidene)nickel(0).

4.1.2 Reduction of Nickel(II)-NHC Complexes

Nickel(0)-NHC complexes can also be prepared via less sensitive nickel(II) precursors. Danopoulos and coworkers applied a two-step synthesis, involving a reductive elimination step, to prepare nickel(0) complexes with different NHCs. One equivalent of [Ni(CH$_3$)$_2$(tmeda)] (tmeda = N,N,N′,N′-tetramethylethylenediamine) was reacted with two equivalents of free NHC in tetrahydrofurane under gradual increase of the reaction temperature from −78 to 25 °C (Scheme 1.3) [18]. The free NHC was either preformed or generated *in situ* by deprotonation of the precursor salt with KO^{t}Bu. This approach led to the isolation of [Ni(IMes)$_2$], [Ni(IPr)$_2$], [Ni(SIPr)$_2$], and [Ni(SI^{t}Bu)$_2$] in higher yields. This alternative methodology appears to be more advantageous in particular for the metallation of NHCs with aliphatic substituents. Better insights into the reaction pathway were obtained by reacting two equivalents of 1,3-diisopropyl-4,5-dimethylimidazolin-2-ylidene (I^{i}Pr$_{Me2}$) with [Ni(CH$_3$)$_2$(tmeda)]. The neutral NHCs displace the neutral, but chelating tmeda ligand in the first step leading to the *cis*-configured bis(NHC) nickel(II) complex *cis*-[Ni(CH$_3$)$_3$(I^{i}Pr$_{Me2}$)$_2$] that was isolated as yellow crystals and also characterized by single crystal X-ray diffraction. In general, the substitution of a chelating ligand (here tmeda) by two monodentate ligands is entropically not favorable, but it highlights the very strong *trans* effect of the methyl ligand. The main contributor to the negative free Gibbs energy of this reaction is therefore the enthalpy term driven by the favorable binding of the stronger donating NHCs to the nickel center. Upon heating, the two anionic methyl ligands in this nickel(II) complex reductively eliminate under formation of ethane among other byproducts to give the nickel(0)-bis(NHC) complex [Ni(I^{i}Pr$_{Me2}$)$_2$] as the final product (Scheme 4.8). Notably, 2-methylimidazolium salts that could be formed by methyl and NHC coupling were not observed as elimination products.

Nickel(0) NHC complexes can also be generated reduction of Ni^{II}-NHC precursors by addition of external reductants. This is an attractive alternative, since the handling of instable, air sensitive, and often toxic nickel(0) precursors can be avoided. As such, Matsubara and coworkers reported a two-step protocol for the synthesis [Ni(IPr)$_2$] that involves the use of nickel(II) acetylacetonate [Ni(acac)$_2$] as a nickel source [19]. The first step simply involves the coordination of the free IPr ligand to a [Ni(acac)$_2$] fragment giving rise to a 5-coordinate [Ni(acac)$_2$(IPr)] complex, which can be isolated as a green compound (Scheme 4.9). Optimization of reaction conditions for the subsequent reduction of the latter were carried out in a typical manner involving screening of various reducing agents and different solvents. The use of NaH in the presence of one additional equivalent of IPr ligand finally afforded the target compound [Ni(IPr)$_2$] in a moderate yield of 45%. Notably, the one-pot and direct reduction of [Ni(acac)$_2$] using NaH in 1,4-dioxane with the presence of

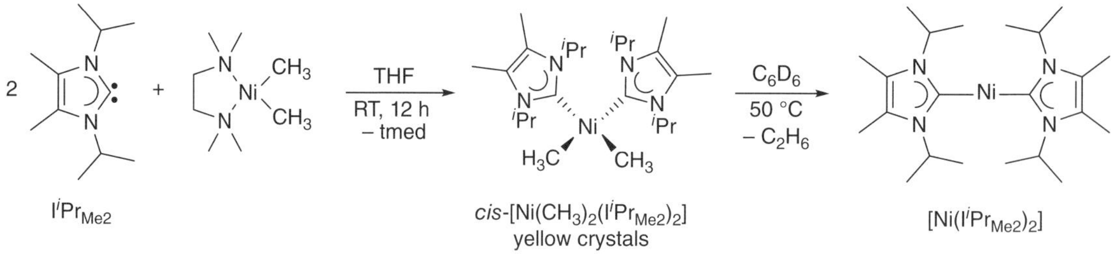

Scheme 4.8
Preparation of a homoleptic nickel(0)-NHC complex by reductive elimination of ethane.

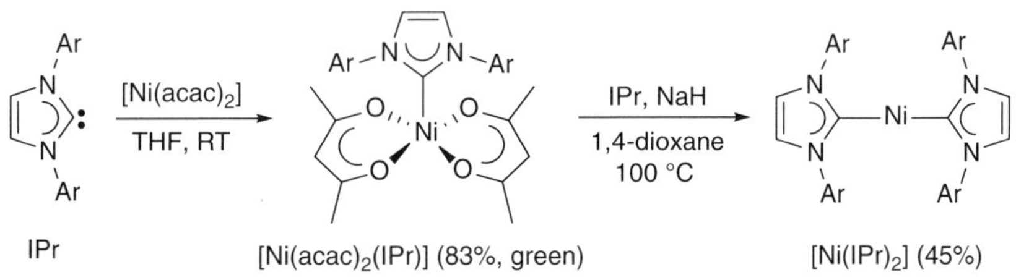

Scheme 4.9
Preparation of [Ni(IPr)$_2$] by NaH reduction of a Ni^{II}-NHC precursor.

Table 4.3 Comparison of the ^{13}C NMR resonances of the carbene atom in homoleptic bis(NHC) nickel(0) complexes measured in C$_6$D$_6$ (ppm).

	[Ni(IMes)$_2$]	[Ni(I^{t}Bu)$_2$]	[Ni(IPr)$_2$]	[Ni(Np$_2$-bimy)$_2$]	[Ni(SI^{t}Pr)$_2$]	Ni(SI^{t}Bu)$_2$]
^{13}C$_{NCN}$	193.2	191.2	193.8	207.6	211.2	211.2

two equivalents of imidazolium salt also gave the desired product even in a slightly better yield of 53%. Examination of its solid state structure showed that the Ni–C$_{carbene}$ bond lengths of 1.856(2) and 1.872(2)Å are slightly longer than those in the mesityl analog [Ni(IMes)$_2$] (see Section 4.1.1), which is most likely due to steric repulsion between the two more bulky IPr ligands.

Table 4.3 shows a comparison of the ^{13}C NMR data for the carbene atom in homoleptic bis(NHC) nickel(0) complexes containing classical imidazolin-2-ylidenes, benzimidazolin-2-ylidenes, and imidazolidin-2-ylidenes. As with the ^{13}C NMR data for the respective free proligands, the chemical shift for the benzannulated derivative is found to be intermediate between those from the unsaturated and saturated NHC analogs. The most downfield signals are exhibited by complexes of saturated NHCs, and coordination to nickel(0) leads to an upfield shift of approximately equal magnitude for all three types of classical NHCs.

Given the increasing interest in Ni complexes as economically more attractive catalysts, one can anticipate further studies in this field. This is particularly true for nickel(0) complexes of carbenes derived from benzimidazole or 1,2,4-triazole, of which currently no detailed structural, spectroscopic, and reactivity data are available.

4.2 Palladium(0)-NHC Complexes

4.2.1 Reactions of Free NHCs

Phosphine Substitution Reactions

Palladium(0) complexes are important catalysts for a wide range of transformations due to the ease with which they can activate substrates via oxidative additions. This step is particularly favorable when the metal center is ligated by strong electron donors, which explains the interest in palladium(0)-NHC complexes. Initial studies focused on the applications of *in situ* generated palladium(0)-NHC species [20], and the first successful attempt to isolate a well-defined complex was reported by the Cloke group, who used an unusual MVS approach already described for a nickel(0)-NHC analog. Here, palladium pellets were evaporated, and co-condensation with excess I^{t}Bu led to the formation of the

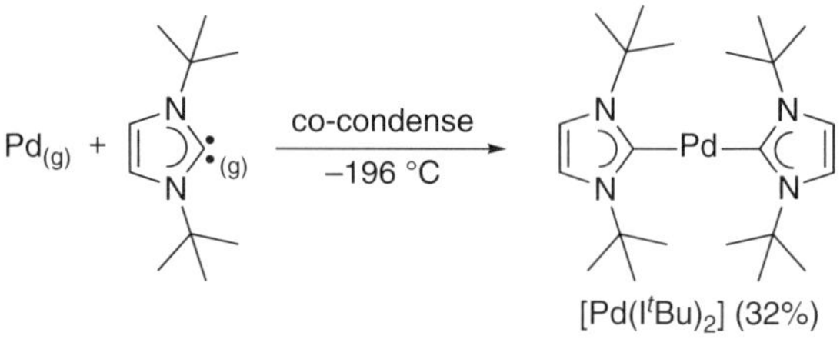

Scheme 4.10
Metal vapor synthesis of [Pd(I^{t}Bu)$_2$].

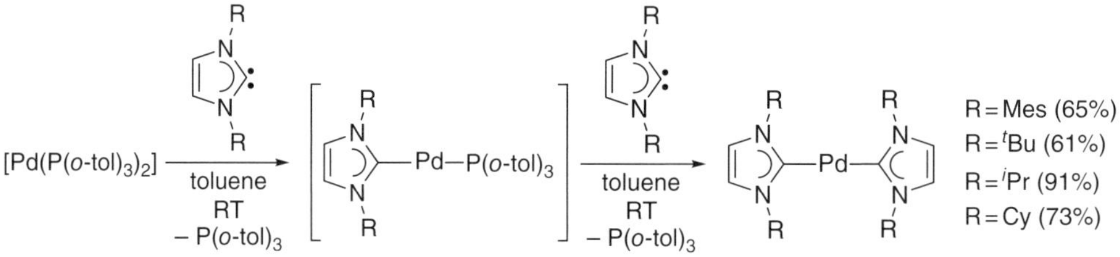

Scheme 4.11
Preparation of the homoleptic, linear palladium(0)-NHC complexes by phosphine displacement.

homoleptic complex [Pd(I^{t}Bu)$_2$], which after a workup was isolated as yellow crystals in a 32% yield (Scheme 4.10) [11].

The involvement of specialized equipment, the limitation to volatile free carbenes and the low yield makes this route unfavorable for general applications. Hencec, simpler and more efficient methods were sought, and the initial alternative for the preparation of homoleptic palladium(0) complexes under standard laboratory techniques involved a ligand exchange of free NHCs with a suitable palladium(0)-phosphine precursor complex. The group of Herrmann tested bis(tri-*ortho*-tolyphosphine)palladium(0), tetrakis(triphenylphosphine)palladium(0) and bis(tricyclohexyl-phosphine)palladium(0) as precursors, but only the former was found to give the target product in good yields after ligand exchange. In the optimized procedure, a slurry of [Pd(P(o-tol)$_3$)$_2$] in toluene was treated with a solution of the free NHC proligand in toluene at room temperature and stirring continued for about 30–40 minutes. The carbenes were used in slight excess. Monitoring of the reaction by NMR spectroscopy revealed a step-wise reaction with the intermediacy of a mixed carbene-phosphine palladium(0) complex, which upon second ligand displacement afforded the desired homoleptic bis(NHC) palladium(0) complex (Scheme 4.11) [21]. The 14-electron complexes were applied as catalyst in the Suzuki-Miyaura cross-coupling reaction, where they showed distinct differences to catalytic systems generated *in situ*, which may point to different mechanisms.

Extension of the carbene scope to a saturated bulky 1,3-diarylimidazolidin-2-ylidene (SIPr) was possible, and careful stoichiometry control also allowed for the isolation of heteroleptic mixed NHC/phosphine complexes. More important, it was also found that well-defined bis(NHC) palladium(0) complexes could still readily undergo NHC-phosphine exchange or ligand redistributions to give mixed complexes (Scheme 4.12) [22]. This observation highlights that the metal-carbene bond, at least in these systems, is more labile than generally assumed.

The application of the phosphine substitution method to obtain a homoleptic palladium(0) complex with the very bulky 1,3-diadamantylimidazolin-2-ylidene ligand,

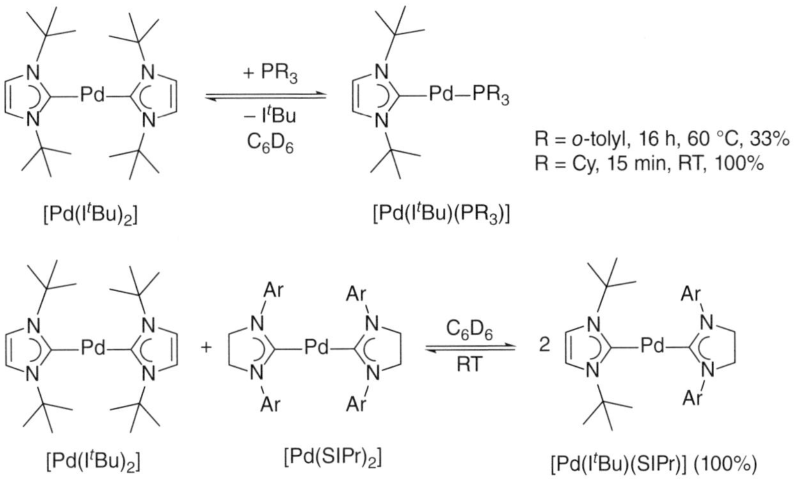

Scheme 4.12
Ligand exchange and redistribution reactions of bis(NHC) palladium(0) complexes.

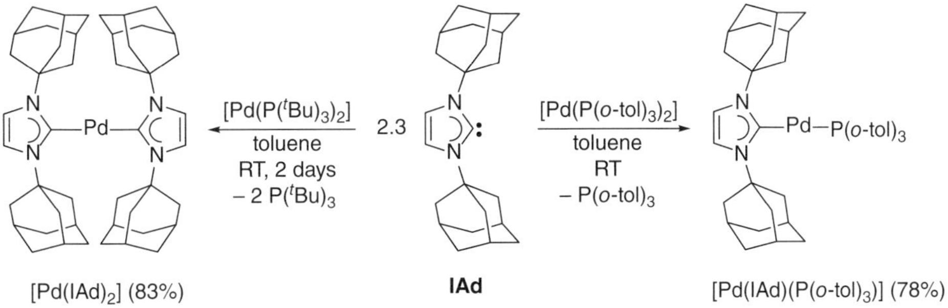

Scheme 4.13
Synthesis of mixed IAd-phosphine and bis(IAd) complexes of palladium(0).

however, did not succeed, but stopped at the mixed carbene-phosphine intermediate [Pd(IAd)(P(*o*-tol)$_3$)] [23]. The low reactivity of this complex for further phosphine substitution may be attributed to both steric and electronic factors. First, the phosphine-metal bond could be enhanced by π-backdonation from the electron-rich palladium(0) center. Second, the approach of the extremely bulky NHC ligand to the already bulky heteroleptic intermediate is anticipated to be difficult. The final finding, that bis(tri-*tert*-butylphosphine)palladium(0) turned out to be a more suitable starting material for obtaining the target complex, seems to corroborate the first electronic factor, since tri-*tert*-butylphosphine is indeed a much poorer π-acceptor compared to the triarylphosphine. Using [Pd(P(*t*Bu)$_3$)$_2$], the bright yellow complex [Pd(IAd)$_2$] was obtained in a good yield of 83% after a prolonged reaction time of two days (Scheme 4.13) [23].

Alkene Substitution Reactions

In addition to two-coordinate, linear 14-electron complexes, palladium(0)-NHC species with formally three ligands have been reported as well. Cavell and coworkers described that the reaction of two equivalents of 1,3,4,5-tetramethylimidazolin-2-ylidene (IMe$_{Me2}$) with mixed alkene-cyclooctadiene palladium(0) complexes of the type [Pd(alkene)(COD)] {alkene = maleic anhydride (MAH), tetracyanoethylene (TCNE)} proceeds with clean cyclooctadiene substitution to afford the 16-electron bis(carbene)(alkene) complexes of the general formula [Pd(alkene)(IMe$_{Me2}$)$_2$] in high yields (Scheme 4.14) [24]. The preference for the replacement of the bidentate COD over the "monodentate" MAH or TCNE ligands indicates that the latter two are more strongly bound to the metal center due to enhanced backdonation as a result of their electron deficiency, which makes them better π-acceptors.

The bis(carbene) complexes show very good stabilities, and decomposition of the solid maleic anhydride complex in air occurred only after several days, while the tetracyanoethylene derivative appears to be indefinitely stable in air. Again, this can be explained by the efficient transfer of electron density from the NHC via the palladium(0) center to the olefin. Evidence for the enhanced backbonding was provided by NMR spectroscopy, where large upfield shifts were observed for the MAH protons ($\Delta\delta_H = 3.71$ ppm) and the C=C carbon atoms ($\Delta\delta_C = 32.2$ ppm) upon coordination. In agreement, IR spectroscopic analyses of these complexes showed a significant shift to smaller wavenumbers for the carbonyl stretches of MAH and cyano stretches of TCNE functions, respectively, in comparison to the free alkenes. As a result the double bond character of the alkene ligand decreases, and both complexes can be regarded as approaching palladium(II) derivatives (Figure 4.3). Nevertheless, these complexes underwent oxidative addition reactions with aryl halides and halogens that are still typical for palladium(0) complexes.

The palladium(0)-NHC complexes described thus far contain either two carbenes or an additional phosphine ligand. Monocarbene-palladium(0) complexes without supporting phosphine ligands are, on the other hand, quite rare. The first examples of such species were reported by the Beller group, who employed a very similar approach by reacting free NHCs with palladium(0)-alkene complexes. The first representative was obtained by mixing two equivalents of IMes with one equivalent of the dinuclear diallylether-palladium(0)

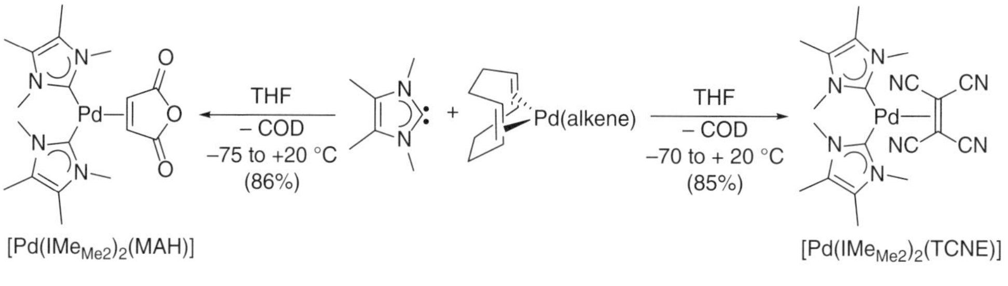

[Pd(IMe$_{Me2}$)$_2$(MAH)]   [Pd(IMe$_{Me2}$)$_2$(TCNE)]

Scheme 4.14
Preparation of three-coordinate bis(NHC)palladium(0) complexes.

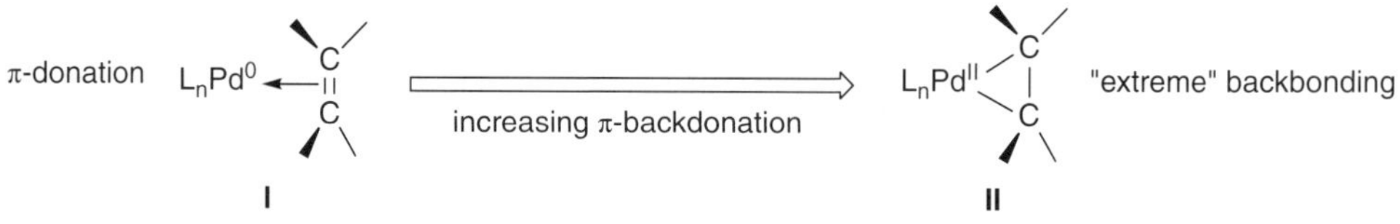

Figure 4.3
Two extremes in the bonding of an olefin to palladium: pure donor bonding (I) and pure acceptor bonding (II).

complex [Pd$_2$(DAE)$_3$] in 1,1,3,3-tetramethyl-1,3-divinyldisiloxane (DVDS) [25]. The resulting complex [Pd(DVDS)(IMes)] contains a palladium(0) center that is ligated by one IMes and one bidentate DVDS ligand resulting in a 16-electron species, which was reported to be highly active in the telomerization reaction of butadiene in methanol (Scheme 4.15).

The group further developed monocarbene-olefin complexes by replacing COD from [Pd(COD)(alkene)] complexes {alkene = *p*-benzoquinone (BQ) and 1,4-naphthoquinone (NQ)} with IMes or IPr (Scheme 4.16) [26]. NMR spectroscopic analyses of the target complexes revealed unsymmetrical olefinic protons and carbon atoms, which led to the conclusion that the complexes exist as [Pd(BQ)(IMes)]$_2$ and [Pd(IMes)(NQ)]$_2$ dimers in solution. A single crystal X-ray diffraction study on [Pd(BQ)(IMes)]$_2$ confirmed the dimeric nature of the complex also in solid state, in which the quinones coordinatively bridge the palladium(0) centers in a very unusual binding mode. In analogy to the MAH or TCNE complexes described earlier, these monocarbene metal adducts are also stable to air and moisture due to the electron-deficient and good π-accepting olefin ligands, which stabilize the complexes via backbonding.

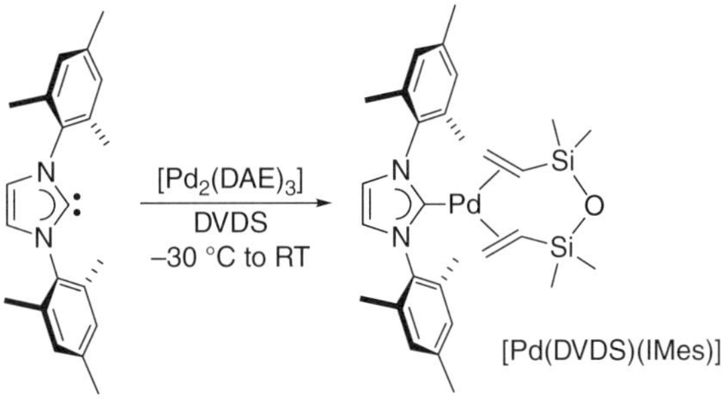

Scheme 4.15
Preparation of the first monocarbene-olefin palladium(0) complex.

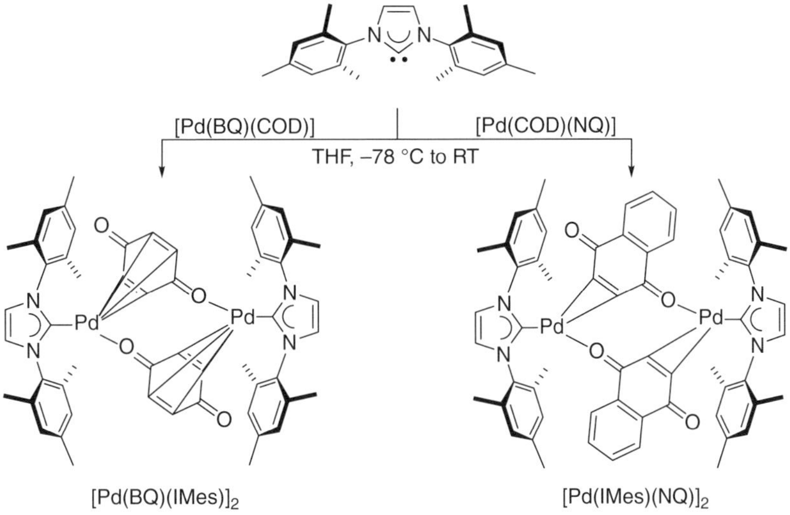

Scheme 4.16
Synthetic pathway to dimeric mono(NHC) complexes of palladium(0).

4.2.2 Reduction of Palladium(II)-NHC Complexes

A major disadvantage of the phosphine or alkene displacement route is the requirement for air sensitive palladium(0)-phosphine complexes and free NHCs, which are either expensive or cumbersome to prepare. The long reaction time required for bulky phosphines is another drawback. A more convenient and improved alternative method, first reported by Caddick and Cloke, involves the reduction of more stable palladium(II)-carbene precursors [27]. The dichlorido-bis(methallyl)dipalladium(II) complex $[Pd(\mu\text{-Cl})(\eta^3\text{-}C_4H_7)]_2$ was found to be a suitable starting material, and its reaction with four equivalents of I^tBu with concurrent reduction using sodium-dimethylmalonate afforded the homoleptic bis(NHC) complex $[Pd(I^tBu)_2]$ in a good yield of 60% (Scheme 4.17). The palladium(II) complex $[PdCl(\eta^3\text{-}C_4H_7)(I^tBu)]$, which has also been isolated as the primary byproduct, was proposed to be the initial complex formed prior to metal-based reduction to palladium(0). The complex can be deliberately formed by bridge-cleavage reaction of $[Pd(\mu\text{-Cl})(\eta^3\text{-}C_4H_7)]_2$ with two equivalents of I^tBu in high yields. Application of this approach to the very bulky IAd ligand afforded $[Pd(IAd)_2]$ in a shorter time of 16 h albeit with a lower yield of 55% [23].

This strategy was further developed into a general synthetic route for the preparation of mixed NHC-phosphine palladium(0) complexes by Nolan and coworkers [28]. It is known that π-allyl complexes of palladium(II) are useful precursors to palladium(0)-phosphine complexes. The reduction is proposed to proceed via attack of a nucleophile on the coordinated allyl group, which is subsequently eliminated after bond formation [29]. By analogy, $[PdCl(\eta^3\text{-}C_4H_7)(IPr)]$ was therefore reacted with potassium *tert*-butoxide in isopropanol in the presence of a phosphine. The procedure is fairly general and affords good yields ranging from 61–92%. Moreover, the synthesis is also applicable to the preparation of homo- and hetero-bis(NHC) complexes. This can be achieved by substituting the phosphine with a free carbene, and yields of 59–87% have been reported (Scheme 4.18).

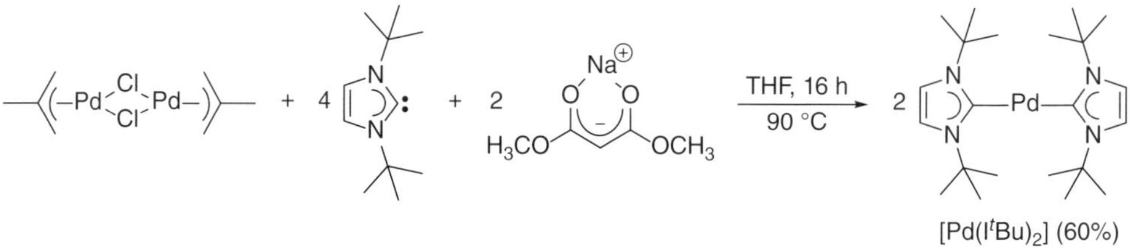

[Pd(I*t*Bu)_2] (60%)

Scheme 4.17
Preparation of homoleptic [Pd(I*t*Bu)_2] via *in situ* reduction of a palladium(II) methallyl-complex with sodium methylmalonate.

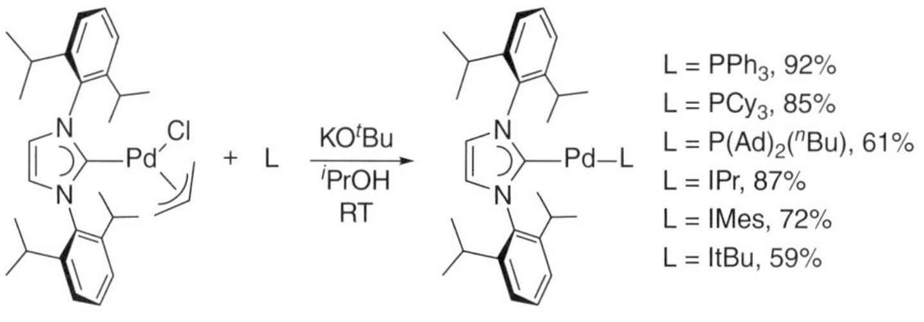

L = PPh_3, 92%
L = PCy_3, 85%
L = P(Ad)_2(*n*Bu), 61%
L = IPr, 87%
L = IMes, 72%
L = I*t*Bu, 59%

Scheme 4.18
Synthesis of homo- and heteroleptic [Pd(IPr)(L)] complexes via *in situ* reduction of palladium(II)-allyl complexes.

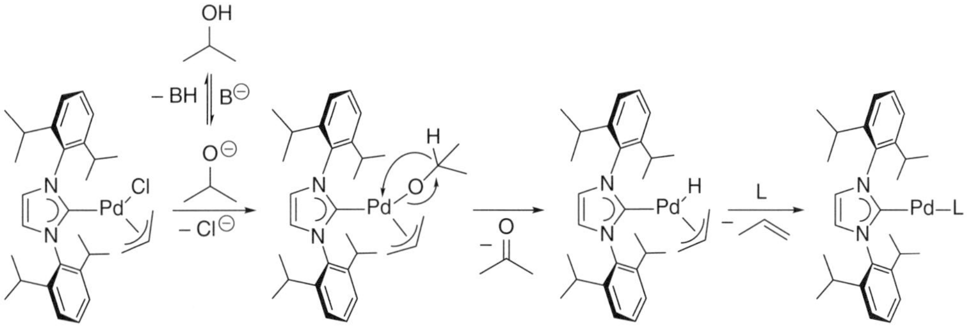

Scheme 4.19
Proposed mechanism for the Pd^{II}/Pd^0 reduction step.

To gain a more in-depth understanding of the Pd^{II}/Pd^0 reduction step, the authors also studied the influence of the base and solvent. It was found that various bases can affect the transformation, but isopropanol was a crucial solvent. The detection of acetone in the mixture at the end of the reaction led to the conclusion that the isopropoxide anion resulting from deprotonation of isopropanol with a base is a likely reducing agent. In the proposed mechanism, the isopropoxide anion substitutes the chlorido ligand. Subsequent β-hydride elimination from the isopropoxido ligand with release of one acetone molecule leads to formation of a palladium(II)-hydrido intermediate. The latter undergoes reductive elimination of 1-propene to give the palladium(0) species, which in the last step is coordinated by the added ligand L (Scheme 4.19).

Another method to prepare palladium(0) NHC complexes via reduction of palladium(II) precursors was reported by Danopoulos and coworkers [30]. They showed that the room temperature reaction of [Pd(CH$_3$)$_2$(tmeda)] with two equivalents of 1,3-bis(2-isopropyl-phenyl)imidazolidin-2-ylidene, generated *in situ* by deprotonation of the respective imidazolinium chloride with KO^{t}Bu, produced the bis(carbene)palladium(0) complex in 74% yield (Scheme 4.20). It was noted that temperature control in this reaction plays a very important role in determining the reaction products. When the reaction was carried out at 0 °C, simple tmeda ligand replacement occurred instead and a square planar, *cis*-configured dimethyl-bis(carbene)palladium(II) complex with the formula *cis*-[Pd(CH$_3$)$_2$(NHC)$_2$] was afforded in 83% yield. Warming up a solution of this dimethyl complex to 30 °C led cleanly to the formation of the linear bis(carbene)palladium(0) complex. The formation of the latter was anticipated to occur via reductive elimination of the two *cis*-standing methyl ligands as ethane. It should be noted that this palladium(0) complex was the first example bearing saturated and less bulky NHCs to be structurally characterized by single crystal X-ray diffraction.

Figure 4.4 depicts solid state molecular structures of hetero- and homoleptic palladium(0)-NHC complexes obtained by single crystal diffraction. The coordination geometry is linear as one would expect for 14-electron d^{10} complexes (left). The Pd–C$_{carbene}$ distance in the mixed NHC/phosphine complex is 2.0292(9) Å, which was surprisingly noted to be 0.03 Å longer than that in the homoleptic bis(NHC) complex [Pd(IPr)$_2$]. Moreover, it is also longer than the Pd–C$_{carbene}$ bond of 2.022(6) Å in the depicted homoleptic complex bearing saturated NHC ligands (right). Notably, the latter was the first structurally characterized palladium(0) complex with imidazolidin-2-ylidene ligands.

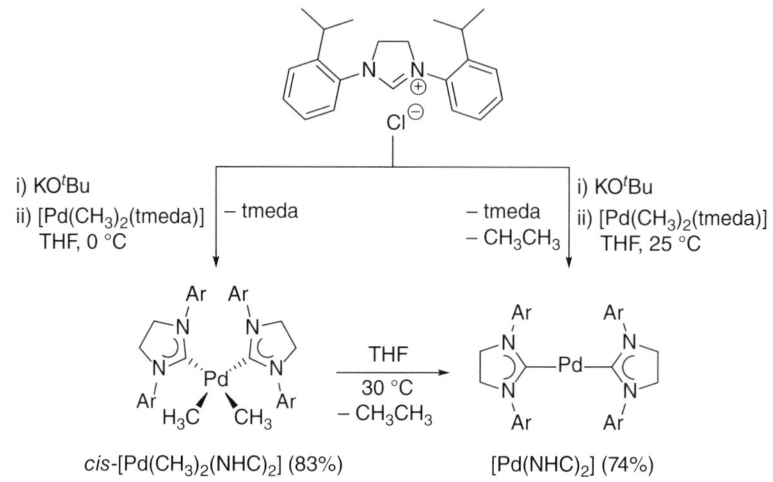

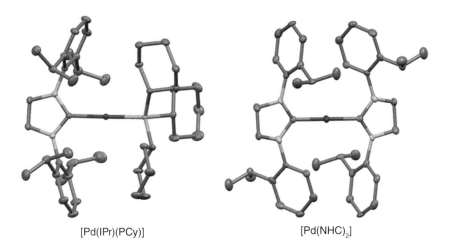

Scheme 4.20
Synthesis of palladium(0)-NHC complex via reductive elimination of ethane from a dimethyl-bis(NHC)palladium(II) precursor.

[Pd(IPr)(PCy)]   [Pd(NHC)₂]

Figure 4.4
Molecular structures of mixed NHC/phosphine and bis(NHC) complexes of palladium(0).

In general, palladium(0) NHC complexes have most commonly been applied as catalysts for coupling reactions of aryl halides (e.g. Suzuki-Miyaura, Mizoroki-Heck, Sonogashira), where they initiate the catalytic cycle by oxidative addition of the aryl halide substrate. The involvement of the Pd^{II}/Pd^0 redox couple is a commonly accepted catalytic cycle. In comparison to their phosphine congeners, addition of excess ligands to stabilize the active palladium(0) species is often not required, which has been attributed to the stronger Pd-NHC bond. However, this statement should not be generalized, but evaluated case by case as different NHCs do possess different electronic properties. This is already apparent from certain examples discussed in earlier sections. In addition to being catalytically active species, palladium(0)-NHC complexes have also shown interesting reactivity toward small

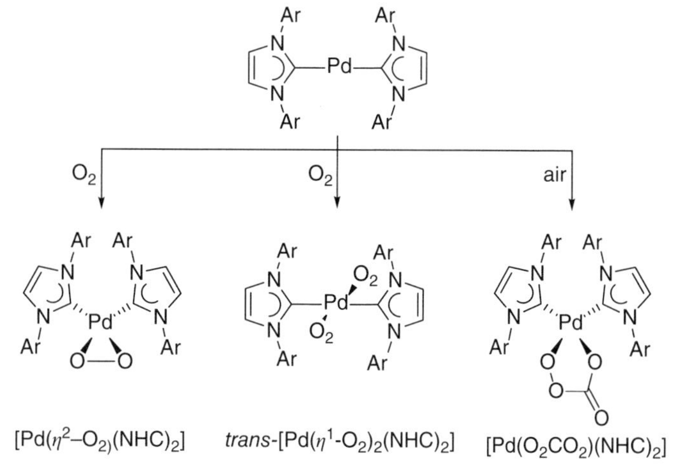

Scheme 4.21
Reaction of [Pd(NHC)$_2$] with oxygen and air.

molecules such as oxygen and carbon dioxide, which gave rise to peroxocarbonato [31], side-on η^2-peroxo [28,31,32] and end-on bis(η^1-peroxo) [33] complexes of palladium(II) (Scheme 4.21).

4.3 Platinum(0)-NHC Complexes

Despite being a rather costly metal, platinum has the distinct advantage that its most abundant isotope ^{195}Pt (~33.8%) is NMR active with a nuclear spin of ½. This allows for an additional method of characterization, and platinum compounds can be studied in greater detail. For example, the ^{195}Pt chemical shift is indicative of the electron density of the complex, and therefore can provide additional information about the oxidation state. Moreover, ligand binding can be evaluated via heteronuclear coupling to the metal center, which gives rise to ligand signals with characteristic platinum satellites. For NHC complexes, one would therefore expect such satellites for the directly bound carbene carbon donor. However, these are sometimes difficult to observe due to the generally low intensity of carbene signals in the ^{13}C NMR spectra of their complexes. The measurement of $^1J_{\text{Pt-C}}$ coupling constants in platinum-carbene complexes requires prolonged acquisition times of samples with well soluble complexes, in order to provide suitably resolved ^{13}C NMR spectra. Most of the platinum(0) NHC complexes are prepared by displacement of bound olefins by either pre-isolated or *in situ* generated free NHCs. Thus, a differentiation between homo- and heteroleptic complexes is made in the following sections.

4.3.1 Homoleptic Complexes

An early example for a homoleptic platinum(0) NHC complex was reported by Arduengo and coworkers, who reacted two equivalents of the free carbene IMes with bis(1,5-cyclooctadiene)platinum(0) in hexane and at room temperature (Scheme 4.22). This reaction gave the bis(IMes) complex [Pt(IMes)$_2$] under ligand displacement in analogy to their synthesis

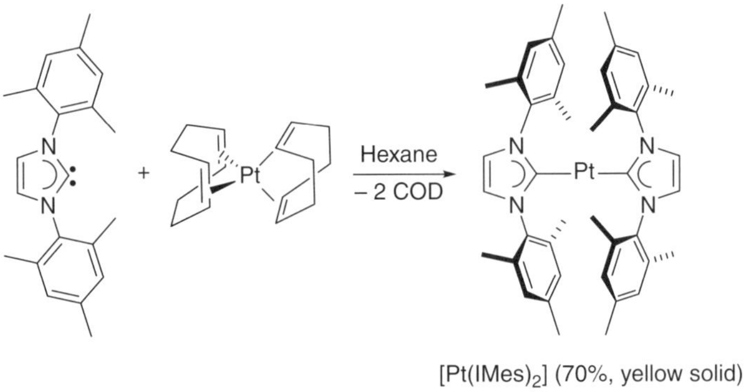

[Pt(IMes)$_2$] (70%, yellow solid)

Scheme 4.22
Preparation of the homoleptic, linear platinum(0)-NHC complex [Pt(IMes)$_2$].

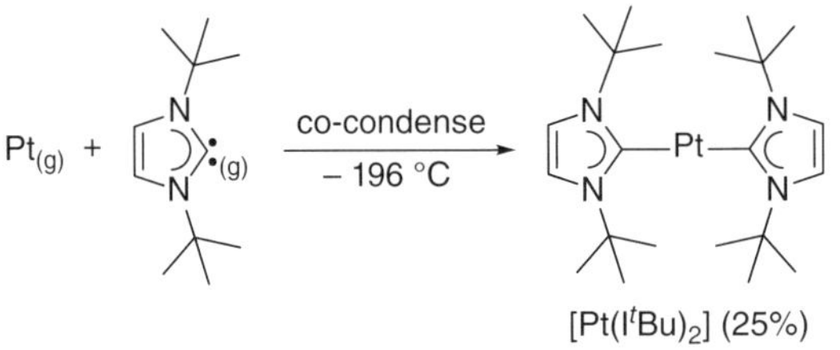

[Pt(I^{t}Bu)$_2$] (25%)

Scheme 4.23
Metal vapor synthesis of [Pt(I^{t}Bu)$_2$].

of the nickel(0) analog [9]. The 14-electron complex precipitated from the reaction mixture and could simply be filtered and subsequently dried to give a yellow powder in 70% yield.

Detailed NMR spectroscopic studies in C$_6$D$_6$ showed an upfield shift of the carbene carbon atom of $\Delta\delta = 22.2$ ppm upon coordination to platinum(0) (^{13}C$_{carbene} = 197.5$ ppm, $^1J_{Pt-C} = 1218$ Hz). The ^{195}Pt nucleus resonates at -5729 ppm, which falls in the range observed for platinum(0) complexes. In addition, a substantial upfield shift of ~0.5 ppm was also observed in the same deuterated solvent for the ring protons H$_{4/5}$ in the complex, which was attributed to an increase of π-electron density, possibly as a result of π-backbonding from the electron-rich metal center into the formally vacant p_π-orbital at the carbene carbon. Single crystals obtained by crystallization from THF and subjected to X-ray diffraction studies also revealed relatively short metal-carbon distances of 1.959(8) and 1.942(8) Å in the linear complex, which together with the NMR spectroscopic observations pointed to a certain degree of $d_\pi \rightarrow p_\pi$ backdonation in the electron-rich, zero-valent bis(IMes) complex. Although, NHCs are primarily strong σ-donors, we have to note that π-backbonding may also occur particularly with low-valent metal centers.

A homoleptic bis(NHC)platinum(0) complex [Pt(I^{t}Bu)$_2$] with the aliphatic imidazolin-2-ylidene I^{t}Bu was also synthesized via the rarely used metal vapor/ligand co-condensation method as outlined in Scheme 4.23 [11]. The bis(1,3-di-*tert*-butylimidazolin-2-ylidene) platinum(0) complex was isolated as dark yellow crystals.

Interestingly, ^{1}H NMR spectroscopic data for this complex did not show the upfield shifts for the H$_{4/5}$ ring protons observed for the [Pt(IMes)$_2$] analog. Furthermore, the ^{13}C$_{carbene}$ signal for the complex, also recorded in C$_6$D$_6$, was detected more upfield with a slightly larger coupling constant ($^{13}C_{carbene} = 193.9$ ppm, $^1J_{Pt-C} = 1257$ Hz), while the ^{195}Pt NMR signal was detected more downfield at -5390 ppm. The Pt-C distances, on the other hand, amount to 1.965(12) Å, which is essentially the same as those observed for the IMes complex within standard deviations. In addition, a gas phase photoelectron spectroscopic study and DFT calculations on [Pt(I^{t}Bu)$_2$] strongly suggest that bonding in this complex occurs predominantly through σ interactions with minimal π contributions [34]. This contrasts with the conclusion reached for the bonding in [Pt(IMes)$_2$]. The supposed decrease in π-acceptor property of the alkyl-substituted I^{t}Bu compared to IMes with aryl wing tips could be rationalized by the strong positive inductive effects of the two *tertiary*-butyl N-substituents, which would increase the energy of the formally vacant p_π orbital at the carbene carbon atom making it less accessible for backdonation. Nevertheless, steric repulsion between the two bulky NHCs may have an influence on the bonding as well, further complicating the interpretation of the bonding situation in terms of pure electronic effects.

4.3.2 Heteroleptic Complexes

Although the above two homoleptic complexes are 14-electron species, they are thermally relatively stable due to shielding of the metal center by the bulky wing tip groups [35]. As such, it was proposed that one of the carbene ligands might be replaced by a more labile ligand, such as an alkene, in order to obtain more reactive complexes for catalysis.

Markó and coworkers prepared a large range of such complexes in generally good yields of >65% using the Karstedt catalyst [Pt$_2$(DVDS)$_3$] as the platinum(0) source [36]. The general procedure involves mixing of the latter with two equivalents of imidazolium salt and sodium hydride or potassium *tert*-butoxide in toluene at room temperature (Scheme 4.24). The strong base was added for the *in situ* generation of the NHC ligand, which upon generation cleaves the bridged diplatinum complex under exclusive liberation of the more labile bridging siloxane ligand.

The products of the general formula [Pt(DVDS)(NHC)] feature a platinum(0) center that is ligated by one carbene and one chelating DVDS ligand. The enthalpically favorable formation of the strong platinum-carbene bond is the main driving force for the reaction, and the complexes were obtained as crystalline solids, that are even stable to air and moisture. However, a slight light-sensitivity of these compounds was noted. The coordination of a strong σ donor enhances the backdonation from platinum to the olefinic units of the bidentate siloxane ligand, which leads to an upfield shift of the olefinic protons. Overall, this increases the stability of the resulting complex relative to the Karstedt catalyst.

The straightforward methodology could also be extended to complexes with saturated and benzannulated carbenes. Here it was noted that saturated NHCs gave the lowest yields of ~50%. This is possibly due to the greater tendency of *in situ* generated saturated NHCs

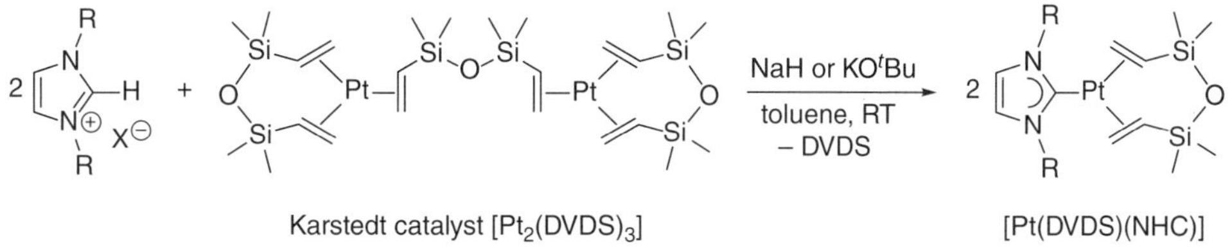

Karstedt catalyst [Pt$_2$(DVDS)$_3$]   [Pt(DVDS)(NHC)]

Scheme 4.24
General scheme for the preparation of mixed NHC/DVDS platinum(0) complexes.

Table 4.4 ^{13}C, ^{195}Pt and $^1J_{Pt-C}$ NMR data for [Pt(DVDS)(NHC)] complexes and [Pt(I^{t}Bu)$_2$].

NHC	^{13}C$_{carbene}$ [ppm]	^{195}Pt [ppm]	$^1J_{Pt-C}$ [Hz]
IMe	179.8	−5343	-
ICy	180.0	−5343	1350
IAd	180.3	−5306	-
I^{t}Bu	181.2	−5333	1361
IMes	184.2	−5339	-
IPr	186.4	−5340	-
^{n}Pr$_2$-bimy	198.3	−5383	1373
Me$_2$-bimy	199.5	−5379	1378
SIMes	211.0	−5365	-
SIPr	213.3	−5361	-
[Pt(I^{t}Bu)$_2$]	193.9	−5390	1257

to dimerize or to decompose via ring-opening reactions prior to complex formation. Most of the complexes showed superior catalytic activity in the hydrosilylation of olefins covering a wide range of substrates compared to their precursor.

Depending on the nature of the carbene used, the ^{13}C$_{carbene}$ signal in these complexes can be found in the range from 179.8–213.3 ppm with $^1J_{Pt-C}$ coupling constants of 1350–1379 Hz (Table 4.4). The downfield shifts are in line with an electron-rich metal center. Furthermore, their ^{195}Pt NMR resonances were observed ranging from −5306 to −5383 ppm. Within both chemical shift data series, no obvious trend can be discerned.

The reason for the absence of any simple trend can be found if one considers the bonding in the monocarbene/DVDS complexes. The platinum resonance is primarily affected by two opposing factors. On one hand, an upfield shift is to be expected as a result of the strong carbene-to-metal σ-donation in line with an electron enriched center. However, this increased electron density is also efficiently backdonated into antibonding π^*-orbitals of the two η^2 olefinic donors of the chelating DVDS ligand, which would lead to a downfield shift due to increased contribution to the paramagnetic term of the shielding constant. These two opposing effects clearly complicate the identification of any simple trend related to the electronic structure of the complex. However, compared to the significantly more upfield ^{195}Pt NMR shifts observed for analogous phosphine systems (approx. −5598 to −5735 ppm) or even the Karsted catalyst (−6156 ppm), one can clearly conclude that the NHC complexes are more electron rich, which enhances metal-to-alkene backdonation.

Structural elucidation of selected complexes in the solid state reveals the expected pseudo-trigonal planar coordination geometry around platinum in all cases, and only two examples are shown in Figure 4.5. The carbene plane is almost perpendicular to this triangular coordination plane, and the bound disiloxane ligand chelates the metal center in a pseudo-chair conformation. Such an arrangement allows for the optimum overlap of orbitals involved in the bonding. As with the key NMR parameters, the Pt-C distances seem to vary unsystematically, but all are in the range expected for single bonds with a mean value of 2.05 Å.

A very similar approach to more reactive platinum(0) NHC complexes was reported by Cavell and Elsevier [37]. Instead of a bidentate dialkene, these authors opted for the use of two monodentate, but electron-deficient dimethylfumarate (DMFU) ligands. These olefins are expected to be strong π-acceptors, thereby enhancing backbonding from the metal

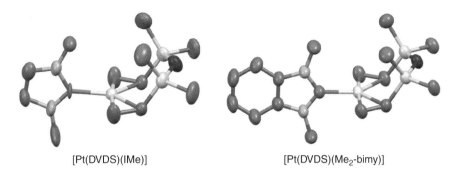

Figure 4.5
Representative molecular structures of [Pt(DVDS)(IMe)] (left) and [Pt(DVDS)(Me$_2$-bimy)] (right).

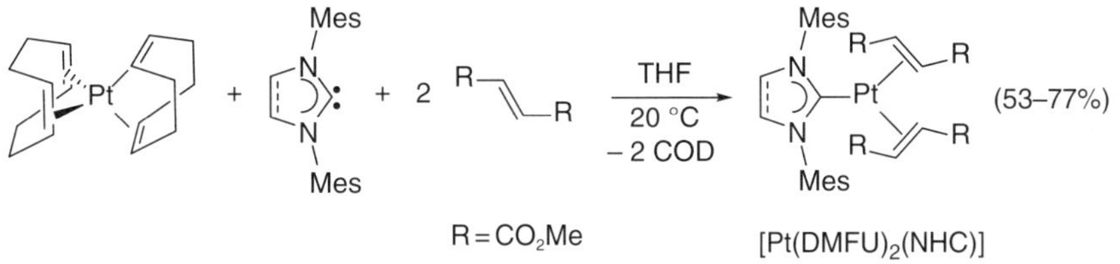

Scheme 4.25
Preparation of NHC-bis(alkene)platinum(0) complexes of via free carbenes.

center. The preparation involves the reaction the free IMes or SIMes with [Pt(COD)$_2$] in a 1:1 ratio in the presence of two equivalents of DMFU. Stirring the reactants in THF at 20 °C afforded the monocarbene-bis(olefin) complexes of the general formula [Pt(DMFU)$_2$(NHC)] with liberation of the COD ligand (Scheme 4.25). The reaction products were isolated as air stable white solids. Notably, they are also stable in solution for extended periods of time and can even be heated in acetone under reflux.

Despite their thermal stability, these complexes readily react with simple imidazolium salts under oxidative addition across the C–H bond to furnish square planar hydrido-hetero-bis(NHC) platinum(II) complexes. Remarkably, this C–H activation can even occur at room temperature. The relatively labile coordination of the monodentate olefin ligands and the high electron density of the metal center contribute to this enhanced reactivity. The latter is apparent from the ^{195}Pt chemical shifts of –5184 and –5200 ppm for the IMes and the SIMes complexes, respectively. These resonances are in the same range observed for other zero-valent platinum-NHC complexes.

Elsevier and coworkers later simplified the synthetic procedures for these two complexes and their 1,3-diphenylimidazolidin-2-ylidene (SIPh) analog by direct reaction of the azolium salts with the platinum source and the olefin in the presence of sodium hydride as the base (Scheme 4.26) [38]. This route eliminates the necessity for the pre-isolation of the free carbene, but requires longer reaction times giving slightly lower yields. In the case of IMes and SIMes complexes, an *in situ* deprotonation of the salt precursors, which gives rise to stable NHCs, is feasible.

However, the same pathway was ruled out for the SIPh analog, since this NHC would rapidly dimerize to form the enetetramine. Here, an alternative route was proposed that involves the intermediacy of the cationic hydrido-platinum(II) complex [PtH(DMFU)

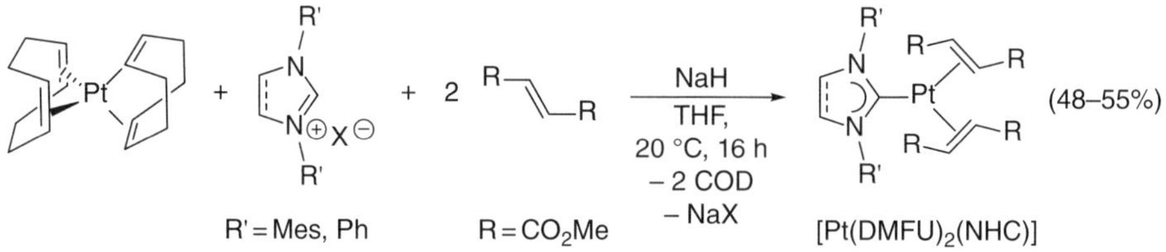

Scheme 4.26
Preparation of NHC-bis(alkene)platinum(0) complexes of via azolium salts.

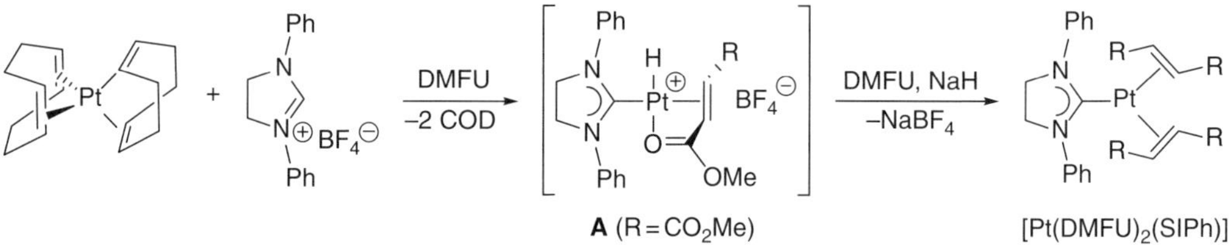

Scheme 4.27
Proposed intermediate A in the formation of [Pt(DMFU)$_2$(SIPh)].

(SIPh)]BF$_4$ (**A**) resulting from oxidative addition of the imidazolinium salt to the zero-valent platinum source. Subsequent "deprotonation" of this complex finally affords the target complex (Scheme 4.27).

Platinum(0) complexes of this type containing NHC/olefin mixed donor sets have been primarily applied as highly active catalysts for the hydrosilylation of olefins. The hydrosilylation reaction can often be carried out in aerobic conditions and may even be accelerated by oxygen. Thus, the air stability of these complexes makes them a suitable choice. In addition, they have been studied for their ability to undergo oxidative addition with small molecules such as dihydrogen or imidazolium salts.

References

[1] M. F. Lappert, P. L. Pye, *J. Chem. Soc. Dalton Trans.* **1977**, 2172.
[2] C. Elschenbroich, (**2006**). *Organometallics*, 3rd ed. Weinheim: Wiley-VCH. ISBN 3-527-29390-6.
[3] (a) L. S. Meriwether, M. L. Fiene, *J. Am. Chem. Soc.* **1959**, *81*, 4200. (b) W. D. Horrocks, R. H. Mann, *Spectrochim. Acta* **1965**, *21*, 399.
[4] (a) B. E. Mann, *Adv. Organomet. Chem.* **1974**, *12*, 135. (b) L. J. Todd, J. R. Wilkinson, *J. Organomet. Chem.* **1974**, *77*, 1.
[5] K. Öfele, W. A. Herrmann, D. Mihalios, M. Elison, E. Herdtweck, W. Scherer, J. Mink, *J. Organomet. Chem.* **1993**, *459*, 177.
[6] (a) G. Kuchenbeiser, M. Soleilhavoup, B. Donnadieu, G. Bertrand, *Chem. Asian J.* **2009**, *4*, 1745. (b) H. V. Huynh, Y. Han, R. Jothibasu, J. A. Yang, *Organometallics* **2009**, *28*, 5395.
[7] W. A. Herrmann, L. J. Goossen, G. R. J. Artus, C. Köcher, *Organometallics* **1997**, *16*, 2472.
[8] R. Dorta, E. D. Stevens, N. M. Scott, C. Costabile, L. Cavallo, C. D. Hoff, S. P. Nolan, *J. Am. Chem. Soc.* **2005**, *127*, 2485.
[9] A. J. Arduengo III, S. F. Gamper, J. C. Calabrese, F. Davidson, *J. Am. Chem. Soc.* **1994**, *116*, 4391.
[10] V. P. W. Böhm, C. W. K. Gstöttmayr, T. Westkamp, W. A. Herrmann, *Angew. Chem. Int. Ed.* **2001**, *40*, 3387.
[11] P. L. Arnold, F. G. N. Cloke, T. Geldbach, P. B. Hitchcock, *Organometallics* **1999**, *18*, 3228.

[12] F. G. N. Cloke,; M. L. H. Green, *J. Chem. Soc., Dalton Trans.* **1981**, 1938.

[13] S. Caddick, F. G. N. Cloke, P. B. Hitchcock, A. K. de K. Lewis, *Angew. Chem. Int. Ed.* **2004**, *43*, 5824.

[14] T. Schaub, U. Radius, *Chem. Eur. J.* **2005**, *11*, 5024.

[15] B. Gehrhus, P. B. Hitchcock, M. F. Lappert, *J. Chem. Soc., Dalton Trans.* **2000**, 3094.

[16] In 2002, Bertrand and coworkers reported nickel(0) complexes of cationic carbon donors formally obtained by mono-deprotonation of dicationic 1,2,4-triazolium salts. Despite the cationic charge of their C-ligands, these complexes can be regarded as close relatives of nickel(0)-NHC complexes. See: C. Buron, L. Stelzig, O. Guerret, H. Gornitzka, V. Romanenko, G. Bertrand, *J. Organomet. Chem.* **2002**, *664*, 70.

[17] R. J. Keaton, J. M. Blacquiere, R. T. Baker, *J. Am. Chem. Soc.* **2007**, *129*, 1844.

[18] A. A. Danopoulos, D. Pugh, *Dalton Trans.* **2008**, 30.

[19] K. Matsubara, S. Miyazaki, Y. Koga, Y. Nibu, T. Hashimura, T. Matsumoto, *Organometallics*, **2008**, *27*, 6020.

[20] D. S. McGuinness, M. J. Green, K. J. Cavell, B. W. Skelton, A. H. White, *J. Organomet. Chem*, **1998**, *565*, 165.

[21] V. P. W. Böhm, C. W. K. Gstöttmayr, T. Weskamp, W. A. Herrmann, *J. Organomet. Chem*, **2000**, *595*, 186.

[22] L. R. Titcomb, S. Caddick, F. G. N. Cloke, D. J. Wilson, D. McKerrecher, *Chem. Commun.* **2001**, 1388.

[23] C. W. K. Gstöttmayr, V. P. W. Böhm, E. Herdtweck, M. Grosche, W. A. Herrmann, *Angew. Chem. Int. Ed.* **2002**, *4*, 1363.

[24] D. S. McGuinness, K. J. Cavell, B. W. Skelton, A. H. White, *Organometallics* **1999**, *18*, 1596.

[25] R. Jackstell, M. G. Andreu, A. C. Frisch, K. Selvakumar, A. Zapf, H. Klein, A. Spannenberg, D. Röttger, O. Briel, R. Karch, M. Beller, *Angew. Chem. Int. Ed.* **2002**, *41*, 986.

[26] (a) K. Selvakumar, A. Zapf, A. Spannenberg, M. Beller, *Chem. Eur. J.* **2002**, *8*, 3901. (b) A. C. Frisch, A. Zapf, O. Briel, B. Kayser, N. Shaikh, M. Beller, *J. Mol. Catal.* **2004**, *214*, 231.

[27] S. Caddick, F. G. N. Cloke, G. K. B. Clentsmith, P. B. Hitchcock, D. McKerrecher, L. R. Titcomb, M. R. V. Williams, *J. Organomet. Chem*, **2001**, *617-618*, 635.

[28] S. Fantasia, S. P. Nolan, *Chem. Eur. J.* **2008**, *14*, 6987.

[29] W. Kuran, A. Musco, *Inorg. Chim. Acta* **1975**, *12*, 187.

[30] N. Stylianides, A. A. Danopoulos, D. Pugh, F. Hancock, A. Zanotti-Gerosa, *Organometallics*, **2007**, *26*, 5627.

[31] M. Yamashita, K. Goto, T. Kawashima, *J. Am. Chem. Soc*, **2005**, *127*, 7294.

[32] M. M. Konnick, I. A. Guzei, S. S. Stahl, *J. Am. Chem. Soc.* **2004**, *126*, 10212.

[33] X. Cai, S. Majumdar, G. C. Fortman, C. S. J. Cazin, A. M. Z. Slawin, C. Lhermitte, R. Prabhakar, M. E. Germain, T. Palluccio, S. P. Nolan, E. V. Rybak-Akimova, M. Temprado, B. Captain, C. D. Hoff, *J. Am. Chem. Soc.* **2011**, *133*, 1290.

[34] J. C. Green, R. G. Scurr, P. L. Arnold, F. G. N. Cloke, *Chem. Commun.* *1997*, **1963**.

[35] M. A. Duin, M. Lutz, A. L. Spek, C. J. Elsevier, *J. Organomet. Chem.*, **2005**, *690*, 5804.

[36] (a) I. E. Markó, S. Stérin, O. Buisine, G. Mignani, P. Branlard, B.Tinant, J.-P. Declercq, *Science* **2002**, *298*, 204. (b) I. E. Markó, S. Stérin, O. Buisine, G. Berthon G. Michaud, B. Tinant, J.-P. Declercq, *Adv. Synth. Catal.* **2004**, *346*, 1429. (c) G. Berthon-Gelloz, O. Buisine, J. F. Brière, G. Michaud, S. Stérin, B. Tinant, J.-P. Declercq, D. Chapon, I. E. Markó, *J. Organomet. Chem.* **2005**, *690*, 6156.

[37] M. A. Duin, N. D. Clement, K. J. Cavell, C. J. Elsevier, *Chem. Commun.* **2003**, 400.

[38] M. A. Duin, M. Lutz, A. L. Spek, C. J. Elsevier, *J. Organomet. Chem.* **2005**, *690*, 5804.

5 Group 10 Metal(II)-NHC Complexes

Group 10 transition metal complexes are widely known for their applications in metal-mediated catalysis. The real catalysts in homogenous transformations are often low-valent and electron-rich metal complexes. This is particular so for the nowadays very well-studied C–C coupling reactions. To further increase the life-time of such catalysts, NHC ligands have been successfully used. Their metal coordination increases the electron density further, and at the same time, complex stability is imparted due to the stronger metal-NHC bonds.

Although a metal(0)-NHC complex can be directly used for such catalytic studies, it is often more convenient to use their metal(II) derivatives as catalyst precursors, which are generally even more stable. Often such complexes can be handled without precautions to exclude air and moisture. The metal(II)-NHC catalyst precursor is then reduced *in situ* to its more active zero-valent complex in the course of the catalytic reaction either by the substrates, additional reducing agents, or by the solvent itself.

Due to this fact, a large number of methodologies have been developed for the preparation of group 10 metal(II)-NHC complexes. This chapter summarizes some of the more common synthetic NHC chemistry related to group 10 metals in their +II oxidation state.

5.1 Nickel(II)-NHC Complexes

In comparison to nickel(0)-NHC complexes, those of the +II oxidation state are much more easily accessible due to their increased stability toward air and moisture. In addition to the curiosity-driven research in the early days, the prospect of obtaining cost saving Ni-NHC catalysts as potential alternatives to the more expensive palladium compounds has significantly fueled the development in this area.

Many nickel-phosphine complexes have found application in catalysis, particularly in aryl-aryl bond formation. For example, precatalysts of the general formula $[NiCl_2P_2]$ ($P_2 = 2\ PPh_3$, $2\ PCy_3$, dppe, dppb, dppf) [1] have been reported to give high conversion in the cross-coupling of aryl chlorides even in absence of an additional reducing agent [2]. As with many transition metals, the NHC chemistry of nickel(II) is often based on their phosphine counterparts.

5.1.1 Cleavage of Enetetramines and the Free Carbene Route

The first example of a nickel(II)-NHC complex was again reported by Lappert and coworkers. Moderate heating of a mixture of tetrahedral $[NiCl_2(PPh_3)_2]$ and a tetramethyl-substituted tetraazafulvene in the presence of $NaBF_4$ in toluene led to both phosphine and chlorido

The Organometallic Chemistry of N-heterocyclic Carbenes, First Edition. Han Vinh Huynh.
© 2017 John Wiley & Sons Ltd. Published 2017 by John Wiley & Sons Ltd.

substitution of the former, and a yellow, square planar compound with a melting point of ~120 °C of the proposed formula [NiCl(SIMe)$_3$]BF$_4$ (SIMe = 1,3-dimethylimidazolidin-2-ylidene) was obtained (Scheme 5.1) [3]. It is interesting to note that the first nickel(II) complex of NHCs is proposed to be a cationic tris(NHC) complex, which is even nowadays still rare.

The most common nickel(II)-NHC coordination compounds are actually neutral and square planar bis(NHC) complexes of the type [NiX$_2$(NHC)$_2$]. The first examples of these were again prepared by the Lappert group using a very similar approach. Phosphine substitution in [NiCl$_2$(PPh$_3$)$_2$] with NHCs derived from the corresponding tetrabenzyl-substituted tetraazafulvalene furnished the desired complex *trans*-[NiCl$_2$(SIBn)$_2$] (Scheme 5.1) [4].

Again, we can note the geometry change from tetrahedral to square planar upon ligand substitution, which also changes the magnetic properties of the respective complexes from para- to diamagnetic. In contrast to the first example described above, no additional chlorido displacement was observed. The reactivity difference between benzyl- versus methyl-substituted enetetramines observed in this reaction is remarkable and may be accounted for by the increased steric bulk of the benzyl-derivative, since these two groups are electronically very similar. Furthermore, it was noted that the colors of the nickel(II) complexes studied ranged from yellow to red, which was rationalized in terms of the strong Ni–C$_{carbene}$ σ-bond and the associated high ligand-field strength of NHCs in general. The latter in turn caused the relative instability of a paramagnetic, tetrahedral (S = 1) complex compared to a spin-paired, d^8 square planar complex. Thus, electronic rather than steric factors are considered controlling in terms of complex geometry [4].

Since the dichlorido-bis(NHC) complex with 1,3-dimethylimidazolidin-2-ylidene ligands could not be prepared *via* direct ligand substitution, an oxidative protocol was developed, which involves oxidation of the previously prepared dicarbonyl-bis(NHC) nickel(0) compound (see Section 4.1.1) with elemental chlorine in dichloromethane at low temperature (Scheme 5.2). Contrary to the direct ligand substitution described earlier, this

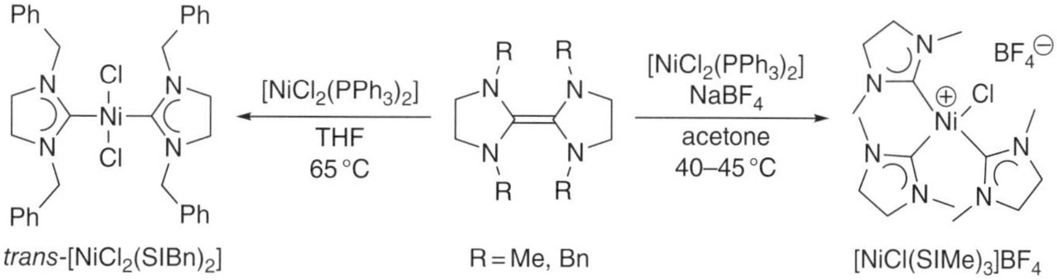

Scheme 5.1
Preparation of the first nickel(II)-NHC complexes by direct ligand substitution using enetetramines.

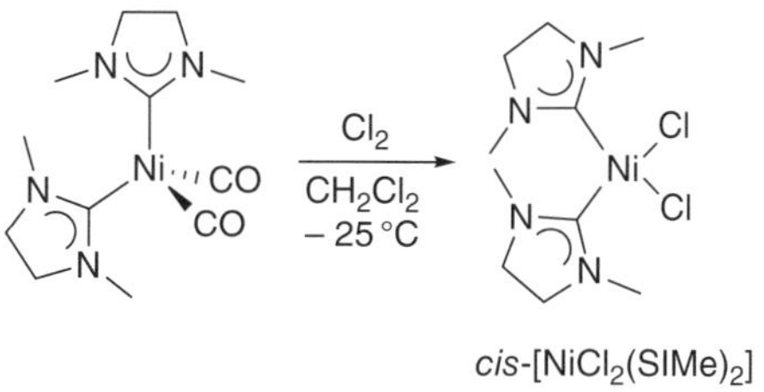

cis-[NiCl$_2$(SIMe)$_2$]

Scheme 5.2
Preparation of a bis(NHC)nickel(II) complex by oxidation of its nickel(0)-NHC precursor.

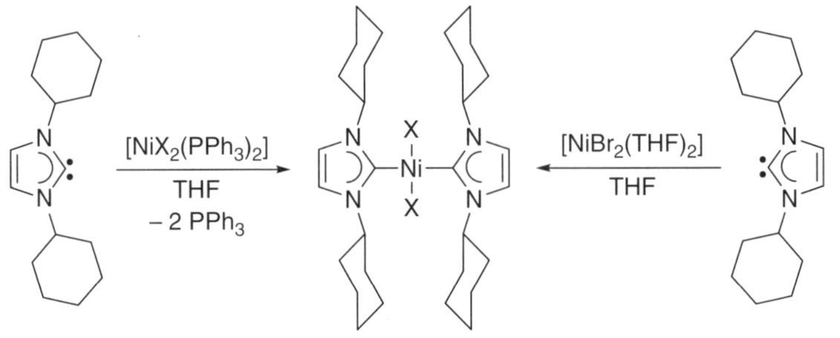

trans-[NiX₂(NHC)₂]; X = Cl, Br

Scheme 5.3
Preparation of the nickel(II)-NHC complexes by direct ligand substitution using free NHCs.

approach afforded the *cis* isomer of the desired dichlorido-bis(NHC)nickel(II) complex *cis*-[NiCl₂(SIMe)₂] [4]. Apparently, preparative routes have an influence on the stereo-chemistry of these simple NHC complexes.

In recent years, nickel(II) complexes of unsaturated imidazolin-2-ylidene ligands have attracted more attention. Since imidazolin-2-ylidenes can be isolated as free carbenes, they can be directly used to replace weaker ligands in the coordination sphere of nickel. This reaction was first studied by Herrmann and coworkers, who reacted two equivalents of the free carbene 1,3-dicyclohexylimidazolin-2-ylidene with [NiCl₂(PPh₃)₂], [NiBr₂(PPh₃)₂] or [NiBr₂(THF)₂] [5]. The action of the free NHC on these nickel(II) precursor complexes led to the displacement of the phosphine or ether ligands, and the *trans*-configured dihalido-bis(NHC) complexes of the general formula *trans*-[NiX₂(ICy)₂] could be obtained in good yields as air stable, red solids (Scheme 5.3).

5.1.2 *In Situ* Deprotonation of Azolium Salts with Basic Metal Salts

The necessity for air sensitive free carbenes or their dimers represents a drawback of the earlier methods. More conveniently, nickel(II)-NHC complexes can also be obtained by *in situ* deprotonation of azolium salts, which are generally air stable, in analogy to the original approach by Wanzlick [6] and Öfele [7]. In an early report, Herrmann and coworkers compared the reactivities of 1,3-dimethylimidazolium iodide versus 1,4-dimethyl-1,2,4-triazolium iodide for this route (Scheme 5.4). Two equivalents of each salt were heated with nickel(II) acetate [Ni(OAc)₂] as the basic metal precursor to yield the bis(NHC) complex. It was found that the 1,2,4-triazolium salts reacted faster and under much milder conditions as compared to the imidazolium counterpart. The former could react in THF as a solvent at 60 °C under ultrasonic irradiation giving the product in 77% yield. The latter on the other hand required much harsher conditions, and only heating the two reactants in the absence of any solvent under vacuum to 150 °C led to the desired complex, but only in 30% yield. The drastic reactivity difference can be rationalized by comparing the acidity of the azo-lium protons at the C2 or C5 position, respectively, which for the 1,2,4-triazolium salt is much greater, thereby facilitating carbene formation by deprotonation.

The general idea of this solventless reaction is that the imidazolium salt melts upon heat-ing and dissolves the Ni(OAc)₂ for reaction. The application of a vacuum ensures constant

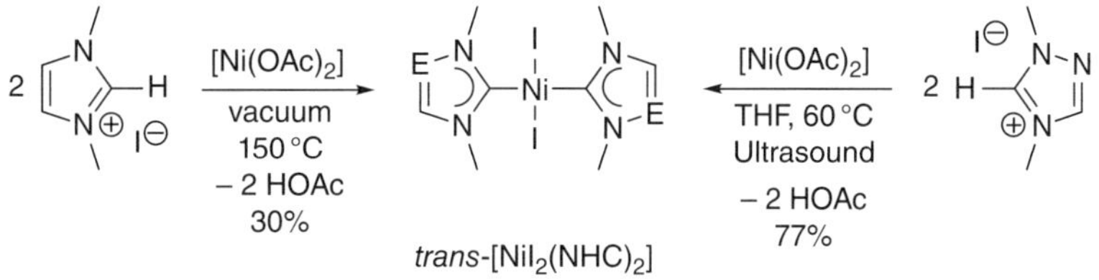

Scheme 5.4
Preparation of bis(NHC) nickel(II) complexes by *in situ* deprotonation of azolium salts.

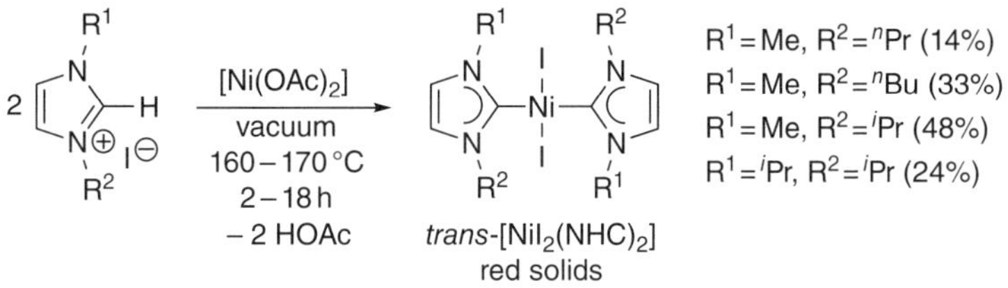

$R^1 = Me, R^2 = {}^nPr$ (14%)
$R^1 = Me, R^2 = {}^nBu$ (33%)
$R^1 = Me, R^2 = {}^iPr$ (48%)
$R^1 = {}^iPr, R^2 = {}^iPr$ (24%)

trans-[NiI$_2$(NHC)$_2$]
red solids

Scheme 5.5
Solventless preparation of bis(NHC) nickel(II) complexes.

removal of the acetic acid (AcOH) byproduct, which should drive the reaction to completion. Since many imidazolium salts are indeed low temperature ionic liquids (ILs), one can expect an extension to a wider range of nickel(II)-complexes. Indeed the reaction scope was easily extended by Wasserscheid and Cavell to a range of complexes with unsymmetrical and symmetrical imidazolin-2-ylidenes; they also simplified the workup by final trituration of the cooled melt with water [8]. Nevertheless, the yields of the red and exclusively *trans*-configured complexes remained low ranging from 14–48% (Scheme 5.5).

A direct application of this reaction for the preparation of benzimidazolin-2-ylidene complexes is not feasible for two reasons. First, the melting points of benzimidazolium salts are generally significantly higher than their imidazolium counterparts. Second, the increased temperatures thus required will also likely decompose the formed complexes. To overcome these shortcomings, Huynh and Hahn reported a modification of the methodology by simply running the reaction in tetraalkylammonium salts, which allows for the preparation of the first benzimidazolin-2-ylidene nickel(II) complexes with relative ease (Scheme 5.6) [9]. The ammonium salt melts at relatively low temperatures under vacuum and acts as an ionic liquid dissolving both benzimidazolium salt and Ni(OAc)$_2$ for a homogeneous reaction. To warrant product purity and to prevent halido ligand scrambling in the complexes, both ammonium and benzimidazolium salts must contain the same halide as counter-anion. Moreover, a medium to high vacuum is crucial to ensure removal of the acetic acid byproduct formed, which drives the reaction to the product side.

The yields obtained in ammonium salts were generally better, but remained relatively low for NHCs with bulkier N-substituents, which is not surprising. Further improvements for the yields of the latter were realized by increasing the reaction temperature and prolonging the reaction time to 12 h. By analogy to the imidazolin-2-ylidene analogs, all complexes were isolated as red solids, and X-ray diffraction on suitable single crystals revealed the *trans*-configuration in all cases (Figure 5.1).

In addition to Ni(OAc)$_2$, the air sensitive and 20-electron sandwich complex nickelocene [NiCp$_2$] or its bis(indenyl)nickel(II) analog [Ni(Ind)$_2$] can also be used as a basic nickel(II)

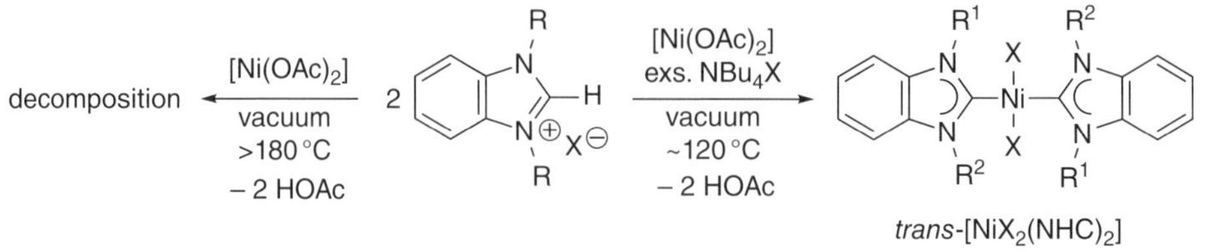

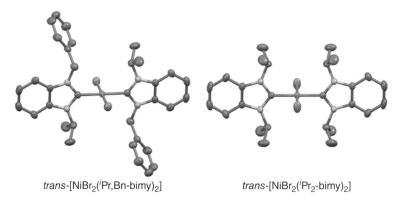

$R^1 = R^2 = Me$, $X = I$ (60%, 2 h, 120 °C)
$R^1 = R^2 = allyl$, $X = Br$ (54%, 2 h, 120 °C)
$R^1 = R^2 = {}^nPr$, $X = Br$ (38%, 2 h, 120 °C)
$R^1 = Me$, $R^2 = allyl$, $X = Br$ (28%, 2 h, 120 °C)
$R^1 = Me$, $R^2 = {}^nPr$, $X = I$ (44%, 2 h, 120 °C)

$R^1 = R^2 = CH_2Ph$, $X = Br$ (71%, 12 h, 130 °C)
$R^1 = R^2 = {}^iPr$, $X = Br$ (78%, 12 h, 130 °C)
$R^1 = R^2 = CHPh_2$, $X = Br$ (34%, 36 h, 170 °C)
$R^1 = R^2 = {}^iBu$, $X = Br$ (41%, 12 h, 130 °C)
$R^1 = CH_2Ph$, $R^2 = {}^iPr$, $X = Br$ (80%, 12 h, 130 °C)
$R^1 = CH_2Ph$, $R^2 = CHPh_2$, $X = Br$ (21%, 12 h, 130 °C)

Scheme 5.6
"Solventless" preparation of dihalido-bis(benzimidazolin-2-ylidene)nickel(II) complexes.

trans-[NiBr₂(*i*Pr,Bn-bimy)₂]   *trans*-[NiBr₂(*i*Pr₂-bimy)₂]

Figure 5.1
Representative molecular structures for *trans*-bis(NHC)nickel(II) complexes.

precursor to access nickel(II) NHC complexes. The first example for this approach was reported by Cowley and coworkers in 2000 [10]. The treatment of nickelocene with one equivalent of 1,3-dimesitylimidazolium chloride affords the 18-electron half-sandwich complex [NiClCp(IMes)], which contains one η^5-Cp and newly introduced NHC and chlorido ligands (Scheme 5.7) [11]. One of the coordinated Cp ligands in nickelocene acts as the base to deprotonate the imidazolium salt eliminating cyclopentadiene (CpH) and generating the free NHC *in situ*, which is immediately trapped by coordination to nickel(II). The chloride counter-anion completes the coordination sphere at the metal. A color change from dark green to red is indicative of reaction progress. The solid state structure of this compound features a pseudo-trigonal planar arrangement of Cp, Cl, and NHC ligands around the metal center. Given that the η^5-Cp-anion donates six electrons and formally occupies three coordination sites, it is probably more accurate to consider this complex as a penta-coordinated species. Compared to nickelocene or the phosphine analog [NiClCp(PPh₃)], the mixed Cp/NHC nickel(II) complex is more stable and can be handled in air [11b]. This not only highlights the generally observed improved stability of 18-electron complexes, but in particular those containing NHC ligands.

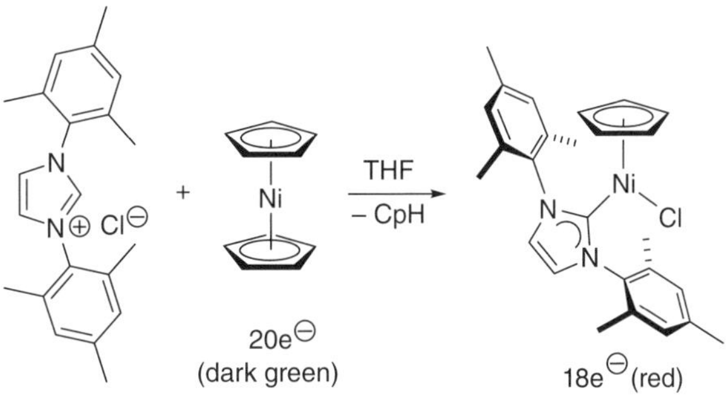

Scheme 5.7
Preparation of nickel(II)-NHC complexes using nickelocene.

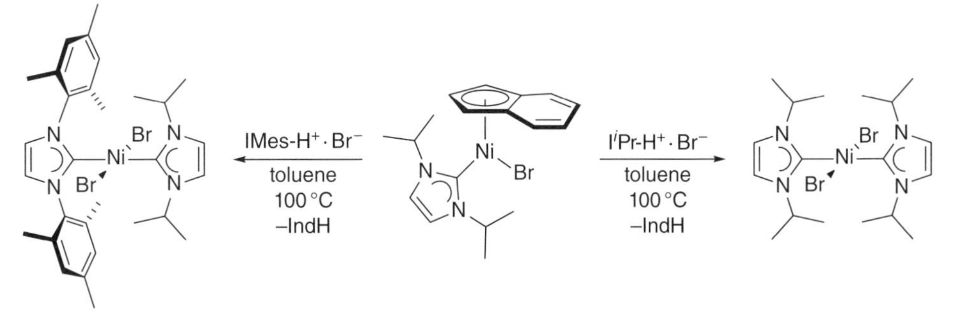

square-planar
trans-[NiBr$_2$(IMes)(I^{i}Pr)] (63%)

"pseudo"-trigonal-planar

square-planar
trans-[NiBr$_2$(I^{i}Pr)$_2$] (65%)

Scheme 5.8
Preparation of homo- and hetero-bis(carbene)nickel(II) complexes.

The nickel(II)-NHC half-sandwich compound is a useful precursor for further transformations either involving metathesis of the chlorido ligand or a second acid-base reaction mediated by the coordinated Cp ligand. This possibility was exploited in the case of the indenyl/NHC nickel(II) [12] complex to prepare homo- or hetero-bis(carbene) complexes by simply treating the half-sandwich complex with a second equivalent of azolium salt. The remaining Ind ligand in the complex is sufficiently basic to deprotonate another azolium salt in an analogous manner described earlier for the Cp analog. Upon elimination of indene (IndH) and coordination of the additional NHC and chlorido ligand, a typical 16-electron, square planar bis(carbene) nickel(II) complex is obtained. When homo-bis(carbene) complexes are desired, this two-step reaction sequence can be combined into a single step protocol by reacting bis(indenyl)nickel(II) directly with at least two equivalents of azolium salts. The two-step separation, however, allows for the controlled preparation of hetero-bis(carbene) complexes, which contain two different NHC ligands (Scheme 5.8) [13].

5.1.3 The Silver-Carbene Transfer Route

Apart from these methodologies, there is also the possibility to access nickel(II) NHC complexes *via* the usually very versatile and mild Ag-carbene transfer route. Surprisingly, this method is less common in nickel(II)-NHC chemistry with simple monodentate NHCs.

$$Ag^+\text{-}NHC \; + \; Ni^{2+} \; \rightleftharpoons \; Ag^+ \; + \; Ni^{2+}\text{-}NHC$$

soft-soft   hard-soft

Scheme 5.9
NHC transfer equilibrium involving silver and nickel ions.

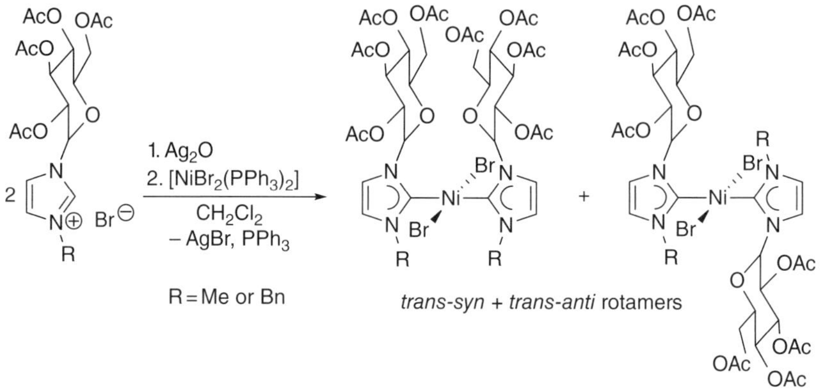

R = Me or Bn   *trans-syn* + *trans-anti* rotamers

Scheme 5.10
Silver-carbene transfer to nickel resulting in a mixture of *trans*-rotamers.

The small number of examples may be reasoned by considering the hard-soft-acid-base (HSAB) concept. NHCs are soft donors, and therefore an equilibrium involving silver-NHC species with soft-soft interactions as opposed to the hard-soft interaction for a nickel-NHC complex should favor the side of the former in the absence of any additional driving force (Scheme 5.9).

A literature survey revealed that transfer of carbenes from silver to nickel is more often conducted with ditopic NHCs featuring harder nitrogen donor functions. The hard nitrogen function in such ligands can assist in the transfer by providing additional driving force by pre-coordination of the hard nitrogen donor to nickel affording cationic bis(chelates) [14].

Successful transfer of NHCs from silver to nickel without the additional coordination of donors has been described for carbohydrate-functionalized [15] and amine-functionalized imidazolin-2-ylidenes [16]. In both cases, the Ag-NHC species were generated by reacting two equivalents of the respective imidazolium bromides with one equivalent of silver(I) oxide. The silver-NHC intermediates generated were not isolated and directly transferred to dihalido-bis(triphenylphosphine)nickel(II) complexes. The carbenes replace the phosphine ligands from the nickel complex and the liberated soft phosphine donors can either form a complex with the soft silver(I) center or simply get oxidized in the course of the reaction [17]. Precipitation of silver halides constitutes an additional driving force. The transfer onto a dibromido-nickel complex using the sugar-derived NHC leads to retention of the bromido ligands at nickel and a mixture of *trans-anti* and *trans-syn* rotamers is obtained, which can be examined by NMR spectroscopy (Scheme 5.10; see Chapter 5.2.5) [15].

In the case of the amine-functionalized NHC, the transfer was conducted using a dichlorido-nickel complex. Here halido ligand scrambling could occur in the final nickel complex. However and in accordance with the HSAB concept, the bromido ligands have greater affinity to the silver ion leading to the formation of a tetrasilver cubane-cluster, while the chlorido ligands are retained at the nickel center (Scheme 5.11).

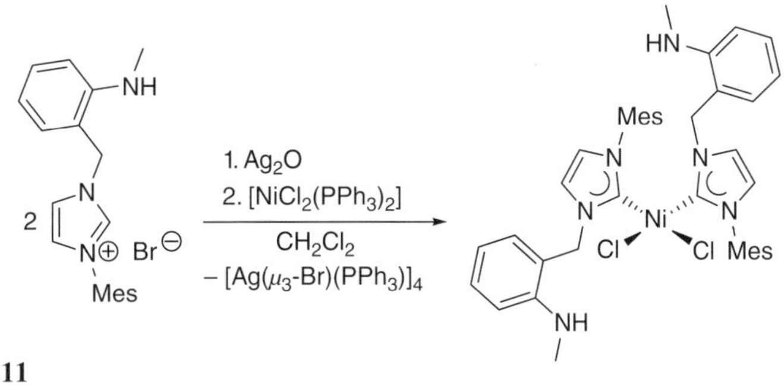

Scheme 5.11
Silver-carbene transfer to nickel resulting in a *cis*-bis(NHC) complex.

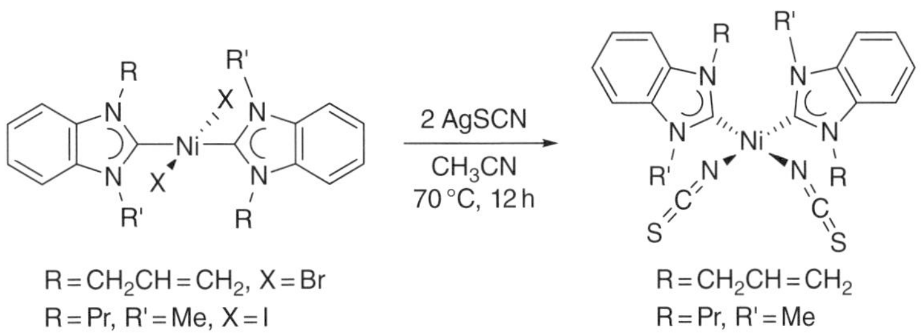

R = CH$_2$CH = CH$_2$, X = Br
R = Pr, R' = Me, X = I

R = CH$_2$CH = CH$_2$
R = Pr, R' = Me

Scheme 5.12
trans-cis isomerization upon halido ligand exchange.

The stereochemistry of the dihalido-bis(NHC)nickel(II) complexes is vastly dominated by *trans* isomers, which are thermodynamically more favorable due to steric reasons. Therefore, the isolation of *cis*-configured complexes is rare. A combination of several factors must be met to obtain *cis* isomers: [18]

1. Steric factor: The NHC ligand cannot be too bulky.
2. Electronic factor: The donating ability and *trans* influence of the co-ligands must be significantly smaller than that of the NHCs.
3. Solvent factor: *cis* complexes have a greater dipole compared to their *trans* isomers and can therefore be better stabilized in polar solvents.

All these factors contribute to a smaller energy gap between *cis* and *trans* isomers, and with their consideration, *cis*-configured bis(NHC)nickel(II) complexes can be obtained. Thus a simple substitution of bromido or iodido with weaker donating isothiocyanato ligands, as shown in Scheme 5.12, leads to the successful isolation of the respective *cis* isomers as evidenced by NMR spectroscopic and X-ray diffraction studies (Figure 5.2) [18].

5.2 Palladium(II)-NHC Complexes

N-heterocyclic carbene complexes of palladium(II) are among the most widely investigated metal-carbenes to date. The reason for their popularity is the capability to act as catalyst precursors for a wide variety of organic transformations. Moreover, most of them are

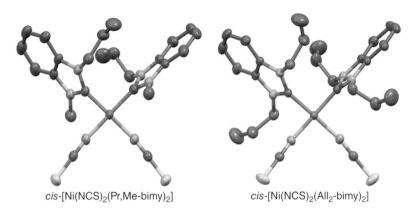

cis-[Ni(NCS)$_2$(Pr,Me-bimy)$_2$]   *cis*-[Ni(NCS)$_2$(All$_2$-bimy)$_2$]

Figure 5.2
Representative molecular structures for *cis*-bis(NHC) nickel(II) complexes.

stable to air and moisture and can be easily prepared. They also exhibit a long shelf-life, and therefore storage does not pose any problem. Due to the great interest in these complexes, a range of methodologies have been developed for their preparation, of which only the most common and convenient ones will be discussed in this section.

5.2.1   Cleavage of Entetramines and the Free Carbene Route

Cleavage of Enetetramines

As with most common NHC complexes, palladium(II) NHC complexes can be prepared by reaction of a suitable palladium(II) source with the free carbene, either preformed or generated *in situ*. Since saturated NHCs readily dimerize, it is also feasible to use their respective electron-rich olefins for such a reaction. Indeed, the first example for a palladium(II)-NHC complex had already been prepared using this approach by Lappert and coworkers in 1972. Heating of the tetraphenylenetetramine with the coordinatively unsaturated dipalladium(II) complex [PtCl$_2$(PEt$_3$)]$_2$ in xylene to 140 °C for one hour led to the kinetically controlled formation of *trans*-[PdCl$_2$(PEt$_3$)(SIPh)], however, only in a low yield of 17% [19]. The authors also noted that this mixed carbene/phosphine complex undergoes an unusual thermal *trans-cis* isomerization (i.e. *transphobia*, see Section 5.2.5), which can be accelerated in a polar solvent such as methanol (Scheme 5.13). Due to the low yields and the cumbersome handling of the materials required, this route is nowadays only of historical interest and seldom applied for the preparation of palladium(II)-NHC complexes.

The Free Carbene Route

About thirty years later, Nolan and coworkers reported the reaction of one equivalent of the preformed bulky IPr carbene with the dichlorido-bis(benzonitrile)palladium(II) complex [PdCl$_2$(NCPh)$_2$] in search for monocarbene complexes [20], which are believed to be catalytically more active than their bis(carbene) counterparts (Scheme 5.14) [21]. In this reaction, the weaker nitrile ligands are displaced by the free carbene, eventually leading to formation of the dimeric monocarbene-palladium complex [PdCl$_2$(IPr)]$_2$, which exhibits

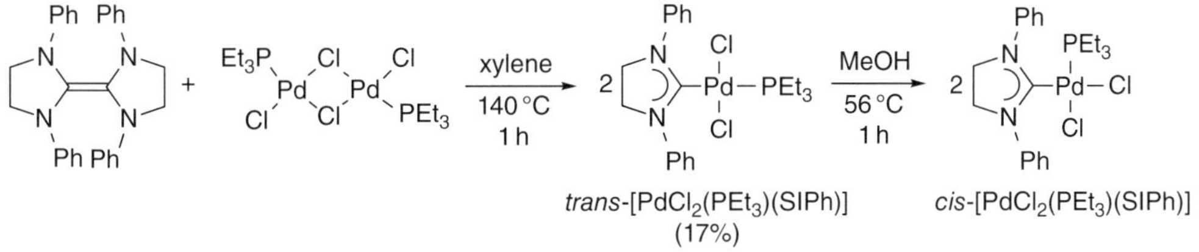

Scheme 5.13
Preparation of the first palladium(II) NHC complex and its thermal *trans-cis* isomerization.

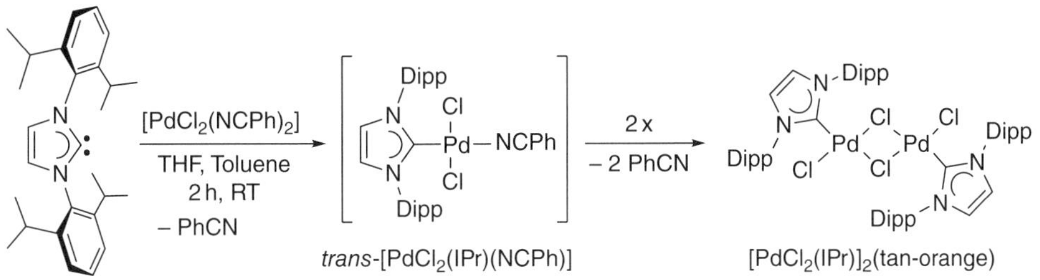

Scheme 5.14
Formation of a dimeric mono-NHC complex *via* "free carbene" route.

high catalytic activity in the amination of aryl chlorides and bromides. Although not explicitly mentioned, one could suppose that the reaction is initiated by an enthalpically driven one-to-one ligand substitution, which gives rise to the square planar intermediate *trans*-[PdCl$_2$(IPr)(NCPh)] with concurrent release of one benzonitrile molecule. The subsequent formation of the final tan-orange dipalladium complex is then accomplished by entropically favorable loss of the weakly bound second nitrile ligand due to the strong *trans* effect of the NHC, which is followed by dimerization. Such mono-nitrile adducts as proposed intermediates are not uncommon species in metal-NHC chemistry, and are indeed often formed by dissolution of dimeric palladium complexes in aceto- or benzonitrile solvents.

X-ray analysis on single crystals of the complex grown from 1,2-dichloroethane/hexanes confirmed the dimeric nature (Figure 5.3). The palladium centers are each coordinated by one terminal chlorido, one bridging chlorido, and one IPr carbene ligand in a distorted square planar fashion. The bulky carbenes are orientated in a favorable *anti*-fashion with respect to the Pd–Pd vector to minimize intramolecular repulsion.

Formally unsaturated dimeric palladium complexes can easily undergo enthalpically driven bridge-cleavage reactions with two equivalents of a free NHC ligand. This protocol has been demonstrated by Caddick and Cloke (see Section 5.2.5) [22], and Nolan and co-workers extended this approach to a family of monocarbene complexes of the general formula [PdCl(η^3-allyl-R)(NHC)] with imidazolin- and imidazolidin-2-ylidenes and differently substituted allyl-type ligands (Scheme 5.15) [23].

The reactions have to be conducted under an inert atmosphere due to the air sensitivity of the free carbenes, but they generally proceed under mild conditions, and the workup is easily achieved by evaporation of the solvent, trituration, and filtration in air. The air and moisture stable target complexes are isolated as yellow-orange solids in very good yields and can be stored in air.

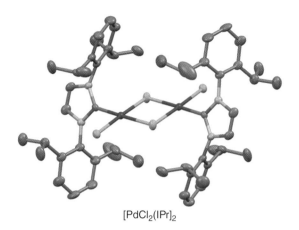

[PdCl₂(IPr)]₂

Figure 5.3
Molecular structure of the dimeric mono-NHC complex [PdCl$_2$(IPr)]$_2$.

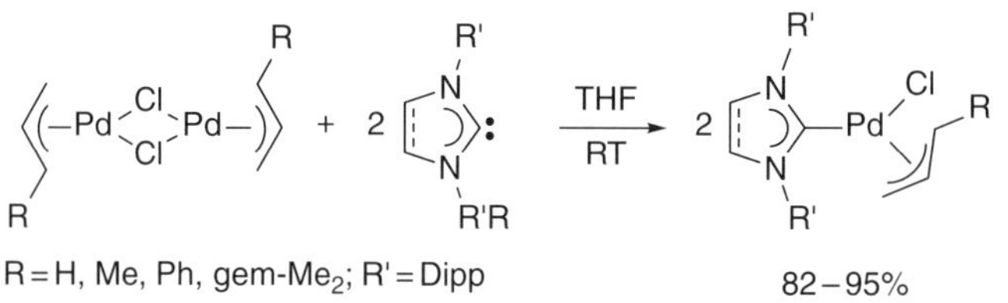

R = H, Me, Ph, gem-Me₂; R' = Dipp

82–95%

Scheme 5.15
Preparation of [PdCl(allyl-R)(NHC)] complexes using free carbenes.

Table 5.1 Comparison of Pd-C$_{allyl-R}$ distances in [PdCl(allyl-R)(IPr)] complexes.

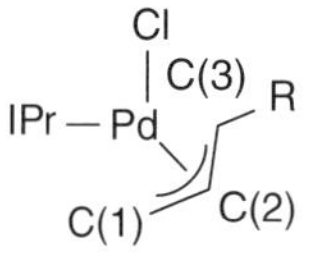	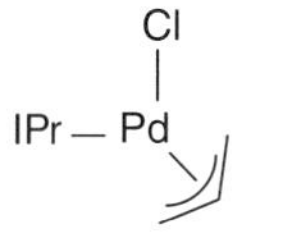	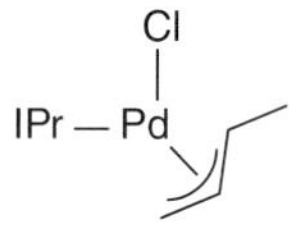	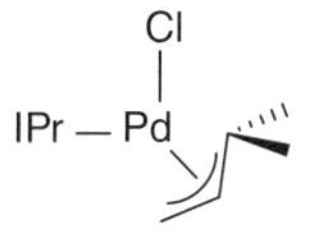	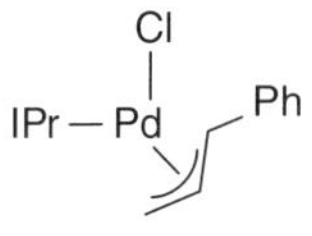
Pd–C(1)	2.098(6)	2.147(18)	2.095(4)	2.082(9)
Pd–C(2)	2.124(7)	2.122(18)	2.137(5)	2.136(10)
Pd–C(3)	2.210(6)	2.209(16)	2.252(5)	2.284(9)

The solid state molecular structures for a range of IPr-allyl complexes were obtained by single crystal X-ray diffraction. Comparison of the pseudo-square planar structures and relevant bond parameters revealed that the R-substituent in the allyl-type ligands are always found *transoid* to the IPr ligand, which can be attributed to the steric bulk of the latter. In addition, it was found that the Pd–C(3) distance increases with substitution as a result of stereoelectronic influences of the R-group. The palladium distances to C(1) and C(2), on the other hand, vary unsystematically (Table 5.1).

5.2.2 *In Situ* Deprotonation of Azolium Salts with External Base

The isolation of air and moisture sensitive free carbenes required in such syntheses can be avoided by *in situ* deprotonation of azolium salts using an external base. The feasibility of this approach has already been demonstrated by Enders and coworkers as early as 1996 in the preparation of the first chiral imidazolinylidene and triazolinylidene palladium(II) complexes [24]. The general procedure involves mixing of the azolium perchlorates with KO^{t}Bu, NaI, and Pd(OAc)$_2$ in tetrahydrofurane. Work up by column chromatography after 1–5 hours of stirring at room temperature gave the palladium(II) carbene complexes in high yields. The formation of mono- versus bis(carbene) complexes can be controlled by stoichiometry (Scheme 5.16). A 2:1 ratio of azolium salt to palladium afforded bis(carbene) complexes that were isolated as a mixture of *trans/cis* and/or *syn/anti* isomers (see Section 4.2.2). A 1:1 ratio of these reagents led as expected to dimers of the monocarbene-palladium complexes, which can undergo facile bridge-cleavage reaction with additional ligands such as amines. The generation of the free carbenes occurs *via* reaction of the azolium perchlorates with the strong KO^{t}Bu base, which releases *tert*-butanol as a byproduct. In addition, the acetato ligands in the initial palladium(II) precursor are replaced by the addition of sodium iodide giving rise to neutral iodido complexes.

While the handling of preformed free carbenes in these procedures is not required, the use of dry solvents is still a necessity due to the moisture sensitivity of the strong base KO^{t}Bu. However, the use of highly sensitive strong bases can be omitted, when sufficiently acidic azolium salts are used. Among the classical carbene precursors, triazolium salts in particular are expected to be the most acidic due to the additional –*I* effect of the third electronegative nitrogen atom in the heterocycle.

Indeed a much simpler protocol has been applied by Gosh and coworkers to access bis(triazolin-5-ylidenes) palladium complexes [25]. Heating two equivalents of triazolium bromide with palladium chloride and an excess of triethylamine in acetonitrile under reflux, already led to the formation of the desired dibromido-bis(NHC) complexes, which have been isolated as mixtures of *trans-anti* and *trans-syn* rotamers in moderate overall yields of 47–50% after column chromatography (Scheme 5.17).

Since triethylamine is a much weaker base compared to the free triazolin-5-ylidene, one can exclude significant amounts of free carbenes in the reaction mixture. However, the

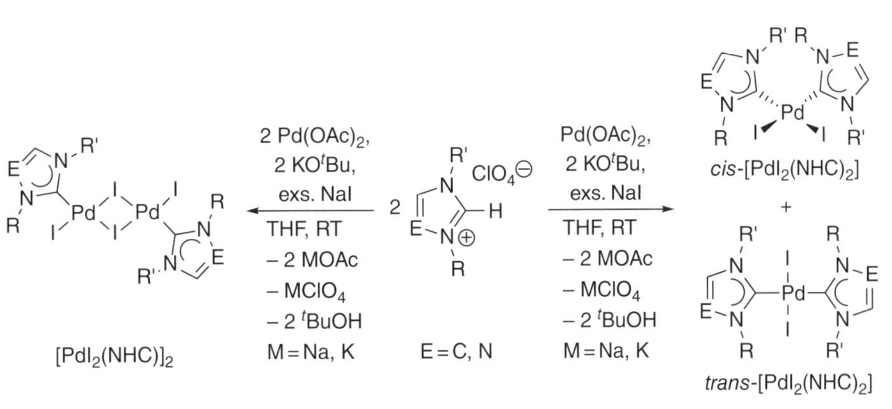

Scheme 5.16
Preparation of mono- and bis(NHC) palladium(II) complexes *via in situ* generation of free carbenes. Only the *anti* rotamers are depicted for clarity (R = Ph, PhCHMe; R′ = PhCHMe).

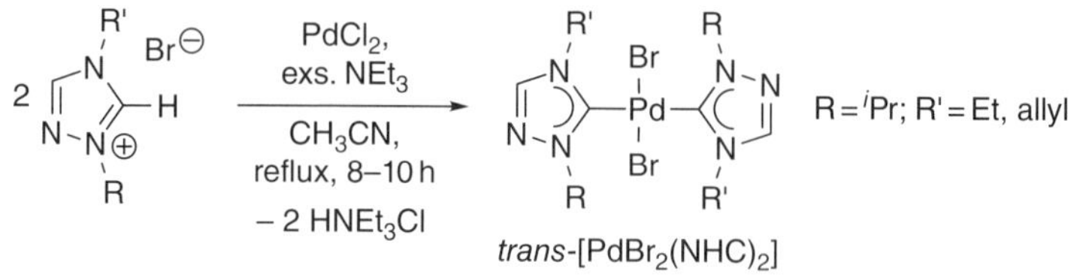

Scheme 5.17
Preparation of bis(NHC) palladium(II) complexes *via in situ* generation of free carbenes with a weak base.

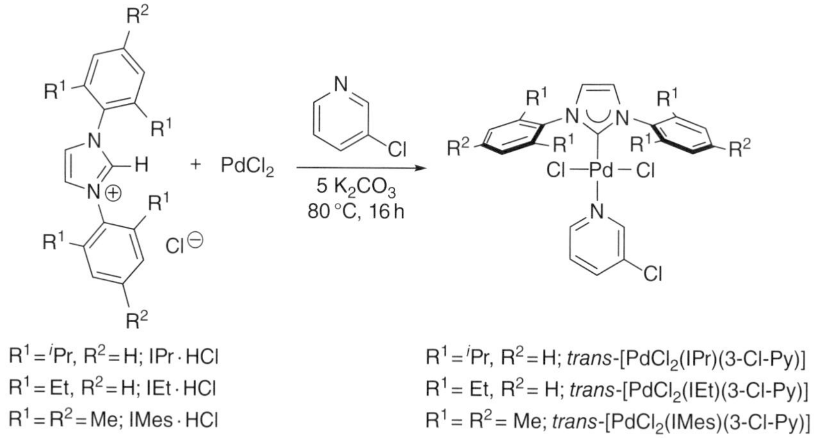

R¹ = ⁱPr, R² = H; IPr·HCl
R¹ = Et, R² = H; IEt·HCl
R¹ = R² = Me; IMes·HCl

R¹ = ⁱPr, R² = H; *trans*-[PdCl₂(IPr)(3-Cl-Py)]
R¹ = Et, R² = H; *trans*-[PdCl₂(IEt)(3-Cl-Py)]
R¹ = R² = Me; *trans*-[PdCl₂(IMes)(3-Cl-Py)]

Scheme 5.18
Preparation of PEPPSI™-type catalysts.

minute amounts of free NHCs generated in this acid/base equilibrium can be immediately trapped by coordination to the palladium(II) Lewis acid, which drives the equilibrium to the product side. The moderate yields may be a result of halide scrambling, since two different halides (Br⁻ and Cl⁻) have been employed in these reactions. The same methodology has also been applied for the preparation of the respective nickel(II) analogs.

In situ deprotonation of imidazolium salts using a relatively weaker base and without the requirement for anhydrous conditions is also possible. Indeed, this is the preferred synthetic route to the commercially available PEPPSI™ (Pyridine-Enhanced Precatalyst Preparation Stabilization and Initiation) type catalysts that were developed by Organ and coworkers. According to the reported original procedure, heating a mixture of PdCl₂, 1.1 equivalents of the imidazolium salt with excess K₂CO₃ in neat 3-chloropyridine afforded excellent yields of >91% of the desired monocarbene complexes of the general formula *trans*-[PdCl₂(NHC)(3-Cl-Py)] (Scheme 5.18).

The 5-fold excess of K₂CO₃ base ensures reasonable deprotonation rates of the imidazolium salt to generate the free carbene, which directly binds to the palladium(II) center. 3-chloropyridine acts as a high boiling solvent, but notably, it is also a source for the fourth co-ligand completing the square planar coordination sphere around the palladium(II) center. It is proposed that this "throw away" ligand dissociates from the metal center after reduction to palladium(0) activating the catalyst for the oxidative addition of incoming substrates. In this example, both the imidazolium salt and the transition metal precursor contain the same halide, and scrambling is avoided.

5.2.3 The "Palladium Acetate" Route

The need for air and moisture sensitive free carbenes, their dimers, or other sensitive reagents for the making of complexes is cumbersome and requires more advanced synthetic skills. More convenient is the direct *in situ* deprotonation of azolium salts in the presence of a basic palladium(II) source without the addition of any external base. Both starting materials are generally air and moisture stable, and the free NHC thus generated is immediately trapped *via* coordination to palladium affording equally stable palladium(II) NHC complexes. Since there is no need to take precautions to exclude air and moisture, such reactions can be conducted with standard glassware and even in wet solvents. The base for the deprotonation of the azolium carbene precursors can be added externally as described before, but it is easiest to use a metal source that already contains the base as a ligand. The compound of choice in palladium(II)-NHC chemistry is palladium(II) acetate [Pd(OAc)$_2$]. The most common red-brown modification of this compound is a trinuclear complex in the solid state and more accurately written as the trimer [Pd(OAc)$_2$]$_3$. Each of the three palladium centers is bridged by the bidentate, μ,κ^2-*O*-acetato, ligands, and the well-defined molecular nature of the complex makes it well soluble in organic solvents. Heating a solution of Pd(OAc)$_2$ with two equivalents of an azolium halide in an aprotic, but polar solvent generally affords the dihalido-bis(carbene)palladium(II) complex in good yields. An early example of this method involving the use of 1,3-dimethylimidazolium iodide is depicted in Scheme 5.19 [26]. A color change of the reaction mixture from red-brown to yellow indicates consumption of Pd(OAc)$_2$ and formation of the diiodido-bis(carbene) complex *cis*-[PdI$_2$(IMe)$_2$]. Acetic acid is formed as the byproduct, and the iodide counter-anions of the carbene precursors end up as additional ligands to complete the square planar coordination sphere around the metal center.

In the given example, the reaction proceeds relatively fast indicating a sufficiently acidic C2-proton of the imidazolium salt. This "Pd(OAc)$_2$ route" has been applied to a wide range of azolium salts leading to the successful preparation of numerous palladium(II) NHC complexes, and has since become a routine preparative method. Although, the *cis* isomer has been solely isolated in the given example, the formation of both *cis* and *trans* isomers in such reactions have been reported (see Section 5.2.5) [24,27].

To ensure reasonable reaction rates, it is important to choose a solvent that can dissolve both azolium salt as well as the Pd(OAc)$_2$. In addition to tetrahydrofurane, more polar solvents such as acetonitrile, or better dimethylsulfoxide, have been successfully used for this

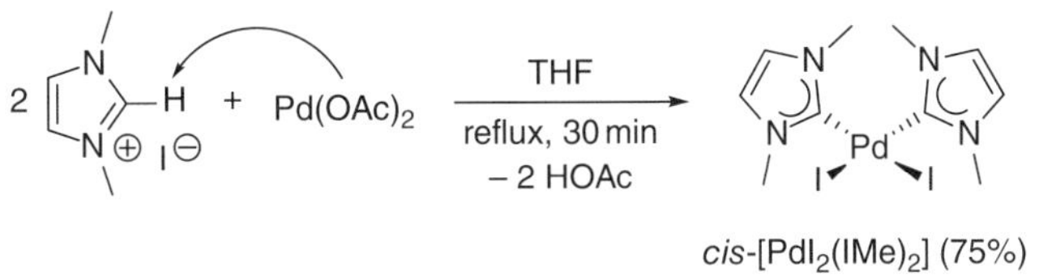

cis-[PdI$_2$(IMe)$_2$] (75%)

Scheme 5.19
Preparation of a bis(NHC) palladium(II) complex by *in situ* deprotonation of an imidazolium salt with Pd(OAc)$_2$.

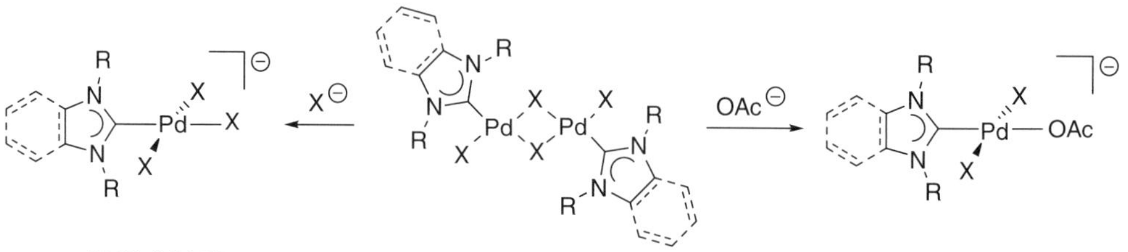

Scheme 5.20
Intermediates in the preparation of bis(NHC) palladium(II) complexes.

reaction. These solvents can dissolve the ionic carbene precursor better and also have higher boiling points. Therefore, reactions can be conducted at higher temperatures, which may be required for the activation of less acidic azolium salts. On the other hand, removal of the high boiling solvent after reaction becomes more difficult, but can be straightforwardly achieved using vacuum distillation. Alcoholic solvents can also dissolve most of the azolium salts; however, their use is usually less practical due to the interference in the deprotonation process.

The formation of the bis(carbene) complex requires successful *in situ* generation and coordination of two carbenes from the deprotonation of two azolium cations. Apparently, this must occur in a stepwise manner, and a monocarbene complex is a likely intermediate. Indeed, various monocarbene-palladium complexes have been identified in incomplete reaction mixtures [28,29]. These include halido-bridged monocarbene-palladium(II) dimers, anionic trihalido(carbene)palladium(II) and acetato-dihalido(carbene)palladium(II) complexes. The latter two result from bridge-cleavage reactions of the former with a halido or acetato ligand present in the mixture (Scheme 5.20).

A detailed study on the course of this reaction was conducted by Herrmann and Gardiner using a methylene-linked diimidazolium salt [30]. Upon mixing of the salt with $Pd(OAc)_2$ a new species was detected that was identified as $(diNHC \cdot H_2)[Pd(OAc)_2X_2]$ (Scheme 5.21). This complex results from a simple bridge-cleavage reaction of the $[Pd(OAc)_2]_3$ trimer by the free halides of the diimidazolium salt, which is driven by enthalpically favorable Pd–X bond formations. Slight heating leads to intermolecular deprotonation of one acidic azolium NCHN moiety by an acetato ligand and generation of the first NHC donor, which immediately coordinates to the metal center. This process concurrently liberates one equivalent of acidic acid. The new zwitterionic complex $[Pd(OAc)X_2(NHC–NHC \cdot H)]$ contains an azolium-functionalized monocarbene ligand, that is stabilized by intramolecular hydrogen bonds with the remaining acetato ligand. Subsequent heating of a solution of this complex to higher temperatures activates intramolecular deprotonation of the remaining azolium moiety, furnishing the desired dihalido-dicarbene complex $[PdX_2(diNHC)]$, which also releases the second molecule of acetic acid. On the other hand, addition of excess halides to the zwitterionic complex leads to ligand substitution, and the new trihalido complex $[PdX_3(NHC–NHC \cdot H)]$ is formed. The deprotonation of the latter with external base at elevated temperatures also yields the dicarbene complex $[PdX_2(diNHC)]$ as the most stable product.

Although the above study has been conducted with a diimidazolium salt, one can anticipate that a similar reaction pathway is operative with ordinary azolium salts, which eventually gives rise to a mixture of *cis* and *trans* bis(NHC) complexes *via* the intermediacy of monocarbene complexes.

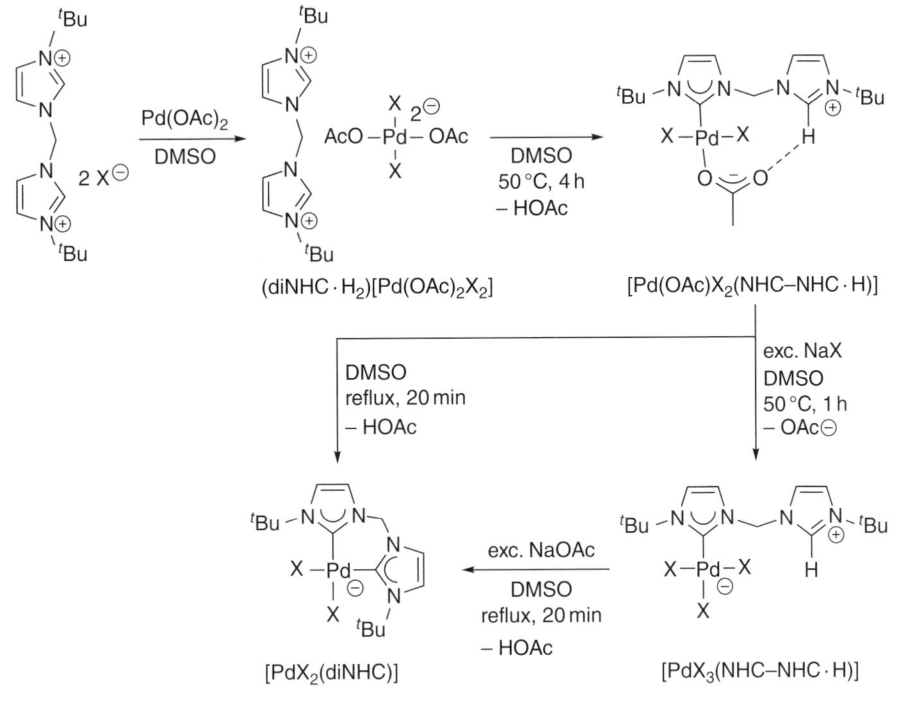

Scheme 5.21
Reaction pathway in the formation of a dihalido(diNHC) complex *via* reaction of a diimidazolium salt with Pd(OAc)$_2$.

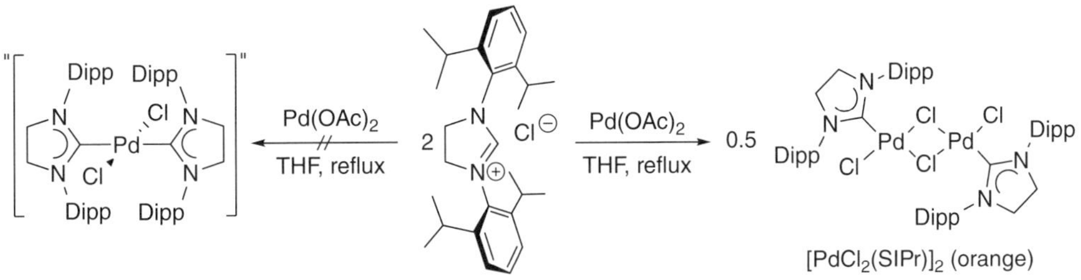

Scheme 5.22
Non-optimized synthesis of the dimeric monocarbene complex [PdCl$_2$(SIPr)]$_2$.

These monocarbene complexes are interesting study objects in their own right, since coordinatively unsaturated palladium complexes with a single NHC or phosphine ligand have been proposed to be the actual catalytic species for a range of transformations [21].

An early example for a deliberate preparation of monocarbene-palladium(II) complexes without the addition of an external base was reported by Andrus and coworkers and simply involves heating an imidazolinium chloride (SIPr·HCl) with Pd(OAc)$_2$ in a 2:1 ratio in tetrahydrofurane (Scheme 5.22) [31]. This reaction gives the desired dimeric monocarbene-palladium(II) complex [PdCl$_2$(SIPr)]$_2$ albeit in a low yield of 37%. Furthermore, 50% of the imidazolinium chloride in this non-optimized reaction has to be sacrificed as a simple source for chlorido ligands to complete the coordination sphere of the dichlorido-bridged dipalladium complex. The isolation of this dimeric complex is notable and confirms the above pathway, since strict stoichiometric considerations would actually favor the formation of the elusive bis(carbene) complex [PdCl₂(SIPr)₂] instead.

The low yield of the dimeric species and the fact that no significant amounts of the bis(NHC) complex was formed can be attributed to two factors. First, the fast deprotonation of the less acidic, saturated imidazolinium salt requires harsher conditions, which were supposedly not effectively met by heating a solution in tetrahydrofurane under reflux. Therefore, only small amounts of NHC ligands could be generated at any given time. Second and more importantly, coordination of a second very bulky SIPr ligand to palladium as required for the formation of the bis(NHC) complex is sterically hindered. In fact, complexes of the type [PdX₂(SIPr)₂] (X = halido) are unknown thus far.

A more economical preparation of similar halido-bridged dipalladium complexes can be accomplished by using a 1:1 ratio of carbene precursor and Pd(OAc)₂ with an excess of a simple and cheaper alkali metal halide. Accordingly, Glorius and coworkers prepared dipalladium complexes of sterically demanding bioxazoline-derived NHCs with restricted flexibility, which have been used as highly active catalysts for the Suzuki-Miyaura reactions of aryl chlorides.

Optimum yields of the dimeric complex [PdCl₂(IBiox7)]₂ were obtained by heating the elaborate imidazolium salt IBiox7·HOTf together with Pd(OAc)₂ and an excess of LiCl in THF to 100 °C for two days in a sealed tube (Scheme 5.23) [32]. Generally, reactions in sealed vessels allow for higher reaction pressures, and temperatures well above the boiling point of the reaction mixtures can be achieved. Nevertheless, more stringent safety precautions must be met to prevent potential explosions. Overall, better stoichiometry control, harsher reaction conditions and the prolonged reaction time led to a substantially improved 91% yield of the target complex.

Huynh and coworkers applied a similar strategy for the preparation of the first benzimidazolin-2-ylidene analogs. The reaction conditions could be further simplified, and moderate heating of 1,3-diisopropylbenzimidazolium bromide with one equivalent Pd(OAc)₂ and four equivalents of sodium bromide in DMSO to 90 °C for one day at atmospheric pressure already afforded the dimeric monocarbene complex [PdBr₂(ⁱPr₂-bimy)]₂ in a very good yield of 93% (Scheme 5.24) [29]. This complex shows high activity in aqueous Suzuki-Miyaura reactions of aryl bromides. Moreover, it has been used as a starting material for the preparation of a wide range of monocarbene/co-ligand and hetero-bis(carbene) palladium(II) complexes of the type *trans*-[PdBr₂(ⁱPr₂-bimy)L] by bridge-cleavage reactions, with which a new and simple ¹³C NMR spectroscopic electronic parameter for the donor strength determination of the ligands L has been developed [33].

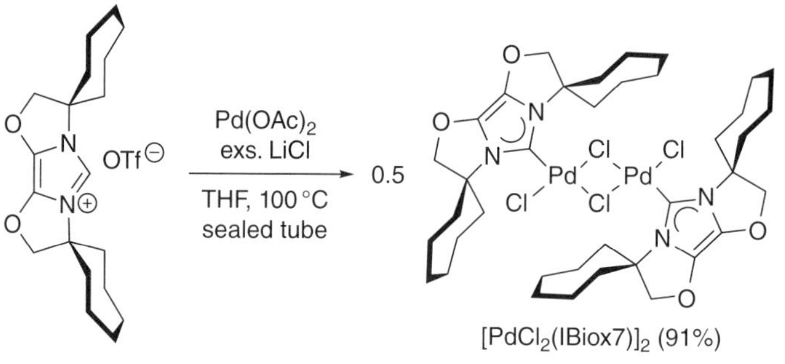

Scheme 5.23
Optimized synthesis of dimeric monocarbene complex [PdCl₂(IBiox7)]₂.

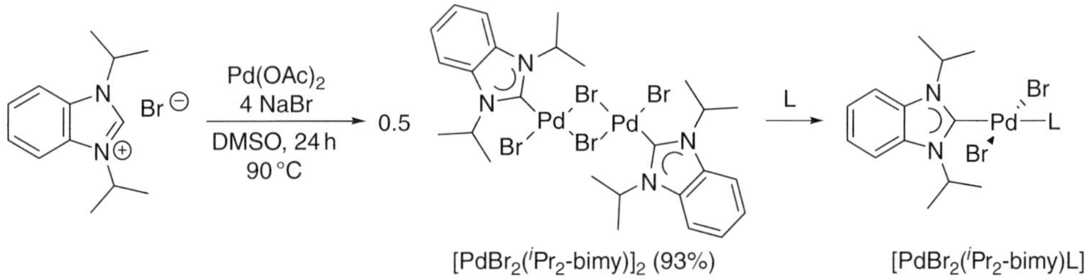

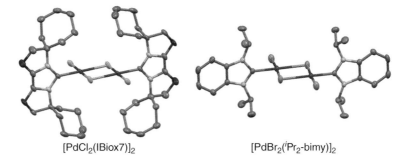

Scheme 5.24
Facile synthesis of the dimeric complex [PdBr$_2$(^{i}Pr$_2$-bimy)]$_2$ and its cleavage with ligands L.

[PdBr$_2$(^{i}Pr$_2$-bimy)]$_2$ (93%)   [PdBr$_2$(^{i}Pr$_2$-bimy)L]

[PdCl$_2$(IBiox7)]$_2$ [PdBr$_2$(^{i}Pr$_2$-bimy)]$_2$

Figure 5.4
Molecular structures of chlorido- and bromido-bridged palladium-NHC dimers.

X-ray diffraction studies on single crystals of both chlorido- and bromido-bridged palladium-monocarbene dimers obtained *via* the "Pd(OAc)$_2$ route" confirmed the dimeric structure. Their molecular structures are depicted in Figure 5.4.

A commonly observed feature for dimeric complexes of this type is the elongated Pd-halido bond *trans* to the carbene, compared to the length of the other two Pd-halido bonds in the complex. This bond elongation is indicative of the strong influence exerted by the NHC donors.

5.2.4 The Silver-Carbene Transfer Route

The second most commonly applied methodology to prepare palladium(II) NHC complexes involves the pre-generation of silver-carbene complexes, which are excellent carbene transfer reagents. In the original procedure reported by Lin and coworkers [34], silver(I) oxide is reacted with two equivalents of 1,3-diethylbenzimidazolium salt to give a silver(I) NHC complex, the structure of which strongly depends on the counter-anion of the initial benzimidazolium salt precursor (Scheme 5.25). For the bromide salt, a dynamic equilibrium between the bis(NHC) complex salt [Ag(Et$_2$-bimy)$_2$][AgBr$_2$] and the neutral monocarbene species [AgBr(Et$_2$-bimy)] was found in solution. The cation and anion in the former complex ion pair are held together by argentophilic interactions, and its formation supposedly through ligand disproportionation of two monocarbene species indicates a certain degree of lability of the silver-NHC bond. In contrast, only the cationic bis(NHC) silver(I) complex [Ag(Et$_2$-bimy)$_2$]PF$_6$ was isolated when the benzimidazolium salt with the non-coordinating PF$_6^-$ anion was employed in the reaction.

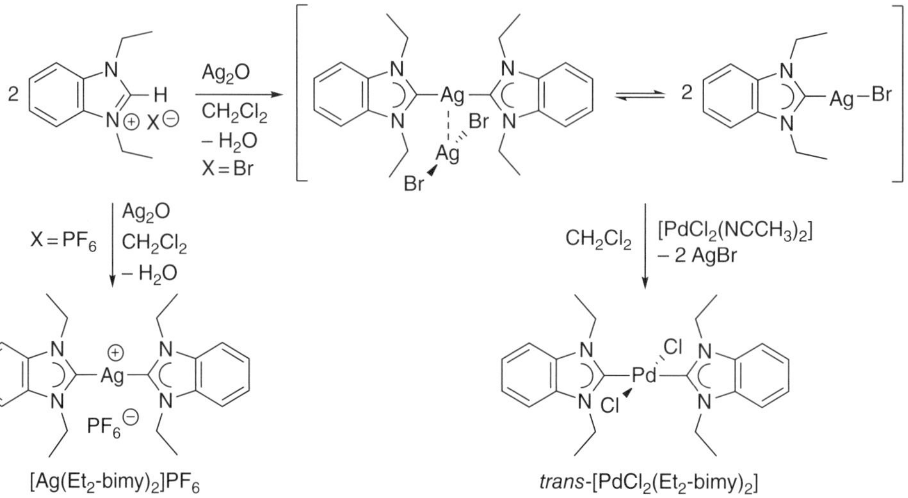

Scheme 5.25
Transfer of the NHC from silver(I) to palladium(II).

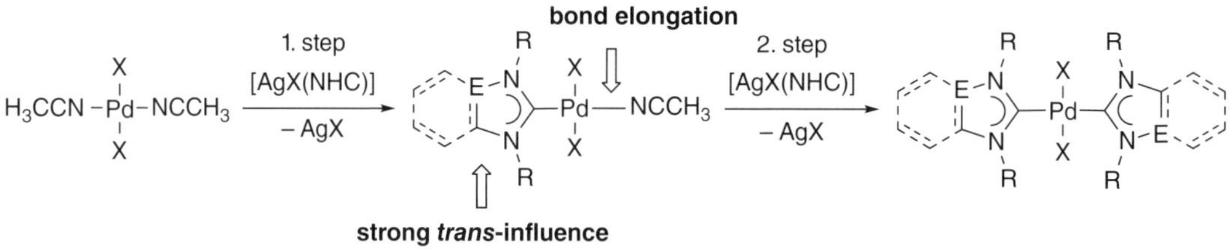

Scheme 5.26
Stepwise silver-NHC transfer to palladium(II).

Addition of the silver-carbenes to bis(acetonitrile)dichloridopalladium(II) leads to a mild transfer of two NHCs to palladium under displacement of the weakly bound acetonitrile ligands. The driving force for the ligand transfer, which affords the dichlorido-bis(NHC)palladium(II) complex *trans*-[PdCl$_2$(Et$_2$-bimy)$_2$], is the precipitation of insoluble silver bromide. A successful carbene transfer from silver to another metal therefore requires the presence of some halide source in the reaction mixture. It is noteworthy, that the resulting metal complex is very likely to retain the lighter halido ligand (here Cl *vs.* Br), due to the increasing preference of silver(I) for halides in the order AgCl<AgBr<AgI.

Palladium(II)-bis(carbene) complexes formed by silver-carbene transfer on [PdX$_2$(NCCH$_3$)$_2$] are usually isolated as *trans*-isomers even if the carbenes are sterically unassuming. This is understandable, since [PdX$_2$(NCCH$_3$)$_2$] complexes are *trans* configured. The retention of the *trans* geometry upon stepwise carbene transfer is apparent when the strong *trans* influence of NHCs is considered. In the first step, one of the weaker acetonitrile ligands is substituted with one NHC giving complexes of the type *trans*-[PdX$_2$(NHC)(NCCH$_3$)]. The remaining acetonitrile ligand in these complexes is even further labilized by the *trans* influence of the NHC ligand, which directs the attack of the second carbene to the *trans* position with respect to the first carbon donor (Scheme 5.26).

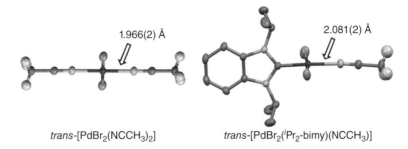

1.966(2) Å   2.081(2) Å

trans-[PdBr₂(NCCH₃)₂]   *trans*-[PdBr₂(ⁱPr₂-bimy)(NCCH₃)]

Figure 5.5
Comparison of palladium-nitrogen bond distances in *trans*-[PdBr₂(NCCH₃)₂] and *trans*-[PdBr₂(ⁱPr₂-bimy) (NCCH₃)]. Hydrogen atoms are only depicted for the acetonitrile ligands for clarity.

Support for this concept can be found by comparison of the Pd–N bond distances in the respective bis(solvate) with those in the mixed monocarbene-acetonitrile adduct. Figure 5.5 depicts the solid state molecular structures for *trans*-dibromido-bis(acetonitrile) palladium(II) [35] and *trans*-dibromido(acetonitrile)(1,3-diisopropylbenzimidazolin-2-ylidene)palladium(II)[29] as a representative along with their key Pd–N bond distances. The latter complex indeed shows a significant Pd–N bond elongation of 11.5 pm compared to the bis(solvate) as a consequence of the strong *trans* influence exhibited by the NHC.

Intuitively, bulky NHC ligands would enhance the discrimination for *trans* complexes even further. Overall, the combination of stereoelectronic effects, the low polarity of dichloromethane as solvent of choice and the low reaction temperature employed, contribute to the very favorable formation of *trans* configured bis(NHC) palladium(II) complexes in the silver-carbene transfer protocol using [PdX₂(NCCH₃)₂] complexes. This is in contrast to the "palladium acetate route" where *cis-trans* mixtures are often obtained due to the greater polarity of the solvents used and the higher reaction temperatures employed.

As shown in the Scheme 5.26, the transfer of only one NHC from silver to palladium gives *trans*-[PdBr₂(NHC)(NCCH₃)] complexes with a weakly bound acetonitrile ligand. These mono-solvates are generally only stable in excess acetonitrile, from which they can be crystallized as mononuclear species. In the absence of this solvent, two processes can be envisioned (Scheme 5.27). Dissociation of the acetonitrile ligand would generate a coordinatively unsaturated tri-coordinated palladium-species, which can stabilize itself *via* dimerization (**I**). Alternatively, a ligand disproportionation process can occur, which affords a bis(NHC) and bis(acetonitrile) complexes (**II**). Both processes can occur simultaneously, however, process **I** is much preferred, since it does not involve breaking of the stable metal-carbene bond.

This can be exploited for the deliberate preparation of palladium dimers. For example, it was found that dissociation of acetonitrile readily occurs by washing powders of *trans*-[PdBr₂(Bh₂-bimy)(NCCH₃)] with diethyl ether yielding the dimeric complex [PdBr₂(Bh₂-bimy)]₂ as an orange powder (Scheme 5.28), which can be crystallized from non-coordinating solvents such as chloroform [36]. The solid state molecular structure of the complex and that of its precursor obtained by single crystal X-ray diffraction analyses are depicted in Figure 5.6. Attempts to prepare the latter complex by the harsher "palladium acetate route" according to Scheme 5.24 failed, which was attributed to the side reactions involving the bulky and more reactive benzhydryl N-substituents. Therefore, this two-step protocol represents a mild and high-yielding, alternative route to valuable palladium(II)-NHC dimers.

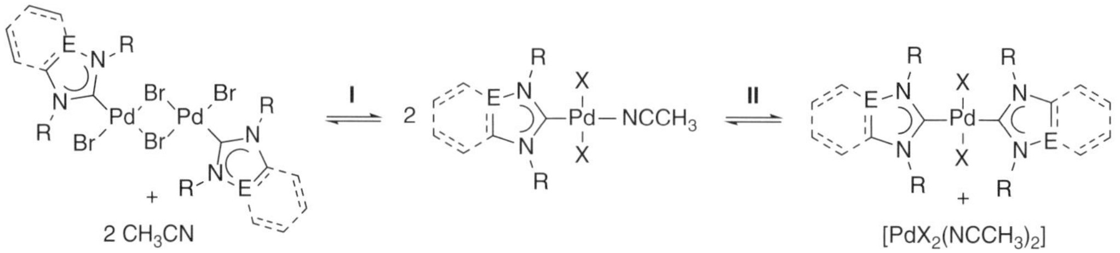

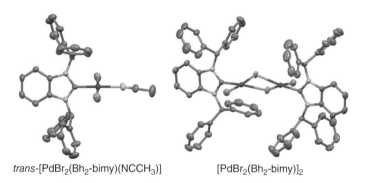

Scheme 5.27
Feasible reactions of *trans*-[PdX_2(NHC)($NCCH_3$)].

[AgBr(Bh_2-bimy)] *trans*-[$PdBr_2$(Bh_2-bimy)($NCCH_3$)] [$PdBr_2$(Bh_2-bimy)]$_2$

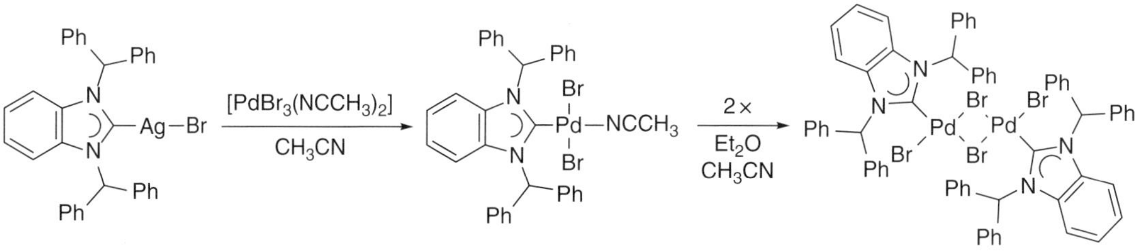

Scheme 5.28
Preparation of a monocarbene-palladium(II) dimer *via* the Ag-NHC transfer route.

trans-[$PdBr_2$(Bh_2-bimy)($NCCH_3$)] [$PdBr_2$(Bh_2-bimy)]$_2$

Figure 5.6
Molecular structures of *trans*-[$PdBr_2$(Bh_2-bimy)($NCCH_3$)] and [$PdBr_2$(Bh_2-bimy)]$_2$.

Not only can these dimers be cleaved with co-ligands such as phosphines or amines, they can also undergo further reactions with additional carbenes to furnish hetero-bis(carbene) palladium(II) complexes. Such complexes contain two different NHC ligands coordinated to the same metal center, and the first deliberate attempt for their preparation was presented by Huynh and coworkers using the silver-NHC transfer route.

The high-yielding and straightforward approach only involves mixing of a dimeric monocarbene complex with a halido-carbene silver(I) complex. The silver-carbene complexes can be pre-prepared, but more convenient is a one-pot procedure by stirring a suspension of the palladium dimer, the azolium halide, and silver(I) oxide in dichloromethane

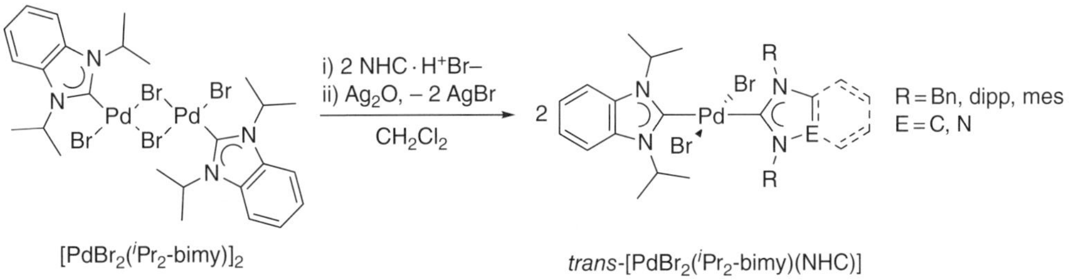

Scheme 5.29
Preparation of hetero-bis(NHC) complexes *via* the Ag-NHC transfer route.

at ambient temperature (Scheme 5.29). It is crucial to employ azolium halides to provide sufficient driving force by silver halide precipitation. Using this method, a host of hetero-bis(carbene) complexes of the general formula *trans*-[PdBr$_2$(^{i}Pr$_2$-bimy)(NHC)] bearing various NHCs derived from imidazole, imidazoline, benzimidazole, 1,2,4- and 1,2,3-triazoles have been prepared for comparison of their donor strengths. The exclusive *trans* geometry of these complexes was confirmed by solution NMR spectroscopic studies as well as single crystal X-ray diffraction. The molecular structures of two representatives are shown in Figure 5.7.

5.2.5 Isomers of Bis(NHC) Palladium(II) Complexes

The first publication on simple dihalido-bis(NHC) complexes with 1,3-dimethylimidazo-lin-2-ylidene ligands prepared by the "Pd(OAc)$_2$ route" reported only the *cis* isomer [26]. A detailed reinvestigation of the analogous reaction with 1,3-dimethylbenzimidazolium iodide revealed the formation of both *cis* and *trans* isomers (Scheme 5.30) [37], although initially only the *cis* isomer has been isolated [38]. The two isomers slightly differ in their colors and show distinct solubilities in organic solvents. The interconversion barrier is relatively high, and therefore *cis* and *trans* isomers can usually be separated [27]. The bright yellow *trans*-[PdI$_2$(Me$_2$-bimy)] complex is readily soluble in halogenated solvents and insoluble in more polar solvents such as dimethylsulfoxide and dimethylformamide, while the orange *cis*-[PdI$_2$(Me$_2$-bimy)] complex, due to its greater polarity, dissolves only in the two latter solvents and acetonitrile. Based on this observation, an improved synthesis for the less polar *trans* form was proposed by employing dimethylsulfoxide as the solvent, where it simply precipitates upon formation [37].

The differentiation and identification of *cis* and *trans* isomers of square planar bis(NHC) metal complexes in solution is best carried out using ^{13}C NMR spectroscopy. In general, *trans* isomers exhibit distinctively more downfield chemical shifts for the carbene carbon atoms compared to their *cis* counterparts. In the given example, the ^{13}C$_{carbene}$ signal for *trans*-[PdI$_2$(Me$_2$-bimy)] was found at 181.0 ppm, while that for its *cis* isomer resonates more upfield at 175.1 ppm. It is important to note that the different solubilities of the two complexes did not allow for the measurement in the same deuterated solvent. Nevertheless, the chemical shift difference observed is much larger than that expected to be induced by solvent effects.

For bis(carbene) complexes of NHCs bearing more elaborated N-substituents, such as longer alkyl chains or benzylic groups, differentiation by ^{1}H NMR spectroscopy is also

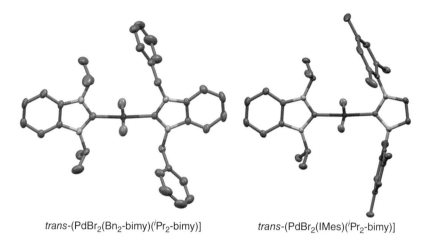

trans-(PdBr$_2$(Bn$_2$-bimy)(^{i}Pr$_2$-bimy)]   *trans*-(PdBr$_2$(IMes)(^{i}Pr$_2$-bimy)]

Figure 5.7
Molecular structures of hetero-bis(carbene) complexes *trans*-[PdBr$_2$(Bn$_2$-bimy)(^{i}Pr$_2$-bimy)] and *trans*-[PdBr$_2$(IMes)(^{i}Pr$_2$-bimy)].

	cis-[PdI$_2$(Me$_2$-bimy)$_2$] 175.1 ppm (NCN, DMSO-d_6)	*trans*-[PdI$_2$(Me$_2$-bimy)$_2$] 181.0 ppm (NCN, CDCl$_3$)
i) THF, RT, overnight:	54%	40%
ii) DMSO, 80 °C, overnight:	20%	77%

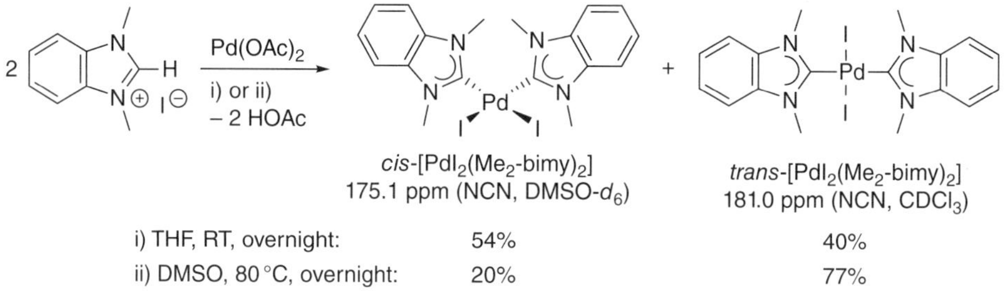

Scheme 5.30
Preparation and ^{13}C$_{carbene}$ shifts of *cis*- and *trans*-[PdI$_2$(Me$_2$-bimy)] complexes.

often possible. NHCs coordinate to square planar metal centers in a way that the carbene plane is perpendicular to the coordination plane with an ideal dihedral angle of 90°. The two N-substituents of a carbene ligand are then located above and below this coordination plane, and a rotation around the metal-carbene bond interchanges their positions. This occurs with ease in *trans* isomers, where the two carbene ligands are further apart (Figure 5.8, left). In addition, free rotations around the N–C and other bonds within the N-substituents are also possible. Overall, the higher degree of rotational freedom in the *trans* isomer leads to equivalence of the wing tip groups. In the case of N-benzyl substituents, only a singlet is observed for the NCH$_2$Ph methylene protons [33,39].

The situation in the respective *cis* isomer is very different. In this case, the two carbene ligands are in close proximity, and rotation around the metal-carbon bond is restricted by the bulk of the N-substituents at room temperature. This steric effect also hampers rotation around the N–C and other bonds. As a result, the NCH$_2$ geminal protons are rotationally arrested and become diastereotopic (Figure 5.8, right). The resulting H$_a$ and H$_b$ protons couple to each other and give rise to two distinct doublets (Figure 5.11). The ^2J(H-H) geminal coupling constants for such methylene protons typically amount to ~14–16 Hz [39,40].

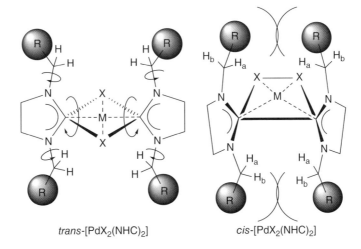

Figure 5.8
Free and restricted bond rotations in *cis* versus *trans* isomers of square planar bis(NHC) metal complexes.

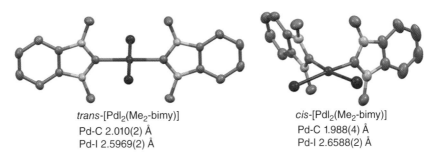

trans-[PdI$_2$(Me$_2$-bimy)]
Pd-C 2.010(2) Å
Pd-I 2.5969(2) Å

cis-[PdI$_2$(Me$_2$-bimy)]
Pd-C 1.988(4) Å
Pd-I 2.6588(2) Å

Figure 5.9
Molecular structures of *trans*- and *cis*-[PdI$_2$(Me$_2$-bimy)] complexes.

In some cases, single crystal X-ray data is also available for both *cis* and *trans* isomers, and a comparison of the key bond parameters around the palladium(II) center reveals that the Pd–C bonds in the *trans* isomer are significantly longer than those in the *cis* form (Figure 5.9). On the other hand, the Pd–I bonds in the latter are significantly elongated due to the strong *trans* influence of the benzimidazolin-2-ylidene ligand. This translates into a greater *trans* effect in catalysis, which facilitates the dissociation of the iodido ligand prior to catalyst initiation. As a result, the *cis* isomer does not show any significant induction period in the Mizoroki-Heck reaction, while the *trans* catalyst precursor shows comparable activity only after an induction period of 60 min.

When unsymmetrically substituted azolium salts are employed in this reaction, an even more complicated mixture is to be expected. In addition to *cis* and *trans* isomers, additional *syn* and *anti* rotation isomers (rotamers) have to be considered, which differ in the relative orientation of equivalent wing tip substituents to each other with respect to the coordination plane (Figure 5.10) [27,28,40]. As the name suggests, isomerization of rotamers occurs *via* simple rotation of the metal-carbene bond.

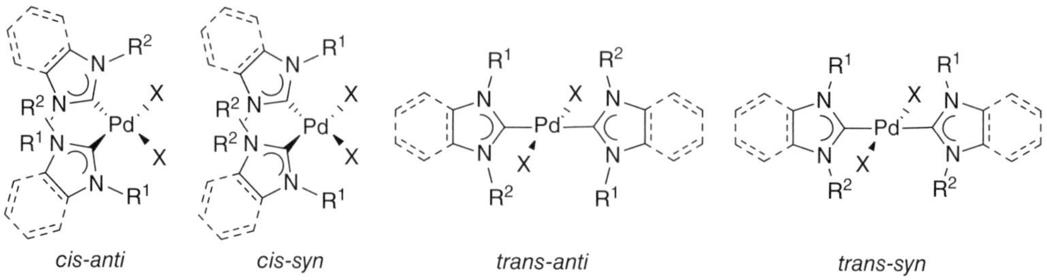

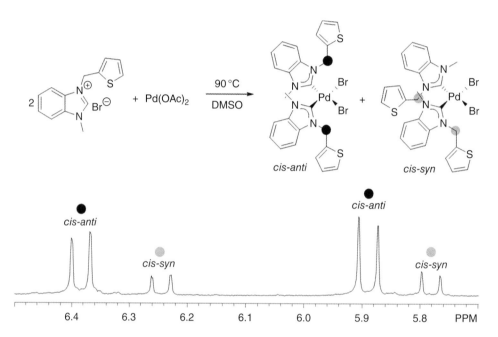

Figure 5.10
Possible isomers of bis(carbene) palladium(II) complexes with unsymmetrical NHC ligands.

Figure 5.11
Diastereotopic ^{1}H NMR signals of NCH$_2$ protons in *cis-anti* and *cis-syn* rotamers.

In the *syn* rotamers, the equivalent substituents are found on the same side of the [PdC$_2$X$_2$] coordination plane, while those in the *anti* form are pointing in opposite directions. Complexes with bulky NHC ligands favor the latter to release intramolecular steric repulsion. This is particularly severe in *cis* rotamers, which explains why *cis-anti* isomers are usually the dominant species [27,40,41]. As elaborated above, the rotational freedom is restricted in *cis* isomers with larger wing tips groups, which leads to diastereotopy of the N-methylene protons. In a *cis*-rotameric mixture of such compounds, four distinct doublets are observed for the *anti* and *syn* forms (Figure 5.11).

The interconversion of *syn* and *anti* complexes can be achieved by a simple rotation of Pd-NHC bond, which usually does not require very high activation energy. This isomerization barrier is especially low in *trans* complexes [27]. This generally precludes separation of *trans* rotamers, and isolation of distinct species has only rarely been reported [27]. In most

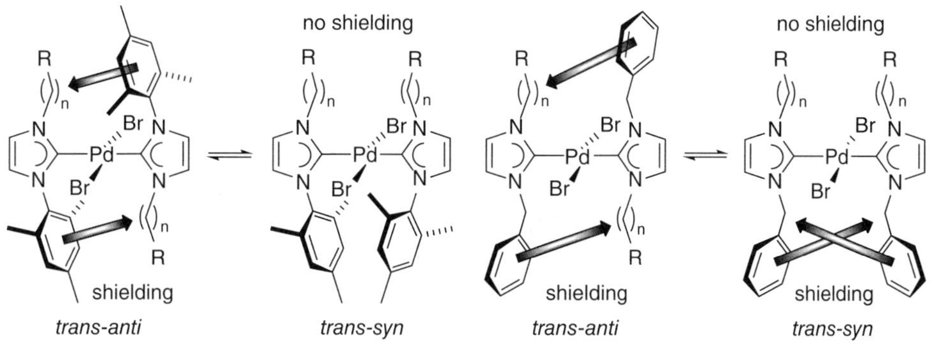

Figure 5.12
Shielding effects in *trans-anti* and *trans-syn* rotamers with aromatic and benzylic substituents.

cases, the two members of a rotameric pair have almost identical physical properties, and are even difficult to differentiate by most analytical methods. NMR spectroscopy, however, is powerful enough to discern them, and two sets of signals are usually observed in the spectra of a rotameric pair. Their ratio can be easily estimated by integration and comparison of the respective proton signals. The assignment of the resonances to the correct rotamer, on the other hand, is generally very difficult and requires additional and more sophisticated methods such as NOE (nuclear Overhauser effect) investigations [27]. This is particularly problematic for NHC ligands with simple N-alkyl substituents. However, one can exploit the anisotropic effect caused by ring currents in aromatic N-substituents for an easy assignment of rotamers in *trans* complexes, where this influence is the strongest [42].

In the *trans-anti* rotamers, the NCH_2 groups are situated in the shielding region of the aromatic mesityl substituent giving rise to distinctively more highfield shifts as compared to those in the *syn* rotamers, which do not experience shielding by the ring current (Figure 5.12). Aromatic substituents can therefore act as built-in spectroscopic probes for the differentiation of *trans* rotamers. This methodology can also be extended to benzyl wing tip groups, which although more flexible still exhibit a sufficiently strong shielding effect *via* their phenyl moieties [43].

The application of this method is demonstrated in Figure 5.13 using a square planar palladium(II) complex with two *trans*-standing cyano-functionalized mesitylimidazolin-2-ylidene ligands. The ¹H NMR spectrum between 5 to 2 ppm displays two sets of triplets for methylene protons in an approximate ratio of 1:2. The two larger triplets at ~4.5 and 2.9 ppm are more highfield compared to the smaller set at ~4.9 and 3.4 ppm, respectively. Notably, only the methylene protons of the cyanoethyl tether in the *trans-anti* rotamer experience the shielding effect of the mesityl substituent. Therefore the more highfield signals at ~4.5 and 2.9 ppm can be assigned to the *anti* rotamer. Overall, one can conclude that the mixture consists of *trans-syn* and *trans-anti* rotamers in approximate 1:2 ratio based on the integrals of the respective signals [42].

Transphobia in Mixed Carbene/Phosphine Complexes

Palladium(II) complexes containing both carbene and phosphine ligands deserve a separate discussion due to their interesting isomerization behavior. Both bis(carbene) and bis(phosphine) complexes of palladium(II) are often found in the preferred *trans* geometry.

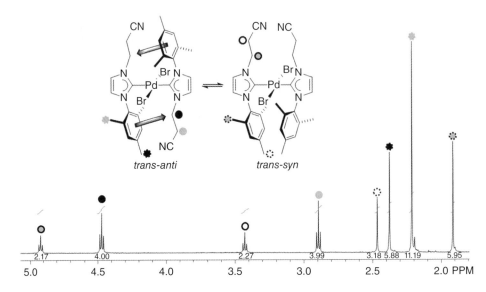

Figure 5.13
Assignments of *trans-anti* and *trans-syn* complexes in a rotameric mixture based on shielding phenomena.

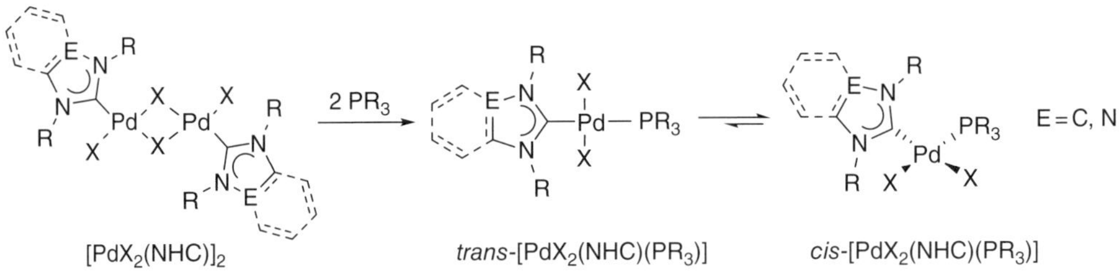

Scheme 5.31
Preparation and isomerization of mixed carbene/phosphine palladium(II) complexes.

A mixed system with both ligands, however, behaves very differently, which is surprising since NHCs have initially been considered phosphine mimics. As already noted by Lappert and coworkers, such complexes isomerize to the *cis* form [19], which appears to be the thermodynamically more favorable isomer due to electronic reasons. Often such complexes are formed by bridge-cleavage reactions of dimeric monocarbene or monophosphine palladium complexes with the other free ligand [44]. This reaction usually affords the *trans* isomer as the kinetically controlled product, which isomerizes to the *cis* form (Scheme 5.31) [29].

The isomerization has been studied in greater detail for the cleavage reaction of [PdBr$_2$(^{i}Pr$_2$-bimy)]$_2$ with triphenylphosphine to yield the complex [PdBr$_2$(^{i}Pr$_2$-bimy)(PPh$_3$)] by time-dependent ^{1}H NMR spectroscopy (Figure 5.14) [29]. Dissolution of the dipalladium complex and addition of the phosphine in CD$_2$Cl$_2$ leads to the almost immediate formation of the *trans* configured mixed-ligand complex, in which the isopropyl-CH protons (CHMe$_2$) resonate as a septet at ~6.30 ppm. With time, the respective signal for the *cis* isomer appears at ~5.84 ppm and grows in intensity, while that for the *trans* complex becomes weaker. After 46 h, this isomerization has almost completed (left plot). Since the

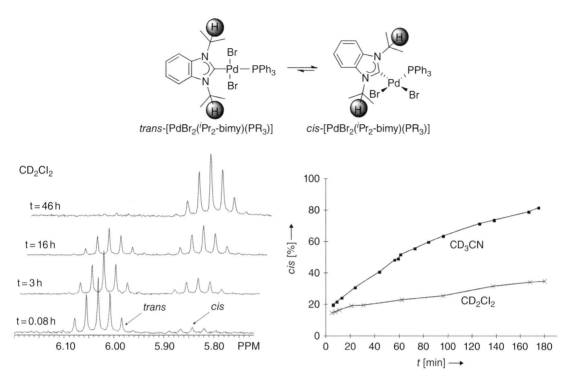

Figure 5.14

Time-dependent ^{1}H NMR spectra showing the *trans-cis* isomerization of [PdBr$_2$(^{i}Pr$_2$-bimy)(PPh$_3$)] in CD$_2$Cl$_2$ (left). Concentration/time diagram (amount of *cis* form [%] vs time [min]) for the same *trans-cis* isomerization in CD$_2$Cl$_2$ and CD$_3$CN (right).

cis form has a greater dipole moment, the isomerization rate can be increased in more polar solvents. Monitoring the same process in more polar CD$_3$CN revealed that after 3 h 80% of the *cis* complex has already formed. In contrast, only less than 30% *cis* isomer was detected in CD$_2$Cl$_2$ after the same time period (right plot).

This phenomenon is not limited to neutral carbenes, but was initially studied by Vicente and coworkers with palladium phosphine complexes bearing anionic carbon donors, such as aryl or aroyl ligands, which can be considered as oxidative addition products of aryl halides to zero-valent palladium-phosphine based catalysts [45]. These authors also proposed the term "*transphobia effect*" for the generally observed difficulty of placing a phosphine ligand *trans* to strong carbon donors in palladium complexes. This "*transphobia effect*" is an electronic effect and occurs in complexes with two strong, but sufficiently different donors. The thermodynamically favorable configuration is eventually achieved, when both strong donors are *trans* to two weaker donors. In our example, this is the *cis* form, where both carbene and phosphine are *trans* to the weaker bromido ligands. The molecular structure determined by single crystal X-ray diffraction for this complex is depicted in Figure 5.15 together with a second example containing a thioether-functionalized imidazolin-2-ylidene ligand [46].

Intuitively, *trans* disposed complexes can be obtained with very bulky ligands [44b,c]. Here the steric factors clearly overrule electronic preferences. Figure 5.16 displays molecular structures of two representative complexes with the bulky I^{t}Bu and SIMes carbene ligands that fulfill this requirement [44c],[47].

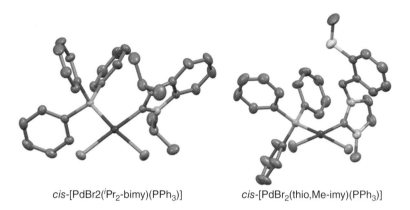

cis-[PdBr2(*i*Pr$_2$-bimy)(PPh$_3$)]   cis-[PdBr$_2$(thio,Me-imy)(PPh$_3$)]

Figure 5.15
Molecular structures of *cis*-configured mixed carbene/phosphine palladium(II) complexes as a result of "*transphobia*."

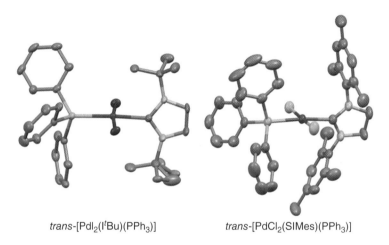

trans-[PdI$_2$(I*t*Bu)(PPh$_3$)]   trans-[PdCl$_2$(SIMes)(PPh$_3$)]

Figure 5.16
Molecular structures of *trans*-configured mixed carbene/phosphine palladium(II) complexes with bulky NHCs.

5.3 Platinum(II)-NHC Complexes

The platinum(II) ion with its d^8 electronic configuration prefers square planar coordination geometries particularly with strong field ligands such as N-heterocyclic carbenes. This usually gives rise to stable and diamagnetic complexes, which can be conveniently studied by NMR spectroscopy. In the resulting spectra, characteristic platinum satellites are expected due to coupling with the NMR active isotope ^{195}Pt. The magnitude of the respective heteronuclear coupling constants often gives insights into the relative position of ligands with respect to each other, and helps in the structural elucidation of the complexes. Despite these apparent advantages, the development of NHC chemistry has been slow compared to that of its lighter, but also cheaper nickel(II) and palladium(II) homologs. The following section provides an overview on early methodologies as well as more recent and more common preparative routes.

5.3.1 Cleavage of Entetramines

The first platinum(II)-NHC complex *trans*-[PtCl$_2$(SIPh)(PPh$_3$)] was reported by Lappert and coworkers in 1971 [48]. The air stable complex was obtained as an yellow solid by the equimolar reaction of the carbene dimer SIPh=SIPh with the dimeric platinum(II)-phosphine complex [PtCl$_2$(PEt$_3$)]$_2$ in boiling xylene (Scheme 5.32). This reaction requires additional activation energy for the breaking of the chlorido-platinum bridges as well as the formal C=C double bond of the carbene dimer, which is provided by the relatively high reaction temperature. The formation of the strong platinum-NHC bond drives the reaction enthalpically. Further heating of the complex just below its melting point led to thermal isomerization to the corresponding *cis* isomer as the thermodynamically more stable species [49]. The electronically favorable *cis* arrangement is a consequence of the *transphobia effect*, which, however, is much less pronounced in platinum complexes compared to their palladium congeners and therefore requires higher activation energy to occur (see Section 5.2.5).

Structural elucidation of both isomers by single crystal X-ray diffraction revealed that the change in the metal-carbene bond is insignificant (Figure 5.17). On the other hand, the metal-phosphine bond distance decreases indicating stronger bond strength in the *cis* isomer. In addition, the platinum-chlorido bonds become substantially elongated [from 2.311(6) Å to 2.362(3) and 2.381(3) Å] upon isomerization to the *cis* form, which is due to the *trans* influences of NHC and phosphine ligands. All these characteristic changes between *trans* and *cis* isomers underline the electronic nature of the *transphobia effect*.

Based on this initial result, the same group conducted a more detailed study to shed light into various factors affecting such *cis-trans* isomerization processes [50]. For this purpose, a range of platinum(II)-carbene complexes bearing various anionic (halido, hydrido, and methyl) and neutral co-ligands (phosphines and arsines) were synthesized by the same methodology and subsequent metathesis reactions (Scheme 5.33).

As expected, more bulky phosphines or carbenes would retard this isomerization, while chlorido and bromido complexes do not behave significantly differently. The isomerization rate can be increased in more polar solvents, which can stabilize the greater dipole in *cis* complexes better. Furthermore, it was found that *cis* complexes are more reactive toward anionic ligand exchange, since these ligands in the latter are subjected to the strong *trans* influence of either the carbene or the phosphine ligand, respectively. The isomeric complexes can also be differentiated by the magnitude of the $^1J_{Pt-P}$ coupling constants, which are larger in the *cis* complexes in line with a stronger platinum-phosphine bond also observed by single crystal X-ray diffraction. Finally, it was noted that the thermodynamically more stable *cis* compounds have higher melting points compared to their *trans* isomers.

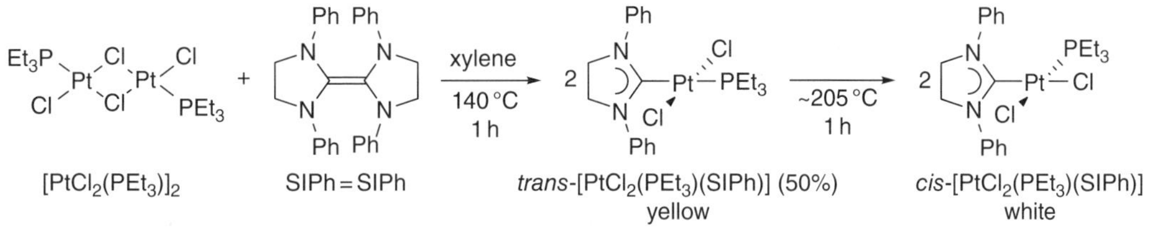

Scheme 5.32
Preparation of the first platinum(II)-NHC complex *via* cleavage of an electron-rich entretramine with a diplatinum complex.

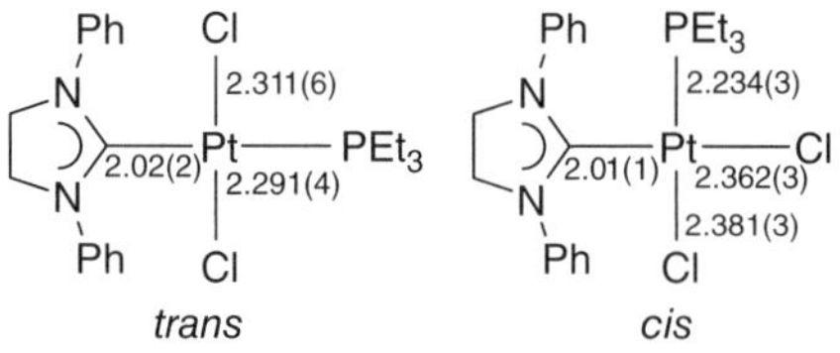

Figure 5.17
Selected bond parameters for *trans* and *cis* isomers of the [PtCl$_2$(SIPh)(PPh$_3$)] complex.

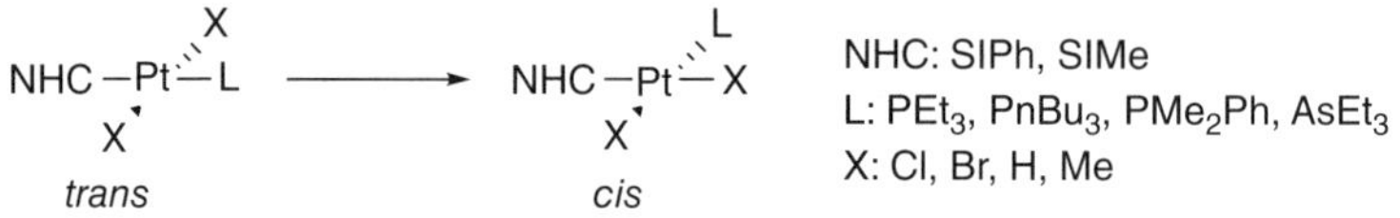

Scheme 5.33
Platinum(II)-carbene complexes studied by Lappert and coworkers.

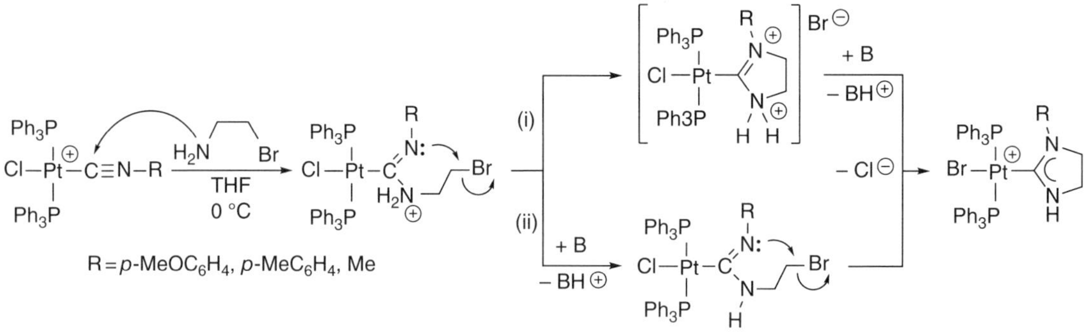

R = *p*-MeOC$_6$H$_4$, *p*-MeC$_6$H$_4$, Me

Scheme 5.34
NHC complex formation by template-directed reaction of platinum-isocyanide adducts with 2-bromoethylamine.

5.3.2 Cyclization of Isocyanide Complexes

More than a decade later, Michelin and coworkers reported the syntheses of related NHC complexes *via* template-directed reaction of platinum-isocyanide adducts with 2-bromoe-thylamine [51] or aziridines [52]. The basis for these reactions is the attack of protic nucleophiles on the metal-activated carbon donor of an isocyanide ligand. In the first case, the nucleophilic attack of the primary amine to a cationic mixed bis(phosphine)/isocyanide complex is proposed to give an intermediate imino-platinum(II) species. Cyclization to give a cationic cyclic diaminocarbene complex then occurs through intramolecular nucleophilic substitution of the bromide, either before (route i) or after (route ii) deprotonation of the ammonium moiety with a base B (Scheme 5.34). The liberated bromide is anticipated to bind to platinum in substitution of the chlorido ligand due to its softer character.

Aziridine also reacts spontaneously with platinum-isocyanide adducts to give the cyclic diaminocarbene complexes in good yield. An example involving a neutral mixed phosphine/isocyanide complex is shown in Scheme 5.35. Formally, one could consider this reaction as an insertion of the coordinated C≡N moiety into a C–N bond of the aziridine, which is thermodynamically driven by the release of the ring strain of the initial three-membered heterocycle.

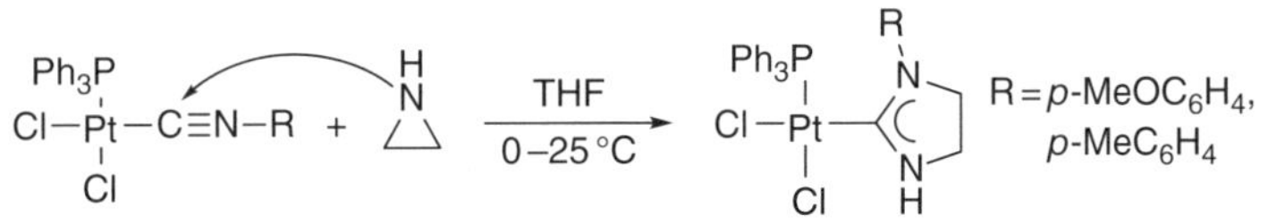

Scheme 5.35
NHC complex formation by template-directed reaction of platinum-isocyanide adducts with aziridine.

In general, complexes containing aromatic isocyanide ligands react faster than their aliphatic analogs. Notably, these reactions lead to complexes containing unusual protic NH,NR-carbenes, that cannot be obtained using azolium precursors. Therefore, subsequent deprotonation of the N–H function followed by reaction with electrophiles (e.g. alkyl halides) furnishes classical and saturated N,N-dialkylimidazolidin-2-ylidene complexes of platinum(II). The methodology is not limited to platinum, but can also be applied to a range of other transition metals. For example, it was noted that palladium complexes react even better. Nevertheless, carbenes derived from imidazole, benzimidazole, or triazole are not obtainable from simple monodentate isocyanide adducts.

5.3.3 The Oxidative Addition Route

A method for the preparation of platinum(II)-imidazolin-2-ylidene complexes involving oxidative addition was described by Cavell and Yates [53]. They drew inspiration from early works by Stone and coworkers who studied the oxidative addition of 2-halo-(benzo) thiazolium and -oxazolium salts with low-valent metal precursors, which gave rise to carbene complexes [54]. In a similar manner, 2-iodo-1,3,4,5-tetramethylimidazolium tetrafluoroborate was heated with a solution of [Pt(PPh$_3$)$_4$] in THF to 65 °C, which resulted in the easy oxidative addition of the imidazolium cation across the C–I bond to the platinum(0) precursor. This reaction gave the cationic platinum(II) complex *trans*-[PtI(timy) (PPh$_3$)$_2$]BF$_4$ in 89% yield (Scheme 5.36), where a new carbene and new anionic iodido ligand were introduced into the coordination sphere of platinum. Solution ^{31}P NMR studies of the white complex revealed a singlet for the phosphine ligands at 13.8 ppm with platinum satellites of $^1J_{Pt-P} = 1234$ Hz, which is indicative of their equivalence and thus a *trans* arrangement. A single crystal X-ray analysis confirmed that this arrangement is retained in the solid state. Although oxidative addition generally occurs in a *cis* manner it is evident that a fast *cis* to *trans* isomerization has taken place.

The oxidative addition of the simple 1,3-dimethylimidazolium tetrafluoroborate salt without C2–halogen bond has also been attempted. Under similar conditions, only 15% of the expected hydrido-platinum complex was isolated, corroborating the more difficult activation of C–H versus C–I bonds (Scheme 5.37). The dissociation of one or two phosphine ligands from tetrakis(triphenylphosphine)platinum(0) as a prerequisite for the reaction to occur, would make the overall process less exothermic (or more endothermic). Therefore, the reaction was repeated with bis(triphenylphosphine)platinum(0) instead. Indeed, the oxidative addition proceeded smoothly giving the carbene-hydrido-bis(phosphine) complex *cis*-[PtH(IMe)(PPh$_3$)$_2$]BF$_4$ in an improved yield of 63%. The *cis* arrangement of the two phosphines in solution was concluded from the ^{31}P NMR spectrum, which depicts two doublets ($^2J_{P-P} = 18$ Hz) at 24.0 and 20.1 ppm with platinum satellites of

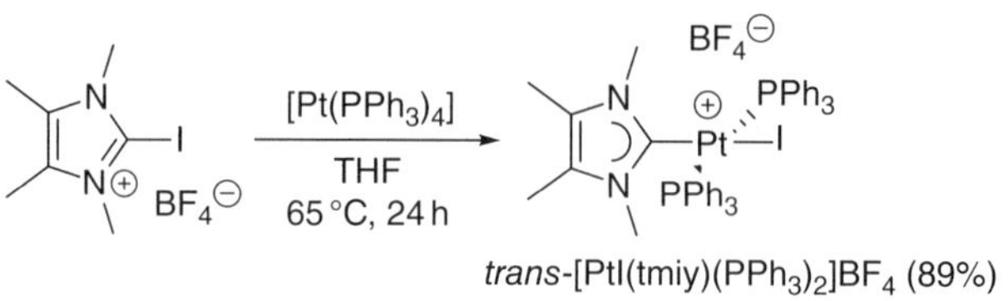

Scheme 5.36
NHC complex formation *via* oxidative addition of an iodo-imidazolium salt.

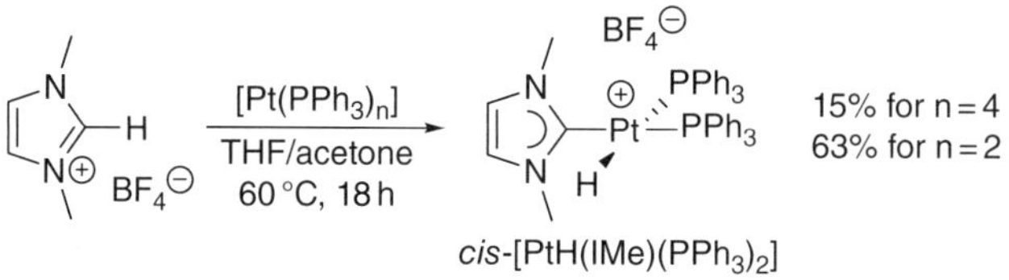

Scheme 5.37
NHC complex formation *via* oxidative addition of a 2H-imidazolium salt.

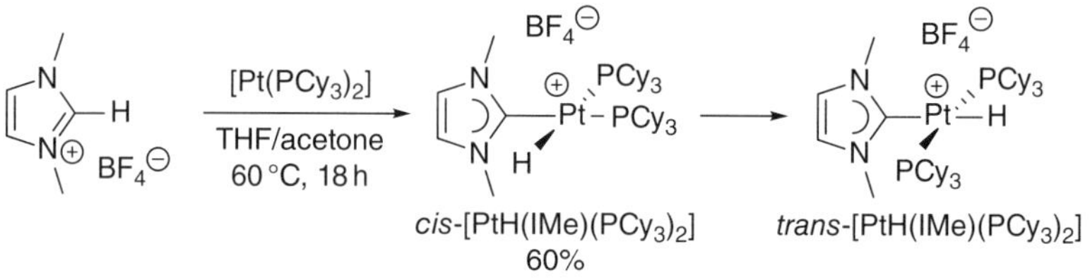

Scheme 5.38
Synthesis and isomerization of the hydrido complex [PtH(IMe)(PCy$_3$)$_2$]BF$_4$.

$^1J_{Pt-P} = 654$ and 663 Hz, respectively. In addition, the ^{1}H NMR spectrum features a doublet of doublets for the hydrido resonance centered at –5.23 ppm ($^2J_{P-H} = 19$ and 176 Hz) with platinum satellites of $^1J_{Pt-H} = 511$ Hz. Supposedly, this was the first example of an C–H bond activation using a platinum(0) complex that does not contain any chelating ligand, which was thought to be essential for such a reaction.

The reaction can also be carried out with the bis(tricyclohexylphosphine)platinum(0) complex indicating that extension of the methodology to trialkylphosphine systems is possible. In fact, the latter are even more electron rich, which is expected to facilitate C–H bond activation. Indeed, oxidative addition occurs with ease, and the hydrido complex *cis*-[PtH(IMe)(PCy$_3$)$_2$]BF$_4$ was isolated in 60% yield (Scheme 5.38). Its ^{1}H NMR spectrum shows characteristics similar to those of the triphenylphosphine analog (e.g. δH = –7.95 ppm, $^2J_{P-H} = 21$ and 160 Hz, $^1J_{Pt-H} = 880$ Hz).

Attempts to grow single crystals from the complex led to the isolation of two different types of crystals, which were analyzed as the *cis* and *trans* geometrical isomers of the desired hydrido complex (Figure 5.18). Again, the slow isomerization that apparently occurred during the period of crystallization can be ascribed to the *transphobia effect*. In addition, one can note that the coordination geometry around the platinum center in the *cis* isomer is significantly more distorted than that in the *trans* form. In this particular case,

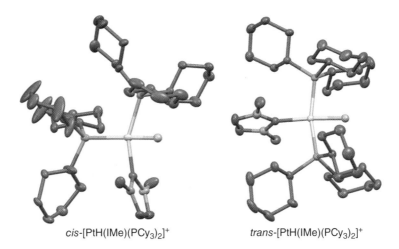

cis-[PtH(IMe)(PCy$_3$)$_2$]$^+$ $trans$-[PtH(IMe)(PCy$_3$)$_2$]$^+$

Figure 5.18
Molecular structures of the cations in cis- and $trans$-[PtH(IMe)(PCy$_3$)$_2$]BF$_4$. The BF$_4^-$ anion, solvent molecules, and hydrogen atoms with exception of the hydrido ligands are omitted for clarity.

isomerization is therefore driven by both steric and electronic factors. In the cis complex, the two very bulky phosphines push each other apart, which results in an enlargement of the P–Pt–P angle from an ideal 90° for a perfect square plane to ~108°. Steric repulsion also leads to an increase of the P–Pt–NHC angle to ~98° bringing the smaller hydrido and carbene ligand closer to each other. Also notable is that the Pt–NHC distance of 2.058(3) Å in the $trans$ isomer is slightly elongated relative to that in the cis complex [2.028(3) Å] due to the higher $trans$ influence of the hydrido compared to the tricyclohexylphosphine ligand.

Instead of platinum(0)-phosphine complexes, more stable platinum(0)-carbene complexes can be employed as the metal source as well. Stirring a mixture of [Pt(DMFU)$_2$(NHC)] (NHC = IMes or SIMes; Chapter 4) with the simple 1,3-dimethylimidazolium iodide salt in THF under reflux afforded the neutral oxidative addition products $trans$-[PtHI(IMe)(NHC)] in quantitative yields with elimination of two equivalents of DMFU [55]. The quantitative yields and the clean reaction are proof for the enhanced electron donation as well as stabilizing character of NHCs relative to phosphines. Remarkably, the reaction can also go to completion at a mild temperature of 20 °C although longer reaction times are a necessity (Scheme 5.39). The two square planar hydrido-hetero-bis(NHC) platinum(II) complexes are the first of their kind to be isolated. The hydrido signals in the two complexes were observed at −14.67 (IMes) and −14.50 ppm (SIMes) with very high coupling constants of $^1J_{Pt-H}$ = 1727 (IMes) and 1738 Hz (SIMes), respectively.

Since the oxidative addition of imidazolium salts is successful, it is apparent that neutral dihalogens, which are easier to activate, will react in an analogous manner. In contrast to the previous case, such a reaction will give rise to two new anionic ligands at the oxidized metal center. Indeed, this was later demonstrated by Marinetti and coworkers, who simply subjected Markó's platinum(0)-NHC complexes (see Section 4.3.2) to elemental iodine (I$_2$) and additional mono- and diphosphine ligands (Scheme 5.40) [56].

The reaction leads to a change of the coordination geometry from pseudo-trigonal planar to essentially square planar with two new iodido ligands. Prior to the addition of the phosphines, the authors proposed an intermediate, in which the initially bidentate DVDS ligand is now only bound via one olefinic moiety making it more prone to ligand displacement

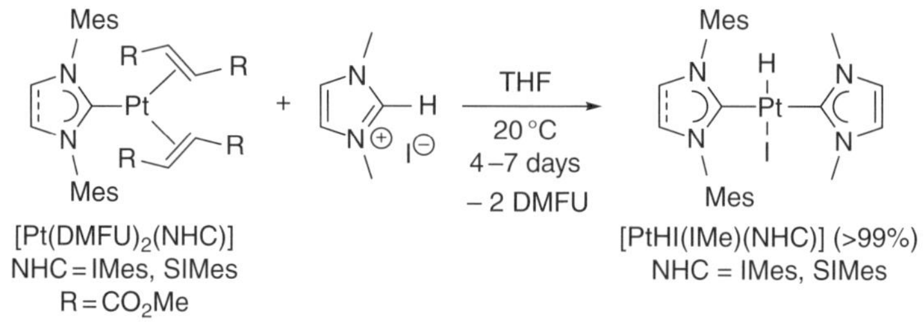

Scheme 5.39
Preparation of hydrido-hetero-bis(NHC) complexes of platinum(II) *via* oxidative addition.

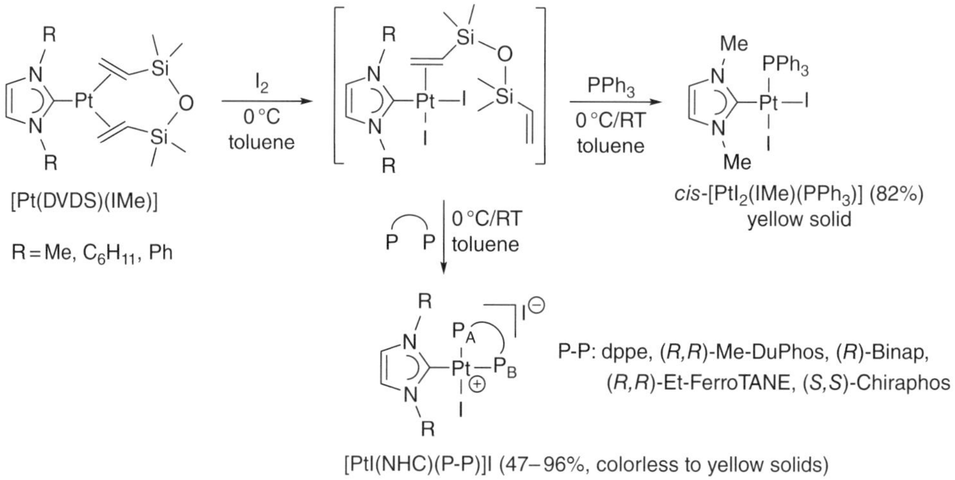

Scheme 5.40
Oxidative addition of iodine to platinum(0)-NHC complexes.

reactions. The subsequent reaction of the IMe representative with triphenylphosphine leads to formation of the neutral platinum(II) complex *cis*-[PtI$_2$(IMe)(PPh$_3$)] incorporating the NHC and phosphine in the thermodynamically favorable *cis* arrangement. The solution ^{31}P NMR spectrum displays a signal at 8.5 ppm for the phosphine with a $^1J_{\text{Pt-P}}$ coupling of 3695 Hz, consistent with values in the range usually observed for platinum(II) complexes (i.e. >3000 Hz), where iodido and phosphine are *trans* to each other. A single crystal X-ray determination of the molecular structure confirms the *cis* configuration of the complex also in the solid state (Figure 5.19, left).

Reaction of the mono-olefin intermediate with a range of mostly chiral *cis*-chelating diphosphines led to the additional displacement of one iodido ligand from the platinum(II) center as well affording cationic complexes of the type [PtI(NHC)(P–P)]I, which were isolated as colorless to pale yellow solids in moderate to excellent yields. These square planar complexes contain two different phosphorus donors, which give rise to an AB spin system as evidenced by two distinct resonances in their ^{31}P NMR spectra. Their $^2J_{\text{P-P}}$ coupling constants range from 6–18 Hz. In addition, they exhibit platinum satellites with $^1J_{\text{P-Pt}}$ coupling constants in two distinct ranges (2140–2293 Hz and 3104–3370 Hz), which allow reliable assignment of the ^{31}P NMR signals to *cis*- and *trans*- phosphorus donors with

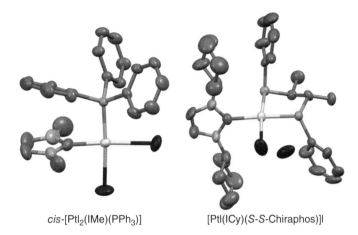

cis-[PtI₂(IMe)(PPh₃)]   [PtI(ICy)(*S-S*-Chiraphos)]I

Figure 5.19
Molecular structure of *cis*-[PtI₂(IMe)(PPh₃)] and [PtI(ICy)(*S-S*-Chiraphos)]I.

respect to the remaining iodido ligand. Thus, the smaller couplings (~2000 Hz) are expected for phosphorus atoms *cis* to the halide ligand. The solid state molecular structure of [PtI(ICy)(*S-S*-Chiraphos)]I as a representative is shown in Figure 5.19 (right).

The cationic mixed-donor platinum(II) complexes have been tested for their suitability to act as precatalysts for the enantioselective cycloisomerization of an allyl propargylamine derivative, where enantiomeric excesses up to 74% could be obtained in this model reaction.

5.3.4 The Free Carbene Route

The obvious reaction of an isolated free NHC with a platinum(II) source for the preparation of simple platinum(II)-NHC complexes has only been studied surprisingly recently. Nolan and coworkers chose the bis(DMSO) complex *cis*-[PtCl₂(DMSO)₂] as a suitable metal source, and its exposure to one equivalent of free NHC led to the displacement of one dimethylsulfoxide ligand and the formation of the mono-NHC complexes *cis*-[PtCl₂(DMSO)(NHC)], where the *cis* geometry is retained (Scheme 5.41) [57]. The synthetic procedure is general and works well for NHCs that can be easily obtained as free base, which include the unsaturated imidazolin-2-ylidenes (IPr, IMes), very bulky saturated imidazolidin-2-ylidenes (SIPr, SIMes) and also Enders' 1,3,4-triphenyltriazolin-5-ylidene (TTP).

All complexes have been isolated as air and moisture stable white solids in good yields in most cases. Lower yields were observed for the saturated NHCs, which are more prone to decomposition by hydrolysis and ring-opening reactions compared to their unsaturated cousins due to their increased basicity and lack of aromaticity.

The NMR spectra of these compounds display the characteristic platinum satellites, which indicate a successful coordination of the carbene to the metal. When compared to the free carbene ligands, duplication of some signals for the substituted N-aryl groups perpendicular to the NHC plane is observed in the ¹H and ¹³C NMR spectra of the complexes. This is corroborative of a hindered rotation around the N–C bound as well as a *cis* arrangement of the carbene and DMSO ligands with respect to each other. As a result, half of the

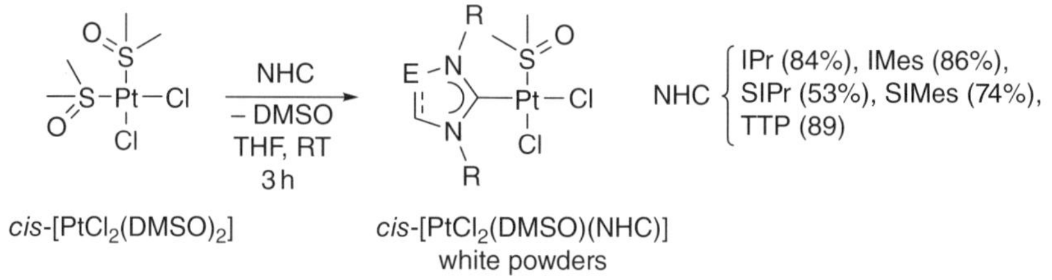

Scheme 5.41
Preparation of monocarbene-DMSO platinum(II) complexes with free NHCs.

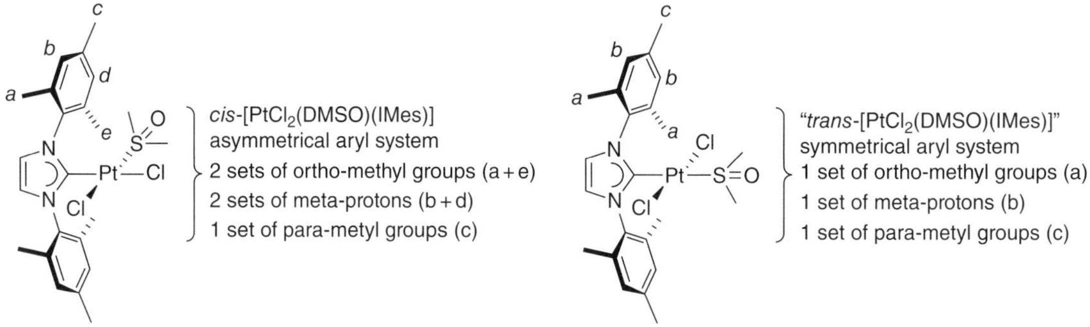

Figure 5.20
Expected sets of NMR signals for the aromatic mesityl substituent of *cis*-[PtCl$_2$(DMSO)(IMes)] and its fictive *trans* isomer.

ortho- and *meta*-substituents of the aromatic rings are pointing toward the *cis*-DMSO, while the other half points to the *cis*-chlorido ligand giving rise to two individual signals. These features can be discussed using the signals observed in ^{1}H NMR spectrum of *cis*-[PtCl$_2$(DMSO)(IMes)] as an example. Two singlets for the *meta*-protons of the mesityl groups, each integrating to 2H, are observed at 7.07 (for *b*) and 7.03 ppm (for *d*, Figure 5.20). The *ortho*-methyl groups also give rise to two signals, each integrating to 6H, at 2.38 and 2.24 ppm for *a* and *e*, respectively. Such an asymmetry was not observed in the unsubstituted and less bulky N-phenyl wing tips of the TTP ligand, where rotation above the N–C bond is therefore not sterically hindered.

In addition to the chemical shifts for the carbene donors, the heteronuclear platinum-carbene coupling constants $^1J_{Pt-C}$ and the ^{195}Pt NMR resonances of the complexes have been measured as well, and are summarized in Table 5.2. It has been suggested that the magnitude of the $^1J_{Pt-C}$ values are indicative of the σ contribution of the respective bond. In the present series, these values decrease in the order IPr > IMes > TTP > SIPr > SIMes indicating that platinum-carbon bonds of saturated NHCs have the smallest σ and in turn the largest π contribution. The ^{195}Pt NMR chemical shifts, on the other hand, are indicative of the electron density at the metal, whereby a more electron-rich metal center would give rise to a more highfield signal.

Apparently, this electron density is influenced by the coordinated ligands, and since the complexes of concern only differ in the NHC ligands, these ^{195}Pt NMR resonances can be used to rank their donating abilities. The δ values for ^{195}Pt NMR signals ranging from

Table 5.2 Selected ^{13}C and ^{195}Pt NMR Data for *cis*-[PtCl$_2$(DMSO)(NHC)] complexes.

Complex	δ_{carbene} [ppm]	δ_{Pt} [ppm]	$^{1}J_{\text{Pt–C}}$ [Hz]
cis-[PtCl$_2$(DMSO)(IMes)]	145.8	1017	1472
cis-[PtCl$_2$(DMSO)(IPr)]	147.4	1023	1479
cis-[PtCl$_2$(DMSO)(SIMes)]	173.8	992	1358
cis-[PtCl$_2$(DMSO)(SIPr)]	174.4	1010	1373
cis-[PtCl$_2$(DMSO)(TTP)]	153.0	1033	1412

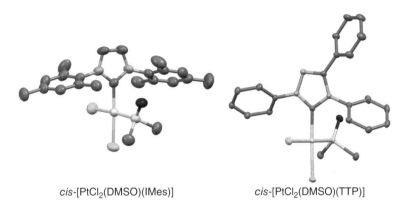

cis-[PtCl$_2$(DMSO)(IMes)]   *cis*-[PtCl$_2$(DMSO)(TTP)]

Figure 5.21
Molecular structures of *cis*-[PtCl$_2$(DMSO)(IMes)] and *cis*-[PtCl$_2$(DMSO)(TTP)].

992–1033 ppm, which are typical for neutral and four-coordinated platinum(II) complexes, decrease in the order TTP > IPr > IMes > SIPr > SIMes. This result is in line with chemical intuition: The TTP ligand with an additional electronegative nitrogen atom in the five-membered ring is expected to be the weakest donor, while the saturated carbenes with two sp^3-hybridized carbon atoms in the backbone transfer the largest amount of electron density to the metal center. Overall and in combination with theoretical calculations, the conclusion can be made that the unsaturated IMes and IPr ligands are both less efficient σ donors and π acceptors compared to their saturated counterparts SIMes and SIPr.

The *cis* arrangement of the square planar complexes found in solution by NMR spectroscopy is retained in the solid state as evidenced by single crystal X-ray diffraction analyses of the SIMes, IMes, and TTP complexes. The molecular structures of the latter two are depicted in Figure 5.21. The platinum–carbon bond lengths in the three complexes are indistinguishable within 3σ and average to 1.981 Å. The platinum-chlorido distances *trans* to the carbene ligand with a mean value of 2.345 Å are also equal within experimental errors. Therefore, the *trans* influence of the three carbenes is essentially of the same strength.

5.3.5 *In Situ* Deprotonation of Azolium Salts with External Base

The methodologies described thus far require either expensive starting materials, such as low-valent platinum(0) complexes, or moisture sensitive free carbenes, which are often cumbersome to handle. Therefore easier and more convenient routes for the preparation of

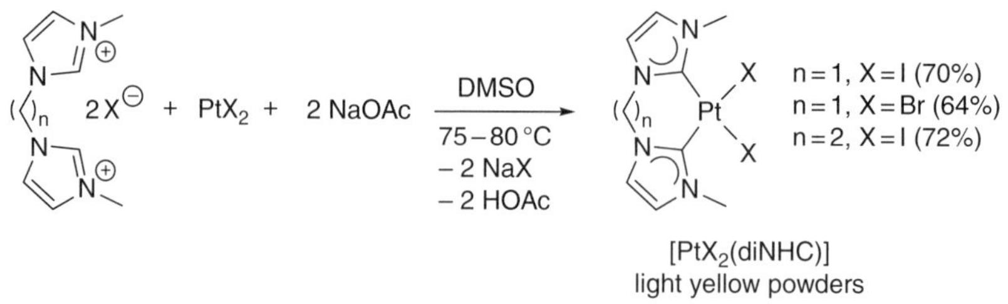

Scheme 5.42
Syntheses of platinum(II)-diNHC complexes *via in situ* deprotonation of diimidazolium salts with NaOAc.

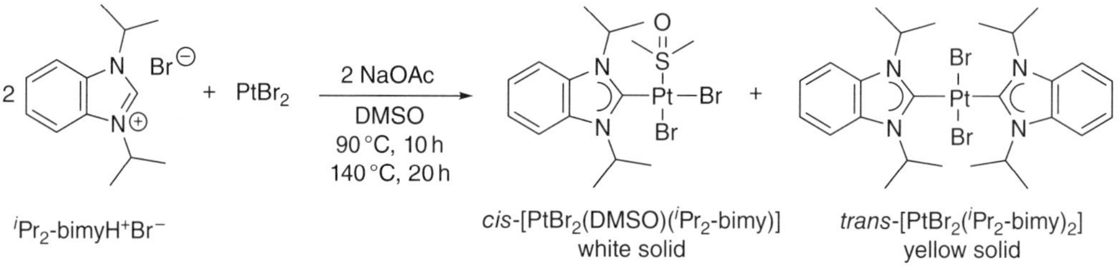

Scheme 5.43
Syntheses of mono- and bis(NHC) complexes *via in situ* deprotonation of a benzimidazolium salt with NaOAc.

platinum(II)-NHC complexes were sought. An obvious alternative to the use of pre-isolated free NHCs is their *in situ* generation in the presence of a suitable platinum source, for the direct formation of the desired complex. Indeed, Strassner and coworkers demonstrated in 2002 that the simple equimolar reaction of an diimidazolium salt with platinum(II) halides in the presence of two equivalents of sodium acetate provides the respective platinum(II)-dicarbene complexes in decent yields (Scheme 5.42) [58]. Notably, the reaction can be carried out in wet DMSO at elevated temperatures, which indicates that the coordination of the *in situ* generated free NHCs to platinum(II) occurs faster than their hydrolysis. The products are isolated as air stable and light yellow powders. In order to avoid halide scrambling, both imidazolium and platinum salts used should contain the same halides.

The diamagnetic nature of the compounds in line with square planar coordination geometry allows for a characterization by solution NMR spectroscopy in DMSO-d_6. The absence of downfield signals for the acidic NCHN azolium protons characteristic of the diimidazolium halides confirm successful dicarbene formation. In addition, the bridging methylene protons become diastereotopic and resonate as two distinct doublets (n=1) or multiplets (n=2) due to the hindered rotation upon formation of the six- (n=1) or seven-membered (n=2) metallacycles. The $^{13}C_{carbene}$ NMR signals are found in the narrow range ~143–145 ppm. Elucidation of the molecular structure for the dibromido-complex by single crystal X-ray diffraction confirmed the square planar geometry of the complex, with a diNHC chelate also found in solution.

Huynh and coworkers extended the methodology for the preparation of mono- and bis(NHC) platinum(II) complexes bearing monodentate benzimidazolin-2-ylidenes, which lack the stabilizing chelate ring [59]. The reaction of the benzimidazolium bromide $^iPr_2$-bimyH$^+$Br$^-$ with PtBr$_2$ in a 2:1 ratio with two equivalents of NaOAc was explored under various conditions and in different solvents (Scheme 5.43). Only the reaction in DMSO at elevated temperatures turned out to be successful possibly pointing to the importance of the solvate *cis*-[PtBr$_2$(DMSO)$_2$] as an intermediate. Yet the conversion was slow and prolonged heating of the mixture was required, which was attributed to the steric bulk and the relatively lower acidity of the C2–H proton in the precursor salt due to the donating nature (+*I* effect) of the two N-isopropyl substituents. Despite the 2:1 stoichiometric ratio, the mono-NHC complex *cis*-[PtBr$_2$(DMSO)(*i*Pr$_2$-bimy)] formed as the major product in acceptable, but moderate yield of 55%, while the bis(NHC) complex *trans*-[PtBr$_2$(*i*Pr$_2$-bimy)$_2$] was isolated in only 3% as a precipitate insoluble in DMSO. The latter is also only sparingly soluble in chlorinated solvents, while the former can be dissolved in most polar organic solvents.

The ^{1}H NMR spectra of both complexes recorded in CDCl$_3$ feature a significant downfield shift of the isopropyl C–H proton (Table 5.3) compared to that in the precursor salt *i*Pr$_2$-bimyH$^+$Br$^-$ (i.e. 5.21 ppm) corroborating rare C–H···Pt anagostic interactions characteristic for square planar d^8 metal complexes of the *i*Pr$_2$-bimy ligand [60]. The carbene signals were found at 153.6 and 176.5 ppm for the mono and bis(NHC) complexes, respectively. The much more downfield signal for the latter is in agreement with its lower Lewis acidic metal center resulting from the simultaneous strong electron donation of two NHC ligands.

The DMSO solvate *cis*-[PtBr$_2$(DMSO)(*i*Pr$_2$-bimy)] is a useful starting point to other heteroleptic mono-NHC complexes as it can undergo facile ligand substitution. Its exposure to one equivalent of triphenylphosphine or pyridine as two commonly used ligands with different *trans* influences was studied by ^{1}H NMR experiments at room temperature (Scheme 5.44). In the first case, the kinetically controlled formation of *trans*-[PtBr$_2$(*i*Pr$_2$-bimy)(PPh$_3$)] was noted, which with time transforms into *cis*-[PtBr$_2$(*i*Pr$_2$-bimy)(PPh$_3$)] due to the *transphobia* of the phosphine ligand. This process is however much slower than the same isomerization observed for the isostructural palladium congeners, and does not even reach completion after 15 h.

In comparison, DMSO substitution with one equivalent pyridine was much slower and required ~18 h to complete. Here, the *trans* isomer formed exclusively as the thermodynamically stable product in agreement with the significantly decreased *trans* influence of pyridine compared to a phosphine.

All four complexes are air stable and can be handled and crystallized without taking precautions to exclude moisture or oxygen. The molecular structures thus determined by X-ray diffraction are depicted in Figure 5.22. Selected spectroscopic key parameters are summarized in Table 5.3. Interestingly, the shortest and longest platinum-carbon bond distances are both found in the *trans* complexes. The shortest and supposedly strongest metal–carbene bond is found in the pyridine adduct, where the weak co-donor and the lack of steric interference allow for an optimal binding of the NHC ligand with the platinum(II) center. The longest platinum-carbon bonds are exhibited by the most electron-rich bis(NHC) complex with *trans* arrangement of the two competing carbene ligands.

Table 5.3 Selected spectroscopic key parameters for platinum(II)-*i*Pr$_2$-bimy complexes.

Complex	$\delta^{13}C_{carbene}$ [ppm]	δH(*i*PrCH) [ppm]	Pt–C [Å]
***cis*-[PtBr$_2$(DMSO)(*i*Pr$_2$-bimy)]**	153.6	6.27	1.979(6)
***cis*-[PtBr$_2$(*i*Pr$_2$-bimy)(PPh$_3$)]**	161.0	6.09	1.973(4)
***trans*-[PtBr$_2$(*i*Pr$_2$-bimy)$_2$]**	176.5	6.52	2.015(4)
***trans*-[PtBr$_2$(*i*Pr$_2$-bimy)(Py)]**	149.7	6.59	1.958(4)

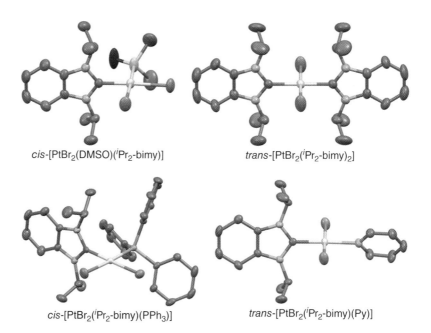

Scheme 5.44
DMSO displacement reactions of *cis*-[PtBr₂(DMSO)(ⁱPr₂-bimy)] with triphenylphosphine and pyridine.

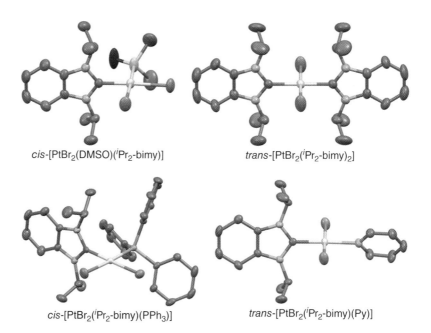

cis-[PtBr₂(DMSO)(ⁱPr₂-bimy)]   trans-[PtBr₂(ⁱPr₂-bimy)₂]

cis-[PtBr₂(ⁱPr₂-bimy)(PPh₃)]   trans-[PtBr₂(ⁱPr₂-bimy)(Py)]

Figure 5.22
Molecular structures of platinum(II)-ⁱPr₂-bimy complexes.

5.3.6 *In Situ* Deprotonation with Basic Platinum Precursors

One of the most common pathways to palladium(II) NHC complexes utilizes commercial palladium(II) acetate, which acts as source for the metal center, but also provides two equivalents of base required for the deprotonation of carbene precursor salts. The analogous

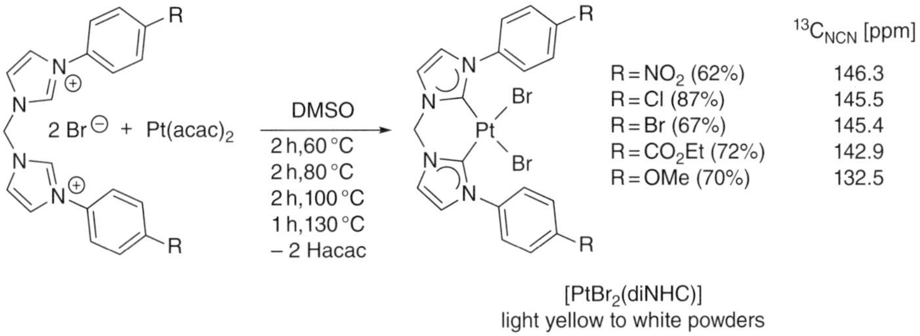

	$^{13}C_{NCN}$ [ppm]
R = NO$_2$ (62%)	146.3
R = Cl (87%)	145.5
R = Br (67%)	145.4
R = CO$_2$Et (72%)	142.9
R = OMe (70%)	132.5

[PtBr$_2$(diNHC)]
light yellow to white powders

Scheme 5.45
Platinum(II)-diNHC chelates by direct metallation of diimidazolium salts with [Pt(acac)$_2$].

platinum(II) compound is, on the other hand, not commercially available, and its use in NHC chemistry is thus uncommon. Platinum(II) acetylacetonate [Pt(acac)$_2$] has been used as an alternative in its place primarily in the preparation of complexes with chelating dicarbene or donor-functionalized carbene complexes. Attempts to prepare monodentate analogs have been reported to be less successful [59,61], and pre-coordination of a donor to platinum(II) appears to be crucial in these procedures.

For example, Strassner and coworkers demonstrated that by heating methylene-bridged diimidazolium bromides with one equivalent [Pt(acac)$_2$] in DMSO, a range of chelating diNHC complexes can be obtained (Scheme 5.45) [62]. However, a strict heating regimen has to be followed for a proper outcome. The reaction commences with stirring the mixture at 60 °C for two hours. It is proposed that during this period, deprotonation of the first imidazolium moiety by one acetylacetonato ligand takes place, which supposedly yields a monocarbene complex with dangling imidazolium tether. Raising the reaction temperature to ~80 °C after this pre-coordination step leads to a clear yellow solution, and further slow increase finally to 130 °C completes generation of the second carbene donor affording the desired platinum(II) diNHC chelates that are isolated as air and moisture stable, white to yellow solids. Overall, two equivalents of acetylacetone are liberated from this acid-base reaction. The terminal aromatic wing tip groups impart an increased stability of the complexes compared to N-alkyl analogs, and these complexes even resist decomposition in strong acidic media.

^{1}H NMR data for solutions of the complexes in DMSO-d_6 revealed diastereotopy of the protons in the methylene linker. This is in line with observations made for palladium analogs and indicates hindered mobility of the six-membered metallacycle as a result of chelate formation. ^{13}C NMR spectroscopic determination of the resonance for the carbene atom shows that this signal is diagnostic of the *para*-group of aromatic N-substituent. Electron-withdrawing groups decrease electron density at the carbene carbon, which leads to a downfield shift. In contrast, electron donating functions would increase the electron density, and thus shielding of the carbene donor is enhanced, which in turn results in an upfield shift. Overall, these chemical shifts were observed in the range of 132.5–146.3 ppm, which is in line with a *cis* orientation enforced by the short methylene spacer of the dicarbene ligand.

The same methodology was extended by Hahn and coworkers to picolyl-functionalized benzimidazolin-2-ylidenes. Again, an elaborate heating sequence was applied in the

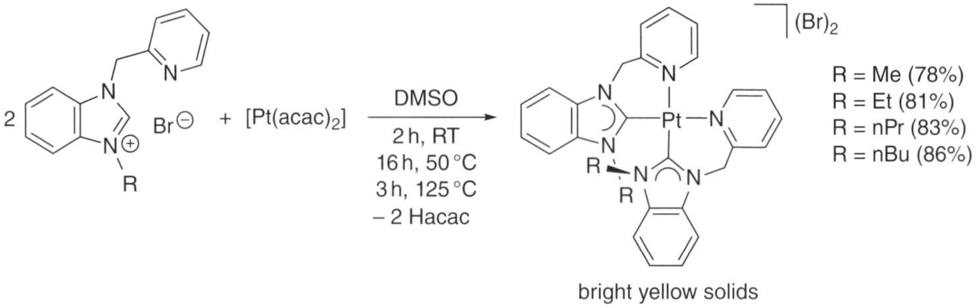

bright yellow solids

Scheme 5.46
Preparation of a bis(chelate) platinum complex *via in situ* deprotonation with [Pt(acac)$_2$].

reaction of the respective benzimidazolium bromides with [Pt(acac)$_2$] in DMSO (Scheme 5.46). Here, a 2:1 ratio of ligand precursor to metal was used, which gave dicationic bis(chelate) complexes that were isolated as bright yellow powders in generally good yields [63].

Solution NMR data recorded in DMSO-d_6 show two doublets for the diasteretopic methylene protons upon coordination, which provides evidence for N-coordination of the picolyl arm. The chemical shifts for the carbene atoms were observed around 149 ppm, which supports their *cis* arrangement in the square planar complexes.

5.3.7 Carbene Transfer Reactions

The first transfer of a NHC to platinum(II) was reported by Liu and coworkers as early as 1998 using pentacarbonyl(imidazolidin-2-ylidene)tungsten(0) complexes as transfer agents [64]. The tungsten complexes themselves were prepared *via* the reaction of an amino-phosphinimine with readily available hexacarbonyltungsten(0) followed by double N-alkylation [65]. Interestingly, heating an equimolar mixture of the tungsten complex with bis(benzonitrile) dichloridoplatinum(II) in chloroform under reflux led to the transfer of both carbene and carbonyl ligand to platinum giving carbonyldichlorido(imidazolidin-2-ylidene)platinum(II) complexes of the general formula *cis*-[PtCl$_2$(CO)(NHC)] in good yields as white to yellow powders (Scheme 5.47). This reactivity is distinct from the carbene transfer to palladium, where carbonyl species have not been observed. In general, palladium(II)-carbonyl complexes exhibit a much lower stability. The fate of the tungsten complex after carbene transfer is unknown and only black decomposition material was noted. The mixed carbonyl/carbene complexes of platinum(II) are air and moisture stable solids exhibiting $^{13}C_{carbene}$ NMR resonances at 166.5 (R = Et; $^1J_{Pt-C}$ = 1125 Hz) and 168.2 ppm (R = Bn; $^1J_{Pt-C}$ = 1134 Hz), respectively. The IR carbonyl stretches observed at 2104 and 2106 cm^{-1} were noted to be essentially identical to that observed for the phosphine analog *cis*-[PtCl$_2$(CO)(PEt$_3$)] (i.e. 2105 cm^{-1}) [66], which highlights some limitations of IR spectroscopy in the comparison of ligand donor strengths.

These complexes are useful complex precursors, and can undergo further carbonyl substitution reactions. Exposure to one equivalent triphenylphosphine leads to the clean formation of mixed carbene/phosphine complexes analogous to Lappert's complexes (see Section 5.3.1) in high yields. The replacement of the carbonyl as a strong π-acceptor ligand with a stronger σ-donor increases the electron density of the complex, and in agreement a

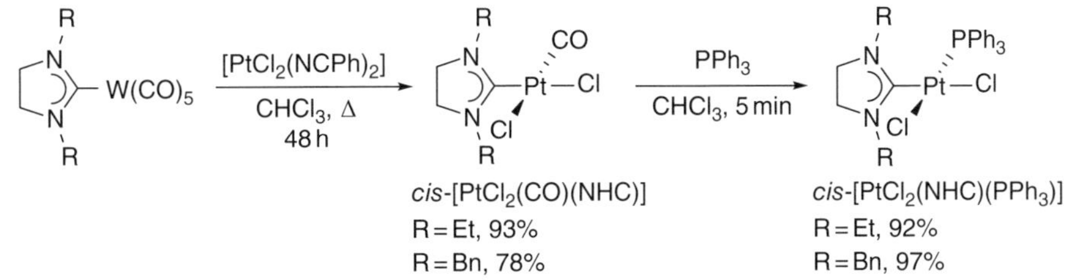

Scheme 5.47
Carbene transfer from tungsten(0) to platinum(II).

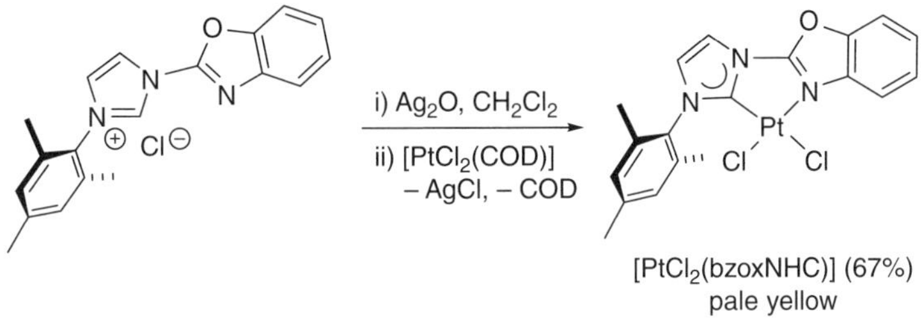

[PtCl$_2$(bzoxNHC)] (67%)
pale yellow

Scheme 5.48
Ag-carbene transfer of an benzoxazole-functionalized NHC to [PtCl$_2$(COD)].

significant downfield chemical shift for the carbene atoms is observed, to 176.7 and 178.1 ppm, respectively, using ^{13}C NMR spectroscopy.

Despite this success, the methodology has received limited application in NHC complex chemistry. Much more attention has been paid to the use of silver-NHC as carbene transfer agents, which was pioneered by Lin and coworkers, due to the ease of their preparation and the milder reaction conditions required [34].

Nevertheless, earlier reports on such an approach in the NHC chemistry of platinum(II) appeared only in 2006 by the groups of Gade and Hahn, who described the finding that silver-carbene transfer reactions of donor-functionalized imidazole- and benzimidazole-derived NHCs could be carried out with different platinum(II) precursors. In the first example, reaction of a benzoxazole-functionalized imidazolium salt with silver(I) oxide and subsequent transfer of the transient Ag-NHC species to [PtCl$_2$(COD)] affords the *cis*-chelate [PtCl$_2$(bzoxNHC)] under replacement of one neutral cyclooctadiene molecule (Scheme 5.48) [67]. The main driving force for this reaction is the very favorable formation and precipitation of silver halide. To avoid halide scrambling, both carbene precursor salt and platinum(II) precursor should ideally contain the same halide.

The second example involves a similar transfer of a phosphine-functionalized benzimidazolin-2-ylidene from silver(I) to *trans*-[PtCl$_2$(NCPh)$_2$] under liberation of two molecules of benzonitrile (Scheme 5.49) [68]. Despite the initial *trans* arrangement of the platinum precursor complex, a *cis* arrangement is enforced in the target complex due to the *cis*-chelating nature of the ditopic NHC ligand.

Notably, these two earlier examples of carbene transfer from silver(I) to platinum(II) employ donor-functionalized NHCs, which can assist in the transfer *via* potential precoordination of the N- or P-donor to the metal, respectively. In addition, the chelate formation of the ditopic ligands provides extra stability to the resulting complex.

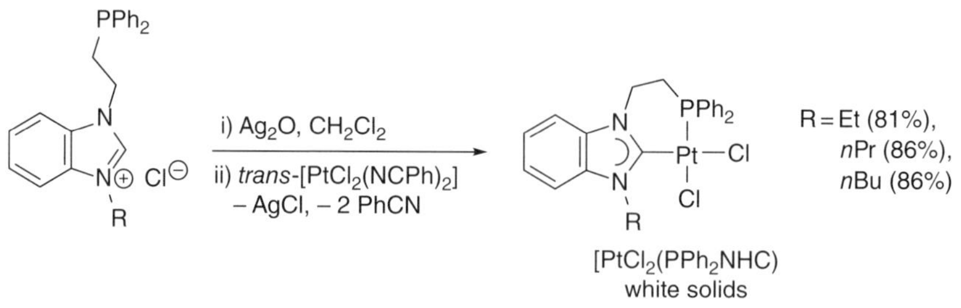

Scheme 5.49
Ag-carbene transfer of an phosphine-functionalized NHC to *trans*-[PtCl$_2$(NCPh)$_2$].

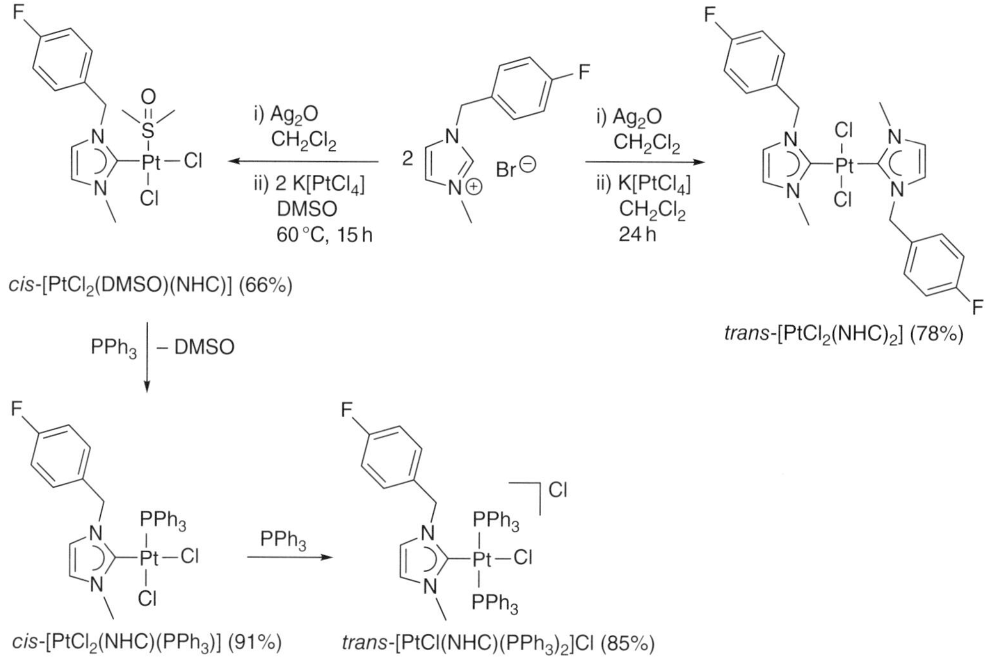

Scheme 5.50
Selectivities in the carbene transfer from silver to platinum and subsequent reactivities.

One of the first studies that demonstrates successful transfer of a simple, monodentate carbene from silver(I) to platinum(II) in the absence of such additional donor interactions appeared in 2007. Rourke and coworkers described that the reaction of a silver(I) complex containing a 1-(*p*-fluorobenzyl)-3-methylimidazolin-2-ylidene ligand with potassium-tetrachloridoplatinate(II) gave either mono- or bis(NHC) complexes of platinum(II) dependent on the solvent used (Scheme 5.50) [69]. In dichloromethane, this reaction leads to the exclusive formation of the *trans*-bis(NHC) complex with two additional chlorido ligands regardless of the initial stoichiometry between transfer agent and platinum source used.

In DMSO a very different outcome was observed, and instead a DMSO solvate of a monocarbene complex was isolated as the major product with a *cis* geometry. In this case, the transfer of two NHCs to the metal center is supposedly hampered by the high affinity

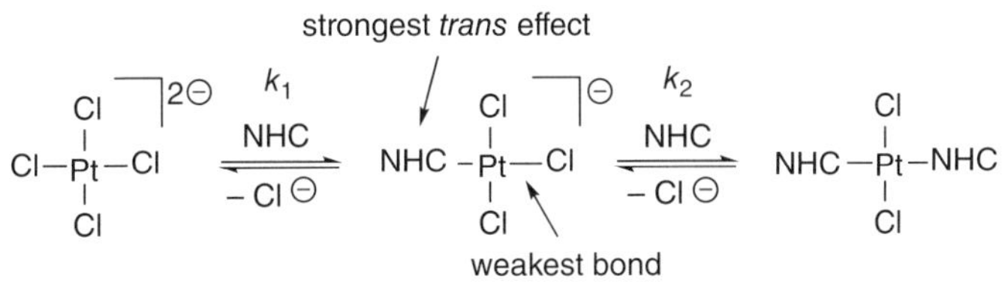

strongest *trans* effect

Scheme 5.51
Stepwise substitution of Pt–Cl bonds with NHC in [PtCl$_4$]$^{2-}$ and exclusive formation of *trans*-[PtCl$_2$(NHC)$_2$].

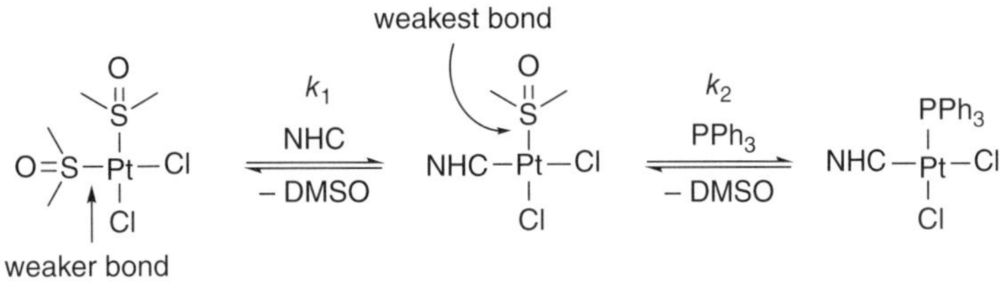

Scheme 5.52
Stepwise substitution of Pt-S bonds with NHC and PPh$_3$ in *cis*-[PtCl$_2$(DMSO)$_2$] and exclusive formation of *cis*-[PtCl$_2$(NHC)(PPh$_3$)].

of platinum to the sulfur donor of the DMSO molecule, which is also the solvent and therefore present in large excess. In general, the transfer is driven by precipitation of silver bromide, which forms preferably over silver chloride due to its greater stability and lower solubility.

The preferred formation of *trans*-[PtCl$_2$(NHC)$_2$] in dichloromethane can be explained by separating the overall reaction into two individual steps with rate constants k_1 and k_2, which are affected by the *trans* effects of all ligands involved (Scheme 5.51). The reaction commences with substitution of one chlorido ligand in tetrachloridoplatinate(II) with one NHC. Since all Pt–Cl bonds are equal in the precursor, this first substitution is random and statistical occurring with a rate of k_1 and giving rise to the formation of the [PtCl$_3$(NHC)]$^-$ complex anion as the intermediate. The introduction of the first NHC ligand leads to a distinction of the three remaining Pt–Cl bonds. The strong *trans* effect of the NHC will substantially weaken the *trans*-disposed Pt–Cl bond, which results in a fast (second) substitution of this chlorido ligand with a rate constant of k_2 and the *trans*-bis(NHC) complex is obtained as the sole product. Due to the strong *trans* effect of NHCs in general, one can expect that the rate constant k_2 is larger than k_1, which explains the exclusive formation of the *trans*-[PtCl$_2$(NHC)$_2$] even when a 1:1 stoichiometry of NHC to platinum metal is applied.

Dissolving the tetrachloroplatinate(II) in DMSO, on the other hand, gives *cis*-[PtCl$_2$(DMSO)$_2$] as the direct precursor containing a very different electronic situation, which is subjected to the incoming NHC ligand. Due to the stronger *trans* effect of the chlorido ligands, the two Pt–DMSO bonds are the weaker ones and much more prone to substitution (Scheme 5.52). Reaction with a single NHC therefore solely affords *cis*-[PtCl$_2$(DMSO)(NHC)]. The remaining DMSO ligand in this complex still has the weakest interaction with the metal center and is thus easily displaced upon addition of a stronger donor, such as triphenylphosphine giving *cis*-[PtCl$_2$(NHC)(PPh$_3$)] as the only product.

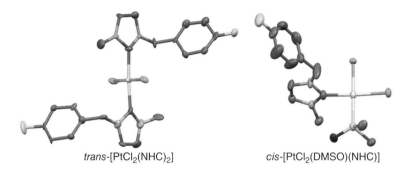

trans-[PtCl$_2$(NHC)$_2$]   *cis*-[PtCl$_2$(DMSO)(NHC)]

Figure 5.23
Molecular structures for *trans*-[PtCl$_2$(NHC)$_2$] and *cis*-[PtCl$_2$(DMSO)(NHC)] prepared by Ag-NHC transfer reactions.

Chlorido substitution is only affected with the introduction of a third strong donor (Scheme 5.50) and gives rise to the *trans*-[PtCl(NHC)(PPh$_3$)$_2$]$^+$ complex cation.

The solid state molecular structures obtained by X-ray diffraction studies on single crystals of the bis(NHC) complex and the DMSO adduct are in agreement with their proposed structures found in solution. The bis(NHC) complex was found to crystallize in a *trans* arrangement, whereas the DMSO adduct adopts the *cis* geometry (Figure 5.23).

References

[1] (a) S. Salto, M. Sakai, N. Miyaura, *Tetrahedron Lett.* **1996**, *37*, 2993. (b) S. Salto, S. Oh-tani, N. Miyaura, *J. Org. Chem.* **1997**, *62*, 8024.

[2] (a) A. F. Indolese, *Tetrahedron Lett.* **1997**, *38*, 3513. (b) V. Percec, G. M. Golding, J. Smidrkal, O. Weichold, *J. Org. Chem.* **2004**, *69*, 3447.

[3] B. Çetinkaya, P. Dixneuf, M. F. Lappert, *J. Chem. Soc., Chem. Commun.* **1973**, 206.

[4] M. F. Lappert, P. L. Pye, *J. Chem. Soc. Dalton Trans.* **1977**, 2172.

[5] W. A. Herrmann, G. Gerstberger, M. Spiegler, *Organometallics* **1997**, *16*, 2209.

[6] H.-W. Wanzlick, H. Schönherr, *Angew. Chem. Int. Ed. Engl.* **1968**, *7*, 141.

[7] K. Öfele, *J. Organomet. Chem.* **1968**, *12*, P42.

[8] D. S. McGuinness, M. Mueller, P. Wasserscheid, K. J. Cavell, B. W. Skelton, A. H. White, U. Englert, *Organometallics* **2002**, *21*, 175.

[9] (a) H. V. Huynh, C. Holtgrewe, T. Pape, L. L. Koh, E. Hahn, *Organometallics* **2006**, *25*, 245. (b) H. V. Huynh, L. R. Wong, P. S. Ng, *Organometallics* **2008**, *27*, 2231.

[10] C. D. Abernethy, A. H. Cowley, R. A. Jones, *J. Organomet. Chem.* **2000**, *596*, 3.

[11] (a) V. Ritleng, E. Brenner, M. J. Chetcuti, *J. Chem. Educ.* **2008**, *85*, 1646. (b) J. Cooke, O. C. Lightbody, *J. Chem. Educ.* **2011**, *88*, 88.

[12] H. M. Sun, Q. Shao, D. M. Hu, W. F. Li, Q. Shen, Y. Zhang, *Organometallics* **2005**, *24*, 331.

[13] Z.-h. Liu, Y.-C. Xu, L.-Z. Xie, H.-M. Sun, Q. Shen, Y. Zhang, *Dalton Trans.* **2011**, *40*, 4697.

[14] (a) X. Wang, S. Liu, G.-X. Jin, *Organometallics*, **2004**, *23*, 6002. (b) G. Huang, H. Sun, X. Qiu, Y. Shen, J. Jiang, L. Wang, *J. Organomet. Chem.* **2011**, *696*, 2949.

[15] T. Shibata, S. Ito, M. Doe, R. Tanaka, H. Hashimoto, I. Kinoshita, S. Yano, T. Nishioka, *Dalton Trans.* **2011**, *40*, 6778.

[16] D. Zhang, S. Zhou, Z. Li, Q. Wang, L. Weng, *Dalton Trans.* **2013**, *42*, 12020.

[17] L. M. Engelhardt, P. C. Healy, V. A. Patrick, A. H. White, *Aust. J. Chem.* **1987**, *40*, 1873.

[18] R. Jothibasu, K.-W. Huang, H. V. Huynh, *Organometallics* **2010**, *29*, 3746.

[19] D. J. Cardin, B. Çetinkaya, E. Çetinkaya, M. F. Lappert, L.J. Manojlović-Muir; K. W. Muir, *J. Organomet. Chem.* **1972**, *44*, C59–C62.

[20] M. S. Viciu, R. M. Kissling, E. D. Stevens, S. P. Nolan, *Org. Lett.* **2002**, *4*, 2229.

[21] U. Christmann, R. Vilar, *Angew. Chem. Int. Ed.* **2005**, *44*, 366.

[22] S. Caddick, F. G. N. Cloke, G. K. B. Clentsmith, P. B. Hitchcock, D. McKerrecher, L. R. Titcomb, M. R. V. Williams, *J. Organomet. Chem*, **2001**, *617–618*, 635.

[23] (a) M. S. Viciu, R. F. Germaneau, O. Navarro-Fernandez, E. D. Stevens, S. P. Nolan, *Organometallics* **2002**, *21*, 5470. (b) N. Marion, O. Navarro, J. Mei, E. D. Stevens, N. M. Scott, S. P. Nolan, *J. Am. Chem. Soc.* **2006**, *128*, 4101.

[24] D. Enders, H. Gielen, G. Raabe, J. Runsink, J. H. Teles, *Chem. Ber.* **1996**, *129*, 1483.

[25] C. Dash, M. M. Shaikh, R. J. Butcher, P. Ghosh, *Dalton Trans.* **2010**, *39*, 2515.

[26] W. A. Herrmann, M. Elison, J. Fischer, C. Köcher, G. R. J. Artus, *Angew. Chem. Int. Ed. Engl.* **1995**, *34*, 2371.

[27] W. A. Herrmann, J. Fischer, K. Öfele, G. R. J. Artus, *J. Organomet. Chem.* **1997**, *530*, 259.

[28] L. Xu, W. Chen, J. Xiao, *Organometallics* **2000**, *19*, 1123.

[29] H. V. Huynh, Y. Han, J. H. H. Ho, G. K. Tan, *Organometallics* **2006**, *25*, 3267.

[30] W. A. Herrmann, J. Schwarz, M. G. Gardiner, *Organometallics* **1999**, *18*, 4082.

[31] Y. Ma, C. Song, W. Jiang, G. Xue, J. F. Cannon, X. Wang, M. B. Andrus, *Org. Lett.* **2003**, *5*, 4635.

[32] G. Altenhoff, R. Goddard, C. W. Lehmann, F. Glorius, *J. Am. Chem. Soc.* **2004**, *126*, 15195.

[33] H. V. Huynh, Y. Han, R. Jothibasu, J. A. Yang, *Organometallics* **2009**, *28*, 5395.

[34] H. M. J. Wang, I. J. B. Lin, *Organometallics* **1998**, *17*, 972.

[35] L. Ademi, E. C. Constable, C. E. Housecroft, M. Neuburger, S. Schaffner, *Acta Cryst.* **2006**, *E62*, m1059.

[36] Y. Han, Y.-T. Hong, H. V. Huynh, *J. Organomet. Chem.* **2008**, *693*, 3159.

[37] H. V. Huynh, J. H. H. Ho, T. C. Neo, L. L. Koh, *J. Organomet. Chem.* **2005**, *690*, 3854.

[38] F. E. Hahn, M. Foth, *J. Organomet. Chem.* **1999**, *585*, 142.

[39] C.-F. Fu, C.-C. Lee, Y.-H. Liu, S.-M. Peng, S. Warsink, C. J. Elsevier, J.-T. Chen, S.-T. Liu, *Inorg. Chem.* **2010**, *49*, 3011.

[40] H. V. Huynh, R. Jothibasu, L. L. Koh, *Organometallics* **2007**, *26*, 6852.

[41] H. V. Huynh, Y. X. Chew, *Inorg. Chim. Acta* **2010**, *363*, 1979.

[42] H. V. Huynh, J. Wu, *J. Organomet. Chem.* **2009**, *694*, 323.

[43] Q. Teng, D. Upmann, S. A. Z. N. Wijaya, H. V. Huynh, *Organometallics* **2014**, *33*, 3373.

[44] (a) W. A. Herrmann, L. J. Gooßen, M. Spiegler, *J. Organomet. Chem.* **1997**, *547*, 357. (b) T. Weskamp, V. P. W. Böhm, W. A. Herrmann, *J. Organomet. Chem.* **1999**, *585*, 348. (c) W. A. Herrmann, V. P. W. Böhm, C. W. K. Gstöttmayr, M. Grosche, C.-P. Reisinger, T. Weskamp, *J. Organomet. Chem.* **2001**, *617–618*, 616.

[45] (a) J. Vicente, J. A. Abad, A. D. Frankland, M. C. Ramírez de Arellano, *Chem. Eur. J.* **1999**, *5*, 3066. (b) J. Vicente, A. Arcas, D. Bautista, P. G. Jones, *Organometallics* **1997**, *16*, 2127. (c) J. Vicente, A. Arcas, M. A. Blasco, J. Lozano, M. C. Ramírez de Arellano, *Organometallics* **1998**, *17*, 5374. (d) J. Vicente, M. T. Chicote, C. Rubio, M. C. Ramírez de Arellano, P. G. Jones, *Organometallics* **1999**, *18*, 2750. (e) J. Vicente, A. Arcas, J. Fernández-Hernández, D. Bautista, *Organometallics* **2001**, *20*, 2767.

[46] (a) H. V. Huynh, C. H. Yeo, G. K. Tan, *Chem. Commun.* **2006**, 3833. (b) H. V. Huynh, C. H. Yeo, Y. X. Chew, *Organometallics* **2010**, *29*, 1479.

[47] R. Sevinçek, H. Türkmen, B. Çetinkaya, S. García-Granda, *Acta Crystallogr.* **2007**, *C63*, m277.

[48] D. J. Cardin, B. Çetinkaya, M. F. Lappert, Lj. Manojlovic-Muir, K. W. Muir, *Chem. Commun.*, **1971**, 400.

[49] D. J. Cardin, B. Çetinkaya, E. Çetinkaya, M. F. Lappert, Lj. Manojlovic-Muir, K. W. Muir, *J. Organomet. Chem.* **1972**, *44*, C59.

[50] B. Çetinkaya, E. Çetinkaya, M. F. Lappert, *J. Chem. Soc., Dalton Trans.* **1973**, 906.

[51] R. A. Michelin, L. Zanotto, D. Braga, P. Sabatino, R. J. Angelici, *Inorg. Chem.* **1988**, *27*, 93.

[52] R. Bertani, M. Mozzon, R. A. Michelin, *Inorg. Chem.* **1988**, *27*, 2809.

[53] D. S. McGuinness, K. J. Cavell, B. F. Yates, B. W. Skelton, A. H. White, *J. Am. Chem. Soc.* **2001**, *123*, 8317.

[54] P. J. Fraser, W. R. Roper, F. G. A. Stone, *J. Chem. Soc., Chem. Commun.* **1974**, 102.

[55] M. A. Duin, N. D. Clement, K. J. Cavell, C. J. Elsevier, *Chem. Commun.* **2003**, 400.

[56] (a) D. Brissy, M. Skander, P. Retailleau, A. Marinetti, *Organometallics* **2007**, *26*, 5782. (b) D. Brissy, M. Skander, P. Retailleau, G. Frison, A. Marinetti, *Organometallics* **2009**, *28*, 140.

[57] S. Fantasia, J. L. Petersen, H. Jacobsen, L. Cavallo, S. P. Nolan, *Organometallics* **2007**, *26*, 5880.

[58] M. Muehlhofer, T. Strassner, E. Herdtweck, W. A. Herrmann, *J. Organomet. Chem.* **2002**, 660, 121.

[59] Y. Han, H. V. Huynh, G. K. Tan, *Organometallics* **2007**, *26*, 4612.

[60] Intramolecular C–H···M *anagostic* interactions are primarily electrostatic in nature and resemble hydrogen bonds, which are 3c-4e$^-$ interactions. Both lead to a significant downfield shift of the respective C–H proton. They should not be confused with *agostic* interactions, which are electron deficient 3c-2e$^-$ interactions characterized by a significant upfield shift (negative ppm) of C–H proton involved (refer to Ref 9b).

[61] J. C. Bernhammer, H. V. Huynh, *Organometallics* **2014**, *33*, 172.

[62] S. Ahrens, E. Herdtweck, S. Goutal, T. Strassner, *Eur. J. Inorg. Chem.* **2006**, 1268.

[63] M. C. Jahnke, T. Pape, F. E. Hahn, *Z. Naturforsch.* **2010**, *65b*, 341.

[64] (a) S.-T. Liu, T.-Y. Hsieh, G.-H. Lee, S.-M. Peng, *Organometallics* **1998**, *17*, 993. (b) R.-Z. Ku, J.-C. Huang, J.-Y. Cho, F.-M. Kiang, K. R. Reddy, Y.-C. Chen, K.-J. Lee, J.-H. Lee, G.-H. Lee, S.-M. Peng, S.-T. Liu, *Organometallics* **1999**, *18*, 2145.

[65] C.-Y. Liu, D.-Y. Chen, G.-H. Lee, S.-M. Peng, S.-T. Liu, *Organometallics* **1996**, *15*, 1055.

[66] C. Y. Mok, S. G. Tan, G. C. Chan, *Inorg. Chim. Acta* **1990**, *176*, 43.

[67] M. Poyatos, A. Maisse-François, S. Bellemin-Laponnaz, L. H. Gade, *Organometallics* **2006**, *25*, 2634.

[68] F. E. Hahn, M. C. Jahnke, T. Pape, *Organometallics* **2006**, *25*, 5927.

[69] C. P. Newman, R. J. Deeth, G. J. Clarkson, J. P. Rourke, *Organometallics* **2007**, *26*, 6225.

6 Group 11 Metal-NHC Complexes

The group 11 elements copper, silver, and gold are commonly known as the coinage metals. N-heterocyclic carbene (NHC) complexes of these metals, in particular those of silver and gold, have received great attention in recent years. Nowadays, they belong to the most widely studied in the organometallic chemistry of NHCs [1]. Historically, those of copper have received relatively less attention, but as more and more research has set focus on them in recent years, the gap between copper and its heavier homologues is getting narrower. A contributing factor is the increasing interest in low cost catalysts. Some coinage metal-NHC complexes are nowadays even commercially available. Among the coinage metal complexes, silver(I) NHC complexes [2] are of prime importance due to their ease of synthesis and their widespread applications as very efficient carbene transfer agents. Using silver(I) NHC complexes a wide range of other transition metal-NHC complexes can be accessed with ease. They have also been studied as antimicrobial agents [3], but the scale of these applications pale in comparison to their use as NHC transfer agents. In contrast, the interest in gold NHC complexes is primarily fueled by their use as efficient catalysts in the activation of unsaturated bonds, for example alkynes, and this development goes hand in hand with the general resurgence of gold catalysis. Many gold NHC compounds have also been studied in bio-related areas for their potential as new metal-based drugs.

Using DFT calculations, Boehme and Frenking compared the bond dissociation energies of metal-NHC bonds in coinage metals [4]. Their results indicated decreasing bond energies in the order $Au^I > Cu^I > Ag^I$. Experimental observations that gold and copper-NHC complexes can be formed via transfer of the carbene from silver(I) to either copper(I) or gold(I) are in line with this prediction. Furthermore, gold NHC complexes are generally very stable, while decomposition has more often been described for their copper analogs.

6.1 Copper(I)-NHC Complexes

6.1.1 The Free Carbene Route

The first copper(I) complex of an N-heterocyclic carbene was reported by Arduengo and coworkers in 1993 shortly after their seminal report on the isolation of the first free NHC [5]. The preparation involves mixing of two equivalents of the free IMes proligand with one equivalent of copper(I) triflate/benzene solvate $[CuO_3SCF_3 \cdot 0.5C_6H_6]$ in tetrahydrofurane at room temperature. Notably, the addition of IMes occurred in two steps with the addition of the second equal portion occurring after five minutes of stirring. The reaction of the copper salt with one molecule of IMes probably leads to the formation of the monocarbene/tetrahydrofurane solvate [Cu(IMes)(THF)]OTf, which gives rise to a

The Organometallic Chemistry of N-heterocyclic Carbenes, First Edition. Han Vinh Huynh.
© 2017 John Wiley & Sons Ltd. Published 2017 by John Wiley & Sons Ltd.

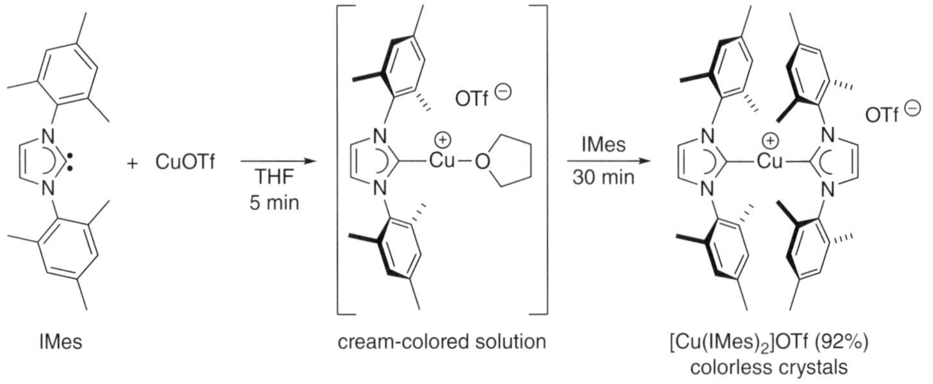

Scheme 6.1
Preparation of the first copper(I)-NHC complex.

cream-colored solution. Exposure of this intermediate to additional IMes rapidly affords the more stable homoleptic bis(NHC)complex $[Cu(IMes)_2]OTf$ that was isolated as colorless crystals after workup (Scheme 6.1), melting in the range of 312–314 °C.

The $^{13}C_{carbene}$ resonance for the IMes ligands in the cationic bis(NHC) complex recorded in THF-d_6 was found at 178.2 ppm, which is significantly upfield shifted from the value for the free IMes molecule (i.e. 219.7 ppm). In addition, all signals for the carbene ring shift downfield, which has been taken as an indication for the increased delocalization and aromaticity compared to the free carbene. Analysis on single crystals, although not of sufficient quality, pointed to the expected linear coordination geometry.

6.1.2 Alkylation of Copper-Azolate Complexes

The first heteroleptic copper(I) NHC complex was reported in 1994 by Raubenheimer and coworkers [6]. Their approach makes use of lithiated azolates as key precursors to carbenes. Treatment of a copper(I) chloride suspension in tetrahydrofurane with lithium-imidazolate or –benzimidazolate at low temperature afforded the respective lithium-cuprates, which were then *N*-methylated with methyltriflate ($CF_3SO_3CH_3$) as a strong electrophile (Scheme 6.2). Upon methylation, the initially anionic imidazolato ligand transforms into a neutral NHC still bound to the copper(I) center. The authors reported an elaborate temperature regime starting at –80 °C slowly warming up to –40 °C, –20 °C and eventually to room temperature. Using this procedure, the first neutral chlorido-monocarbene copper(I) complexes were obtained in moderate yields of 38% and 48%, respectively.

Compared to the cationic and homoleptic bis(NHC) complex reported by Arduengo and coworkers, these two examples decompose slowly in air and have to be stored under inert conditions. The imidazolin-2-ylidene complex [CuCl(IMe)] was reported to undergo color change from light brown to green or blue on exposure to the atmosphere. The lack of stability in these complexes is probably due to the limited steric bulk of the carbene ligands employed. The ^{13}C NMR signal for the carbene donor in [CuCl(IMe)] was found at 177.0 ppm in CD_2Cl_2, while the benzimidazolin-2-ylidene analog turned out to be insufficiently soluble for such analysis.

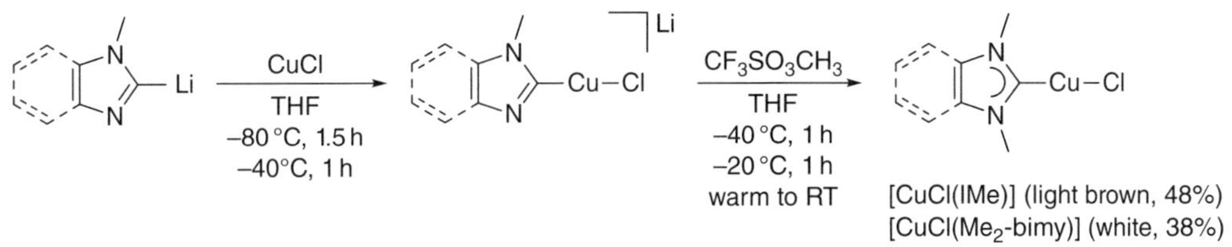

Scheme 6.2
Preparation of the first heteroleptic copper(I)-NHC complexes.

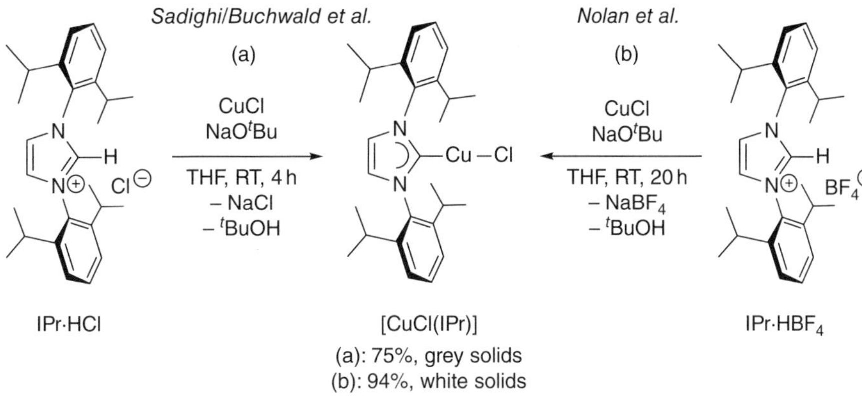

Scheme 6.3
Preparation of the neutral heteroleptic complex [CuCl(IPr)] via *in situ* deprotonation of an imidazolium salt.

While this metal-template-assisted approach was also successfully applied to the preparation of analogous copper(I) complexes of less common NSHCs derived from thiazole and benzothiazole, it has generally received little attention in metal-NHC chemistry. Nevertheless, it is certainly of historical interest. The requirement for the preparation of lithiated azolates using highly reactive alkyl lithium reagents, the subsequent use of strong electrophiles for *N*-alkylation and the generally low yields were possibly contributing factors that made this route less popular.

6.1.3 *In Situ* Deprotonation with External Base

A more straightforward access to this type of heteroleptic complexes was almost concurrently developed a decade later by Sadighi/Buchwald [7] and Nolan and coworkers [8]. The former were in search of a copper catalyst for the conjugate reduction of α,ß-unsaturated carbonyl compounds. They opted for a neutral heteroleptic complex with the IPr ligand and prepared the complex by mixing approximately equimolar amounts of copper(I) chloride, the imidazolium salt IPr·HCl, and sodium *tert*-butoxide in dry tetrahydrofurane [7]. Stirring the mixture at room temperature for four hours followed by filtration to remove inorganic salts and evaporation of the solvent, afforded the desired complex [CuCl(IPr)] as a gray solid in a decent yield of 75% (Scheme 6.3, a). The strong sodium *tert*-butoxide base deprotonates the imidazolium salt to produce the carbene *in situ*, which immediately binds to the copper(I) center giving the target product. In addition, this reaction also produces sodium chloride and *tert*-butanol as stable byproducts, which thermodynamically drive the reaction.

In contrast to Raubenheimer's complexes, [CuCl(IPr)] shows a greater stability, which can be attributed to the better steric protection by the substantially more bulky IPr ligand. Compared to the methyl groups in [CuCl(IMe)], the two aromatic wing tip groups in this complex not only increase the steric demand, but also exhibit a deshielding (inductive) effect, which leads to a downfield shift of the $^{13}C_{carbene}$ signal to 180.2 ppm.

At the time of its first report, the [CuCl(IPr)] catalyst precursor was found to provide the highest activity in the conjugate reduction of α,ß-unsaturated carbonyl compounds. The actual catalytically active species was proposed to be the hydrido complex [CuH(IPr)], which forms on exposure to sodium *tert*-butoxide and a silane *via* ligand substitution and subsequent σ-bond metathesis.

The group of Nolan used IPr·BF$_4$ instead and prolonged the reaction time to 20 hours, which gave the product [CuCl(IPr)] in a better yield of 94% with elimination of NaBF$_4$ (Scheme 6.3, b) [8]. The authors later also extended their methodology to other carbenes and prepared the complexes [CuCl(ICy)] (76%) with aliphatic N-substituents and [CuCl(SIMes)] (70%) containing a saturated backbone from the respective azolium tetra-fluoroborates in good yields [9]. Single crystals for X-ray diffraction could be obtained from the three complexes, and their molecular structures depicted in Figure 6.1 revealed the strongest copper-carbene bond for the SIMes complex, whereas the weakest bond was observed for the aliphatically-substituted ICy complex.

The well-defined copper-NHC complexes were found to be better catalyst precursors for the hydrosilylation of various ketones including hindered or functionalized alkyl, cyclic, bicyclic, aromatic, and heteroaromatic ketones, compared to those generated *in situ* from the respective salts, base, and copper source.

Another example for a homoleptic bis(NHC) complex was surprisingly reported only 13 years after Arduengo's first example by Nolan and coworkers [10]. They used

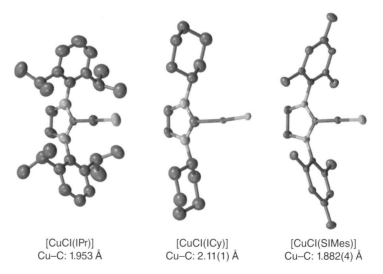

[CuCl(IPr)]
Cu–C: 1.953 Å

[CuCl(ICy)]
Cu–C: 2.11(1) Å

[CuCl(SIMes)]
Cu–C: 1.882(4) Å

Figure 6.1
Molecular structures of chlorido-NHC copper(I) complexes. Hydrogen atoms are omitted for clarity.

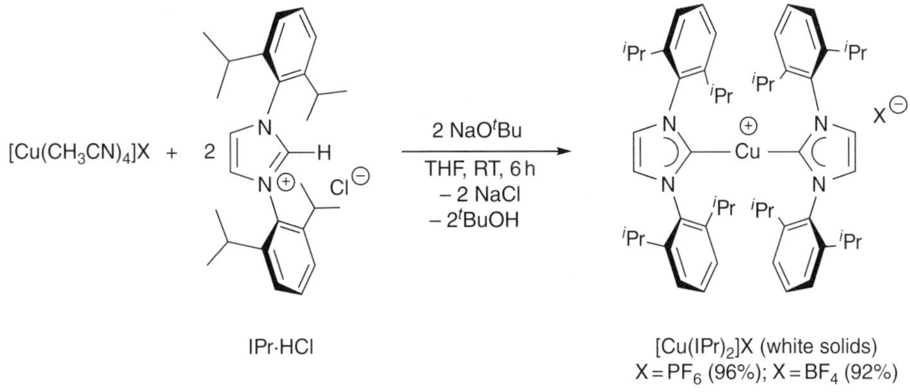

Scheme 6.4
Homoleptic copper(I) bis(IPr) complexes via *in situ* deprotonation.

tetrakis(acetonitrile)copper(I) pentafluorophosphate or tetrafluoroborate salts as precursor instead. The reaction of these salts with free IPr in a 1:2 ratio in tetrahydrofurane at room temperature gave high yields of the desired complexes of the formula $[Cu(IPr)_2]X$ ($X = PF_6$, BF_4). The authors then reported that the use of the imidazolium salt in conjunction with equimolar sodium *tert*-butoxide as a strong base in analogy to their previous approach to neutral heteroleptic complexes of the type [CuCl(NHC)] could be applied for the preparation of cationic bis(NHC) complexes as well in place of the free carbenes (Scheme 6.4).

This *in situ* deprotonation protocol does not require pre-isolation of the free carbene and has simplified the syntheses. At the same time, it also gave rise to very good yields of the target compounds, which were isolated as white air and moisture stable solids.

As expected, the $^{13}C_{carbene}$ NMR signals of the complexes were found in the same range (i.e. 178.4 ppm for PF_6 and 177.4 ppm for BF_4) as that reported for Arduengo's complex, therefore pointing to a very similar structure. Evidence for the identity of the complexes as linear bis(NHC) complexes were finally obtained from single crystal X-ray diffraction. Since both complexes are as expected isostructural and differ solely in their counter-anion, only the molecular structure of $[Cu(IPr)_2]PF_6$ is shown in Figure 6.2 (left). The Cu–C bond distances of 1.938(5) and 1.939(18) Å found for the two complexes are identical within standard deviation, but the angles of intersection between the two carbene planes (torsion angles) as a result of steric repulsion are rather different, with values of 41° (PF_6) and 50° (BF_4).

The authors extended the methodology and prepared a range of cationic bis(NHC) copper(I) complexes bearing other classical NHCs (Table 6.1) [11]. Most complexes were found to be stable to air and moisture with the exception of the I′Bu complexes, which slowly decomposed over a period of several weeks if not kept under argon.

For most cases, the use of a slight excess of the base was beneficial in the synthesis to ensure complete conversion of the azolium salt to the carbene. However, partial decomposition was observed for ICy and I′Bu complexes, and in these cases a near stoichiometric ratios of the base with respect to the imidazolium salt turned out to be more practical.

Comparison of the ^{13}C NMR data for the carbene donors did not revealed any particular trend that can be traced back to the counter-anion of the complexes. The change from PF_6^- to BF_4^- anion either did not affect the chemical shift or only led to an unsystematic change. However, the chemical shifts for the saturated carbenes generally appear at a much more

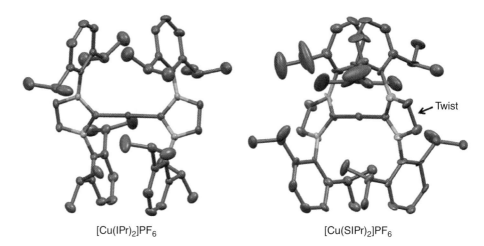

[Cu(IPr)₂]PF₆   [Cu(SIPr)₂]PF₆

Figure 6.2
Molecular structures of the cations [Cu(IPr)₂]⁺ (left) and [Cu(SIPr)₂]⁺ (right). The twist in the saturated backbone of SIPr is highlighted. The PF_6^- anions and hydrogen atoms are omitted for clarity.

Table 6.1 Preparation of bis(NHC) complexes of copper(I) via *in situ* deprotonation and selected spectroscopic and structural parameters.

$$[Cu(CH_3CN)_4]X + 2\ NHC \cdot HBF_4 \xrightarrow[\text{THF, RT, } t]{\text{n NaO}^t\text{Bu}} [Cu(NHC)_2]X + 2\ NaBF_4 + 2\ ^t\text{BuOH}$$

Complex	n(Base) [equiv]	t [h]	Yield [%]	$^{13}C_{NCN}$ [ppm]	Torsion angles[c] [°]	Cu–C [Å]
[Cu(SIPr)₂]PF₆	2.6	15	68	199.8[a]	28	2.000(15), 1.910(15)
[Cu(SIPr)₂]BF₄	2.6	6	63	201.4[a]	-	-
[Cu(IMes)₂]PF₆	2.0	15	91	178.8[b]	67	1.871(3)
[Cu(IMes)₂]BF₄	2.6	15	74	178.8[b]	54	1.884(2)
[Cu(SIMes)₂]PF₆	2.6	15	86	201.4[b]	28	1.899(3)
[Cu(SIMes)₂]BF₄	2.6	15	86	201.3[a]	30	1.896(4)
[Cu(ICy)₂]PF₆	2.0	15	80	174.2[b]	-	-
[Cu(ICy)₂]BF₄	2.2	15	93	173.8[b]	-	-
[Cu(IAd)₂]PF₆	2.6	15	84	169.8[a]	81	1.938(5)
[Cu(IAd)₂]BF₄	2.6	15	95	169.9[a]	86	1.933(5)
[Cu(I^{t}Bu)₂]PF₆	2.2	6	99	171.7[b]	-	-
[Cu(I^{t}Bu)₂]BF₄	2.2	6	100	171.2[b]	82	1.909(2)

[a] measured in CDCl₃.
[b] measured in acetone-d_6.
[c] angle between the two carbene (NCN) planes.

downfield region in agreement with observations made for the respective free carbenes (Table 6.1).

The solid state molecular structures could be obtained for most homoleptic complexes, and the copper–carbon bond lengths range from 1.87–2.00Å (Table 6.1). Generally, an increase of these bonds was observed with increasing steric bulk of the carbenes, which is intuitive. In addition, the torsion angles between the two carbene planes also increase with greater steric demand to reduce steric repulsion of the wing tip groups. At such the IAd and I*t*Bu complexes exhibit the longer metal–carbon distances as well as the largest torsion angles.

The complex [Cu(SIPr)$_2$]PF$_6$ appears to be an exception, due to its longest Cu–C bond and yet rather small torsion angle. This complex is also the only one that shows a twist of the saturated *sp*^{3}-hybridized C4/5 carbene backbones of 16° and 9° (Figure 6.2, right). This backbone twist supposedly can also reduce the steric repulsion of the wing tip groups resulting in a rather small torsion angle between the carbenes.

The cationic and homoleptic bis(NHC) copper(I) complexes discussed above have been used as catalyst precursors for the hydrosilylation of ketones with triethylsilane. Similar to Sadighi and Buchwald's mechanistic suggestion, it was proposed that action of sodium *tert*-butoxide on these complexes eventually led to monocarbene/hydrido species as actual catalysts. Comparative studies revealed a superior performance of these cationic species compared to neutral and heteroleptic copper(I)-NHC complexes.

6.1.4 *In Situ* Deprotonation with Basic Copper Precursors

Danopoulos and coworkers were the first to explore the preparation of copper-NHC complexes from imidazolium salts without addition of any external base, in 2001 [12]. They studied the reaction of N-pyridyl- and picolyl-functionalized imidazolium bromides, which also bear aromatic (Mes or Dipp) wing tip groups, with copper(I) oxide (Cu$_2$O) in dichloromethane (Scheme 6.5).

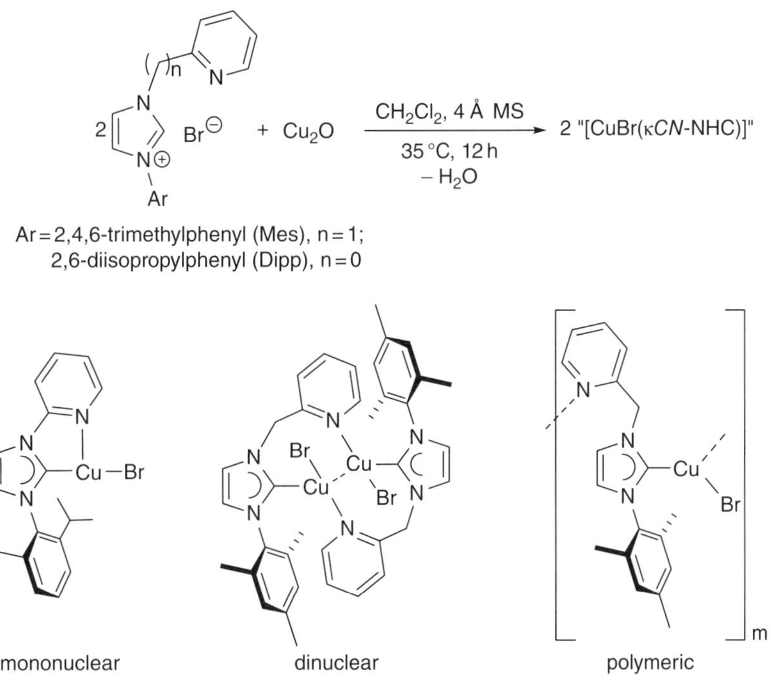

Scheme 6.5
In situ deprotonation of donor-functionalized imidazolium salts with Cu$_2$O.

The approach is analogous to the well-known preparation of silver-NHCs via silver(I) oxide. Generally, each equivalent of the basic Cu_2O is sufficient to deprotonate two equivalents of imidazolium salt under liberation of one molecule of water. However, a slight excess was added to ensure complete deprotonation. In addition, it was found that addition of molecular sieves (4Å) has a beneficial effect on the reaction rate, since the water is readily absorbed, and through this removal the equilibrium is driven to the product side.

The ^1H NMR spectra of the resulting products of the general formula [CuBr(κ^2-C,N-NHC)] show only broad signals even at lower temperatures, which hampered a detailed assignment and thus a proper characterization of the exact solution species. This observation was attributed to non-rigidity in solution and equilibria involving monomer, dimer, and possibly oligomeric species. ^{13}C NMR data were not reported in this study.

Nevertheless, single crystals could be obtained, which X-ray diffraction indeed revealed to be different species. For the pyridyl-functionalized NHC ligand, a tricoordinate, mononuclear copper(I) complex was identified from crystals obtained from a saturated solution in mixed CDCl$_3$/diethyl ether (**I** in Figure 6.3). The organometallic ligand coordinates to the metal center in a κ^2-C,N chelating fashion with a short copper–carbon distance of 1.880(6)Å and a bite angle of 76.5°. One terminal bromido ligand completes the T-shape coordination geometry at copper(I).

The picolyl-functionalized ligand, on the other hand, gave rise to a complex that crystallized in different forms dependent on the solvent mixture used. From a saturated solution in mixed dichloromethane/diethylether, crystals of a centrosymmetric dinuclear complex were obtained upon cooling (**II** in Figure 6.3). The functionalized NHC acts as a bridging ligand instead of chelating a single metal center. The geometry around each copper center is most appropriately described as a distorted trigonal pyramid where the bromido ligand occupies the apical position. A weak metal-metal interaction with a Cu···Cu distance of 2.655Å was observed. In addition, π–π stacking interactions are observed between the mesityl wing tip and the pyridine moiety of the picolyl arm. The copper–carbon bond length amounting to 1.931Å is in the range characteristic for Cu–C single bonds, but markedly longer than that for the mononuclear complex described above.

Crystallization of the same compound from a concentrated solution in CDCl$_3$ afforded a polymeric form instead (**III** in Figure 6.3). The picolyl-functionalized carbene ligand is

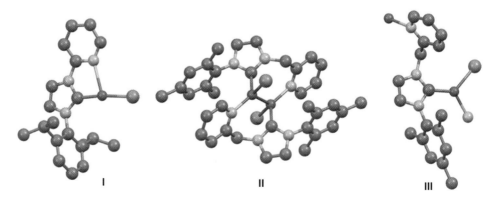

Figure 6.3
Molecular structures for pyridyl- and picolyl-functionalized NHC copper(I) complexes. Only the monomeric building block for the polymeric chain III is depicted. Solvent molecules and hydrogen atoms are omitted for clarity.

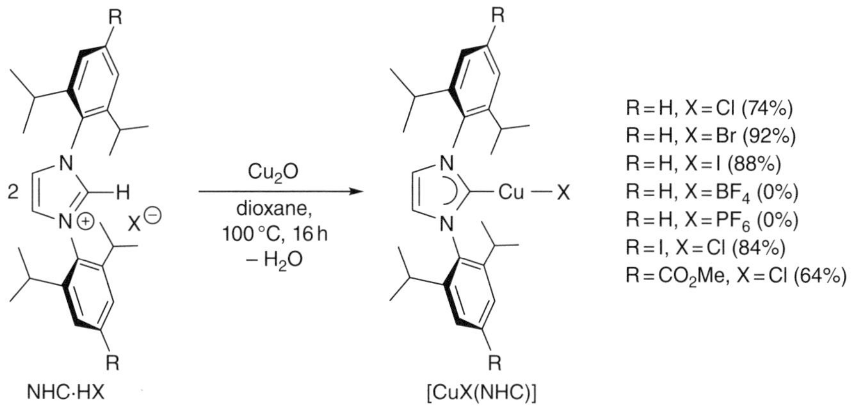

R = H, X = Cl (74%)
R = H, X = Br (92%)
R = H, X = I (88%)
R = H, X = BF$_4$ (0%)
R = H, X = PF$_6$ (0%)
R = I, X = Cl (84%)
R = CO$_2$Me, X = Cl (64%)

Scheme 6.6
Preparation of halido-monocarbene copper(I) complexes using Cu$_2$O.

also bridging, but copper-copper interactions are absent. The copper center is found in trigonal planar environment with a copper–carbon distance of 1.914(4) Å, which similar to the dinuclear species.

Surprisingly, the adaptation of this methodology to heteroleptic copper(I) complexes with simple monodentate NHCs took almost another decade and was only realized in 2010. Son and coworkers reported that heating a mixture copper(I) oxide with various 1,3-diarylimidazolium halides to 100 °C in dioxane leads to the formation of carbene complexes of the general formula [CuX(NHC)] in good yields (Scheme 6.6) [13]. Although 0.5 equiv of Cu$_2$O with respect to the imidazolium salt is sufficient, the amount of the oxide was increased to 0.75 equiv for a better conversion. The desired complexes can be separated from the virtually insoluble metal oxide precursor by a simple filtration step. Removal of all volatiles from the filtrate afforded a residue, which could be washed with water and dried under vacuum to give the copper-NHC complexes as almost colorless solids.

Notably, when imidazolium salts with weakly coordinating anions such as BF$_4^-$ and PF$_6^-$ were used, no heteroleptic complexes were isolated. Therefore, the formation of copper–halido bonds was identified as an important driving force for this reaction.

In the same year, Cazin and coworkers reported the application of essentially the same methodology to the preparation of [CuCl(NHC)] complexes. In addition to 1,3-diarylimidazolium salts, this work also included 1,3-dialkyl derivatives and saturated imidazolinium salts (Scheme 6.7). Here, toluene was chosen as the solvent with a higher boiling point, and moderate to excellent isolated yields were obtained for most complexes after heating the reactants under reflux for 24 hours. Using this approach, the preparation of complex [CuCl(IPr)] could also be efficiently scaled up to 25 g, which demonstrates its viability. However, a complex containing the saturated 1,3-dicyclohexylimidazolidin-2-ylidene could not be obtained. Instead, oxidation to the respective 1,3-dicyclohexylimidazolidin-2-one was observed in this particular case.

Furthermore, it was found that the reactions involving NHC with aromatic wing tips could also be conducted in water as a low cost and non-hazardous alternative solvent, which helps in the dissolution of the azolium ligand precursors. On the other hand, NHC complexes with aliphatic N-substituents could not be obtained in aqueous media.

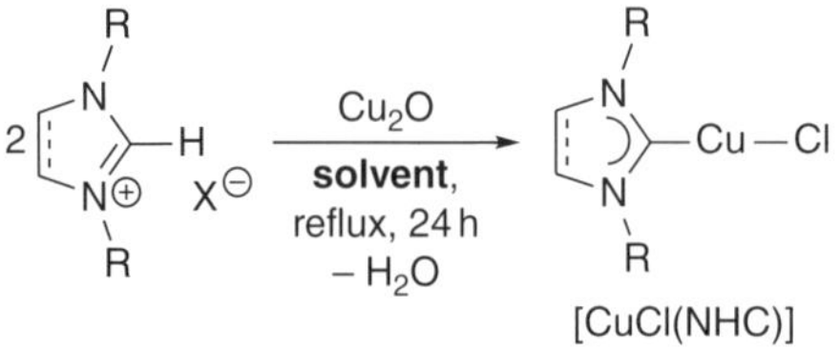

	toluene		water	
NHC	**%Conv.***	**%Yield**	**%Conv.***	**%Yield**
IMes	100	86	100	98
SIMes	100	71	100	99
IPr	88	78	96	94
SIPr	99	88	78	72
ICy	88	70	0	0
SICy	55	0	7	0

*determined by ^{1}H NMR spectroscopy

Scheme 6.7
Preparation of chlorido-monocarbene copper(I) complexes using Cu$_2$O.

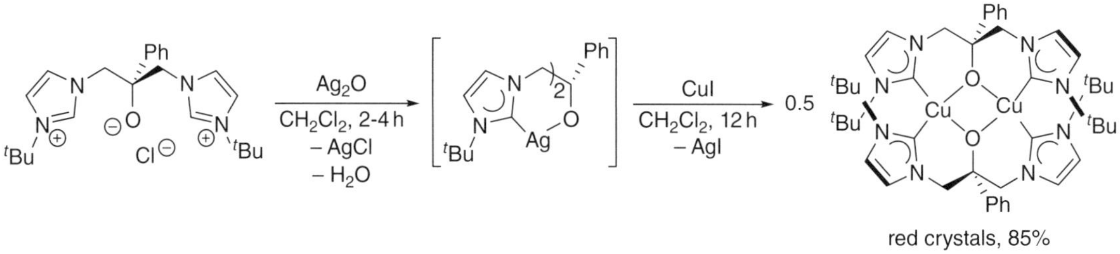

red crystals, 85%

Scheme 6.8
Synthesis of the dinuclear, alkoxido-bridged copper(I)-NHC complex [Cu$_2$(μ,κ^3C,O,C-diNHC)$_2$] via silver-NHC transfer.

6.1.5   The Silver Carbene Transfer Route

Although rarely exercised, copper(I) NHC complexes have also been prepared by transfer of the carbene from silver(I). Arnold and coworkers reported the first example, which involved a more elaborated dinucleating dicarbene ligand [14]. For this, they prepared an alkoxide-bridged diimidazolium monochloride salt, which upon reaction with silver(I) oxide undergoes double deprotonation and formation of an unusual, neutral, halido-free silver(I) diNHC complex, in which alkoxido coordination was proposed (Scheme 6.8). However, the exact structure is difficult to be elucidated in detail due to the fluxionality of silver(I) NHC complexes in general. Precipitation of silver chloride drives this initial reaction.

Subsequent addition of copper(I) iodide leads to the transfer of the tridentate diNHC ligand from silver(I) to copper(I). Coordination of the hard alkoxido donor to the copper as opposed to the softer silver center possibly assists in this process. However, the main driving force is the instantaneous formation of very stable silver(I) iodide. The transmetallation eventually affords a dinuclear copper(I) complex [Cu$_2$(μ,κ^3-C,O,C-diNHC)$_2$], which contains two dicarbene-alkoxido ligand strands. The compound was obtained as red crystals, and their analysis by X-ray diffraction reveals almost square planar copper centers coordinated by two carbene and two bridging alkoxide donors from two different diNHC ligands (Figure 6.4, left). The notable square planar coordination geometry is rarely observed for copper(I). The average Cu–C$_{carbene}$ distance amounts to 1.963(4) Å, and the two metal centers are separated by a distance of 3.1122(9) Å.

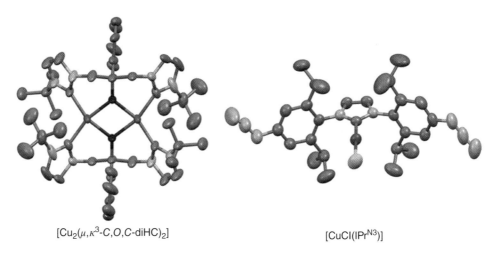

[Cu₂(μ,κ³-*C,O,C*-diHC)₂]   [CuCl(IPr^N3)]

Figure 6.4
Dinuclear and mononuclear copper(I) NHC complexes obtained *via* silver-NHC transfer reactions. Hydrogen atoms are omitted for clarity.

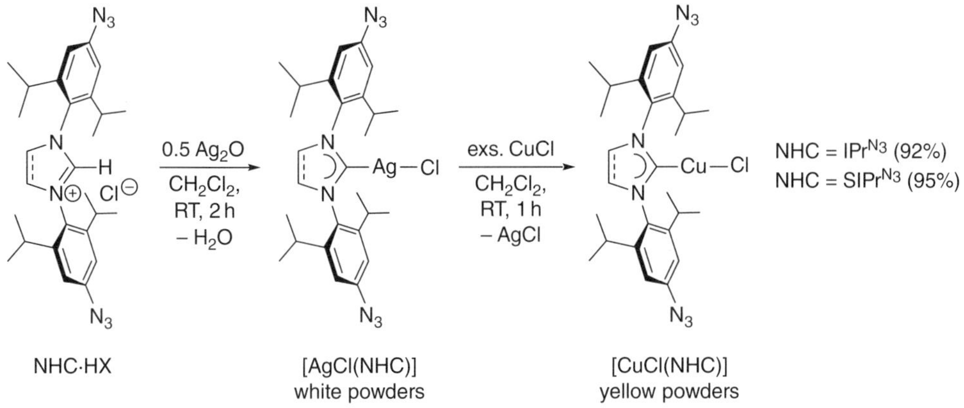

Scheme 6.9
Preparation of [CuCl(NHC)] complexes via silver-NHC transfer.

The transmetallation of simpler monodentate NHCs from silver that gave rise to well-defined and well-characterized copper(I) NHC complexes was surprisingly only realized more than a decade later by Cisnetti and Gaultier [15]. In this work, bulky, azide-functionalized IPr and SIPr carbene precursors were routinely subjected to an half-equivalent of silver(I) oxide to give the silver(I) carbene species, which upon addition of an excess of copper(I) chloride smoothly transferred the NHC to the latter (Scheme 6.9). Expectedly, silver(I) chloride precipitates as the favorable byproduct, and the copper complexes of the formula [CuCl(NHC)] were obtained in high yields of >90% as yellow powders, which were noted to be bench-stable for weeks.

The $^{13}C_{carbene}$ signals recorded in DMSO-d_6 for the unsaturated imidazolin-2-ylidene ligand (IPr^{N3}) comes to resonance at 178.7 ppm, while that for its saturated imidazolidin-2-ylidene (SIPr^{N3}) counterpart was found more downfield at 201.3 ppm, as expected.

Crystals of [CuCl(IPr^{N3})] were obtained by layering a solution in dichloromethane with *n*-pentane and cooling the enclosed mixture to 4 °C for three days. The analysis of these by X-ray diffraction confirmed the proposed composition of a neutral, linear chloride-mono-carbene complex (Figure 6.4, right). The azide-functions in these NHCs were introduced for post-functionalizations to afford 1,2,3-triazole groups via Copper-Catalyzed Azide-Alkyne Cycloaddition (CuAAC).

6.1.6 The Copper Powder Route

The latest method to access copper(I) NHC complexes is also the simplest, and was initially developed by Chen and coworkers for complexes of donor-functionalized NHC ligands [16]. The extension to simple monodentate dentate NHCs was only reported by the same group in 2015 [17]. In this approach, the azolium salt is typically mixed with five equivalents of copper powder in acetonitrile and heated under air to 55 °C for 24 h (Scheme 6.10).

With bulky imidazolium or imidazolinium halides, this reaction affords neutral, hetero-leptic halido-monocarbene complexes of the type [CuX(NHC)] in very good yields. The use of bulky imidazolium or imidazolinium hexafluorophosphate salts, on the other hand, gives cationic, homoleptic bis(NHC) complexes in much lower yields. Yet much higher temperatures and longer reaction times are required to affect their formation.

It is crucial to conduct the reaction under air, since oxygen is required for the oxidation of elemental copper to copper(I). Supposedly, oxygen is reduced to oxide in this process, which can then act as a base for the generation of the carbene ligands with the elimination of water. Azolium halides or donor-functionalized azolium salts were observed to be much more reactive than salts lacking in either halides or additional donors. This positive effect could be explained on two grounds. First, the pre-coordination of halido ligands or other donors to copper is expected to increase its electron density, which in turn should enhance the rate of oxidation (Scheme 6.11, i). Second, the halide or additional donors can also form hydrogen bonds to the C2-H proton of the azolium moiety, and through this polarization the acidity of this group is enhanced facilitating carbene formation (Scheme 6.11, ii).

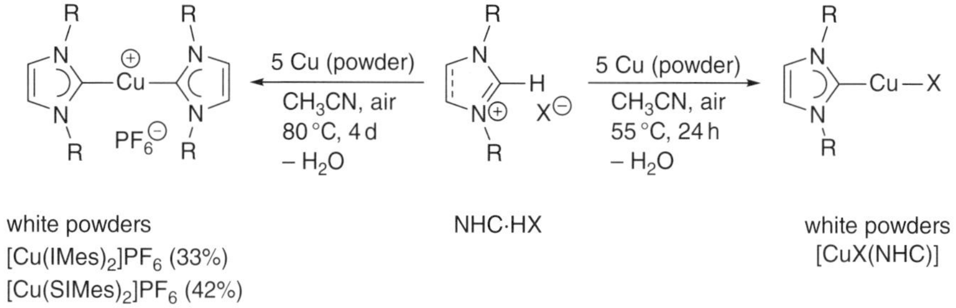

NHC	X	Yield [%]
IPr	Cl	90
IPr	Br	99
IPr	I	79
SIPr	Cl	88
SIPr	I	83
IMes	Cl	49

Scheme 6.10
Preparation of copper(I) NHC complexes using copper powder under oxidative conditions.

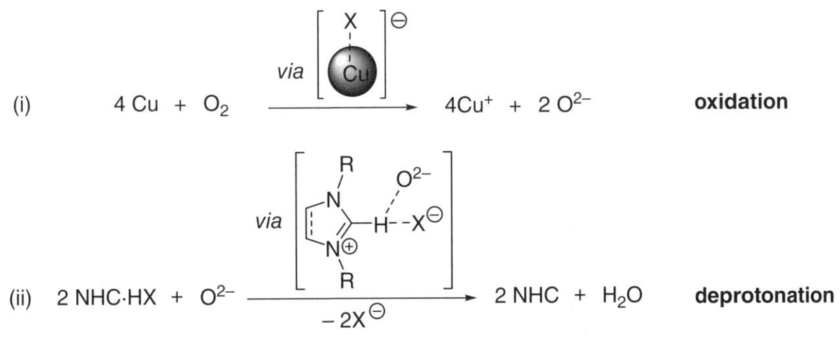

Scheme 6.11
Possible activation processes for the oxidation and deprotonation steps by halides or additional donors.

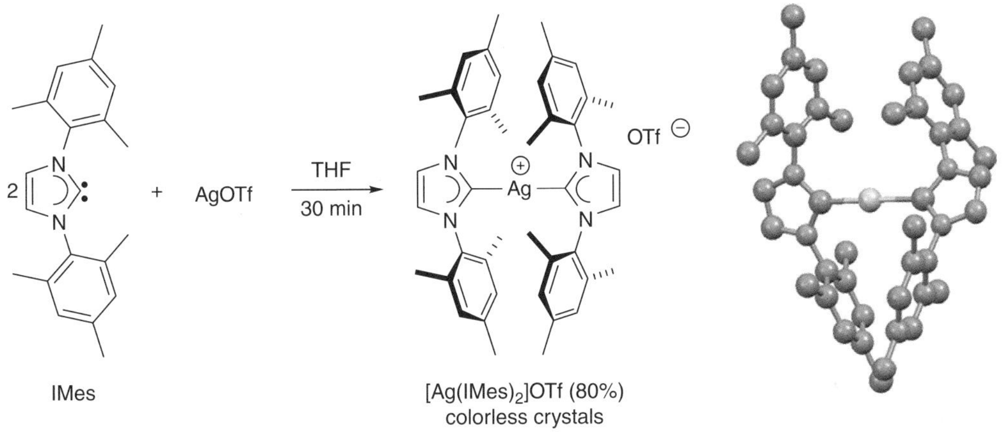

IMes

[Ag(IMes)$_2$]OTf (80%)
colorless crystals

Scheme 6.12
Preparation of the first silver(I)-NHC complex and the solid state molecular structure of its cation. Hydrogen atoms are omitted for clarity.

6.2 Silver(I)-NHC Complexes

6.2.1 The Free Carbene Route

Simultaneous with the first copper(I) NHC complex, Arduengo and coworkers also reported the first example of a silver(I) NHC complex in 1993 [5]. Mixing two equivalents of IMes with silver(I) triflate in tetrahydrofurane and stirring for 30 minutes at ambient temperature straightforwardly led to the formation of the homoleptic bis(NHC) complex [Ag(IMes)$_2$] OTf (Scheme 6.12). The initial cloudy reaction mixture containing the product was first warmed and then filtered. Cooling of the filtrate to –25 °C afforded the complex as colorless crystals in 80% yield.

The complex was characterized by multinuclear NMR spectroscopies even including [109]Ag, which proton coupled NMR spectrum shows a quintet at 642.4 ppm downfield of 5 M aqueous silver(I) nitrate solution due to coupling with the four equivalent C4/5 protons in the imidazolinylidene backbone. Since silver has with [109]Ag and [107]Ag two NMR receptive nuclei with nuclear spins of ½ in a roughly 48:52 ratio, diagnostic coupling pattern with these can also be anticipated in the ^{13}C NMR spectrum. Indeed, the ^{13}C$_{carbene}$ signal is detected as two distinct doublets centered at 183.6 ppm with coupling constants of $^1J(^{13}$C-^{107}Ag$) = 188.0$ Hz and $^1J(^{13}$C-^{109}Ag$) = 208.6$ Hz. Evidence for such couplings is found in the ratio of these constants, which approximately reflects the ratio of gyromagnetic ratios γ for the two silver nuclei involved, that is, $\gamma(^{109}$Ag$)/\gamma(^{107}$Ag$) = 1.15$. The observation of such ^{13}C$_{carbene}$ to silver coupling is quite rare in silver-NHC chemistry, which we nowadays know is characterized by structural fluxionality. Therefore, this first example of a silver-NHC complex does not show any sign of ligand exchange behavior, which can possibly be attributed to the steric protection of the metal center by two bulky IMes ligands.

Elucidation of the solid state structure by single crystal X-ray diffraction reveals the linear two-coordinate molecular structure of the complex also found in solution (Scheme 6.12). The silver-carbene bond distances are 2.067 and 2.078 Å. The two carbene planes are twisted by an angle of 39.7° from coplanarity to relieve steric congestion imposed by the two bulky mesityl N-substituents.

Heteroleptic silver(I) NHC complexes have also been prepared by the "free carbene" route. For example, Chung treated 1,3,4,5-tetramethylimidazolin-2-ylidene (IMe$_{Me2}$) with one equivalent of silver(1,3-diphenyl-1,3-propanedionate) in tetrahydrofuran at room temperature [18]. This reaction is complete after one hour, affording a neutral, three-coordinated silver(I) complex [Ag(dppd)(IMe$_{Me2}$)] (dppd = 1,3-diphenyl-1,3-propanedionato) as an air-stable, pale yellow solid in 80% yield (Scheme 6.13). In contrast to the complex [Ag(IMes)$_2$]OTf, only a singlet was detected at 179.4 ppm in CD$_2$Cl$_2$ that was attributed to the carbene resonance. Here, coupling to the two NMR active silver isotopes could not be observed. The trigonal planar silver(I) center is bound by one carbene and a chelating diketonato ligand.

Linear heteroleptic silver(I) complexes containing one carbene and a second monodentate, anionic ligand are also accessible via this route. Nolan and coworkers reported the successful preparation of [AgCl(NHC)] complexes by mixing two simple imidazolin-2-ylidenes with isopropyl and methyl wing tip groups with excess silver(I) chloride in tetrahydrofurane (Scheme 6.14) [19]. This reaction is much slower than that with the silver(I) diketonate and requires longer reaction times, which is most likely due to the very poor solubility of the silver halide in general. Nevertheless, the neutral products are well soluble in organic solvents and can be separated from the heterogeneous mixture by a simple filtration step. Concentration of the filtrate and precipitation by addition of non-polar hydrocarbon solvents afforded the target complexes as white and yellow solids, respectively. It is notable that the [AgCl(IMe$_{Me2}$)] complex that bears a tetramethyl-substituted NHC and obtained in much lower yield, shows a greater stability to air, compared to its congener [AgCl(I^{i}Pr$_{Me2}$)] with two isopropyl wing tip groups that was obtained in a greater yield.

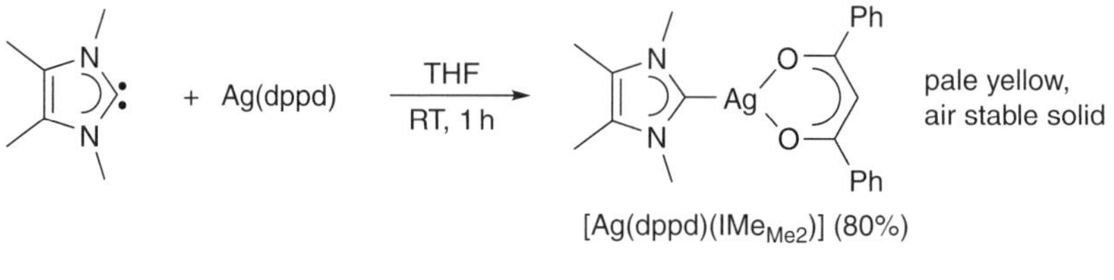

[Ag(dppd)(IMe$_{Me2}$)] (80%)

Scheme 6.13
Preparation of a mixed diketonato/carbene silver(I) complex.

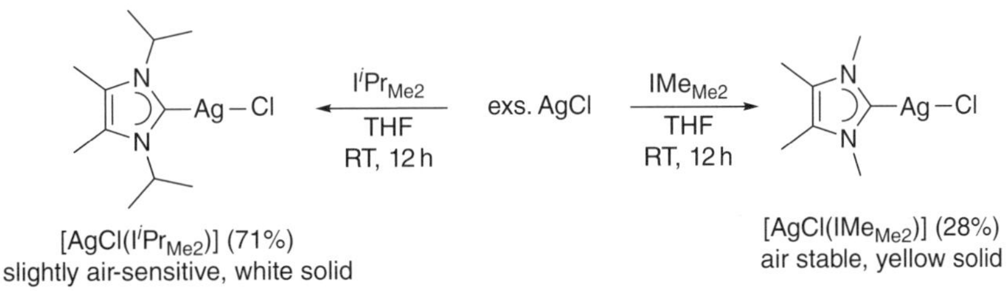

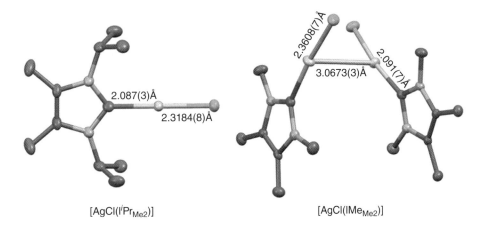

Scheme 6.14
Preparation of [AgCl(NHC)] complexes using free carbenes.

Figure 6.5
Solid state molecular structures of [AgCl(I^{i}Pr$_{Me2}$)] and [AgCl(IMe$_{Me2}$)]. Hydrogen atoms are omitted for clarity.

Attribution of the increased sensitivity to either steric or electronic considerations is non-trivial, since I^{i}Pr$_{Me2}$ is more bulky, but also more electron donating than IMe$_{Me2}$.

The ^{13}C NMR resonance for the carbene carbon in the more bulky I^{i}Pr$_{Me2}$ complex was detected as broad singlet at 172.3 ppm. The fact that heteronuclear couplings to silver could not be resolved is attributed to a dynamic behavior of the complex. In contrast, the ^{13}C$_{carbene}$ signal for the IMe$_{Me2}$ complex was found as a sharp singlet, slightly more downfield at 177.6 ppm. We can attribute the downfield shift to a decrease in the shielding effect of N-substituent, which is in agreement with a reduction of the positive inductive effect going from N-isopropyl to N-methyl. The sharpness of the signal indicates that fluxionality via ligand redistribution to give [Ag(NHC)$_2$]$^+$ cations is even faster due to a reduced steric bulk of the IMe$_{Me2}$.

The differences in the solid state molecular structures between these two complexes are also striking. As depicted in Figure 6.5, the [AgCl(I^{i}Pr$_{Me2}$)] complex crystallizes as a well-defined mononuclear species, while [AgCl(IMe$_{Me2}$)] dimerizes in the solid state. The two complex units are held together by argentophilic interactions with an intermetal separation of 3.0673(3) Å. Notably, the dimerization also leads to a lengthening of the silver–chlorido bonds, while the silver–carbene distances are essentially the same in both complexes. It is likely that the formation of dimers via metallophilic interactions could ease ligand exchange, therefore leading to the greater fluxionality observed in solution by NMR spectroscopic means.

6.2.2   *In Situ* Deprotonation with Basic Silver Precursors

The requirement for air sensitive free carbenes is an apparent drawback in the "free carbene route," which is nowadays rarely being used for the preparation of silver(I) NHC complexes. The vast majority of these compounds are prepared by the more convenient *in situ* deprotonation by direct exposure of azolium salts to basic silver precursors, which does not require stringent inert conditions and additional bases. In general, there are a couple of basic silver salts available for this purpose. These include silver(I) acetate (AgOAc), oxide (Ag_2O) and carbonate (Ag_2CO_3).

One of first examples of this approach was documented by Bertrand and coworkers in 1997. They heated a dicationic 1,2,4-trialkyltriazolium ditriflate salt with two equivalents of silver(I) acetate in tetrahydrofurane under reflux for two hours and obtained a unusual one-dimensional silver-dicarbene organometallic polymer (Scheme 6.15) [20]. The silver acetate provides the base for the deprotonation of the triazolium dication, which furnishes an interesting Janus-type diylidene. Through *in situ* coordination of this unusual ligand to silver ions, a light-sensitive organometallic polymer is formed. The $^{13}C_{carbene}$ signal of the polymer appears at 189.2 ppm without coupling to silver, which indicates rapid metal exchange even in the polymeric structure.

One year later, Lin and Wang reported their serendipitously discovered way of making silver(I) NHC complexes and the use of these as carbene transfer agents. In an attempt to prepare a gold(I) NHC complex, the graduate student mistakenly used silver(I) oxide as a base for the deprotonation of the azolium salt, which intuitively may lead to a mixture of different metal complexes. On repeating the reaction in the absence of gold, the authors found that silver(I) NHC complexes can be formed in excellent yields by exposure of azolium salts to silver(I) oxide. This facile preparation of silver(I) NHC complexes has undoubtedly become the most popular and widely used methodology nowadays. In the original procedure, 1,3-diethylbenzimidazolium salts were used to generate the carbene complexes in dichloromethane at ambient temperature (Scheme 6.16). Silver(I) oxide is sufficiently basic to deprotonate two equivalents of most classical azolium salts under elimination of one molecule of water to give the respective carbenes, that can immediately coordinate to the silver ions present in the reaction mixture [21]. Addition of molecular sieves to the reaction mixture to absorb the produced water was reported to have some positive effects on yield and purity of the resulting complexes [22]. The insolubility of the black silver(I) oxide in virtually all organic solvents provides a means for monitoring

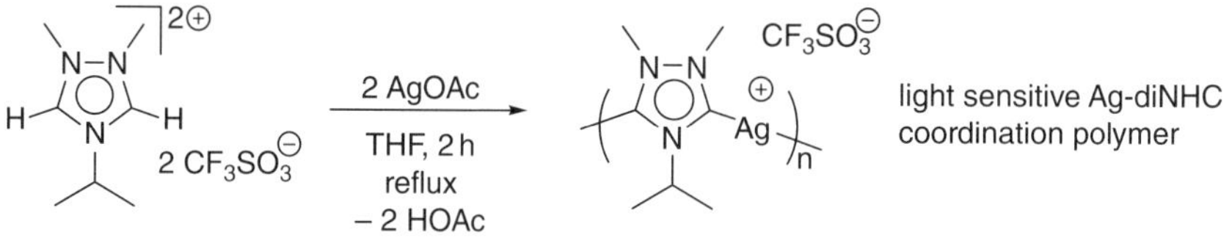

Scheme 6.15
Preparation of an organometallic silver-carbene polymer with bridging 1,2,4-triazolidin-3,5-diylidene ligands.

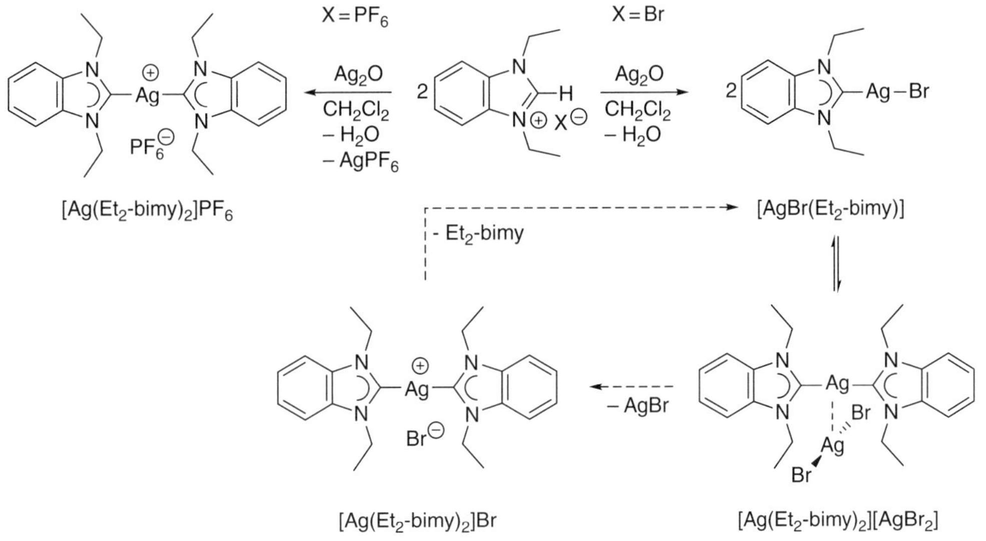

Scheme 6.16
Formation of silver-NHC complexes via silver(I) oxide route and their dynamic behavior.

the reaction progress by observation of its gradual disappearance. As mentioned above, the silver–carbene bond is the weakest among the coinage metals and also subject to dynamic exchange, especially when the steric bulk is less severe. The nature of the anion of the azolium salt also plays an important role. Coordinating anions, such as halides (e.g. Cl^-, Br^-, I^-), form stronger hydrogen bonds and increase the acidity of the C2–H azolium proton as opposed to weakly or non-coordinating anions (e.g. BF_4^-, PF_6^-). In addition, halides usually facilitate the dynamic exchange in the resulting complexes via their reversible binding to the silver center. Weakly or non-coordinating anions, on the other hand, can be used to prevent such dynamic behavior. However, only cationic bis(NHC) complexes are accessible with the latter.

This is demonstrated in Scheme 6.16, where the reaction of two equivalents of benzimidazolium hexafluorophosphate with silver(I) oxide cleanly affords the cationic complex [Ag(Et$_2$-bimy)]PF$_6$ [23]. The absence of any coordinating anions in this system prevents any dynamic ligand exchange behavior. Evidence for this was obtained by ^{13}C NMR spectroscopy, where heteronuclear couplings of the carbene donor centered at ~188 ppm to the two silver isotopes were observed ($^1J(^{13}C\text{-}^{107}Ag) = 180\,Hz$; $^1J(^{13}C\text{-}^{109}Ag) = 204\,Hz$). Furthermore, molar conductivity measurements of 10^{-3} to 10^{-4} molar complex solutions in acetonitrile gave a value of $146\,\Omega^{-1}mol^{-1}cm^2$, which is in the range typically observed for 1:1 electrolytes.

A similar reaction involving the benzimidazolium bromide supposedly leads to the neutral bromido complex [AgBr(Et$_2$-bimy)], which, however, is in equilibrium with the complex ion pair [Ag(Et$_2$-bimy)$_2$][AgBr$_2$] as a consequence of polymerization isomerization. In agreement, the ^{13}C$_{carbene}$ signal was detected as a sharp singlet at 188.9 ppm without any coupling to silver, indicating a fast exchange on the NMR timescale. The much lower molar conductivity of $74\,\Omega^{-1}mol^{-1}cm^2$ measured in equimolar solutions in acetonitrile indicates a large contribution of the neutral [AgBr(Et$_2$-bimy)] complex corroborating the aforementioned dynamic solution behavior.

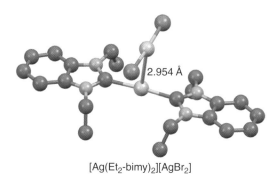

[Ag(Et$_2$-bimy)$_2$][AgBr$_2$]

Figure 6.6
Molecular structure of [Ag(Et$_2$-bimy)$_2$][AgBr$_2$]. Hydrogen atoms are omitted for clarity.

Colorless crystals of [Ag(Et$_2$-bimy)$_2$][AgBr$_2$] were obtained by recrystallization of the product from dichloromethane/hexane. Its molecular structure confirms the presence of an ion pair consisting of a linear [Ag(Et$_2$-bimy)$_2$]$^+$ cation and a linear [AgBr$_2$]$^-$ anion that are held together by argentophilic interactions at an inter-silver distance of 2.954 Å (Figure 6.6). The two complex ions are staggered by an angle of ~85° with averaged Ag–C and Ag–Br bonds of 2.063 Å and 2.488 Å, respectively. The former are in the same range as those observed for Arduengo's [Ag(IMes)]OTf complex [5].

Notably, these are only two major structural forms of many possible isostoichiometric complexes. It is apparent that these forms are dominantly influenced by the concentration and polarity of the solvents used.

To complicate the solution dynamics further, one could also anticipate irreversible precipitation of silver halide from the complex ion pair, which eventually leads to the bis(NHC) complex [Ag(Et$_2$-bimy)$_2$]Br (Scheme 6.16, dotted arrows). The latter complex could in turn undergo ionization isomerization *via* inner and outer sphere exchange, which liberates a free NHC and furnishes back the initial neutral [AgBr(Et$_2$-bimy)] complex closing the "decomposition cycle". In the presence of moisture the liberated carbene is expected to undergo hydrolysis under ring-opening or protonation. Prolonged reaction or storage of halido-monocarbene silver(I) complexes in solution will therefore likely lead to decomposition under formation of silver halide, substituted formamides, or reformation of azolium salts.

Another example that highlights such counter-anion dependent behavior is depicted in Scheme 6.17. The reaction 1,3-di(2-pyridyl)imidazolium bromide with silver(I) oxide under conventional conditions only led to the isolation of the N-(pyridin-2-yl)-N'-(2-(pyridin-2-ylamino)vinyl)formamide [24], possibly a secondary ring-opened product from the hydrolysis of the intermediate free NHC formed by displacement involving a bromido ligand (route 1).

However, isolation of the desired bis(NHC) complex [Ag(IPy)$_2$]BF$_4$ as an air stable white powder is successful when the non-coordinating tetrafluoroborate anion is employed instead (route 2). Here, hydrolysis is inhibited despite the requirement for additional sodium hydroxide to accelerate the deprotonation of the less acidic imidazolium tetrafluoroborate cation [25]. The carbene signal was detected at 182.7 ppm in CD$_3$CN, but coupling to silver was not reported. The insignificant changes of pyridine protons upon complex formation indicate that coordination of the pyridine-units is unlikely.

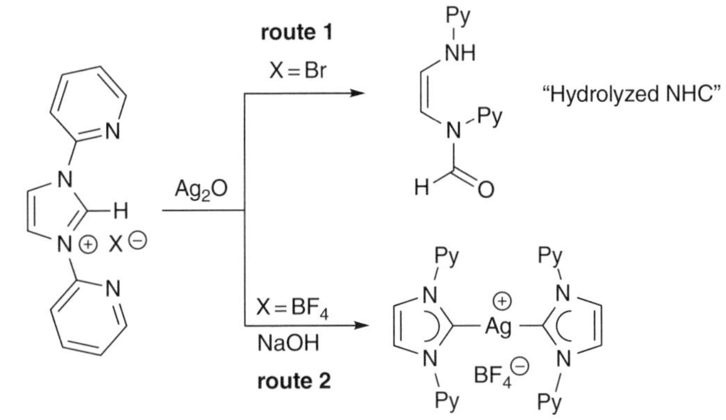

Scheme 6.17
Anion dependence in the preparation of silver-NHC complexes.

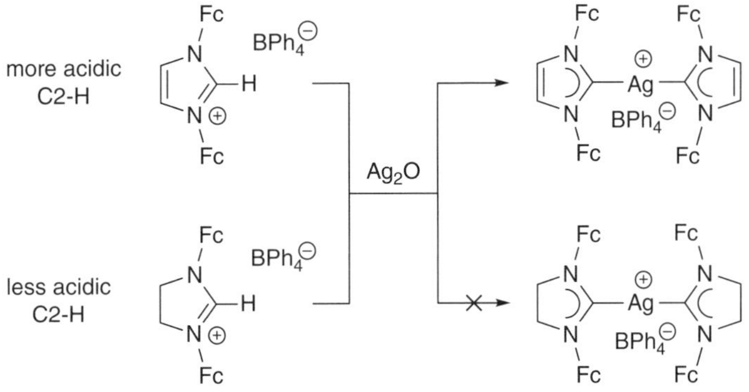

Scheme 6.18
Comparison of the reactivity of diferrocenylazolium salts with saturated and unsaturated backbones in the silver(I) oxide route.

The acidity of carbene precursors is also greatly influenced by the nature of their heterocyclic backbones including the electronic effects of any substituents. Due to their saturated backbone, imidazolinium salts are the weakest proton acids among the four types of classical NHC precursors of our concern. Therefore, it is not surprising that attempts to prepare silver-imidazolidin-2-ylidene complexes using the silver(I) oxide route are often less straightforward.

This has been described in a comparative study involving 1,3-diferrocenylazolium tetraphenylborates [26]. It was observed that the C4/5 saturated imidazolinium salt could not be converted into the bis(NHC) silver complex using silver(I) oxide, which has been attributed to the lower acidity of the C2–H proton (Scheme 6.18). The analogous but more acidic C4/5 unsaturated imidazolium salt, on the other hand, smoothly reacted under the same conditions to give the expected bis(NHC) complex $[Ag(IFc)_2]BPh_4$ in 63% yield, ruling out that steric bulk of the ferrocenyl-wing tips was the determining factor for the reactivity difference. The chemical shift for the carbene donor was observed at 180.6 ppm in CD_2Cl_2 with heteronuclear coupling constants of $^1J(^{13}C\text{-}^{107}Ag) = 189\,Hz$ and $^1J(^{13}C\text{-}^{109}Ag) = 219\,Hz$, which are in the range observed for other bis(NHC) silver complexes. In addition, the ^{109}Ag

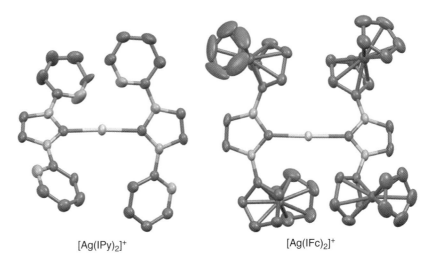

$[Ag(IPy)_2]^+$ $[Ag(IFc)_2]^+$

Figure 6.7
Molecular structures for the complex cations $[Ag(IPy)_2]^+$ and $[Ag(IFc)_2]^+$. Hydrogen atoms are omitted for clarity.

nuclei was also reported to resonate at 727 ppm downfield from 5 M aqueous silver(I) nitrate solution, which is 84 ppm downfield of Arduengo's $[Ag(IMes)_2]$OTf complex.

Analyses of single crystals $[Ag(IPy)_2]BF_4$ and $[Ag(IFc)_2]BPh_4$ by X-ray diffraction revealed a linear molecular complex geometry (Figure 6.7). Their very similar metal-carbene bond distances of 2.117(5) and 2.106(6) Å for the former and 2.082(4) and 2.092(4) Å for the latter are also comparable to the $[Ag(IMes)_2]$OTf complex.

Despite the aforementioned limitations, these processes are usually not very rapid, and the resulting complexes can be isolated as reasonably stable species in their solid form. Often they are also just prepared *in situ* and directly transferred to other metal centers, which makes the "silver(I) oxide route" very versatile and applicable to many different metals.

The third basic silver salt that has found application in the making of silver-NHC complexes is silver carbonate (Ag_2CO_3). An early example of such a protocol was reported by Danopoulos and coworkers using three N-functionalized imidazolium bromides (Scheme 6.19) [22]. They noted that carbonate-facilitated reactions usually require longer times (2 days!) for completion compared those employing silver(I) oxide. In addition, the reaction mixture in dichloromethane (or dichloroethane) had to be heated under reflux. One obvious explanation is the reduced basicity of the carbonate compared to that of the oxide dianion. As with Lin's route, a stoichiometrically balanced reaction only requires half an equivalent of the basic silver precursor for each mol of azolium salt. In practice, a slight excess is employed to ensure complete deprotonation, which would result in the formation of carbonic acid (H_2CO_3). The acid is unstable and decomposes to water and carbon dioxide. The formation of these two very stable compounds drives the reaction to the product side of the equation.

The appearance of the NMR spectra of the complexes shows concentration dependence, with more concentrated samples showing broader signals. Moreover, the carbene signals could not be resolved for the majority of complexes. Both observations point to a high degree of fluxionality is solution.

[AgBr(NHC)] [Ag(μ-Br)(NHC)]$_2$ [Ag(NHC)$_2$][AgBr$_2$]

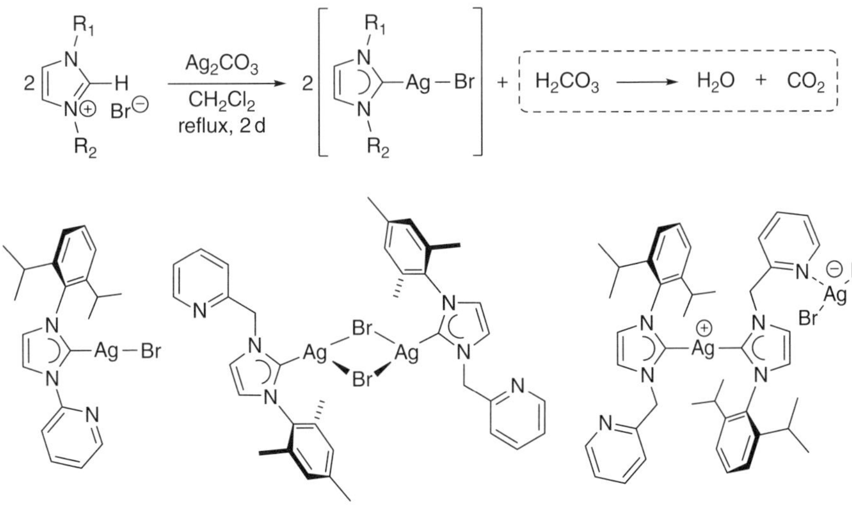

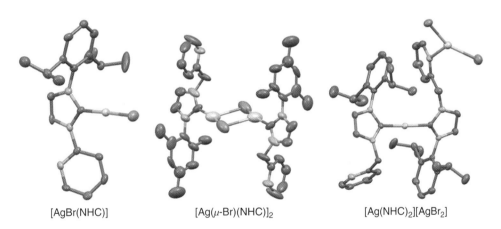

Scheme 6.19
Preparation silver(I)-NHC complexes using silver(I) carbonate.

[AgBr(NHC)] [Ag(μ-Br)(NHC)]$_2$ [Ag(NHC)$_2$][AgBr$_2$]

Figure 6.8
Different types of molecular structures of silver(I)-NHC complexes. Hydrogen atoms are omitted for clarity.

In agreement with these observations, three different types of molecular structures differing in nuclearity, charge and inner coordination sphere were determined by single crystal X-ray diffraction (Figure 6.8). These include the neutral monocarbene complex [AgBr(NHC)], the bromido-bridged dimeric monocarbene complex [Ag(μ-Br)(NHC)]$_2$ and the auto-ionized complex salt [Ag(NHC)$_2$][AgBr$_2$]. None of these show a κ^2-*C,N* chelating mode of the N-functionalized carbene, and only the last shows coordination of one nitrogen donor, however only to the dibromideargentate(I) unit.

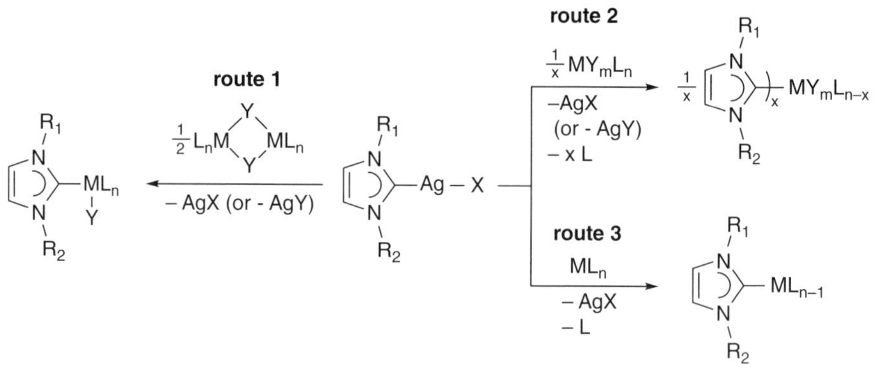

Scheme 6.20
Carbene transfer from silver-halido complexes to three different types of general metal precursors.

6.2.3 Silver-Carbene Transfer Reactions

Although some silver(I) NHC complexes have found applications in catalysis and as anti-microbial agents, the vast majority are prepared for transmetallation to furnish primarily transition metal complexes. Therefore, it is important to discuss some of the characteristic features involved in a typical silver-carbene transfer reaction. Depending on the nature of the silver-NHC transfer agent as well as the receiving metal complex precursors, a few cases have to be differentiated.

If the silver-NHC complex contains a halido ligand, which in its simplest case would be represented by the neutral [AgX(NHC)] species, then the transfer to another metal is usually straightforward. Precipitation of simple silver halides is warranted as the major and strong driving force. Here, three simple reaction outcomes can be anticipated that are based on the use of different receiving metal precursors. Dinuclear metal complexes with bridging μ-halido ligands Y are coordinatively unsaturated and thus ideal precursors. The carbene transferred from the silver center can engage in a simple bridge-cleavage reaction, which results in stable and coordinatively saturated monocarbene complexes without the need for any additional ligand dissociation (Scheme 6.20, route 1). In general, this type of reaction is both enthalpically and entropically favorable. In order to avoid potential halido-scrambling, it is preferred to use the same halido ligand in both the silver and the receiving metal complex precursor, that is, X=Y. If this condition cannot be fulfilled, that is, X≠Y, the carbene transfer will still occur. However, it is important to note that the silver ion precipitates primarily with the heavier halide, which is consistent with the increasing stability of the silver halides in the order AgCl<AgBr<AgI. As a consequence, the lighter halido ligand is usually retained at the metal center M. Nevertheless, minor amounts of halido-scrambled byproducts can be expected, because an exclusive selectivity cannot be ensured as all silver halides are generally poorly soluble.

The transfer from x equivalents [AgX(NHC)] to a coordinatively saturated halido-complex [MY$_m$L$_n$] (Y=terminal halido ligand) can also lead to the same kind of halido-scrambling, and the same considerations described above are valid. More importantly, such a reaction would usually lead to a replacement of the most labile neutral ligands L by the same number (x) of stronger NHCs resulting in di- or poly(NHC) complexes depending on the stoichiometry used (Scheme 6.20, route 2). The formation of new and stronger metal-NHC bonds in the new metal-NHC complex contribute to a negative reaction enthalpy, which together with the precipitation of silver halide salt as the byproduct makes the reaction very favorable.

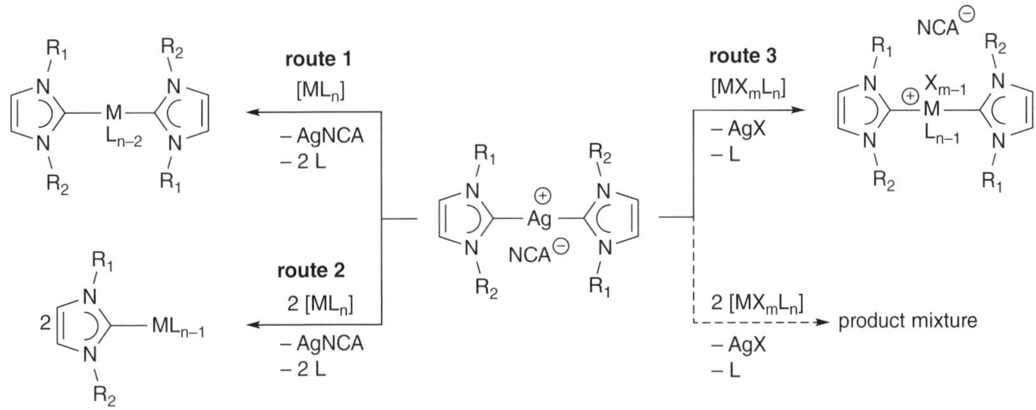

Scheme 6.21
Carbene transfer from bis(NHC) silver complexes to different types of general metal precursors.

The last example simply involves a 1:1 replacement of a neutral ligand L with the carbene (Scheme 6.20, route 3), which occurs with the same driving force as route 2. Halido scrambling is of no concern as no halido ligands are involved in this reaction. Such a reaction would yield a monocarbene complex.

A different reaction outcome is to expected when the transfer agent is a bis(NHC) silver species that only contains weakly or non-coordinating anions (NCA, mostly BF_4^- or PF_6^-). For this case too, a few scenarios have to be differentiated. The first involves a 1:1 transfer to a coordinatively saturated metal complex without any halido ligands (Scheme 6.21, route 1). Again, one driving force that transfers both NHCs to the new metal center is primarily based on the new stronger bond of the NHC to the metal compared to that to silver. Also, two monodentate ligands L initially coordinated to the metal have to dissociate, and a new bis(carbene) metal complex is obtained. The former silver metal center is released as a simple AgNCA salt (e.g. $AgPF_6$) of better solubility than silver halides, thereby providing less contribution to the thermodynamics of this transfer process. On the other hand, the overall process can be estimated to be entropically favored by considering the overall number of species after reaction. Simple stoichiometry control by changing the ratio of transfer agent to receiving metal species to 1:2 could lead to the selective formation of monocarbene complexes under otherwise identical reaction conditions (Scheme 6.21, route 2).

If the receiving metal complex precursor is neutral and contains m halido ligands X, a halido abstraction by the silver(I) ion can be anticipated. In a 1:1 reaction, this would result in the loss of one halido and one neutral ligand at the new metal center, which additionally gains two new NHC ligands. Overall, a new cationic bis(NHC) complex with retention of the coordination number is obtained. The non-coordinating anion balances the positive charge of the metal complex (Scheme 6.21, route 3).

An attempt to prepare monocarbene complexes by a simple change of the stoichiometry of the same reactants to 1:2 is not straightforward and not advisable. The presence of only one silver(I) ion can only lead to abstraction of one halido ligand from the first equivalent of receiving metal precursor, while the second equivalent would retain its halido ligand. Thus a complicated product mixture of NHC complexes can be foreseen upon mixing of the reactants (Scheme 6.21, dotted line).

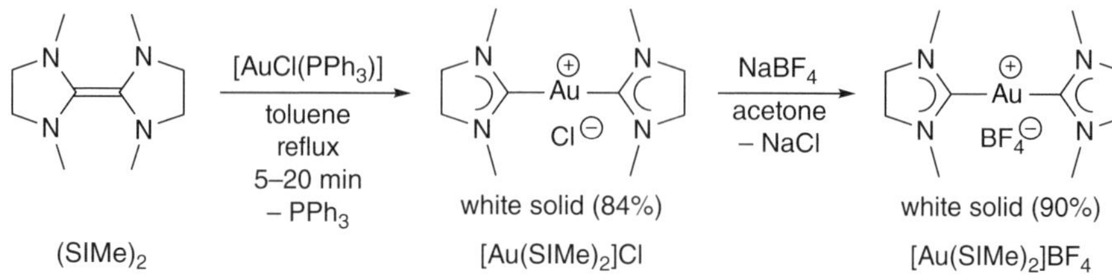

Scheme 6.22
Preparation of the first gold(I) NHC complex.

6.3 Gold(I)-NHC Complexes

6.3.1 Cleavage of Enetetramines

The first gold(I) NHC complex was supposedly reported by Lappert and coworkers in 1974 [27,28]. The authors described that heating a solution of the electron-rich tetramethylene-tetramine with chlorido-triphenylphosphinegold(I) in toluene gave a pale yellow solution, from which upon further heating the cationic bis(carbene)gold(I) complex [Au(SIMe)$_2$]Cl forms as a white precipitate in 84% (Scheme 6.22). The two imidazolidin-2-ylidene ligands substitute both the chlorido as well as the phosphine ligands from the gold(I) precursor complex. The resulting hygroscopic complex is air and moisture stable, but decomposes upon heating to 270 °C. Anion metathesis with sodium tetrafluoroborate affords [Au(SIMe)$_2$] BF$_4$, which shows a reduced solubility in organic solvents, but also an increased thermal stability with a melting point of around 218–225 °C.

Although historically important, the cleavage of enetetramines has only limited scope in terms of ligand diversity. It has been succeeded by simpler methodologies and does not play any significant role in the preparation of gold NHC complexes nowadays.

6.3.2 Protonation/Alkylation of Azolato Complexes

Fifteen years later, Burini and coworkers reported the preparation of gold(I) imidazolin-2-ylidene complexes [29]. These were prepared from the reaction of the respective lithium-N-benzylimidazolate salt with gold(I) complex precursors bearing either phosphine or organic sulfide ligands, that is, [AuCl(PPh$_3$)] or [AuCl(SMe$_2$)], at low temperature in tetrahydrofurane (Scheme 6.23). The initial product is an unusual cyclic trigold(I) NHC complex. Due to its *C,N* bridging mode, the initial imidazolato ligand can be formally regarded as a cyclic carbene donor.

Subsequent treatment with hydrochloric acid involving several workup steps leads to protonation of the anionic nitrogen moiety with concurrent disaggregation, which eventually afforded a cationic monogold(I) bis(imidazolin-2-ylidene) complex with an unusual NH,NR-NHC ligand.

Raubenheimer and coworkers extended the methodology and obtained a range of imidazolin- and benzimidazolin-2-ylidene complexes without the intermediacy of trigold species [30]. The low temperature reaction of 2-lithio-1-methylimidazolate or –benzimidazolate, in turn generated by deprotonation of the 1-methylazoles using *n*-butyllithium, with the [AuCl(THT)] precursor complex led directly to the formation of the linear diazolato gold(I) complexes Li[Au(Me-azoyl)$_2$] (Scheme 6.24). These aurate complexes could be regarded

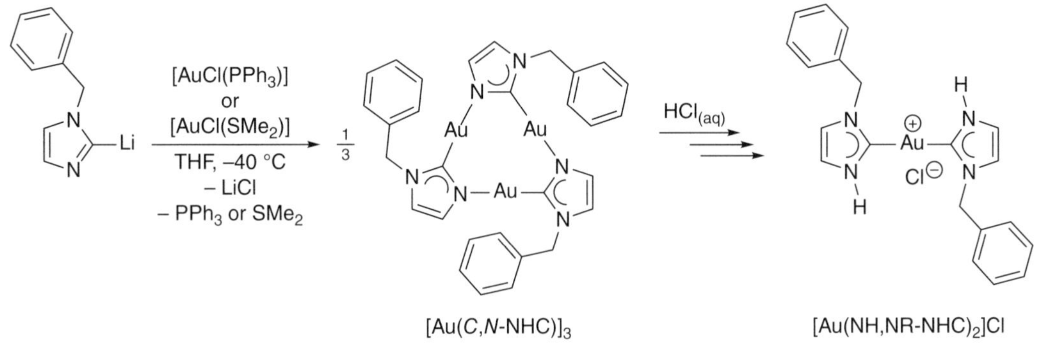

Scheme 6.23
Preparation of a cyclic trigold(I)-NHC and a bis(NHC) gold(I) complex using lithium-imidazolate.

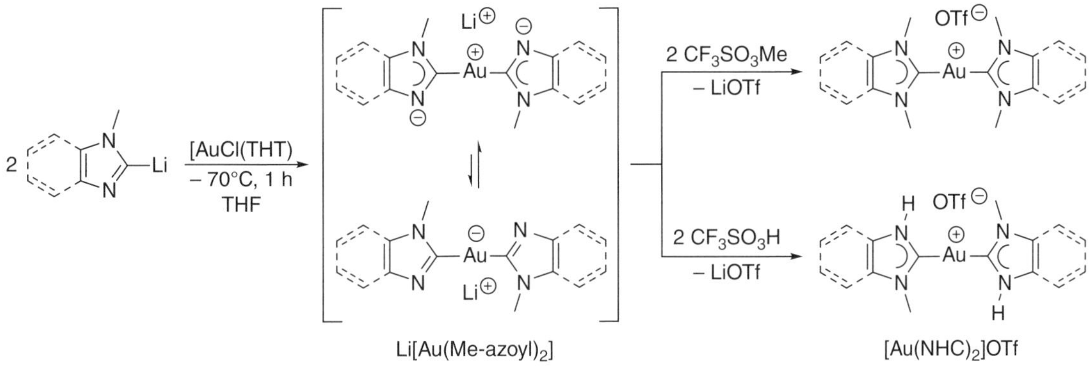

Scheme 6.24
Raubenheimer's approach to bis(NHC) gold(I) complexes.

as complexes of "naked and anionic" NHC ligands as the ground state resonance structure should have the negative charge primarily residing on the electronegative, unsubstituted (thus "naked") nitrogen atom leading to a neutral carbon donor. Indication for the carbene character of the carbon donor atom was provided by a significant downfield shift of the respective ^{13}C NMR signal from 138.2 ppm in the free 1-methylimidazole to 185.8 ppm in the imidazole-derived Li[Au(Me-azoyl)$_2$] complex. Accordingly, the unsubstituted nitrogen atom is the nucleophilic site of the molecule and prone to electrophilic attack. Indeed, subsequent *N*-protonation or *N*-methylation with two equivalents of trifluoromethanesulfonic acid (CF$_3$SO$_3$H) or methyl trifluoromethanesulfonate (CF$_3$SO$_3$Me) produced the cationic bis(carbene) complexes [Au(NHC)$_2$]OTf in acceptable yields as colorless solids after workup (Scheme 6.24). At room temperature, these complexes are thermally stable, and they show good solubility in more polar organic solvents such as tetrahydrofurane, dichloromethane, and acetone. For the imidazole-based compounds, changes in ^{13}C NMR shift of the carbon donor upon N-protonation (i.e. 182.8 ppm) or N-methylation (i.e. 185.7 ppm) were noted to be very small compared to the anionic complex precursor. This further strengthens the notion that the latter contains "anionic" carbene ligands. The respective signals for the benzimidazolin-2-ylidene analogs were found to be more downfield by

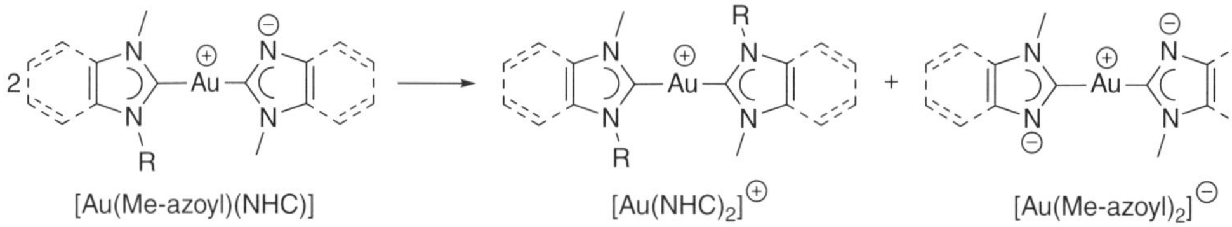

Scheme 6.25
Proposed homoleptic rearrangement reaction of a heteroleptic gold(I) complex.

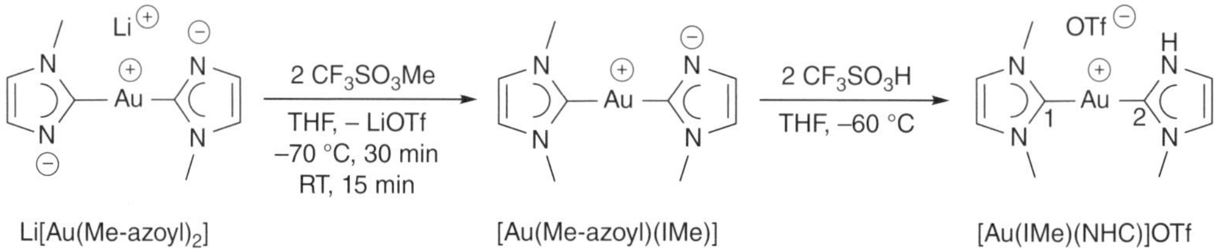

Scheme 6.26
Preparation of the first gold(I) hetero-bis(carbene) complex.

~5–7 ppm, that is, 189.3 and 191.3 ppm, respectively. This is in line with the negative inductive effect as a result of benzannulation.

Notably, the reaction with only one equivalent of acid or methylating agent in an attempt to obtain a heteroleptic complex [Au(Me-azoyl)(NHC)] also afforded the aforementioned homoleptic bis(carbene) complexes, but in lower yields as anticipated. This observation has been attributed to a rearrangement reaction of the initially formed neutral heteroleptic complex into the two homoleptic complex ions [Au(NHC)$_2$]$^+$ and [Au(Me-azoyl)$_2$]$^-$ under autoionization (Scheme 6.25). Such homoleptic rearrangement reactions have been later observed in other types of gold(I) NHC complexes as well and are ligand dependent (see Section 6.3.5).

Nevertheless, this process is sufficiently slow for capture of the neutral mixed-ligand complex to be possible. This was demonstrated with the imidazole-based systems, where a 2-step reaction involving initial *N*-methylation followed by *N*-protonation produced the first gold(I) hetero-bis(carbene) complex [Au(IMe)(NHC)]OTf, which contains two different NHC ligands coordinated to the same metal center (Scheme 6.26). As expected, this complex exhibits two different $^{13}C_{carbene}$ NMR signals. By comparison with the homoleptic species, the more downfield signal at 185.3 ppm has been assigned to the IMe ligand (C$_1$ in Scheme 6.26), while the higher field resonance at 183.0 ppm was attributed to the protic NHC ligand (C$_2$ in Scheme 6.26).

Similar reactions occurred with alternative gold(I) complex precursors such as [AuCl(PPh$_3$)], [Au(C$_6$F$_5$)(THT)] and AuCN, which gave rise to a range of homoleptic and heteroleptic gold(I) NHC complexes.

6.3.3 The Free Carbene Route

The use of free carbenes for the preparation of gold(I) NHC complexes is relatively rare. An early example was reported by Fehlhammer and coworkers in 2001 [31], who employed monoanionic bis(imidazolin-2-ylidene-1-yl)borate dicarbene ligands for this purpose.

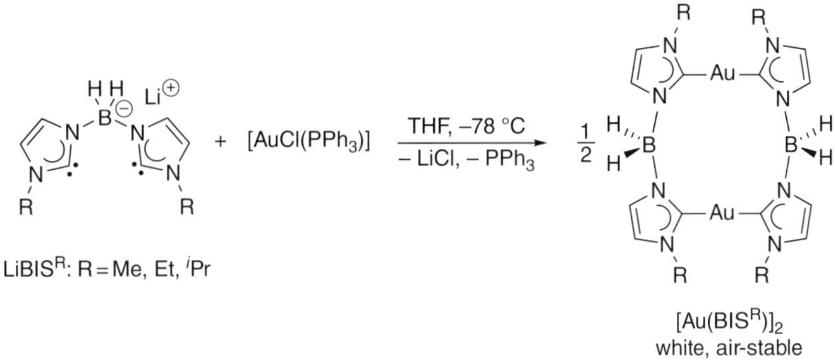

Scheme 6.27
Formation of neutral bis(NHC)digold complexes using free dicarbenes.

The latter were in turn prepared by deprotonation of the respective diimidazolium salts with *n*BuLi in tetrahydrofuran at low temperature. Equimolar reaction of these diNHCs with [AuCl(PPh$_3$)] at −78 °C in tetrahydrofuran led to the formation of the gold(I) complexes as white and analytically pure precipitates (Scheme 6.27). These products were identified as neutral dinuclear gold species, where the dicarbene coordinates in a bridging μ_2-fashion. Each gold(I) center is bound by two carbene donors from different diNHC ligands resulting in a cationic bis(NHC) complex fragment, which charge is balanced by the anionic borate bridge.

Both LiCl salt formation as well as the enthalpically favorable formation of strong gold NHC bonds contributes to the driving force of this reaction. The digold(I)-NHC complexes exhibit ^{13}C$_{carbene}$ signals ranging from 187.5–188.2 ppm in CDCl$_3$ that are in the typical range for bis(NHC)gold(I) complexes and downfield shifted by ~50 ppm from the respective signals in their imidazolium iodide precursors. Also notable are the downfield shifts of the ^{11}B NMR signals (i.e. −6.94 to −8.14 ppm) by >1 ppm upon complexation compared to those observed in the salts (i.e. −9.02 to 9.59 ppm).

Diffusion of *n*-pentane to a solution of the N-ethyl digold(I) complex in tetrahydrofuran gave single crystals that were subjected to an X-ray diffraction study. The analysis reveals the structure of a 12-membered dimetallacycle in a twisted boat-like conformation with a transannular Au···Au separation of 3.3610(7) Å, which is indicative of only a very weak aurophilic interaction (Figure 6.9). The average gold–carbene distance in this homoleptic dinuclear bis(chelate) complex amounts to 2.021 Å.

The reaction of free monodentate NHCs with gold(I) sources was reported almost simultaneously only a few years later in 2005 by the groups of Roesky [32] and Nolan [33]. The latter study noted that direct exposure of free NHCs to AuCl is undesirable and only leads to poor yields and complex product mixtures containing metallic gold and mono- as well as bis(carbene) complexes.

Ligand displacement reactions involving gold(I) complexes are more promising, and indeed, a range of saturated and unsaturated free NHCs react with chlorido(dimethylsulfide) gold(I) to form the desired monocarbene complexes of the general formula [AuCl(NHC)] under liberation of dimethylsulfide (Scheme 6.28, route 1). In this reaction, the stronger carbene donor simply replaces the weak and neutral thioether ligand, and a stronger gold-carbon bond is exothermically formed. The yields ranging from 58 to 81% were moderate to good. Interestingly, the reaction of IMes with [AuCl(SMe$_2$)] was found to proceed less smoothly giving rise to a similar outcome as with simple gold(I) chloride.

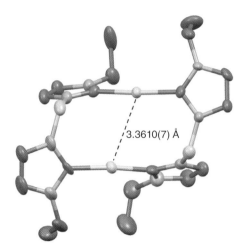

Figure 6.9
Molecular structure of the digold(I) complex [Au(BIS^{Et})]$_2$. Hydrogen atoms are omitted for clarity.

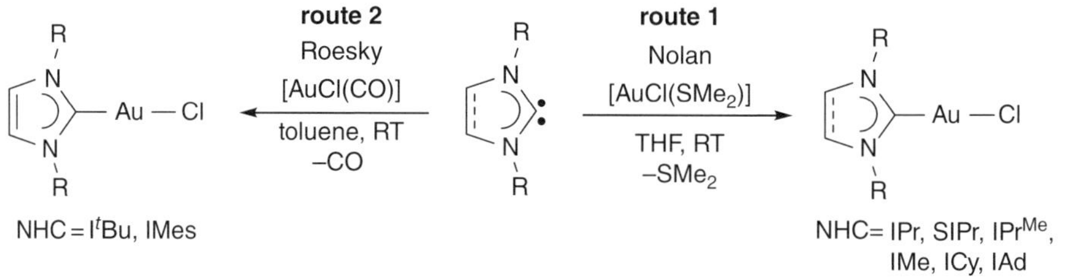

Scheme 6.28
Preparation mono-NHC gold(I) complexes via free carbenes.

On the other hand, Roesky and coworkers described that both IMes and I^{t}Bu react in similar fashion with [AuCl(CO)] in toluene at room temperature under CO substitution to give the respective [AuCl(NHC)] complexes in 67% and 83% yields, respectively (Scheme 6.28, route 2) [32].

All these neutral chlorido-monocarbenegold(I) complexes are thermally stable and isolated as white solids, which are also stable to air and moisture. Since the two-coordinated complexes contain d^{10} metal centers, they are expected to be diamagnetic and linear in geometry throughout. As with the other coinage metal centers, coordination of the free carbene to gold(I) leads to a pronounced upfield shift of the ^{13}C$_{carbene}$ NMR signal, while a significant downfield shift for the C4/5-proton resonances were observed.

The carbene ligands can also be generated *in situ* and immediately trapped by coordination to gold(I). Such a reaction has already been reported in 2002 by Grützmacher and coworkers [34]. Their initial attempt to prepare a tropylidene substituted NHC in its free form by deprotonation of the respective imidazolium chloride only led to the isolation of a 1,2-disubstituted imidazole as an undesired rearrangement product from the transient carbene. In an attempt to trap the latter *in situ*, deprotonation of the imidazolium salt was repeated in the presence of chloridotriphenyl-phosphinegold(I) [AuCl(PPh$_3$)] (Scheme 6.29). This afforded the first mixed carbene/phosphine gold(I) complex [Au(NHC)(PPh$_3$)]Cl in 61% yield. The displacement of the chlorido ligand by

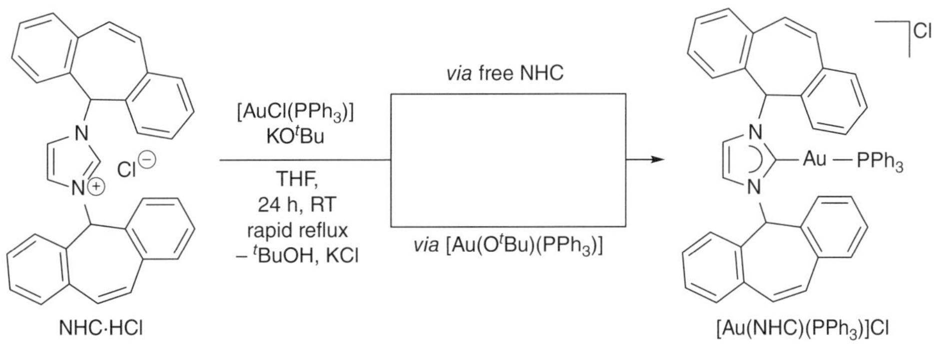

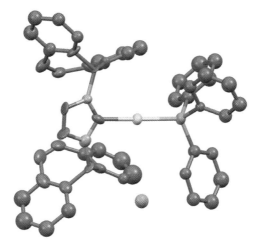

Scheme 6.29
Synthesis of the first mixed phosphine/NHC gold(I) complex by *in situ* deprotonation.

Figure 6.10
Solid state molecular structure of the first mixed carbene/phosphine gold(I) complex. Hydrogen atoms are omitted for clarity.

the generated free carbene is supposedly facilitated by the strong *trans* effect of the triphenylphosphine ligand, which weakens the gold–chlorido bond.

Mechanistic details of the reaction are not known, and an alternative pathway via salt metathesis can also be envisioned. This route would involve substitution of the chlorido ligand with the alkoxide leading to the neutral *tert*-butylalkoxidophosphine complex [Au(O^{t}Bu)(PPh$_3$)] and elimination of KCl salt. Deprotonation of the imidazolium salt with this basic gold phosphine complex then affords the final product.

The ^{13}C NMR spectrum of the cationic complex recorded in CDCl$_3$ displays a singlet for the carbenic carbon at 207.3 ppm, which is substantially downfield shifted from those of the neutral [AuCl(NHC)] counterparts. This downfield shift can certainly be attributed to presence of the phosphine ligand, which is a stronger donor compared to the chlorido ligand.

Layering a solution of the complex in dichloromethane with *n*-hexane gave suitable single crystals for X-ray diffraction. The expected linear molecular structure is depicted in Figure 6.10. The gold–carbene distance of 2.111(7) Å is slightly longer than those observed for cationic bis(carbene) complexes, that is, ~1.99–2.03 Å (Fehlhammer's [Au(BIS^{Et})]$_2$ complex; see Section 6.3.3). This bond elongation can likely be attributed to the presence of two *trans*-standing bulky ligands.

6.3.4 The Silver-Carbene Transfer Route

In 1998, Ivan J. B. Lin and coworkers introduced a new methodology to prepare gold(I) NHC complexes via the transfer of the carbene from silver(I) NHC complexes [21]. Since then, it has become the most popular way to access gold NHC complexes due its ease and versatility, which generally does not require the exclusion of air and moisture. The syntheses can thus be conducted in simple glassware. The basis for this approach is the relatively weak and dynamic silver-carbene bond, which on the other hand is still sufficiently strong to allow for isolation of the respective silver-NHC complexes. Exposure of the latter to neutral gold(I) complex precursors of the type [AuX(L)], for example [AuCl(SMe$_2$)] in Scheme 6.30, results in the transfer of the carbene from silver(I) to gold(I) and a much stronger metal-carbene bond is formed, which contributes favorably to the reaction enthalpy. Simultaneously, the initial and weaker co-ligand L is displaced. Perhaps contributing even more significantly to the driving force is the formation of insoluble silver halides.

Two distinct routes are commonly observed. Route 1 involves the equimolar transfer from a neutral halido-complex "[AgX(NHC)]" or its auto-ionized isomer "½ [Ag(NHC)$_2$] [AgX$_2$]." If the halido ligand X is heavier than the chlorido ligand, then it is preferably— however not exclusively—associated with the silver ion precipitating as the more favorable and heavier silver halide, that is, AgBr or AgI versus AgCl. In route 2, a cationic bis(NHC) silver(I) complex [Ag(NHC)$_2$]Y with a non-coordinating counter-anion Y$^-$ (e.g. BF$_4^-$ or PF$_6^-$) is used. This generally leads to the formation of isoelectronic and isostructural bis(NHC)gold(I) complexes of the formula [Au(NHC)$_2$]Y, and the silver ion abstracts the halido ligand (here the chlorido ligand) from the gold(I) center.

In Lin's seminal first example, these two routes were explored with 1,3-diethylbenzimidazolium salts bearing either a bromide or hexafluorophosphate counter-anion. These reactions led to the isolation of the two linear complexes [AuCl(Et$_2$-bimy)] and [Au(Et$_2$-bimy)$_2$] PF$_6$ in very good yields (Scheme 6.31). Notably, the former can also be easily converted into the latter by reaction with an additional equivalent of hexafluorobenzimidazolium salt and a suitable base (here K$_2$CO$_3$) in acetone. The same reaction was also applied to the 1,3-dimethyl analog [35].

Detailed NMR spectroscopic investigations revealed that the cationic bis(NHC) complexes exhibit markedly more downfield resonances (~190 ppm) than their neutral

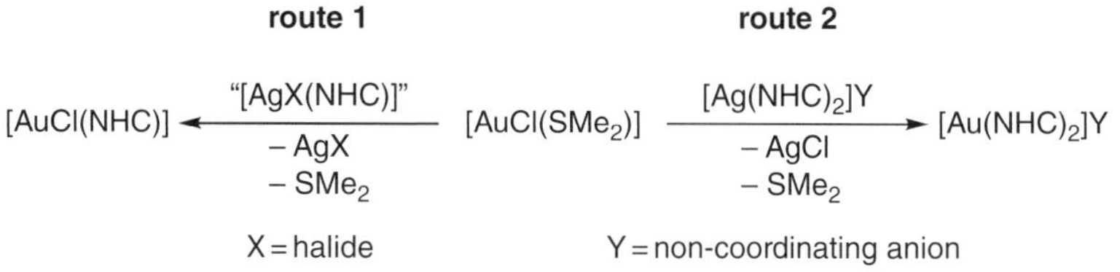

route 1   route 2

[AuCl(NHC)] ⟵ "[AgX(NHC)]"   [AuCl(SMe$_2$)]   [Ag(NHC)$_2$]Y ⟶ [Au(NHC)$_2$]Y
        − AgX   − AgCl
        − SMe$_2$ − SMe$_2$

X = halide   Y = non-coordinating anion

Scheme 6.30
General scheme for the preparation gold NHC complexes via silver-carbene transfer.

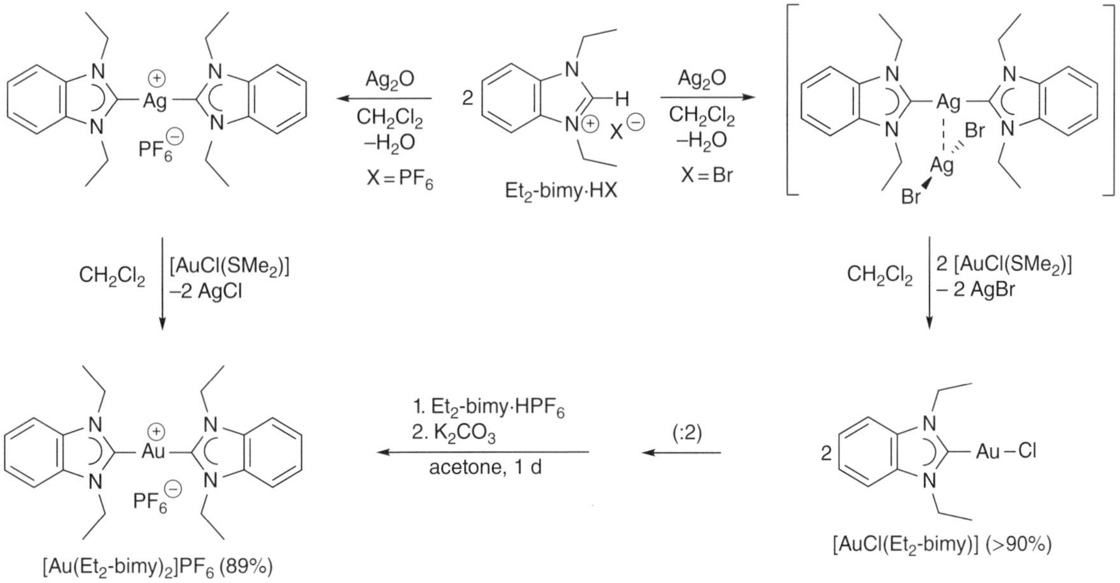

Scheme 6.31
Preparation of mono- and bis(NHC) complexes of gold(I) via silver-carbene transfer.

chlorido-monocarbene counterparts (~178 ppm). As commonly observed, the $^{13}C_{carbene}$ resonance is dependent on the Lewis acidity of the metal center, which in turn can be influenced by the nature of the *trans*-standing co-ligand (see Table 6.3).

Single crystals of the chlorido complexes were grown from their solutions in dichloromethane/hexane. Their molecular structures obtained by X-ray diffraction analyses revealed interesting structural differences (Figure 6.11). The solid state structure of complex [AuCl(Me$_2$-bimy)] with sterically unassuming *N*-methyl groups exhibits intermolecular Au^I··· Au^I aurophilic interactions between two molecules with an intergold distance of 3.166(1) Å. Notably, the subtle change from a *N*-methyl to an *N*-ethyl wing tip is already sufficient to completely remove aurophilicity in the latter. Thus the solid state packing of complex [AuCl(Et$_2$-bimy)] was found to consist of discrete and independent molecules.

The aurophilic interactions do not have any influences on the gold–carbene bond distance, as these bond lengths of ~2 Å in both complexes are essentially identical within experimental error. Likewise, the gold-chlorido distances were found to be very similar.

The presence of aurophilicity, on the other hand, affects the luminescent properties of the complexes to a great extent. The emission spectra recorded of crystalline samples of both complexes show significant differences. Only that for the [AuCl(Me$_2$-bimy)] complex exhibits emission events due to intergold Au^I–Au^I (^{3}MC) transitions, while only intraligand transitions can be noted for the [AuCl(Et$_2$-bimy)] complex.

Using the silver-carbene transfer method, gold(I) complexes of the most common NHC ligands derived from imidazole (e.g. IMes, etc.), triazole (i.e. tazy), benzimidazole (i.e. bimy) and imidazoline (e.g. SIMes, etc.) have been prepared in generally very good yields. The complexes are usually air and moisture stable white solids, and selected key analytical data for some representatives are summarized in Table 6.2. Despite small changes that can be anticipated due to solvent effects, the following trends can be identified by comparison of their $^{13}C_{carbene}$ signals.

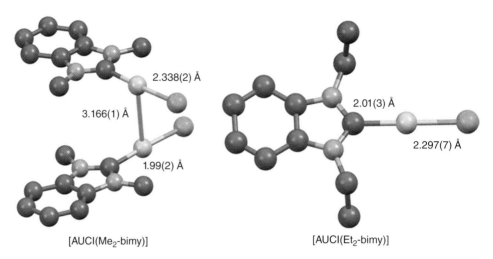

[AuCl(Me₂-bimy)]   [AuCl(Et₂-bimy)]

Figure 6.11
Molecular structures of benzimidazolin-2-ylidene-chlorido gold(I) complexes. Hydrogen atoms are omitted for clarity.

Table 6.2   Selected data for some monocarbene-chlorido complexes of gold(I).

Complex	Solvent-d_n	$^{13}C_{carbene}$ [ppm]	Au–NHC [Å]	Au–Cl [Å]
[AuCl(IMe)] [36]	DMSO-d_6	169	1.98(1)	2.288(3)
[AuCl(IMe^{Me})] [33]	CD$_2$Cl$_2$	168.4	1.987(8)	2.288(2)
[AuCl(ICy)] [33]	CD$_2$Cl$_2$	168.0	1.99(1), 2.00(1)	2.306(3), 2.281(3)
[AuCl(IAd)] [33]	CD$_2$Cl$_2$	166.3	1.989(2)	2.2761(6)
[AuCl(I^{t}Bu)] [37]	CDCl$_3$	168.2	2.018(3)	2.2742(7)
[AuCl(IMes)] [33]	CDCl$_3$	173.4	1.998(5)	2.276(1)
[AuCl(IPr)] [33]	CD$_2$Cl$_2$	175.1	1.942(3)	2.270(1)
[AuCl(IPr^{Me})] [33]	CDCl$_3$	166.3	1.996(9)	2.279(2)
[AuCl(Me$_2$-tazy)] [36]	DMSO-d_6	173	–	–
[AuCl(Me$_2$-bimy)] [35]	CDCl$_3$	179	1.99(1)	2.338(2)
[AuCl(SIMes)] [33]	CDCl$_3$	195.0	1.983(4)	2.277(1)
[AuCl(SIPr)] [33]	CDCl$_3$	196.1	1.979(3)	2.276(1)

1. Within the same type of NHCs, those with aromatic substituents have a more downfield shift, while those with aliphatic substituents resonate at upper field.
2. Across different types of NHCs there is a gradual downfield shift in the order imidazolin-2-ylidene < triazolin-2-ylidene < benzimidazolin-2-ylidene < imidazolidin-2-ylidene.

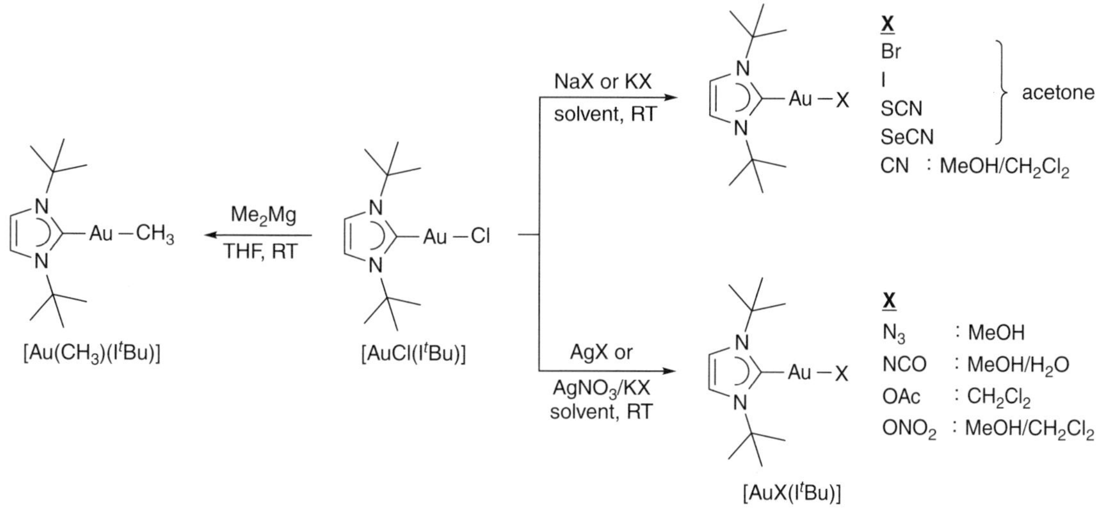

Scheme 6.32
Preparation of neutral monocarbene gold(I) complexes with various anionic co-ligands.

3. The presence of positively inductive methyl groups in 4/5-positions in imidazolin-2-
   ylidenes leads to an upfield shift.

Overall, the chemical shift of the carbene donor is influenced by the backbone and its substituents. Electron releasing groups (e.g. alkyl) have a shielding effect, while electron-withdrawing groups (e.g. aryl, heteroatoms) are deshielding.

Complexes of the general formula [AuCl(NHC)] are very useful complex precursors and offer access to a wide range of other neutral or cationic gold(I) NHC complexes via simple co-ligand exchange. As depicted in Scheme 6.31, cationic bis(NHC) complexes can also be prepared from such complexes.

More straightforward is the preparation of gold(I) NHC complexes bearing simple anionic co-ligands other than chlorido. The team of Murray Baker provided a detailed study on structures and properties of various complexes of the type [AuX(I^{t}Bu)] [37]. The parent compound [AuCl(I^{t}Bu)] was easily prepared using the silver-carbene transfer route (Table 6.2).

Subsequent salt metathesis reactions with various sources of halide, pseudohalides, and dimethylmagnesium afforded a range of neutral gold(I) I^{t}Bu complexes bearing various anionic co-ligands (Scheme 6.32). In cases, where the nucleophile is weak, preceding chlorido abstraction mediated by silver(I) nitrate is required.

All the complexes are soluble in medium to high polarity organic solvents such as ethyl acetate, dichloromethane and methanol, and the methyl complex [Au(CH$_3$)(I^{t}Bu)] was also soluble in hexanes. The good solubility is a consequence of the two *tertiary*-butyl wing tip groups on the NHC. Notably, the thiocyanato, selenocyanato, isocyanato, azido, nitrato, and methyl complexes were the first examples in gold NHC chemistry.

Comparison of the solution NMR data of these complexes revealed that the chemical shift of the carbene carbon is very sensitive to the Lewis acidity of the metal center, which can be tuned by different co-ligands. The greater the Lewis acidity, the more upfield is the carbene resonance. Therefore the net donating abilities of these anionic co-ligands can be compared by ranking the chemical shifts they induce at the carbene carbon (Table 6.3). In general, it was

Table 6.3 $^{13}C_{carbene}$ resonances of [AuX(I^{t}Bu)] complexes recorded in CDCl$_3$.

X	$^{13}C_{carbene}$ [ppm]	Au–C [Å]	X	$^{13}C_{carbene}$ [ppm]	Au–C [Å]
ONO$_2$	156.3	1.973(4)	Br	172.4	1.999(4)
OAc	161.3	1.982(3)	*SCN*	174.7	2.027(4)
NCS	163.8	-	*SeCN*	177.3	2.049(7)
NCO	166.2 (br)	1.986(3)	I	179.9	2.014
N$_3$	166.4	1.993(4)	CN	181.5	2.013(6)
Cl	168.2	2.018(3)	CH$_3$	198.7	2.049(4)

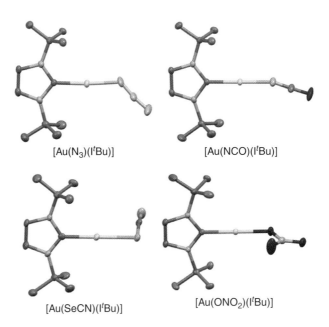

[Au(N$_3$)(I^{t}Bu)] [Au(NCO)(I^{t}Bu)]

[Au(SeCN)(I^{t}Bu)] [Au(ONO$_2$)(I^{t}Bu)]

Figure 6.12
Molecular structures of selected neutral gold(I) I^{t}Bu complexes with pseudohalido ligands.
Hydrogen atoms are omitted for clarity.

found that N donors are stronger than O donors, but inferior to halido and carbon donors, that is, O < N < halido < C. The donor strength of *S*-thiocyanato and *Se*-selenocyanato ligands are in between that of the bromido and iodido ligands. Notably, two inseparable linkage isomers, that is, [Au(*SCN*)(I^{t}Bu] and [Au(*NCS*)(I^{t}Bu], were observed in the thiocyanato case in a ~10:1 ratio. The less Lewis acidic major isomer exhibits binding of the sulfur donor to the gold(I) metal center due to the preferable soft-soft interaction.

Single crystals of most complexes were obtained and their molecular structures determined. A few representatives are depicted in Figure 6.12. All complexes exhibit the expected linear coordination geometry in the solid state. The triatomic co-ligands OC*N$^-$* (158°), N$_3^-$ (128°), NC*S$^-$* (102°) and NC*Se$^-$* (97°) all coordinate in an increasingly bent fashion with decreasing Au-X-E angles (E = adjacent atom), which is due to the additional lone pairs at the donor atoms X. Notably, the bent angle correlates with the donor strength of the respective triatomic ligand as well, which reflects its ability to stabilize the negative charge at the donor atom. Overall, it was found that the gold–carbene distance roughly

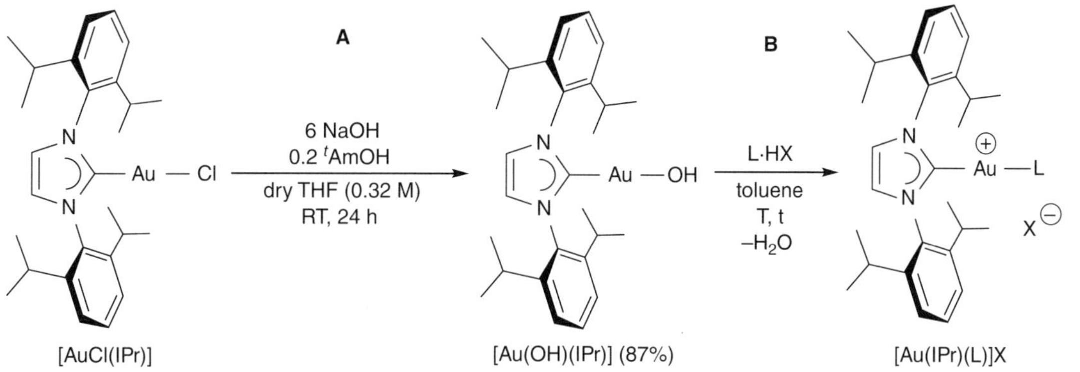

Scheme 6.33
Preparation of [Au(OH)(IPr)] and cationic complexes of the type [Au(IPr)(L)]X, where L is NHC or phosphine.

increases with increasing donor strength of the co-ligand (Table 6.3). This is intuitive since a stronger donating strength is generally associated with a stronger *trans*-influence. However, these bond parameters are not accurate enough to trace smaller differences in electronic properties as additional effects, for example packing effects, and intermolecular interactions also play a major role in solid state structures. For example, this can be clearly seen by comparing the metal-carbon distances in chlorido versus bromido and iodido versus cyanido complexes. The significant deviations from the NCS⁻ and NCSe⁻ ligands are likely due to their very different structural features.

[AuCl(NHC)] complexes have also been used to prepare complexes of the type [Au(OH)(NHC)] [38] as useful synthetic equivalents for the [Au(NHC)]⁺ synthon. The very basic hydroxido complexes can deprotonate other substrates *in situ*, which proves useful in the preparation of other gold–carbene derivatives and in catalysis. The optimized preparation of the latter involves stirring a mixture of [AuCl(IPr)], six equivalents of freshly ground sodium hydroxide and 0.2 equivalents of *tertiary* amyl alcohol in dry tetrahydrofurane at room temperature and under air for 24 hours (Scheme 6.33, **A**). The reaction is sufficiently robust and can be scaled up to produce 20 g of the product. More detailed investigations points to the *tert*-amylato complex [Au(O^{t}Am)(IPr)] as a primary intermediate. The *tert*-amylato ligand is formed by reaction of the respective alcohol with sodium hydroxide. Reaction of this complex with water gives the desired hydroxido product, while *tert*-amyl alcohol is regenerated. However, it must be noted that only NHCs with very bulky aromatic wing tips such as IPr and SIPr can stabilize the gold(I)-hydroxido moiety. Attempts to prepare analogs with *N*-alkyl carbenes and even the slightly less bulky IMes or SIMes were met with limited success and led to product mixtures of highly moisture sensitive compounds, whereas [Au(OH)(IPr)] is air and moisture stable.

Upon chlorido-hydroxido exchange, an upfield shift of the $^{13}C_{carbene}$ resonance from 175.1 to 171.9 ppm was observed. This is in line with the observations made for the [AuX(I^{t}Bu)] complexes, where O donors have been determined to be generally weaker donors compared to halido ligands.

The basicity of the [Au(OH)(IPr)] complex (pK_a ~29–31) has also been exploited to prepare a library of hetero-bis(carbene) and mixed carbene/phosphine complexes of gold(I) containing two different kinds of NHC ligands or one NHC and one phosphine, respectively (Scheme 6.33, B) [39]. For this purpose, the hydroxido complex is directly treated

Table 6.4 Optimized conditions and $^{13}C_{carbene}$ signals for [Au(IPr)(L)]X complexes.

[Au(IPr)(L)]X	T [°]	t [h]	Yield [%]	$^{13}C_{IPr}$ [ppm][a]	$^{13}C_{NHC2}$ [ppm][a]
[Au(IPr)(BMIM)]BF$_4$	25	14	93	188.0	181.1
[Au(IPr)(BMIM)]PF$_6$	25	14	95	188.0	181.3
[Au(IPr)(BMIM)]Cl	25	14	94	188.0	180.6
[Au(IPr)(ICy)]BF$_4$	25	24	91	187.0	178.5
[Au(IPr)(I^{t}Bu)]BF$_4$	80	24	95	185.5	179.6
[Au(IPr)(IMes)]BF$_4$	70	48	97	186.1	183.1
[Au(IPr)(SIMes)]BF$_4$	70	48	59	186.7[b]	205.2[b]
[Au(IPr)(IPr)]BF$_4$	70	48	95	184.2	184.2
[Au(IPr)(P^{t}Bu$_3$)]BF$_4$	25	14	89	191.9 (111.9 Hz)[c]	-
[Au(IPr)(P^{n}Bu$_3$)]BF$_4$	25	14	90	191.6 (122.0 Hz)[c]	-
[Au(IPr)(PPh$_3$)]BF$_4$	25	14	91	188.2 (126.4 Hz)[c]	-

BMIM = 1-butyl,3-methylimidazolin-2-ylidene.
[a] Measured in CDCl$_3$.
[b] Measured in CD$_2$Cl$_2$.
[c] ^{13}C-^{31}P coupling constant in parentheses.

with an azolium or phosphonium salt, which leads to an *in situ* deprotonation of the latter followed by coordination of the generated ligand to the gold(I) center. Concurrently, one molecule of water is liberated and a cationic complex is formed, which is charge-balanced by the counter-anion of the initial salt precursors. The reaction is general and proceeds particularly well for phosphines and unsaturated NHCs with aromatic or aliphatic *N*-substituents (Table 6.4). Hetero-bis(carbene) complexes of saturated NHCs are more difficult to prepare, due to the increased tendency of imidazolidin-2-ylidenes to undergo hydrolytic ring-opening reactions. Thus, the complex [Au(IPr)(SIPr)]BF$_4$ complex could not be prepared, while [Au(IPr)(SIMes)]BF$_4$ was only obtained in a significantly lower yield compared to other analogs. Azolium salts with very bulky wing tips are generally less suitable for this reaction, as the IPr spectator already bears a significant steric bulk. Approach and deprotonation of a second very bulky azolium salt, such as IAd·HBF$_4$, is thus hampered preventing the formation of the elusive [Au(IPr)(IAd)]BF$_4$ complex.

Once obtained, the hetero-bis(NHC) and mixed NHC/phosphine complexes are stable and can be straightforwardly characterized. Table 6.4 summarizes the optimized conditions and selected ^{13}C NMR data of the complexes prepared. ^{13}C NMR measurements indicate that all the cationic complexes exhibit more downfield carbene chemical shifts compared to the neutral mono-IPr complexes bearing chorido or hydroxido ligands. Despite the cationic charge, the metal centers in the former are thus less Lewis acidic due to the donation from two strong donors. However, within these complexes a differentiation of the donating abilities between the varying second donors on grounds of the $^{13}C_{carbene}$ signal of the IPr ligand is not possible. Steric repulsion between two bulky ligands appears to diminish the electronic contributions to the metal-ligand bonds. For the BMIM complexes, the influence of the counter-anion on these shifts was also examined. Interestingly, the IPr carbene signals are not at all affected by the different anions, while the BMIM carbene resonances show small variations with a maximum chemical shift difference of $\Delta\delta = 0.7$ ppm. The notable difference between the two carbenes is difficult to explain.

The solid state molecular structures of all cationic complexes could be obtained by single crystal X-ray diffraction, and a few representatives are depicted in Figure 6.13. All complexes exhibit the expected linear coordination geometry. The gold-IPr distances

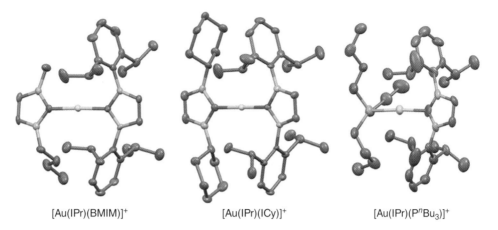

[Au(IPr)(BMIM)]⁺   [Au(IPr)(ICy)]⁺   [Au(IPr)(PⁿBu₃)]⁺

Figure 6.13
Solid state molecular structures of $[Au(IPr)(BMIM)]^+$, $[Au(IPr)(ICy)]^+$, and $[Au(IPr)(P^nBu_3)]^+$. Hydrogen atoms are omitted for clarity.

vary significantly from 2.003(7) to 2.08(1) Å depending on the steric hindrance and the electronic effect of the co-ligands. In hetero-bis(NHC) complexes, the twist angles between the two carbene planes range from 10° to 56° and was found to increase with the steric bulk of the second NHC. The largest angle was found for the SIMes complex, which is also the one most difficult to prepare.

By analogy to the above described methodology, mixed acetato-carbene gold(I) complexes can also be used as direct precursors to hetero-bis(NHC) complexes. This was first demonstrated by Huynh and coworkers using the 1,3-diisopropylbenzimidazolin-2-ylidene ($^iPr_2$-bimy) ligand. The preparation of the $[Au(OAc)(^iPr_2$-bimy)] complex is straightforward and only involves treatment of $[AuCl(^iPr_2$-bimy)] with silver acetate in dichloromethane, which affords the product as a white solid in 75% yield driven by the precipitation of silver chloride (Scheme 6.34, **A**) [40]. The compound is only moderately stable under aerobic and humid conditions, where it slowly decomposes to purple gold. Nevertheless, it can be kept as a properly dried solid under inert atmosphere for at least two weeks.

Reaction of the acetato complex with various azolium tetrafluoroborate or hexafluorophosphate salts derived from imidazole, triazole, benzimidazole, and even benzothiazole afforded the desired hetero-bis(NHC) complexes in high yields (Scheme 6.34, **B**). The reaction has also been extended to a dicationic triazolium bis(tetrafluoroborate) salt in a 2:1 ratio, which affords the hetero-tetrakis(NHC) digold complex $[Au_2(^iPr_2$-bimy$)_2(\mu$-ditz$)](BF_4)_2$ that contains the bridging Janus-type dicarbene ligand 1,2,4-triazolidine-3,5-diylidene (ditz).

In these reactions, the acetato ligand acts as the base to deprotonate the carbene precursor salts under formation of acetic acid. Dependent on the solubility of the azolium salts dichloromethane, acetone, or acetonitrile can be used as the reaction medium. The optimized conditions for the preparation of a series of hetero-bis(NHC) complexes and the hetero-tetrakis(NHC) complex obtained via this route are summarized in Table 6.5.

The NMR spectra of all compounds show the presence of two types of carbenes in the expected ratio. Of greater interest are the $^{13}C_{carbene}$ NMR resonances of the complexes that are collected in Table 6.6 together with two analogs bearing the non-classical Indy and FPyr [41] carbene ligands. Other than the IPr complexes discussed earlier, this series of $^iPr_2$-bimy hetero-bis(NHC) complexes is sterically less demanding. Therefore, the electronic contribution of the individual *trans*-co-ligands to the $^{13}C_{carbene}$ chemical shift of

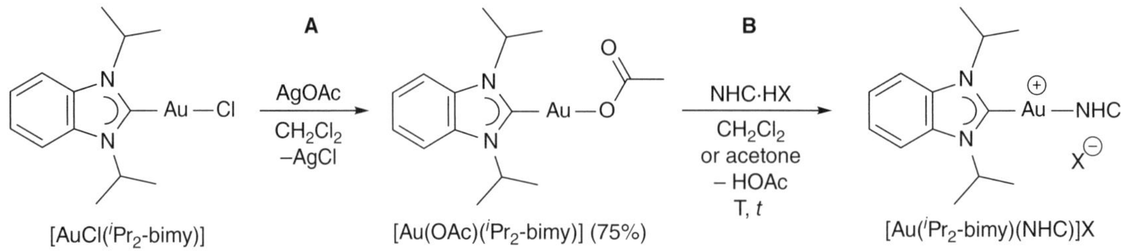

Scheme 6.34
Preparation an acetato-carbene gold complex as a direct precursor to hetero-bis(NHC) complexes.

Table 6.5   Optimized conditions for the Synthesis of [Au(^{i}Pr$_2$-bimy)(NHC)]X complexes.

Complex	Solvent	T [°]	t [h]	Yield [%]
[Au(^{i}Pr$_2$-bimy)(Bn$_2$-tazy)]BF$_4$	CH$_2$Cl$_2$	AT	5	80
[Au(^{i}Pr$_2$-bimy)(Bn$_2$-bimy)]BF$_4$	CH$_2$Cl$_2$	AT	24	82
[Au(^{i}Pr$_2$-bimy)(Bn-btzy)]BF$_4$	CH$_2$Cl$_2$	AT	16	76
[Au(^{i}Pr$_2$-bimy)(IPr)]BF$_4$	Acetone	80	48	88
[Au(^{i}Pr$_2$-bimy)(IPr)]PF$_6$	Acetone	80	48	86
[Au$_2$(^{i}Pr$_2$-bimy)$_2$(μ-ditz)](BF$_4$)$_2$	CH$_3$CN	AT	3	48

Bn$_2$-tazy = 1,4-dibenzyl-1,2,4-triazolin-5-ylidene; Bn$_2$-bimy = 1,3-dibenzylbenzimidazolin-2-ylidene;
Bn-btzy = 3-benzylbenzothiazolin-2-ylidene; ditz = 1,2,4-triazolidine-3,5-diylidene.

Table 6.6 ^{13}C$_{carbene}$ signals of hetero-bis(NHC) and -tetrakis(NHC) gold(I) complexes.

Complex	^{13}C(^{i}Pr$_2$-bimy) [ppm][a]	^{13}C$_{NHC2}$ [ppm][a]
[Au(^{i}Pr$_2$-bimy)(Bn$_2$-tazy)]BF$_4$	186.6	186.7
[Au(^{i}Pr$_2$-bimy)(Bn$_2$-bimy)]BF$_4$	187.6	191.5
[Au(^{i}Pr$_2$-bimy)(Bn-btzy)]BF$_4$	186.2	214.8
[Au(^{i}Pr$_2$-bimy)(IPr)]BF$_4$	186.7	187.3
[Au(^{i}Pr$_2$-bimy)(IPr)]PF$_6$	186.7	187.3
[Au$_2$(^{i}Pr$_2$-bimy)$_2$(μ-ditz)](BF$_4$)$_2$	185.6	189.4
[Au(^{i}Pr$_2$-bimy)(Indy)]BF$_4$	190.8	182.4
[Au(^{i}Pr$_2$-bimy)(FPyr)]PF$_6$ [41]	192.5	179.6

Indy = 6,7,8,9-tetrahydropyridazino[1,2-a] indazolin-3-ylidene; FPyr = 1,2,3,4,6,7,8,9-octahydropyrida-zino
[1,2-a]-indazolin-11-ylidene.
[a] Measured in CDCl$_3$ and internally referenced to the solvent signal at 77.7 ppm relative to TMS.

the ^{i}Pr$_2$-bimy reporter carbene is less diluted and thus more reflective of their donating abilities. This study also reveals that there was no detectable influence of BF$_4^-$ versus PF$_6^-$ counter-anions on the chemical shift of the carbene atoms of both ligands.

Figure 6.14 shows a plot of the seven NHCs and the ^{13}C$_{carbene}$ chemical shift values they induce in the ^{i}Pr$_2$-bimy ligand. The scale indicates that the latter ligand can be employed as a ^{13}C NMR spectroscopic probe to determine the donor strengths of the *trans*-standing co-ligands, whereby stronger donating NHCs would cause a more downfield shift. Thus, the weakest triazolidinediylidene (ditz) ligand leads to the most upfield ^{i}Pr$_2$-bimy carbene resonance, while the strongest pyrazole-derived ligand (FPyr) gives rise to the most

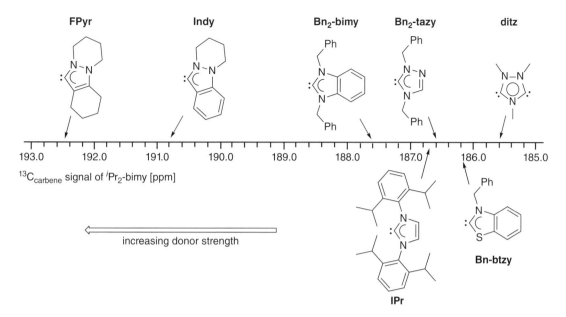

Figure 6.14
^{13}C NMR based donor strength scale obtained from [Au(^{i}Pr$_2$-bimy)(NHC)]BF$_4$ complexes.

downfield signal. Formal "oxidation" of FPyr by removal of four hydrogen atoms affords the Indy ligand, which is weaker donating, as expected. Nevertheless, both of these non-classical carbene donors lack a second electron-withdrawing nitrogen atom making them stronger donating compared to their classical counterparts.

The ^{13}C$_{carbene}$ NMR data obtained for these linear hetero-bis(NHC) gold(I) complexes [Au(^{i}Pr$_2$-bimy)(NHC)]BF$_4$ have also been also subjected to a linear regression with those found for the square planar hetero-bis(NHC) palladium(II) complexes *trans*-[PdBr$_2$(^{i}Pr$_2$-bimy)(NHC)] bearing the same set of carbenes. Notably, a very good regression coefficient ($R^2 = 0.98$) was obtained for correlation between the ^{13}C NMR values of these two metals, indicating a very strong linear regression. Inter–conversion between the gold(I) and palladium(II) systems can be easily achieved by applying the equation $[Pd] = 1.19[Au] - 45.0$, where [Pd] and [Au] are the ^{13}C$_{carbene}$ chemical shifts values in ppm for the ^{i}Pr$_2$-bimy reporter ligand (Figure 6.15).

The introduction of gold(I) as an alternative metal center broadens the scope of the ^{13}C NMR based electronic parameter as a greater choice of suitable complex probes is available. It was also found that the solubility of gold(I) hetero-bis(NHC) complexes in CDCl$_3$ is significantly better than that of their palladium(II) congeners, which allows for a much faster detection of the ^{i}Pr$_2$-bimy ^{13}C NMR signals. Nevertheless, the ^{13}C$_{carbene}$ values detected for the gold(I) complexes fall in a narrower region compared to those in palladium(II) analogs, which may suggest a reduced resolution of the gold system with respect to donor strength evaluation of *trans*-standing NHC ligands.

The solid state molecular structures for some complexes could be obtained by single crystal X-ray analysis, and two representatives are depicted in Figure 6.16. While the gold–carbene distances are found in the expected range, it is interesting to note that the less sterically encumbered complex [Au(^{i}Pr$_2$-bimy)(Bn$_2$-tazy)]BF$_4$ exhibits a larger twist angle between the two carbene planes of 44° as compared to that found for the IPr complex [Au(^{i}Pr$_2$-bimy)(IPr)]BF$_4$ of only 11°.

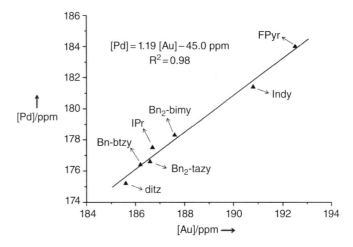

Figure 6.15
Correlation of ${}^{i}Pr_2$-bimy ${}^{13}C_{carbene}$ NMR values of $[Au({}^{i}Pr_2$-bimy)(NHC)]BF$_4$ and *trans*-[PdBr$_2$(${}^{i}Pr_2$-bimy)(NHC)] complexes.

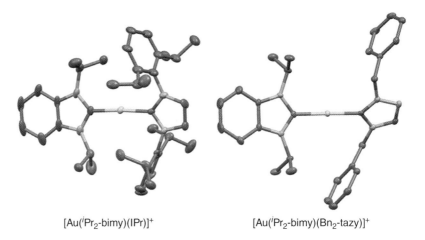

$[Au({}^{i}Pr_2$-bimy)(IPr)]$^+$ $[Au({}^{i}Pr_2$-bimy)(Bn$_2$-tazy)]$^+$

Figure 6.16
Solid state molecular structures of hetero-bis(NHC) complex cations $[Au({}^{i}Pr_2$-bimy)(IPr)]$^+$ and $[Au({}^{i}Pr_2$-bimy)(Bn$_2$-tazy)]$^+$. Hydrogen atoms are omitted for clarity.

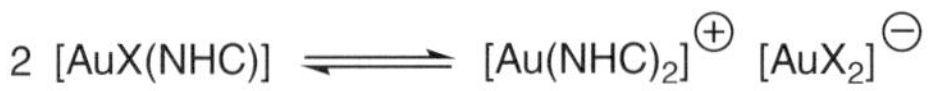

$$2\ [AuX(NHC)] \rightleftharpoons [Au(NHC)_2]^{\oplus}\ [AuX_2]^{\ominus}$$

Scheme 6.35
Autoionization of neutral monocarbene gold complexes to give complex salts.

6.3.5   Ligand Redistribution and Autoionization of Gold(I) NHC Complexes

In general, neutral monocarbene gold(I) complexes with an anionic ligand X of the formula [AuX(NHC)] are observed to undergo autoionization (or ligand disproportionation) to give ion pairs consisting of bis(NHC)gold(I) cations [Au(NHC)$_2$]$^+$ and aurate(I) anions [AuX$_2$]$^-$ in the gas phase (Scheme 6.35).

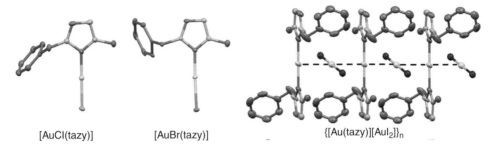

[AuCl(tazy)]   [AuBr(tazy)]   {[Au(tazy)][AuI$_2$]}$_n$

Figure 6.17
Molecular structures of halido-tazy gold(I) complexes in the solid state. Hydrogen atoms are omitted for clarity.

Therefore ions associated with mono-NHC-gold fragments are rarely observed in the mass spectra of such complexes. Instead, [Au(NHC)$_2$]$^+$ cations are observed in almost all cases in the positive mode mass spectra. This behavior is reminiscent of silver-NHC species, which show coordination isomerization not only in gas phase, but also in solution and upon crystallization. In contrast, NMR and UV spectroscopic analyses provide evidence that [AuX(NHC)] complexes retain their linear and neutral structure in solution. In certain rare cases, autoionization has been reported to occur upon crystallization of halido complexes. The tendency for this behavior appears to depend on the donating ability of the carbene and/or the halido ligand. For example, exceptionally strong donors such as the *non-classical* NHCs CAAC [42] and Indy [43] have been reported to induce such disproportionation reactions. Among the halido ligands, iodido complexes appear to be more prone to disproportionate, which is consistent with the larger stability constant for the [AuI$_2$]$^-$ complex anion ($K = 23.8$) compared to that of [AuCl$_2$]$^-$ ($K = 20.2$) and [AuBr$_2$]$^-$ ($K = 20.6$), respectively [44].

For example, this was even noted in halido-gold complexes [AuX(tazy)] (X = Cl, Br, I) bearing a relatively weakly donating 1,2,4-triazolin-5-ylidene (tazy) ligand with non-bulky methyl and benzyl wing tips [45]. In solution, all three heteroleptic complexes are neutral and linear. While the chlorido and bromido complexes remain discrete molecules in the solid state, the iodido complex autoionizes and forms a polymeric chain of ion pairs held together by aurophilic interactions with an intergold distance of ~3.316 Å (Figure 6.17).

Interestingly, the autoionization is reversible. Dissolving crystals of the polymerization isomer [Au(tazy)$_2$][AuI$_2$] leads to the reformation of the neutral and mononuclear species [AuI(tazy)] in accordance with Scheme 6.35. In addition to NMR and UV spectroscopy, the presence of the latter in solution was also proven by subsequent oxidative addition of iodine, which solely afforded the monocarbene complex [AuI$_3$(tazy)].

Ligand redistribution can also occur in cationic hetero-bis(NHC) gold(I) complexes, which are generally stable as solids [40]. However, detailed NMR studies revealed that some of them decompose slowly in solution to other NHC complexes and the respective azolium salts. It was found that irreversible ligand redistribution occurs, which leads to the formation of homo-bis(NHC) complexes as outlined in Scheme 6.36 (see also Scheme 6.25). Compared to the similar process observed in neutral monocarbene complexes, the charge remains unchanged, and thus the term "autoionization" is less appropriate for hetero-bis(NHC) complexes.

Interestingly, such reactivity has only been observed for complexes with sterically unassuming carbenes of medium to weak donating abilities, that is, bimy, tazy, btzy, and ditz.

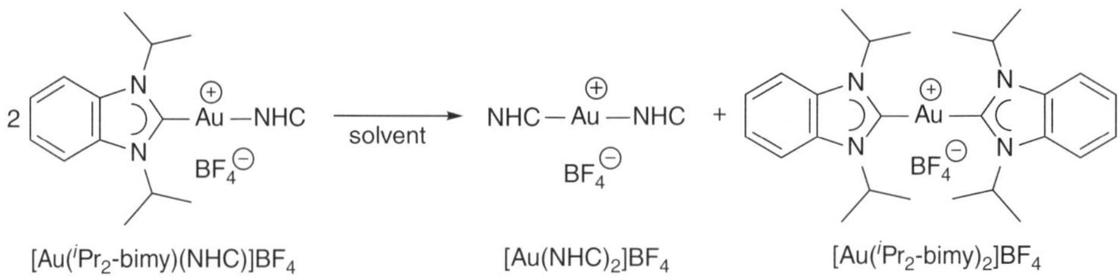

Scheme 6.36
Ligand redistribution reactions of hetero-bis(NHC) gold(I) complexes.

The reason for this process is believed to be the presence of two or more different NHCs with markedly different donating abilities and *trans* influences, resulting in Au–C$_{carbene}$ bonds of rather different strengths. The thermodynamically stable rearrangement products, on the other hand, are homoleptic complexes with only a single type of Au–C$_{carbene}$ bonds of equal strength, which makes them more stable compared to their kinetically controlled heteroleptic counterparts. The detection of azolium salts as hydrolysis compounds of free NHCs points to a dissociative rearrangement pathway involving de- and re-coordination of carbene ligands, but the exact mechanism remains to be explored in future.

In contrast, no homoleptic rearrangement was observed for complexes bearing the IPr, FPyr, and Indy ligands (Figure 6.14). It is intuitive to conclude that the steric bulk of the IPr ligand prevents formation of bis(IPr) complexes. The non-classical carbenes FPyr and Indy, on the other hand, are not bulky but very strong donors. The increased electron density in these two complexes may result in stronger Au–carbene bonds due to enhanced π-back donation, which are non-negligible in d^{10} complexes, hindering further ligand redistribution.

6.4   Gold(III)-NHC Complexes

In addition to the abundance of gold(I) NHC complexes, gold(III) complexes of carbenes have also received increasing attention. In contrast to its lighter homologues copper and silver, the + III oxidation state is easily accessible for gold, and the majority of the respective NHC complexes are even very stable to air and moisture making their handling an ease.

Pioneering work was reported by Raubenheimer and coworkers in 1997, who adapted a methodology already known for gold(III) complexes bearing acyclic carbenes [46]. The first gold(III) NHC complexes were thus prepared by direct oxidative addition of dihalogens (Cl$_2$, Br$_2$, I$_2$) to the cationic bis(carbene) gold(I) complex [Au(IMe)$_2$]CF$_3$SO$_3$ in dichloromethane (Scheme 6.37). Even nowadays, this pathway remains the most convenient and common route to gold(III) NHC complexes. The oxidants were introduced at low temperatures followed by slow warming to room temperature, which eventually gave the desired gold(III) complexes [AuX$_2$(IMe)$_2$]CF$_3$SO$_3$ (X = Cl, Br, I) in moderate to good yields ranging from 63 to 90%. The two-electron oxidation of the electron-rich linear d^{10} system with the strong oxidants leads to *trans*-configured square planar d^8 complexes bearing two new halido ligands in addition to the two carbenes already present in the gold(I) precursor. The *trans* arrangement was confirmed by a single crystal X-ray analysis on [AuCl$_2$(IMe)$_2$] CF$_3$SO$_3$ as a representative. All three bis(carbene)dihalido complexes are air-stable solids

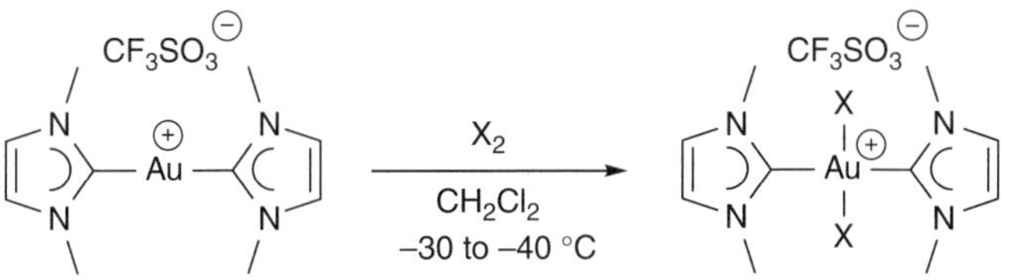

X	Yield [%]	Colour
Cl	90	white
Br	63	yellow
I	84	orange

Scheme 6.37
Bis(NHC) gold(III) complexes via oxidative addition of dihalogens.

with increasing coloration going from chlorido, to bromide, to iodido. Moreover, they exhibit good solubilities in tetrahydrofuran, acetone, and dichloromethane.

Upon oxidation, most of the ^1H NMR signals of the carbene ligands shift downfield. More diagnostic is the significant upfield shift of $\Delta\delta = 31$–40 ppm to 145.6 (I), 154.1 (Br) and 154.5 (Cl) ppm, respectively, observed for the ^{13}C NMR resonances of the carbene atoms. This is a common phenomenon generally observed for such Au^I to Au^{III} conversions and can be explained by the greater Lewis acidity of the gold(III) center bound by the respective carbene donor. Within this series, an upfield trend of the respective ^{13}C NMR signals was observed in the order Cl < Br < I, whereby the *cis*-standing iodido ligands have the greatest shielding effect on the carbene atoms due to their largest electron clouds (see Table 6.8).

In this work, attempts were also made to prepare monocarbene gold(III) complexes, which led to the isolation of a less common trichlorido-thiazolinylidene gold(III) complex. Ten years later, Nolan and coworkers expanded the range of gold(III) NHC complexes based on this discovery by preparation and detailed characterization of various monocarbene-tribromido gold(III) complexes containing the more common NHCs derived from imidazole and imidazoline. Initial attempts to gain access to such compounds involved reaction of the free carbenes with simple gold(III) salts containing the tetrachloridoaurate(III) complex anion $[AuCl_4]^-$. Due to the strongly oxidizing nature of the precursor, only gold(I)-NHC complexes were obtained along with metallic gold and decomposition products of free carbenes. A better strategy involves oxidative treatment of bromido-carbene gold(I) complexes in dichloromethane with liquid bromine, which was chosen over chlorine gas for the ease of handling (Scheme 6.38).

The bromido complex precursors were prepared as described by halido exchange with excess lithium bromide. With this method seven tribromido-monocarbene complexes were prepared in very good yields and isolated as yellow to orange solids that are all stable to air and moisture (Table 6.7). Most of the complexes could easily be obtained by reactions at room temperature, but interestingly, complexes with IMes and SIMes ligands had to be prepared at −78 °C to avoid decomposition.

As for the cationic bis(NHC) counterparts, significant upfield shifts for the $^{13}C_{carbene}$ signals of $\Delta\delta = 24$–38 ppm were observed upon oxidative addition, which is in line with greater Lewis acidic gold(III) centers. For the neutral tribromido gold(III) complexes, these were found in the range 132.9–146.2 ppm for the imidazolin-2-ylidene series and around 173 ppm for the imidazolidin-2-ylidene analogs.

The molecular structures of the thermally stable complexes were also confirmed by X-ray analysis, and three representatives are shown in Figure 6.18. As expected for four-coordinated d^8 metal complexes, the geometry was found to be square planar. The gold centers are each coordinated by one NHC and three anionic bromido ligands, whereby the

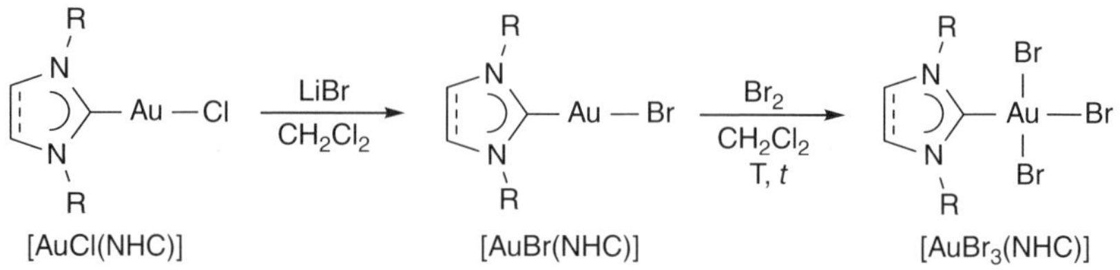

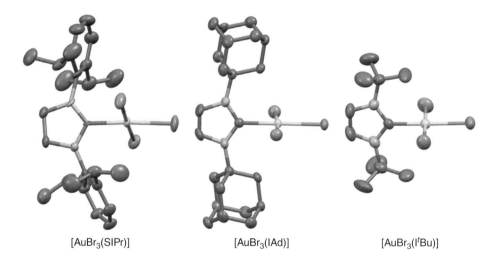

Scheme 6.38
Preparation of [AuBr$_3$(NHC)] complexes.

Table 6.7 Selected reaction conditions and yields for [AuBr$_3$(NHC)] complexes.

NHC	T [°C]	t [min]	Yield [%]
IPr	25	60	84
IMes	–78	20	94
SIPr	25	60	97
SIMes	–78	20	94
ICy	25	60	89
IAd	25	60	95
I^{t}Bu	25	60	93

[AuBr$_3$(SIPr)]   [AuBr$_3$(IAd)]   [AuBr$_3$(I^{t}Bu)]

Figure 6.18
Solid state molecular structures of selected tribromido-NHC gold(III) complexes. Hydrogen atoms are omitted for clarity.

carbene plane is situated almost perpendicular to the [AuBr$_3$C] coordination plane. All gold–carbene distances lie in the range of ~2.01–2.05 Å, which is not significantly different from those observed for the neutral gold(I) complexes.

Oxidative addition of dihalogens has also been conducted on cationic hetero-bis(NHC) gold(I) complexes to give the respective hetero-bis(NHC) gold(III) analogs. This was conducted by Huynh and coworkers to provide additional evidence for the existence of

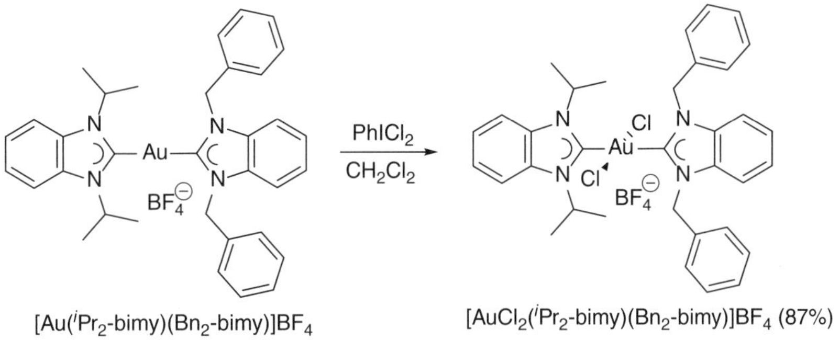

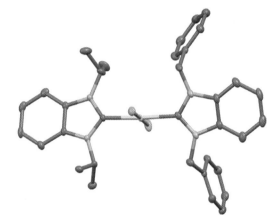

Scheme 6.39
Preparation of the hetero-bis(NHC) gold(III) complex [AuCl$_2$(^{i}Pr$_2$-bimy)(Bn$_2$-bimy)]BF$_4$.

Figure 6.19
Molecular structure of the hetero-bis(NHC) gold(III) complex cation [AuCl$_2$(^{i}Pr$_2$-bimy)(Bn$_2$-bimy)]$^+$. Hydrogen bonds are omitted for clarity.

hetero-bis(NHC) complexes of gold(I), which underwent slow ligand redistribution in solution [40]. For example, the mixed carbene complex [Au(^{i}Pr$_2$-bimy)(Bn$_2$-bimy)]BF$_4$ was treated in dichloromethane with iodobenzene dichloride (PhICl$_2$) as a more convenient source for chlorine. The reaction takes place at ambient temperature and gave the gold(III) complex [AuCl$_2$(^{i}Pr$_2$-bimy)(Bn$_2$-bimy)]BF$_4$ in a high yield of 87% (Scheme 6.39). Unlike its gold(I) precursor, the complex is stable and does not undergo ligand redistribution in solution even on prolonged standing for days. The increased Lewis acidity of gold(III) versus gold(I) likely leads to stronger gold(III)–carbene bonds, stabilizing the hetero-bis(NHC) gold(III) complex and thus preventing ligand disproportionation.

Unambiguous confirmation for the identity of the hetero-bis(carbene) gold(III) complex was provided by X-ray crystallography diffraction on single crystals. The molecular structure depicted in Figure 6.19 clearly shows two different carbenes coordinated to the square planar gold(III) center, with the coordination site completed by the two new chlorido ligands.

Oxidative additions involving the three common halogens on mono- and bis(carbene) gold(I) complexes bearing the identical ligand were conducted in order to shed more light

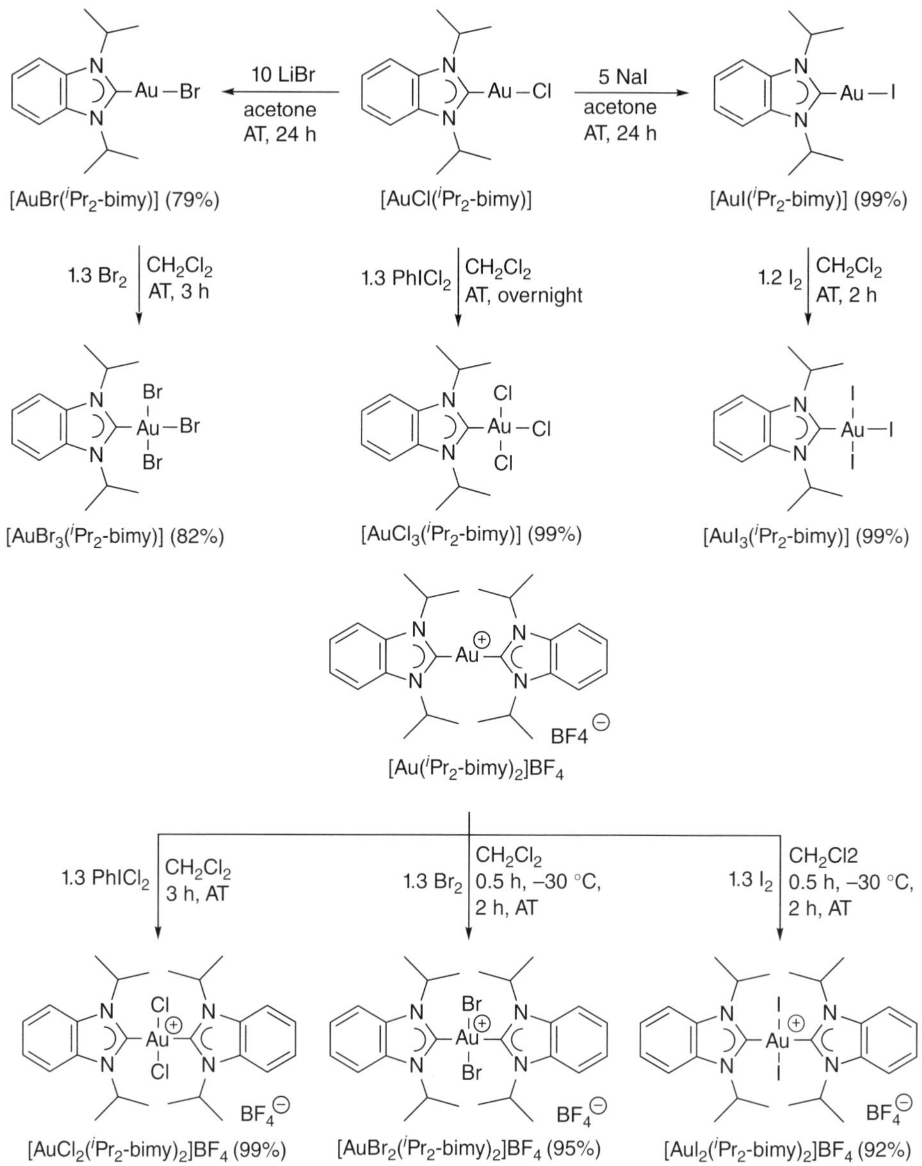

Scheme 6.40
Preparation of a complete series of halido-^{i}Pr$_2$-bimy gold(I) and gold(III) complexes.

on the structural and electronic trends observed in gold(I) and gold(III) NHC complexes [47]. All complexes for this purpose were prepared from the parent complex [AuCl(^{i}Pr$_2$-bimy)], which in turn was obtained by silver-carbene transfer as described in the gold(I) section. Simple halido exchange afforded the bromido and iodido analogs [AuBr(^{i}Pr$_2$-bimy)] and [AuI(^{i}Pr$_2$-bimy)], respectively (Scheme 6.40). In addition, the bis(NHC) complex [Au(^{i}Pr$_2$-bimy)$_2$]BF$_4$ was prepared by treatment of the parent chlorido complex with additional benzimidazolium salt and potassium carbonate. All four gold(I) complexes were subjected to sources of chlorine, bromine, and iodine for oxidative additions. Overall, a complete

Table 6.8 Comparison of the diagnostic ^{1}H and ^{13}C NMR data in ppm.[a]

Complex	CH(CH$_3$)$_2$	^{13}C$_{carbene}$ ($\Delta\delta$)[d][ppm]
[Au^{I}Cl(^{i}Pr$_2$-bimy)] **(1)**	5.48	175.8
[Au^{I}Br(^{i}Pr$_2$-bimy)] **(2)**	5.49	180.0
[Au^{I}I(^{i}Pr$_2$-bimy)] **(3)**	5.51	186.4
[Au^I(^{i}Pr$_2$-bimy)$_2$]BF$_4$ **(4)**	5.39	187.4
[Au^{III}Cl$_3$(^{i}Pr$_2$-bimy)] **(5)**	5.51	150.8 (25.0)
[Au^{III}Br$_3$(^{i}Pr$_2$-bimy)] **(6)**	5.40	148.8 (31.2)
[Au^{III}I$_3$(^{i}Pr$_2$-bimy)] **(7)**	5.21	146.1 (40.3)
[Au^{III}Cl$_2$(^{i}Pr$_2$-bimy)$_2$]BF$_4$ **(8)**[b]	5.38	159.2 (28.2)
[Au^{III}Br$_2$(^{i}Pr$_2$-bimy)$_2$]BF$_4$ **(9)**[b]	5.24	156.3 (31.1)
[Au^{III}I$_2$(^{i}Pr$_2$-bimy)$_2$]BF$_4$ **(10)**[c]	5.06	150.9 (36.5)

[a] measured in CDCl$_3$ and referenced to the solvent signal at 77.7 ppm.
[b] measured in DMSO-d_6.
[c] measured in CD$_2$Cl$_2$.
[d] $\Delta\delta$ is the chemical shift difference to the respective precursor gold(I) complex.

series of ten mono- and bis(NHC) complexes of gold(I) and gold(III) with all three types of common halido ligands were obtained, which were studied by NMR and UV-Vis spectroscopies, ESI mass spectrometry, cyclic voltammetry, and X-ray single crystal diffraction. All complexes are air and moisture stable and well soluble in common organic solvents with exception from the non-polar ones, such as hexane, diethylether, and toluene.

Overall, it was found that the ^{13}C$_{carbene}$ signal of the ^{i}Pr$_2$-bimy ligand is the most useful diagnostic tool for the differentiation of all gold complexes. It is highly sensitive to the oxidation state of the metal, the donor strength of the *trans*-standing halido co-ligands in gold(I) complexes, while it is also sensitive to the shielding effect of the two *cis*-standing halido moieties in square planar gold(III) complexes. Table 6.8 summarizes key NMR spectroscopic data for all complexes.

In general, the following trends are observed:

1. In [Au^{I}X(^{i}Pr$_2$-bimy)] complexes a gradual *downfield* shift of the ^{13}C$_{carbene}$ signal is observed in the order Cl<Br<I reflecting the increasing donor ability of the *trans* halido ligands.
2. In the square planar complexes [Au^{III}X$_3$(^{i}Pr$_2$-bimy)] and *trans*-[Au^{III}X$_2$(^{i}Pr$_2$-bimy)$_2$]BF$_4$ a gradual *upfield* shift of the ^{13}C$_{carbene}$ signal in the order Cl>Br>I is noted reflecting the increasing electron density (i.e. shielding effect) of the *cis*-halido ligands.
3. The ^{13}C$_{carbene}$ signals of bis(NHC) Au^I/Au^{III} complexes are always more *downfield* than those of their mono-NHC counterparts due to an increased electron density (i.e. a decreased Lewis acidic metal center) resulting from the coordination of a second NHC.
4. A significant *upfield* shift of the ^{13}C$_{carbene}$ signal is observed upon oxidative addition to an gold(I) precursor complex, which is due to increased Lewis acidity of the resulting gold(III) center. The chemical shift difference ($\Delta\delta$) between the resulting gold(III) complex and its gold(I) precursor increases in the order Cl<Br<I.

UV–Vis absorption analyses of all gold(III) complexes revealed a LMCT process from halido co-ligands to gold(III) centers, the absorption energy of which is found to be correlated to the respective halido and the number of carbene ligands. This is most striking for

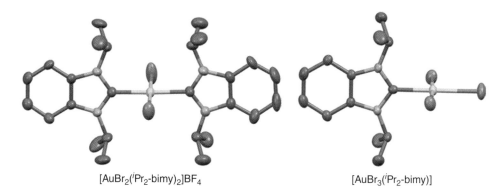

[AuBr₂(ⁱPr₂-bimy)₂]BF₄   [AuBr₃(ⁱPr₂-bimy)]

Figure 6.20
Solid state molecular structures of [AuBr$_2$(^{i}Pr$_2$-bimy)$_2$]$^+$ and [AuBr$_3$(^{i}Pr$_2$-bimy)]. Hydrogen atoms are omitted for clarity.

[Au^{III}X$_3$(^{i}Pr$_2$-bimy)] complexes where a bathochromic shift from 301 nm (X=Cl) to 351 nm (X=Br) and finally to 405 nm (X=I) was observed. This red shift indicates an increasingly easier LMCT associated with higher-energy HOMOs, which in turn are induced by an increase in electron donating ability in the order Cl<Br<I.

In addition, cyclic voltammetric studies showed irreversible Au^{III}/Au^{I} reductions for all gold(III) complexes, where the potentials are found to be sensitive to the number of carbene ligands. The cationic bis(NHC) complexes exhibit potentials of approximately –1.09 V, while the neutral monocarbene counterparts showed the redox process at more negative potential ranging from –1.12 to –1.16 V. Moreover, a quasi-reversible redox process of the Au^{I}/Au^0 couple without a clear halido-dependent trend was observed for all gold complexes between –0.72 and –0.85 V.

The three complexes of each series, [Au^{I}X(^{i}Pr$_2$-bimy)], [Au^{III}X$_3$(^{i}Pr$_2$-bimy)] and [Au^{III}X$_2$(^{i}Pr$_2$-bimy)$_2$]BF$_4$ (X=Cl, Br, I), are essentially isostructural in both solution and solid state. Thus, only two gold(III) representatives are shown in Figure 6.20. Notably, the gold–carbene distances in all complexes including those in the homoleptic [Au^{I}(^{i}Pr$_2$-bimy)$_2$]BF$_4$ complex differ insignificantly and range from 1.973–2.048 Å regardless of the oxidation state and the number of carbenes.

References

[1] J. C. Y. Lin, R. T. W. Huang, C. S. Lee, A. Bhattacharyya, W. S. Hwang, I. J. B. Lin, *Chem. Rev.* **2009**, *109*, 3561.
[2] (a) I. J. B Lin, C. S. Vasam, *Comments Inorg. Chem.* **2004**, *23*, 534. (b) J. C. Garrison, W. J. Youngs, *Chem. Rev.* **2005**, *105*, 3978. (c) I. J. B. Lin, C. S. Vasam, *Coord. Chem. Rev.* **2007**, *251*, 642.
[3] A. Kascatan-Nebioglu, M. J. Panzner, C. A. Tessier, C. L. Cannon, W. J. Youngs, *Coord. Chem. Rev.* **2007**, *251*, 884.
[4] C. Boehme, G. Frenking, *Organometallics* **1998**, *17*, 5801.
[5] A. J. Arduengo, H. V. Rasika Dias, J. C. Calabrese, F. Davidson, *Organometallics* **1993**, *12*, 3405.
[6] (a) H. G. Raubenheimer, S. Cronje, P. H. van Rooyen, P. J. Olivier, J. G. Toerien, *Angew. Chem. Int. Ed. Engl.* **1994**, *33*, 672. (b) H. G. Raubenheimer, S. Cronje, P. J. Olivier, *J. Chem. Soc., Dalton Trans.* **1995**, 313.
[7] V. Jurkauskas, J. P. Sadighi, S. L. Buchwald, *Org. Lett.* **2003**, *14*, 2417.

[8] H. Kaur, F. K. Zinn, E. D. Stevens, S. P. Nolan, *Organometallics* **2004**, *23*, 1157.

[9] S. Díez-González, H. Kaur, F. K. Zinn, E. D. Stevens, S. P. Nolan, *J. Org. Chem.* **2005**, *70*, 4784.

[10] S. Díez-González, N. M. Scott, S. P. Nolan, *Organometallics* **2006**, *25*, 2355.

[11] S. Díez-González, E. D. Stevens, N. M. Scott, J. L. Petersen, S. P. Nolan, *Chem. Eur. J.* **2008**, *14*, 158.

[12] A. A. D. Tulloch, A. A. Danopoulos, S. Kleinhenz, M. E. Light, M. B. Hursthouse, G. Eastham, *Organometallics* **2001**, *20*, 2027.

[13] J. Chun, H. S. Lee, I. G. Jung, S. W. Lee, H. J. Kim, S. U. Son, *Organometallics* **2010**, *29*, 1518.

[14] P. L. Arnold, A. C. Scarisbrick, A. J. Blake, C. Wilson, *Chem. Commun.* **2001**, 2340.

[15] C. Gibard, D. Avignant, F. Cisnetti, A. Gautier, *Organometallics* **2012**, *31*, 7902.

[16] B. Liu, Q. Xia, W. Chen, *Angew. Chem. Int. Ed.* **2009**, *48*, 5513.

[17] B. Liu, X. Ma, F. Wu, W. Chen, *Dalton Trans.* **2015**, *44*, 1836.

[18] M. C. Chung, *Bull. Korean Chem. Soc.* **2002**, *23*, 921.

[19] P. de Frémont, N. M. Scott, E. D. Stevens, T. Ramnial, O. C. Lightbody, C. L. B. Macdonald, J. A. C. Clyburne, C. D. Abernethy, S. P. Nolan, *Organometallics* **2005**, *24*, 6301.

[20] O. Guerret, S. Solé, H. Gornitzka, G. Trinquier, G. Bertrand, *J. Am. Chem. Soc.* **1997**, *119*, 6668.

[21] H. M. J. Wang, I. J. B. Lin, *Organometallics* **1998**, *17*, 972.

[22] A. A. D. Tulloch, A. A. Danopoulos, S. Winston, S. Kleinhenz, G. Eastham, *J. Chem. Soc., Dalton Trans.* **2000**, 4499.

[23] The original synthesis was conducted under phase-transfer conditions employing NaOH/ NBu$_4$PF$_6$ for the deprotonation of the benzimidazolium salt in addition to Ag$_2$O. Nowadays, we know such conditions are not a necessity for the preparation of bis(NHC) silver complexes.

[24] J. C. C. Chen, I. J. B. Lin, *Organometallics* **2000**, *19*, 5113.

[25] V. J. Catalano, M. A. Malwitz, A. O. Etogo, *Inorg. Chem.* **2004**, *43*, 5714.

[26] B. Bildstein, M. Malaun, H. Kopacka, K. Wurst, M. Mitterböck, K.-H. Ongania, G. Opromolla, P. Zanello, *Organometallics* **1999**, *18*, 4325.

[27] B. Çetinkaya, P. Dixneuf, M. F. Lappert, *J. Chem. Soc., Dalton Trans.* **1974**, 1827.

[28] In the same year Fehlhammer and coworkers reported gold(I) and gold(III) complexes bearing a N,O-NHC ligands by spontaneous intramolecular cyclization of functionalized isocyanide complexes: W. P. Fehlhammer, K. Bartel, *Angew. Chem. Int. Ed. Engl.* **1974**, *13*, 599.

[29] F. Bonati, A. Burini, B. R. Pietroni, B. Bovio, *J. Organomet. Chem.* **1989**, *375*, 147.

[30] H. G. Raubenheimer, L. Lindeque, S. Cronje, *J. Organomet. Chem.* **1996**, *511*, 177.

[31] R. Fränkel, J. Kniczek, W. Ponikwar, H. Nöth, K. Polborn, W. P. Fehlhammer, *Inorg. Chim. Acta* **2001**, *312*, 23.

[32] S. Singh, S. S. Kumar, V. Jancik, H. W. Roesky, H.-G. Schmidt, M. Noltemeyer, *Eur. J. Inorg. Chem.* **2005**, 3057.

[33] P. de Frémont, N. M. Scott, E. D. Stevens, S. P. Nolan, *Organometallics* **2005**, *24*, 2411.

[34] C. Böhler, D. Stein, N. Donati, H. Grützmacher, *New J. Chem.* **2002**, *26*, 1291.

[35] H. M. J. Wang, C. Y. L. Chen, I. J. B. Lin, *Organometallics* **1999**, *18*, 1216.

[36] H. M. J. Wang, C. S. Vasam, T. Y. R. Tsai, S.-H. Chen, A. H. H. Chang, I. J. B. Lin, *Organometallics* **2005**, *24*, 486.

[37] M. V. Baker, P. J. Barnard, S. K. Brayshaw, J. L. Hickey, B. W. Skelton, A. H. White, *Dalton Trans.* **2005**, 37.

[38] F. Nahra, S. R. Patrick, A. Collado, S. P. Nolan, *Polyhedron* **2014**, 84, 59 and references therein.

[39] S. Gaillard, P. Nun, A. M. Z. Slawin, S. P. Nolan, *Organometallics* **2010**, *29*, 5402.

[40] S. Guo, H. Sivaram, D. Yuan, H. V. Huynh, *Organometallics* **2013**, *32*, 3685.

[41] H. Sivaram, J. Tan, H. V. Huynh, *Organometallics*, **2012**, *31*, 5875.

[42] G. D. Frey, R. D. Dewhurst, S. Kousar, B. Donnadieu, G. Bertrand, *J. Organomet. Chem.* **2008**, *693*, 1674.

[43] H. Sivaram, R. Jothibasu, H. V. Huynh, *Organometallics* **2012**, *31*, 1195.

[44] R. Roulet, N. Q. Lan, W. R. Mason, G. P. Jr. Fenske, *Helv. Chim. Acta* **1973**, *56*, 2405.

[45] S. Guo, J. C. Bernhammer, H. V. Huynh, *Dalton Trans.* **2015**, *44*, 15157.

[46] H. G. Raubenheimer, P. J. Olivier, L. Lindeque, M. Desmet, J. Hrušak, G. J. Kruger, *J. Organomet. Chem.* **1997**, *544*, 91.

[47] H. V. Huynh, S. Guo, W. Wu, *Organometallics* **2013**, *32*, 4591.

7 Ruthenium, Rhodium, and Iridium Metal-NHC Complexes

In addition to group 10 and 11 transition metals, NHC complexes of ruthenium, rhodium, and iridium are also very common. Interest in them is primarily based on the search for highly active catalysts for various organic transformations. Compared to many other transition metals, these three are known to easily activate C–H bonds. In relation to this, they tend to form more stable hydrido complexes, which is of great interest for the realization of various important catalytic redox processes. Among them, ruthenium stands out due to its relatively lower cost, while still exhibiting high catalytic activities, particularly for olefin metathesis. Many ruthenium complexes are also explored for their luminescent properties. Rhodium, on the other hand, has the advantage that it is naturally composed of only one isotope, that is, ^{103}Rh (100%), which is at the same time also NMR active with a nuclear spin of ½. This allows for much more detailed and elaborate studies in terms of characterizations providing deeper mechanistic insights. The organometallic chemistry of iridium is very similar to that of rhodium, and thus almost the same synthetic strategies can be applied. Despite this similarity, NHC complexes of each metal have their own uniqueness in terms of properties and catalytic activities. Iridium-NHC complexes are particularly active in hydrogenations. Moreover, many iridium compounds offer unique photophysical properties and are thus explored for their application as phosphors.

7.1 Ruthenium(II)-NHC Complexes

7.1.1 Cleavage of Enetetramines and the Free Carbene Route

Cleavage of Entetramines

One of the earliest successful and deliberate attempts in making a ruthenium NHC complex was communicated by Lappert and coworkers as early as 1976 [1]. Their prime focus was the reactivity study of electron-rich carbene dimers with transition metals. Thus heating dichlorido-tris(triphenylphosphine)ruthenium(II) with an excess of such entetramines (2.5 equiv) containing N-methyl or N-ethyl substituents in xylene (140 °C) or methylcyclohexane (100 °C), respectively, led to the isolation of two early examples of ruthenium(II)-NHC complexes as yellow solids (Scheme 7.1).

In this reaction, the three phosphine ligands in the five-coordinated metal precursor are displaced by four NHCs, again highlighting the stronger metal-carbene bonds, as compared to metal-phosphine. The two isolated mononuclear and neutral complexes are six-coordinated and contain four imidazolidin-2-ylidenes (SIMe and SIEt) and two

The Organometallic Chemistry of N-heterocyclic Carbenes, First Edition. Han Vinh Huynh.
© 2017 John Wiley & Sons Ltd. Published 2017 by John Wiley & Sons Ltd.

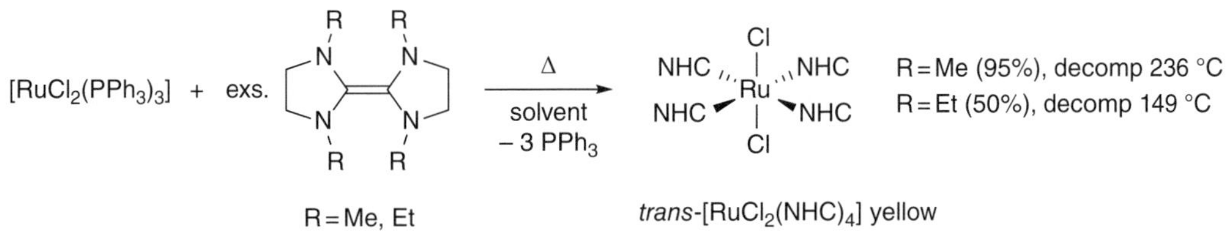

Scheme 7.1
Early preparation of Ru^{II}-NHC complexes via cleavage of NHC dimers.

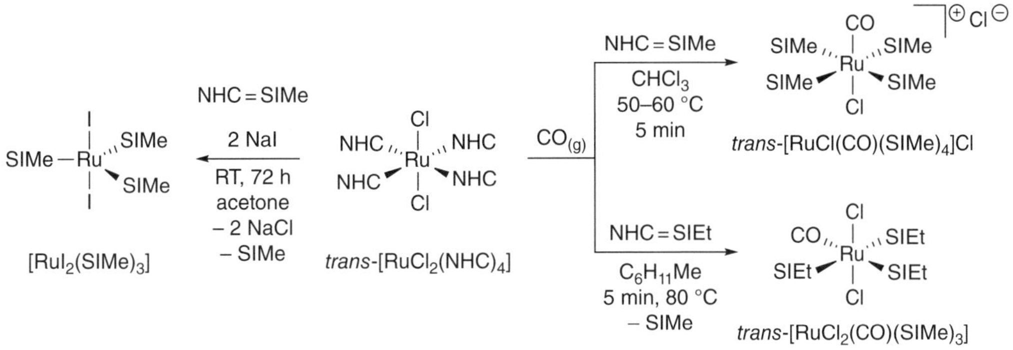

Scheme 7.2
Ligand substitution reactions of dichlorido-tetrakis(NHC) ruthenium(II) complexes.

chlorido ligands arranged in a *trans* fashion. Due to the different types of ligands, the overall coordination geometry of the d^6 metal complexes *trans*-[RuCl$_2$(NHC)$_4$] (NHC = SIMe or SIEt) can be best described as a tetragonally distorted octahedron. Determination of the solid state molecular structure of the SIEt complex by single crystal X-ray diffraction confirmed the equatorial positions of the four carbenes. Despite the simple structure, the ^{1}H NMR spectra of both complexes are unexpectedly complicated with multiple signals for the N-substituents. In particular, the N-ethyl CH$_2$ protons of the SIEt ligands in the complex are diastereotopic, which points to a hindered rotation around the metal-carbon bonds. Insufficient solubility hampered resolution of the ^{13}C$_{carbene}$ signal for the SIMe complex, but its SIEt analog shows only a single resonance for the carbene atoms at 227.8 ppm in C$_6$D$_6$, indicating that the molecular structure found in the solid state is retained upon dissolution. The very downfield chemical shift is also in line with a very electron-rich metal center as a result from the coordination of four saturated NHCs.

This fact is further substantiated by observations made when the complexes were subjected to further ligand displacement reactions [2]. For example, *trans*-[RuCl$_2$(SIMe)$_4$] reacts with sodium iodide not just under simple chlorido substitution. The introduction of two stronger donating iodido ligands would lead to a substantial increase in electron density, which is not preferred in this case. To circumvent this, one carbene ligand is eliminated from the coordination sphere to give an extremely air sensitive five-coordinate tris(NHC) complex of the formula [RuI$_2$(SIMe)$_3$] (Scheme 7.2). Even at –95 °C, solution ^{13}C NMR spectroscopy in CD$_2$Cl$_2$ showed only one singlet at 216.3 ppm for the carbene donors, which is consistent with a trigonal bipyramidal structure.

When exposed to carbon monoxide as a very good π-acceptor ligand, the two parent *trans*-complexes showed very different behavior. For *trans*-[RuCl$_2$(SIMe)$_4$], a chlorido-carbonyl replacement was observed giving the cationic complex *trans*-[RuCl(CO)(SIMe)$_4$] Cl as the major product. In *trans*-[RuCl$_2$(SIEt)$_4$], on the other hand, carbene displacement took place yielding the *trans*-[RuCl$_2$(CO)(SIEt)$_3$] complex. The different reactivities were explained on grounds of steric rather than electronic factors.

The cleavage of enetetramines as a methodology to access NHC complexes has been applied to many transition metals, often leading to the first NHC complexes. Nevertheless, it has not become popular, due to the perceived difficulty and limited modularity. For example electron-rich olefins can only be obtained from saturated and benzannulated systems with small wing tip groups. It has thus been succeeded by a range of other more popular and easier methods.

The Free Carbene Route

The most obvious route to ruthenium NHC complexes is the reaction of a free NHC ligand with a selected ruthenium source. This reaction is highly related to the aforementioned cleavage of electron-rich olefins as the latter are in fact dimers of imidazolidin-2-ylidenes. As with group 10 metals, much of the early ruthenium NHC chemistry was driven by the use of NHCs as phosphine mimics. Grubbs' discovery that complexes of the type [RuCl$_2$(=CHAr)(PR$_3$)] (i.e. Grubbs I; R = Cy, Ph; Ar = aryl) are highly active and yet easy to handle olefin metathesis catalysts provided additional momentum for Ru–NHC research.

In search for even better olefin metathesis catalysts, Herrmann and coworkers were the first to react Grubbs I catalysts of the general formula [RuCl$_2$(=CHPh)(PR$_3$)$_2$] with free NHCs [3]. They chose imidazolin-2-ylidenes with aliphatic wing tip groups for their study, and observed that both phosphines could be easily replaced by these NHCs giving rise to new bis(NHC) complexes of the formula [RuCl$_2$(=CHPh)(NHC)$_2$] in an enthalpically driven process (Scheme 7.3, route A). The new complexes were isolated in high yields of 80–90% as air stable solids. Notably, the substitution of aromatic and aliphatic phosphines goes to completion, which indicates that the chosen NHCs are even stronger donors than the strongest trialkylphosphines, such as, for example, tricyclohexylphosphine.

These complexes feature two very different types of carbenes, which allows a direct spectroscopic and structural comparison. The carbon donor of the "Schrock alkylidene"

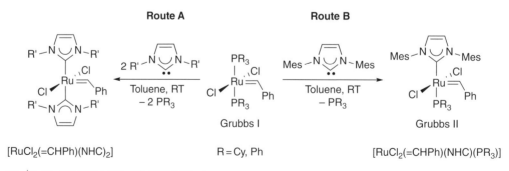

Scheme 7.3
Replacement of phosphines in Grubbs I catalysts with NHCs.

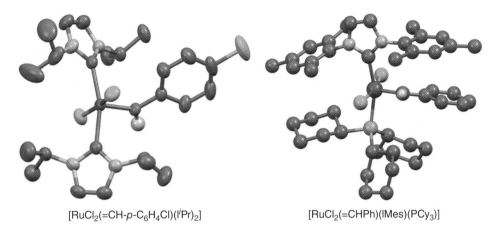

[RuCl$_2$(=CH-*p*-C$_6$H$_4$Cl)(I^{i}Pr)$_2$]   [RuCl$_2$(=CHPh)(IMes)(PCy$_3$)]

Figure 7.1
Molecular structures of Ru–NHC metathesis catalysts.

ligand, usually drawn with a "double bond" to the metal, exhibits a substantially more downfield chemical shift in CD$_2$Cl$_2$ with values of 294–296 ppm in line with a substantial amount of backbonding. The ^{13}C$_{carbene}$ signal for the different NHCs, on the other hand, resonate more upfield in the range of 183–187 ppm. Furthermore, the solid state molecular structure for the analog [RuCl$_2$(=CH-*p*-C$_6$H$_4$Cl)(I^{i}Pr)$_2$] bearing a 4-chloro-substituted Schrock alkylidene could be obtained. Evaluation of the key bond parameters around the metal revealed a much shorter ruthenium-alkylidene (double) bond of 1.821(3) Å, compared to the two ruthenium NHC (single) bonds of 2.107(3) and 2.115(3) Å, respectively (Figure 7.1, left).

A catalytic study using the new bis(NHC) complexes revealed their superiority in Ring-opening Metathesis Polymerization (ROMP) of 2-norbornene or cyclooctadiene and Ring-closing Metathesis (RCM) of 1,7-octadiene compared to the parent complex [RuCl$_2$(=CHPh)(PCy$_3$)$_2$]. Moreover, a strong influence of the different NHCs on the overall activity was noted. This initial study prompted a flurry of research activities concerned with the fine-tuning and optimization of Grubbs type catalysts.

Almost simultaneously and independently, Nolan and coworkers reported the same phosphine replacement reaction, but involving the imidazolin-2-ylidene ligand IMes that contains aromatic N-substituents [4]. Interestingly, this reaction gave a slightly different outcome, according to which only one phosphine can be replaced even when a 10-fold excess of free carbene is used (Scheme 7.3, route B). This interesting observation is likely linked to the decreased donating ability of NHCs with aromatic N-substituents compared to their N-alkyl counterparts. The purple-brown mixed phosphine/NHC complexes [RuCl$_2$(=CHPh)(IMes)(PR$_3$)] (R = Cy, Ph) thus obtained also show much higher thermal stability compared to their bis(phosphine) precursors. The [RuCl$_2$(=CHPh)(IMes)(PCy$_3$)] complex showed the highest stability and showed no sign of decomposition even after prolonged heating at 60 °C for 14 days. Evaluation of its molecular structure in the solid state also corroborated the very different nature of the alkylidene and NHC ligands mentioned above. In addition, one can also note the increased steric bulk of the IMes ligand (Figure 7.1, right).

In this work, the catalytic activity of the [RuCl$_2$(=CHPh)(IMes)(PCy$_3$)] complex for the standard RCM of diethyldiallylmalonate was also found to be better than that of its Grubbs

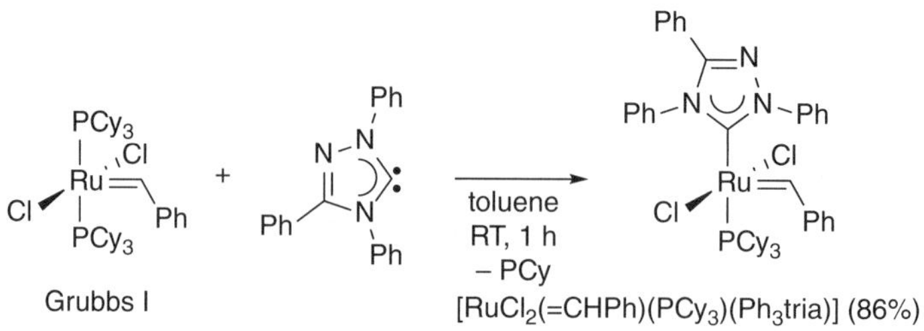

Scheme 7.4
Preparation of a triazolin-5-ylidene containing Grubbs catalyst.

I precursor. Furthermore, the inactive [RuCl$_2$(=CHPh)(PPh$_3$)$_2$] complex could be converted into the highly active species [RuCl$_2$(=CHPh)(IMes)(PPh$_3$)] by simple exchange of one triphenylphosphine with IMes.

Grubbs and coworkers provided more details and observed that the same [RuCl$_2$(=CHPh) (IMes)(PCy$_3$)] complex is actually less active in RCM reactions than its bis(phosphine) precursor at room temperature. However, it possesses a longer life-time and at an elevated temperature of 40 °C, it becomes highly active and can catalyze a larger substrate scope [5].

In addition to imidazolin-2-ylidene complexes, the "free carbene route" for the making of Ru–NHC complexes is also applicable to 1,2,4-triazolin-5-ylidenes. For example, Fürstner and coworkers reported the reaction of the Grubbs I catalyst with the commercially available Enders' carbene (Ph$_3$tria), which led to the displacement of one tricyclohexylphosphine ligand to afford the [RuCl$_2$(=CHPh)(PCy$_3$)(Ph$_3$tria)] complex in a good yield as a brown solid (Scheme 7.4) [6].

Compared to the imidazolin-2-ylidene analogs, this complex shows a decreased stability and slowly decomposes in solution. Notably, the Grubbs I precursor and the respective triazolium salt were detected among the decomposition products, which points to a weaker metal-NHC bond. Carbene dissociation followed by hydrolysis and ligand exchange reactions would explain the observation of these two species.

Moreover, Grubbs and coworkers noted that the [RuCl$_2$(=CHPh)(PCy$_3$)(Ph$_3$tria)] complex exists as a mixture of two conformational isomers differing solely in the orientations of the unsymmetrical triazolin-5-ylidene and/or the Schrock alkylidene moiety [7]. This conclusion was made on grounds of ^{1}H NMR spectroscopy, where two sets of signals were observed in an approximate 60:40 ratio. Likewise, the ^{31}P NMR spectrum shows one singlet for each isomer at ~24 ppm, while two distinct and closely spaced ^{13}C$_{carbene}$ signals were noted for each type of carbene present in the complex. Those for the alkylidenes appear at >300 ppm, while the triazolin-5-ylidene resonances are substantially more upfield at ~192 ppm. Their identity as isomers is further supported by high-resolution mass spectrometry data, where only one molecular ion peak was observed.

Despite exhibiting a high activity for certain substrates in olefin metathesis reactions, the short life-time of this triazolinylidene ruthenium complex in solution is a considerable limitation to its practical use.

Besides providing access to second generation Grubbs catalysts, the "free carbene route" has also been applied to a range of other ruthenium(II)-NHC complexes. Complexes bearing aromatic π-donors are exceptionally widespread and easily synthesized. One common precursor is the dimeric dichlorido-*p*-cymene ruthenium(II) complex [RuCl$_2$(*p*-cymene)]$_2$ containing a neutral 6π-electron donor. In comparison to the reactions discussed above, no

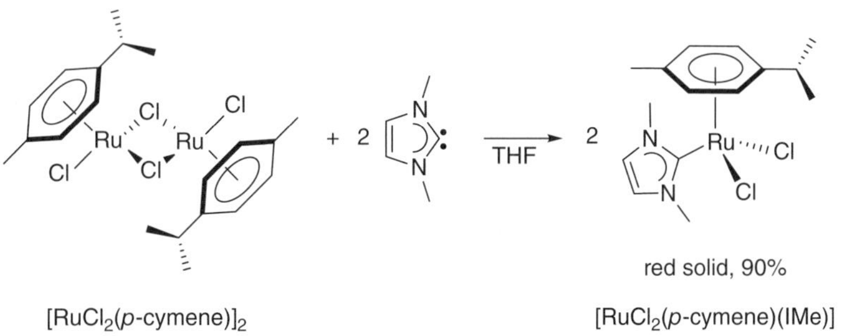

Scheme 7.5
Preparation of a ruthenium(II) half-sandwich complex using the free IMe ligand.

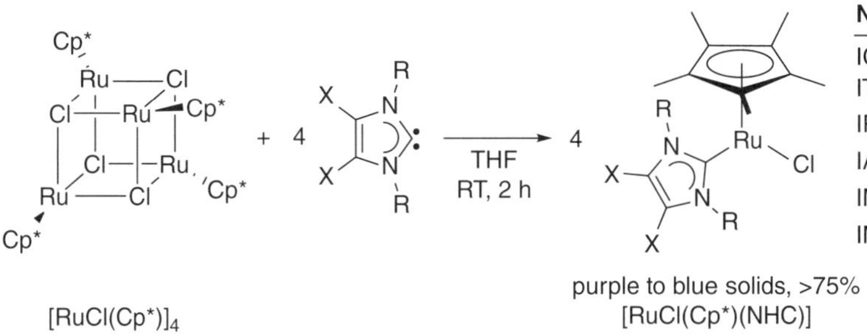

NHC	R	X
ICy	cyclohexyl	H
ITol	4-tolyl	H
IPh^{Cl}	4-Cl-phenyl	H
IAd	adamantyl	H
IMes	mesityl	H
IMesCl	mesityl	Cl

purple to blue solids, >75%
[RuCl(Cp*)(NHC)]

Scheme 7.6
Preparation of coordinatively unsaturated Cp*-NHC Ru^{II} complexes via free NHCs.

ligand needs to be displaced. Instead, the ruthenium-chlorido bridges in the dimer are enthalpically cleaved by addition of two equivalents of free NHC. Conveniently, the reaction can be visually traced by a color change from orange-red to deep red. An early example involving the small IMe ligand was reported by Herrmann and coworkers in 1996, which provided the half-sandwich complex [RuCl$_2$(p-cymene)(IMe)] in a high yield of 90% as a red solid (Scheme 7.5) [8].

The neutral compound features one η^6-p-cymene, two terminal chlorido and one IMe ligands arranged in a three-legged piano-stool geometry with the facial planar p-cymene representing the "seat," while the remaining three monodentate ligands form the three "legs." It dissolves very well in chlorinated solvents, but is hardly soluble in non-polar solvents such as diethylether and alkanes. ^{13}C NMR spectroscopic analysis in CDCl$_3$ reveals the chemical shift for the carbene donor atom at 173.2 ppm. Surprisingly, the 18-electron complex was reported to be air and moisture sensitive, although later analogs bearing other NHCs are mostly stable.

Another useful ruthenium(II) precursor is the tetrameric [RuCl(Cp*)]$_4$ complex, which carries an anionic Cp* as a 6π-electron donor and exhibits a cubane-like structure with μ_3-bridging chlorido ligands. Nolan and coworkers first described its reaction with a range of bulky free imidazolin-2-ylidenes [4,9]. In a typical procedure, the ruthenium tetramer was treated with four equivalents of free NHC at room temperature in tetrahydrofurane for ~2 h (Scheme 7.6). This bridge-cleavage reaction afforded coordinatively unsaturated 16-electron complexes, which were isolated as purple to blue solids.

Table 7.1 Enthalpies and Relative Bond Dissociation Energies for the cleavage of the parent ruthenium(II)-tetramer.

$$[RuCl(Cp^*)]_{(s)} \ + \ 4 \ L_{(soln)} \ \xrightarrow[\ 30\,^{\circ}C\ ]{THF} \ 4[RuCl(Cp^*)(L)]_{(soln)}$$

Ligand L	$-\Delta H_{rxn}$ [kcal mol^{-1}]	Relative Ru–L BDE [kcal/mol^{-1}]
ICy	85.0(0.2)	21.2
ITol	75.3(0.4)	18.8
IPh^{Cl}	74.3(0.3)	18.6
IAd	27.4(0.4)	6.8
IMes	62.6(0.2)	15.6
IMesCl	48.5(0.4)	12.1
PCy$_3$	41.9(0.2)	10.5
P^{i}Pr$_3$	37.4(0.3)	9.4

The direct reactions involving pre-isolated carbenes were found to be rapid, clean, and quantitative, which meant that they were suitable for calorimetric measurements for the determination and comparison of reaction enthalpies. The bond dissociation energy (BDE) for the respective ruthenium–ligand bonds can be estimated by division of the reaction enthalpies by four.

Table 7.1 shows a summary of the values obtained for imidazolin-2-ylidenes, which are compared with those for two common trialkylphosphine ligands. With the exception of IAd, all NHCs bind more strongly to the metal center than their trialkylphosphine counterparts, and so an easy ligand replacement of phosphine with NHC can be anticipated. Generally, N-alkyl-substituted NHCs are expected to be stronger donors than N-aryl analogs. The BDE for ICy underlines this view. The reason for the unexpectedly weaker Ru–IAd bond has been ascribed to the enhanced steric bulk of the NHC preventing ideal orbital overlap. Furthermore, the data suggests that NHCs with additional electron-withdrawing groups, such as chloro substituents in both backbones and wing tips, would decrease BDE values, which is in line with their lower donating abilities.

Overall, the magnitude of the enthalpy of reaction involving nucleophilic carbene ligands and other ligands is a function of a combined stereoelectronic properties of the ligand, where both steric and electronic effects influence the overall availability of the carbene lone pair.

The steric bulk of the Cp* ligand, together with the enhanced spatial demands of most NHCs investigated herein, allows for the formation of coordinatively unsaturated complexes. For most complexes, the unsaturation is maintained in the solid state as evidenced by X-ray diffraction studies on single crystals. The molecular structures of two representatives bearing the IMesCl and ITol carbene ligands are depicted in Figure 7.2.

Notably, the ICy complex exceptionally crystallizes as a dimeric species of the formula $[Ru(\mu\text{-}Cl)(Cp^*)(ICy)]_2$, where the two ruthenium centers are connected via bridging chlorido ligands (not shown). This observation supports the view that cyclohexyl groups are bulky and yet sufficiently flexible to break the unsaturation by accommodating bridging chlorido ligands upon crystallization.

The unsaturation also leads to an increased reactivity. While the ICy and IMes complexes are moderately stable as solids, their deep blue solutions in toluene rapidly undergo

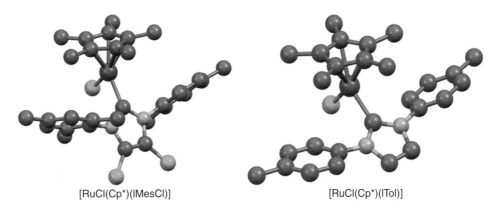

[RuCl(Cp*)(IMesCl)]   [RuCl(Cp*)(ITol)]

Figure 7.2
Two representative molecular structures for coordinatively unsaturated [RuCl(Cp*)(NHC)] complexes.

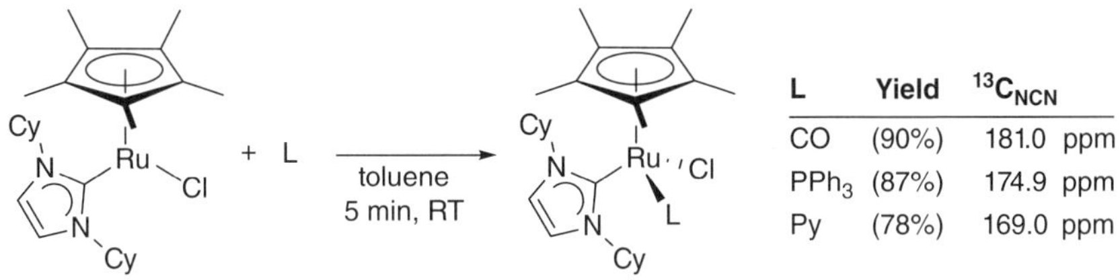

L	Yield	$^{13}C_{NCN}$
CO	(90%)	181.0 ppm
PPh$_3$	(87%)	174.9 ppm
Py	(78%)	169.0 ppm

[RuCl(Cp*)(ICy)]
16-electron, $^{13}C_{NCN}$ = 196.6 ppm

[RuCl(Cp*)(ICy)(L)]
18-electron

Scheme 7.7
Preparation of more stable 18-electron [RuCl(Cp*)(NHC)(L)] complexes.

color change to brown when exposed to air. Their high reactivity was exploited in the catalytic dimerization of terminal alkynes. Both complexes gave rise to the most active catalyst system for alkyne dimerization when they were reported [10].

More stable 18-electron adducts of the type [RuCl(Cp*)(ICy)(L)] are easily formed with additional ligands, such as carbon monoxide, triphenylphosphine and pyridine in high yields (Scheme 7.7). ^{13}C NMR spectroscopic evaluation of the carbene donor in the adducts and comparison with parent compound indicates a gradual upfield shift with decreasing δ values going from CO < PPh$_3$ < Py. Interestingly, this trend appears to follow the decreasing π-acceptor properties of the respective co-ligands.

7.1.2 *In Situ* α-Elimination

The troublesome handling of air sensitive free carbenes in the direct synthesis of ruthenium NHC complexes can be circumvented by *in situ* preparation of the free ligand in the presence of suitable metal precursors. One elegant method exploits the α-elimination of small molecules from the *sp^3*-hybridized C2-carbon in imidazolidines reminiscent to the preparation of Enders' free triazolin-5-ylidene. Again this research was largely driven by the

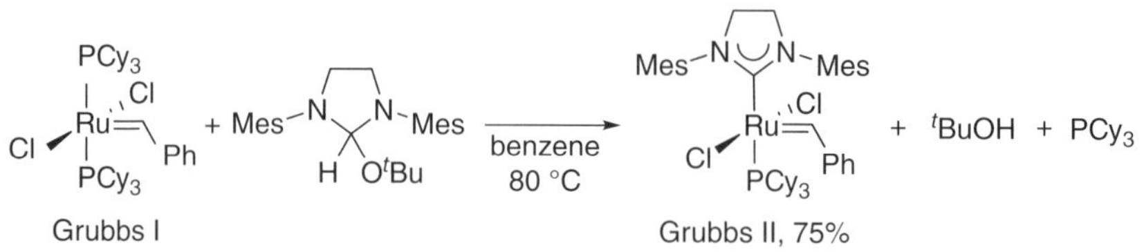

Scheme 7.8
Preparation of a Grubbs II catalyst via *in situ* α-elimination.

search for better and more stable ruthenium based olefin metathesis catalysts. The group of Grubbs first described that moderate heating of a mixture of the Grubbs I catalyst with 2-*tert*-butoxy-1,3-dimesitylimidazolidine in benzene to 80 °C led to the generation of the respective stereotypical Grubbs II catalyst [11]. In this process, the free imidazolidin-2-ylidene was generated by α-elimination of *tert*-butanol from the respective 2-alkoxy precursor and immediate phosphine replacement reaction afforded the stable Grubbs II catalyst bearing a mixed phosphine/NHC donor set (Scheme 7.8).

Purification was simple and involved removal of all volatiles and subsequent washing with methanol. Notably, the formation of the carbene complex occurred only upon heating, while at ambient temperature no reaction took place. Overall, the reaction is both enthalpically and entropically favorable, since a stronger ruthenium–carbene bond is formed with the release of two small molecules.

Apart from the more downfield carbene signal characteristic for saturated NHCs, the NMR spectroscopic features of these new complexes are similar to their earlier analogs containing unsaturated imidazolin-2-ylidenes prepared using pre-isolated NHCs. The thermal stabilities are comparable, but the increased donating ability of saturated NHCs led to a significantly improved catalytic performance and a higher substrate scope that could compete with or outperform highly active but much more sensitive molybdenum based catalysts.

7.1.3   *In Situ* Deprotonation of Azolium Salts

Intermediate free NHCs for the preparation of ruthenium NHC complexes can also be generated by *in situ* deprotonation either via (i) addition of an external base or (ii) with a suitably basic ruthenium complex precursor. One of the first examples of ruthenium-1,2,4-triazolin-5-ylidene complexes reported by Enders and coworkers was prepared using route (i). Accordingly, stirring a mixture of the triazolium perchlorate, the ruthenium-cymene dimer and excess triethylamine at room temperature for 3 h was already sufficient to affect ruthenation of the NHC, and the resulting complexes [RuCl(κ^2-C,C'-Ph,R-tria)(*p*-cymene)] were isolated as air and moisture stable yellow solids (Scheme 7.9) [12]. In this case, triethylamine is already basic enough to deprotonate the acidic C2–H proton of carbene precursor. Notably, the mild base also assists in the *ortho*-metallation of the N-phenyl substituent giving rise to chelating κ^2-C,C' mixed NHC/aryl ligands with the replacement of one chlorido ligand from the metal center.

The cyclometallated complexes are *pseudo*-tetrahedral in structure with four different donors, which makes them inherently chiral-at-ruthenium. Predetermination of chirality can be achieved by using enantiopure ligands with a chiral wing tip group, where formation

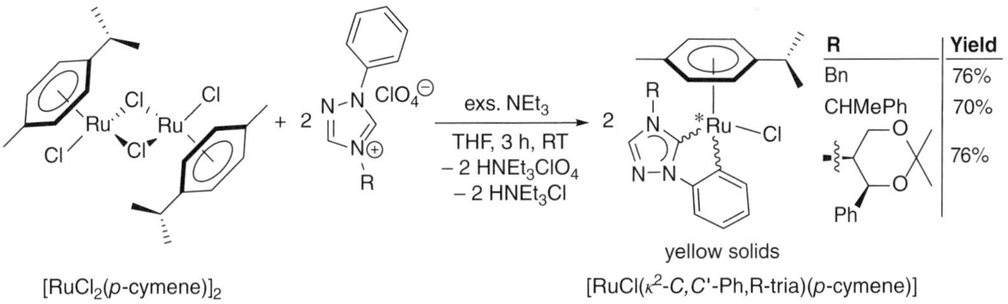

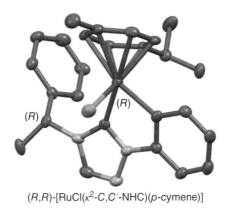

Scheme 7.9
Preparation of cyclometallated ruthenium(II)-*p*-cymene NHC complexes via *in situ* deprotonation with external base.

Figure 7.3
Molecular structure of the major diastereomer of a chiral-at-metal complex.

of two possible diastereomers can be anticipated. For example, the use of the (*R*)-phenyle-thyl wing tip group leads to the preferred formation of the (R_{Ru},R_{NHC}) diastereomer in >87% *de* (*de*=diastereomeric excess), while the more elaborate (*S,S*)-2,2-dimethyl-4-phenyl-1,3-dioxan-5-yl substituted NHC affords the respective (R_{Ru}) diastereomer in >95% *de*. The two diastereomers of each complex could be separated by column chromatography and isolated in >96% *de*. ^{13}C NMR data recorded in CDCl$_3$ revealed that all carbene donors resonate in the narrow range of 189.5–191.3 ppm, while the anionic and more electron-rich σ-aryl donor atoms appear at higher field ranging from 159.9–160.3 ppm. Figure 7.3 depicts the molecular structure of the major (R_{Ru}) diastereomer of the (*R*)-phenylethyl substituted chiral-at-metal complex established by single crystal X-ray diffraction.

An example for route (ii) employing basic ruthenium precursors was reported by Herrmann and coworkers. One such compound is the dimeric [RuCp*(OMe)]$_2$ that contains two ruthenium centers bridged by two basic methoxido ligands. On treatment with two equivalents of azolium salts, *in situ* deprotonation of the cyclic ammonium salts takes place, and two molecules of methanol are liberated [10]. Compared to a simple bridge-cleavage reaction involving free NHCs discussed above, this methodology includes an additional acid-base reaction step and thus requires additional activation energy, which is achieved by moderate heating to 45 °C (Scheme 7.10).

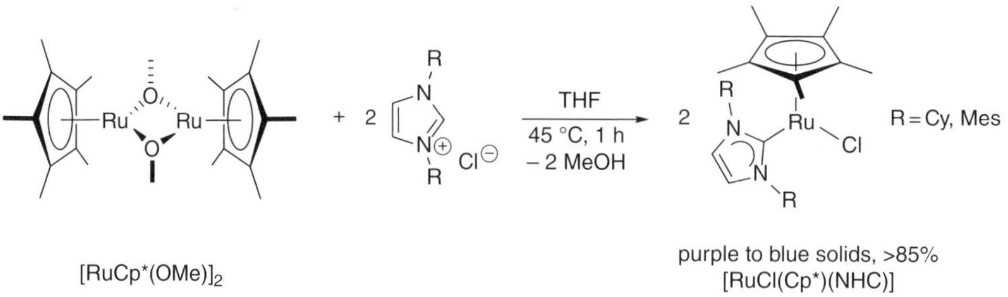

Scheme 7.10
In situ deprotonation of azolium salts using a methoxido-bridged Ru-dimer.

It is important to note that azolium halides should be used in this protocol to provide for an additional halido ligand. Even so, the resulting 16-electron complexes are electronically and coordinatively unsaturated and therefore more reactive. The use of carbene precursors with non-coordinating anions is thus anticipated to give even more unstable 14-electron compounds of the formula [RuCp*(NHC)].

7.1.4   The Silver-Carbene Transfer Route

Nowadays, the most widely used methodology to prepare ruthenium NHC complexes is the silver-carbene transfer route. Its popularity is rooted in its easiness and effectiveness, which does not require stringent exclusion of air and moisture. In most cases, the procedures can be carried out in normal glassware using solvents straight from the bottle. The first step obviously involves preparation of the silver-carbene species, which usually occurs by reaction of suitable azolium salts with silver(I) oxide or another basic silver precursor. The silver-NHC can be isolated or directly reacted with a suitable ruthenium source. An early deliberate attempt at this approach was reported by Joó and coworkers, who treated [RuCl$_2$(*p*-cymene)]$_2$ with the silver-carbene complex [Ag(Bu,Me-imy)$_2$][AgCl$_2$] obtained from the reaction of 1-buty-3-methylimidazolium chloride with silver(I) oxide [13]. The NHC ligand is transferred from the silver(I) to the ruthenium(II) metal center and silver chloride precipitates driving the reaction forward. Concurrently, the initial two chlorido bridges are cleaved, and two equivalents of the neutral and mononuclear complex [RuCl$_2$(Bu,Me-imy)(*p*-cymene)] are obtained (Scheme 7.11).

The complex was isolated as an orange powder and its ^{13}C NMR spectrum recorded in CD$_2$Cl$_2$ shows a downfield signal at 173.7 ppm indicative of the metallated carbene donor. Interestingly, it was noted that the complex dissolves well in water under chlorido displacement giving rise to the cationic aqua complex [RuCl(Bu,Me-imy)(*p*-cymene)(OH$_2$)]Cl (Scheme 7.12).

The ease of chlorido replacement can be exploited to prepare mixed NHC/phosphine complexes. Thus, exposure of the neutral dichlorido complex to an aqueous solution of the unusual phosphine 1,3,5-triaza-7-phosphaadamantane (pta) instantaneously afforded the yellow complex [RuCl(Bu,Me-imy)(*p*-cymene)(pta)]Cl. The ligand exchange is accompanied by an upfield shift of the carbene carbon to 165.4 ppm in line with the greater Lewis acidity of cationic complex. The latter complex showed significantly increased catalytic

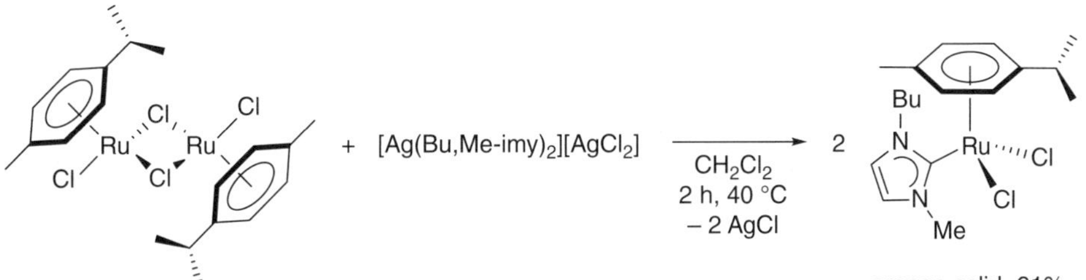

orange solid, 61%

$[RuCl_2(p\text{-cymene})]_2$ $[RuCl_2(Bu,Me\text{-imy})(p\text{-cymene})]$

Scheme 7.11
Preparation of a Ru–NHC half-sandwich complex via silver-NHC transfer.

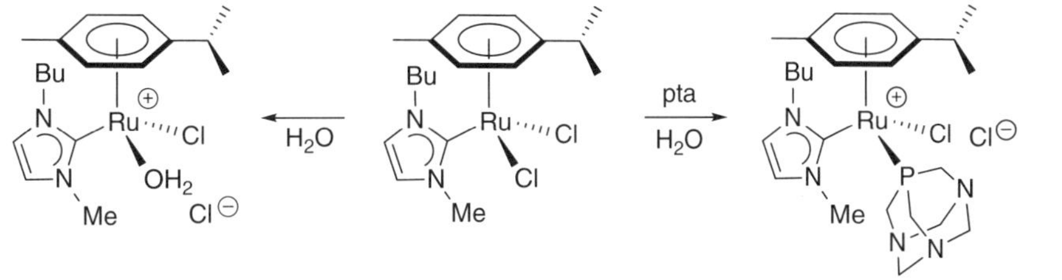

$[RuCl(Bu,Me\text{-imy})(p\text{-cymene})(OH_2)]Cl$ $[RuCl(Bu,Me\text{-imy})(p\text{-cymene})(pta)]Cl$

Scheme 7.12
Chlorido displacement reactions with aqua and phosphine ligands.

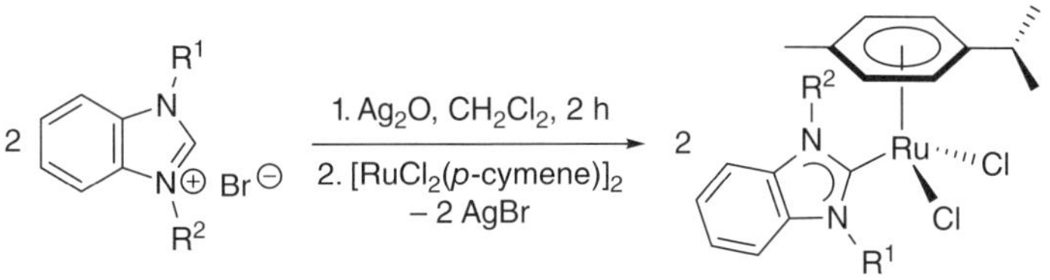

R^1	R^2	Yield	$^{13}C_{carbene}$
Bn	Bn	81%	191.7 ppm
^{i}Pr	Bn	70%	189.7 ppm
^{i}Pr	^{i}Pr	58%	187.6 ppm

$[RuCl_2(R^1,R^2\text{-bimy})(p\text{-cymene})]$, red-orange solids

Scheme 7.13
Ruthenium(II)-NHC half-sandwich complexes via *in situ* silver-NHC transfer.

activity in hydrogenation reactions of olefins, aldehydes and ketones using 10 bar dihydrogen pressure in phosphate buffer (pH 6.90) compared to its neutral parent.

The methodology can also be applied to yield benzimidazolin-2-ylidene complexes. *In situ* formation of silver-carbene species by treatment of benzimidazolium salts with silver(I) oxide followed by addition of the ruthenium-*p*-cymene dimer without prior isolation of the former also leads to the formation of the respective half-sandwich complexes $[RuCl_2(bimy)(p\text{-cymene})]$, which have been isolated as red-orange solids after column chromatography in moderate to good yields (Scheme 7.13) [14].

The overall yields are strongly affected by the stereoelectronic properties of the carbene precursors as highlighted by the preparation of the complexes bearing 1,3-dibenzyl- (Bn_2-bimy), 1-isopropyl,3-benzyl- (^{i}Pr,Bn-bimy) and 1,3-diisopropylbenzimidazolin-2-ylidene

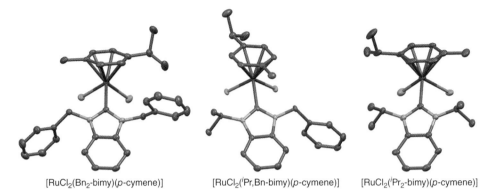

[RuCl₂(Bn₂-bimy)(*p*-cymene)]   [RuCl₂(*i*Pr,Bn-bimy)(*p*-cymene)]   [RuCl₂(*i*Pr₂-bimy)(*p*-cymene)]

Figure 7.4
Molecular structures of [RuCl₂(bimy)(*p*-cymene)] complexes.

(*i*Pr₂-bimy) complexes. The 1,3-dibenzyl-substituted compound is the most acidic and least bulky among the three salts. Stepwise substitution of benzyl with isopropyl groups leads to a decrease in acidity of the C2–H proton due to the increase of +*I* effects from branched alkyl groups. At the same time, the steric crowding around the pre-carbene atom increase as well. Both effects hamper the formation of the intermediate silver-carbenes and their subsequent transfer to the ruthenium center. Thus, a drop in yield was observed in the formation of NHC complexes in the order Bn₂-bimy (81%) > Bn,*i*Pr-bimy (70%) > *i*Pr₂-bimy (58%) under otherwise identical reaction conditions.

The variation of the wing tip groups of the NHCs also affects the chemical shift of the carbene donors recorded in CDCl₃, where an upfield shift was noted with increasing +*I* effect going from benzyl to isopropyl groups (Scheme 7.13). The electron donation (inductive effects) of branched alkyl substituents has intuitively a shielding effect on the carbene carbon.

The solid state molecular structures for all members of this series have been established. (Figure 7.4) Overall, the bond parameters were found to be very similar. In particular, the Ru-centroid distances to the aryl ligand of about 1.7Å were found to be essentially identical within error margin. The Ru-C_carbene bonds of 2.069–2.090 Å are also in a very narrow range. Complexes of this type have been successfully applied as catalysts for alcohol activation in the making of secondary and tertiary amines via the proposed hydrogen borrowing pathway.

7.2 Rhodium(I)- and Rhodium(III)-NHC Complexes

7.2.1 Cleavage of Enetetramines and the Free Carbene Route

Cleavage of Enetetramines

The first rhodium NHC complexes were isolated by Lappert and coworkers as stable intermediates in the metal catalyzed dismutation of electron-rich entetramines (Scheme 7.14) [15]. This is particularly interesting, since cross-over experiments involving such olefins did not give the mixed products, which led to the initial dismissal of the "Wanzlick equilibrium [16]."

The proposed rhodium-carbene intermediates can be formed on treatment of the olefins with either Wilkinson's catalyst [RhCl(PPh₃)₃] or the rhodium analog of Vaska's complex

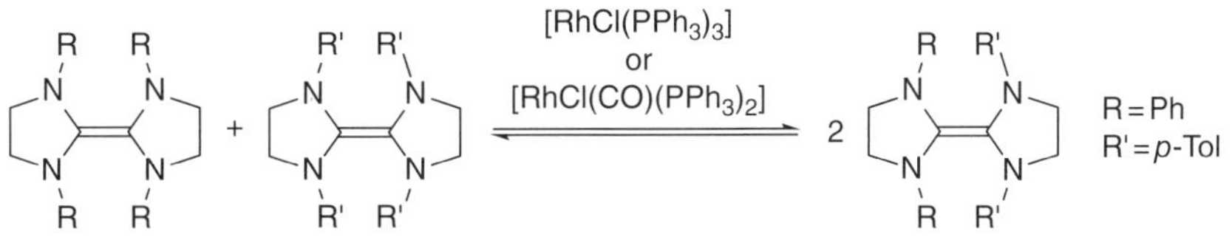

Scheme 7.14
Rhodium-catalyzed "Wanzlick equilibrium."

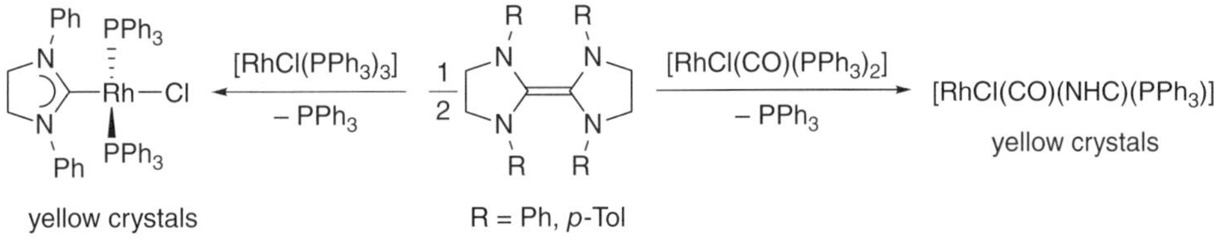

Scheme 7.15
Preparation of the first rhodium NHC complexes.

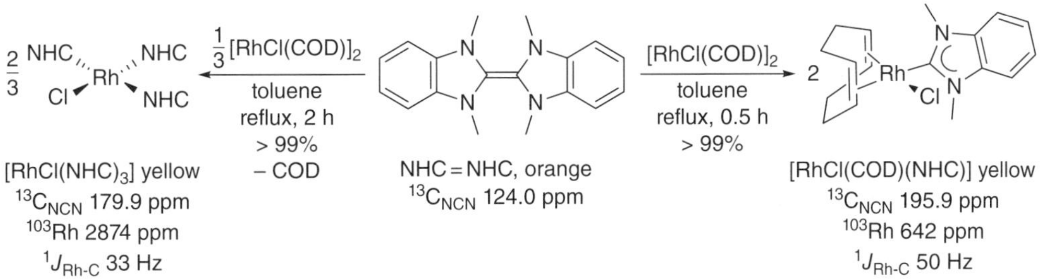

Scheme 7.16
Rhodium(I)-benzimidazolin-2-ylidene complexes via NHC dimer cleavage.

trans-[RhCl(CO)(PPh$_3$)$_2$] (Scheme 7.15). Formally, the C=C double bond is cleaved *in situ* generating two equivalents of imidazolidin-2-ylidene, which displaces one triphenylphosphine from the coordination sphere. The new square planar rhodium(I) carbene complexes were isolated as orange to yellow crystalline solids. While ^{31}P NMR data suggested a *trans* arrangement in the bis(phosphine) complexes, the detailed arrangement of the four different ligands in the carbonyl complexes was not elaborated on.

The methodology has also been applied to the making of benzimidazolin-2-ylidene complexes of rhodium(I). Accordingly, the tetraazafulvalene from the 1,3-dimethyl-substituted benzannulated carbene reacted with one equivalent of the dimeric chlorido-cyclooctadiene-rhodium(I) complex in toluene under reflux to give the monocarbene complex [RhCl(COD)(Me$_2$-bimy)] quantitatively in a bridge-cleavage reaction (Scheme 7.16) [17]. Reducing the amount of the metal precursor to one third and prolonging the reaction time under otherwise identical conditions led to the additional replacement of the cyclooctadiene ligand and incorporation of three carbenes per metal center instead. Both complexes were isolated as air stable yellow solids.

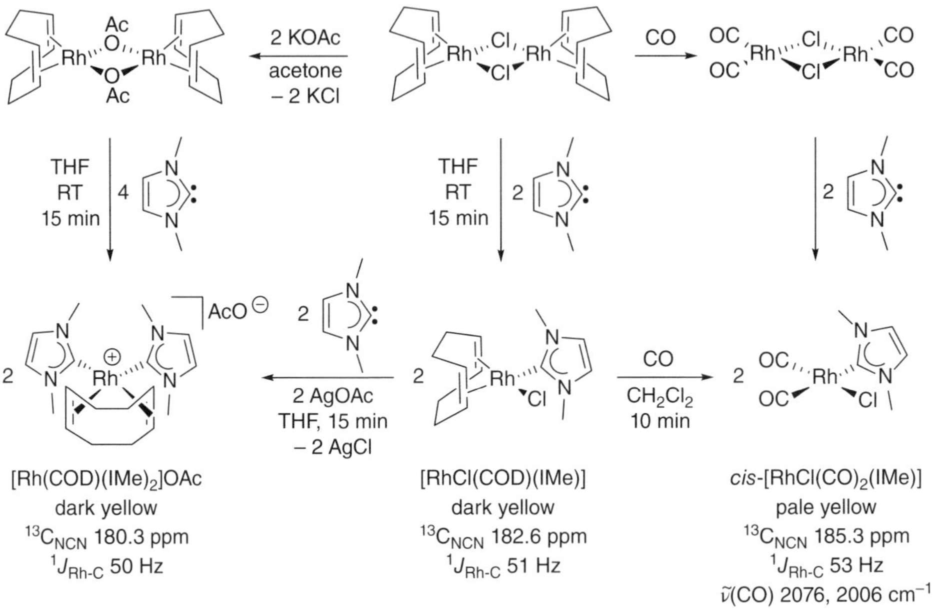

Scheme 7.17
Preparation of rhodium(I)-IMe complexes via the free carbene route.

^{13}C NMR spectroscopic analyses revealed a substantial downfield shift of the carbenoid atom upon olefin cleavage. As expected, the two structurally very different complexes have very different NMR signatures. This is in particular notable in very different chemical shifts for the ^{103}Rh nuclei.

The Free Carbene Route

Carbenes derived from imidazoles can be used directly to prepare rhodium complexes. A straightforward method is to use free NHCs to cleave coordinatively unsaturated complexes such as the dimeric rhodium(I) precursor [RhCl(COD)]$_2$. Herrmann and coworkers were among the first to use this approach to prepare rhodium(I) complexes with unsaturated NHC ligands [8]. The reaction with 1,3-dimethylimidazolin-2-ylidene (IMe) in tetrahydrofurane proceeds enthalpically, already driven rapidly at room temperature, and its progress is indicated by a color change of the reaction solution from yellow to orange (Scheme 7.17).

The target compound [RhCl(COD)(IMe)] is isolated as a dark yellow to orange solid after purification by recrystallization from dichloromethane/n-pentane in an excellent yield of 91%. The complex is very well soluble in common polar organic solvents and thermally stable in the presence of both oxygen and moisture. No decomposition was noted even when a solution in toluene was heated to reflux for several days. The ^{103}Rh–^{13}C heteronuclear coupling constant measured in CDCl$_3$ is essentially of the same magnitude as that of its benzimidazolin-2-ylidene analog, indicating a very similar metal-ligand bonding situation. The resonance of the carbene donor, on the other hand, appears as expected at higher field as a consequence of the unsaturated backbone.

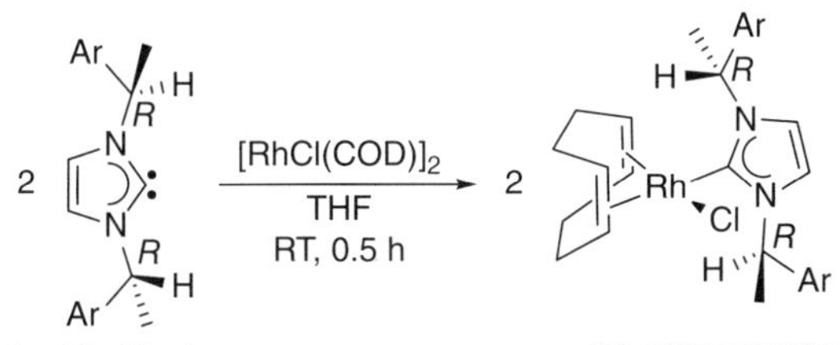

Scheme 7.18
Rhodium(I)-imidazolin-2-ylidene complexes using free NHCs.

Replacement of the 1,5-cyclooctadiene ligand in [RhCl(COD)(IMe)] occurs with ease when carbon monoxide is bubbled through a solution of the complex in dichloromethane. The resulting pale yellow dicarbonyl complex *cis*-[RhCl(CO)$_2$(IMe)] was found to be less stable to air and moisture than its precursor. Alternatively, it can also be quantitatively obtained by bridge-cleavage of the bis[(μ-chlorido)dicarbonylrhodium(I)] dimer with two equivalents of the free IMe ligand. This ligand substitution leads to a downfield shift of the carbene carbon, which points to an increased electron density of the new complex. In addition, the IR data recorded in tetrahydrofurane indicate a substantial amount of rhodium-to-carbonyl backdonation due to the coordination of the strongly donating IMe ligand. The magnitude of this interaction induced by different NHC ligands forms the basis of the currently most popular method for the donor strength determination of NHCs and related ligands.

Another example involves the reaction of excess IMe with bis[μ-acetato(1,5-cyclooctadiene)rhodium(I)]. This dimeric precursor contains weaker acetato bridges and thus additional replacement of the anionic ligand takes place giving rise to the formation of the yellow and cationic bis(carbene) complex [Rh(COD)(IMe)$_2$]OAc. The complex salt is poorly soluble in common organic solvents. Its ^{13}C$_{carbene}$ NMR signal is shifted upfield in comparison to the neutral monocarbene complex, which is in line with the increased Lewis acidity of the complex cation despite the coordination of two NHCs. Moreover, solution NMR studies did not indicate ligand dissociation even at high temperatures. Another route to the same complex involves chlorido replacement of [RhCl(COD)(IMe)] with an additional equivalent of IMe. Subsequent anion substitution of the resulting chloride with silver(I) acetate leads to the formation of [Rh(COD)(IMe)$_2$]OAc with concurrent precipitation of silver(I) chloride.

The bridge-cleavage reaction of [RhCl(COD)]$_2$ with free NHCs is general and has also been conducted with NHCs bearing chiral N-substituents as exemplified in Scheme 7.18 [18]. The products have been isolated as yellow microcrystalline solids and their yields as well as key NMR spectroscopic data are also highlighted in this scheme.

The free carbene route in rhodium chemistry involving the common imidazolin-2-ylidene with aromatic wing tips IMes has been reported by the group of Nolan [19]. In this work, the dimeric complex [RhCl(COE)$_2$]$_2$, which contains the weaker binding mono-olefin cyclooctene (COE), was treated with four equivalents of the IMes ligand, possibly in an attempt to prepare complexes of the type [RhCl(COE)(NHC)$_2$]. Surprisingly, the target complex was not obtained as the final product. Instead, the cyclometallated rhodium(III) complex [RhClH(κ^2-*C,C*-IMes)(IMes)] was isolated in high yield as a result of an intramolecular C–H activation of one methyl group of a mesityl substituent (Scheme 7.19). The ease of the C–H activation under mild conditions is notable. Supposedly, coordination of

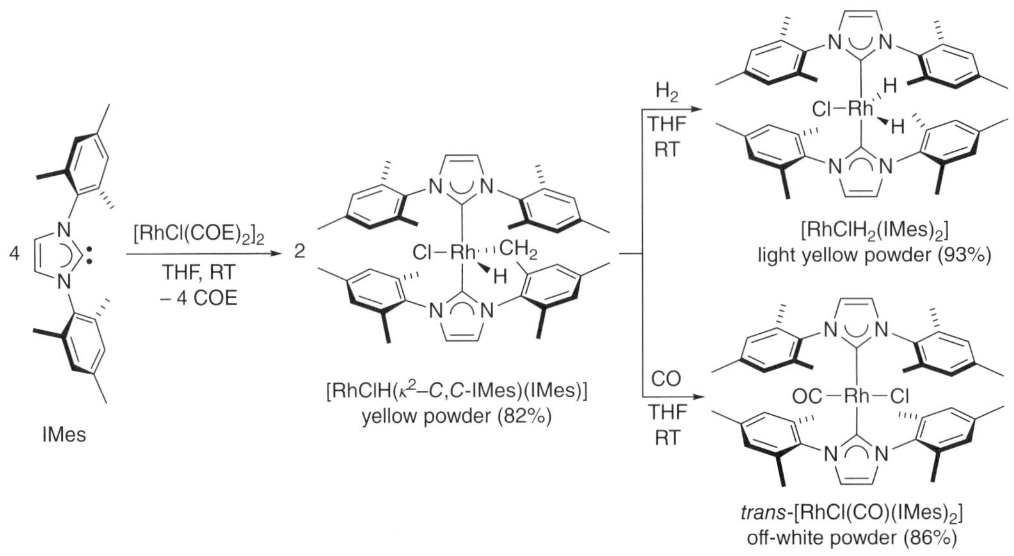

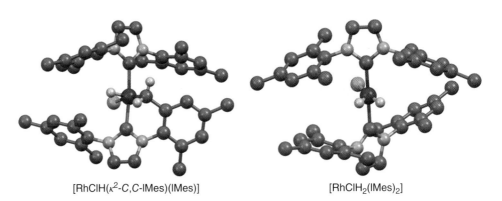

Scheme 7.19
Formation of a cyclometallated IMes complex and its reactivity.

[RhClH(κ^2-*C*,*C*-IMes)(IMes)]   [RhClH$_2$(IMes)$_2$]

Figure 7.5
Molecular structures of rhodium(III)-hydrido complexes with IMes ligands.

two IMes ligands to the metal center occurs first leading to the initially targeted rhodium(I) bis(IMes) complex. However, this intermediate species is very electron rich and an *in situ* intramolecular oxidative addition of one methyl substituent occurs with dissociation of the labile olefin.

The hydrido ligand of the five-coordinated complex resonates as a broad singlet at -27.30 ppm. ^{1}H NMR spectroscopy also revealed the presence of one classical IMes and one second cyclometallated IMes chelating the rhodium center in a κ^2-*C*,*C* mode. The solid state molecular structure obtained by single crystal X-ray diffraction is in line with that proposed by solution studies (Figure 7.5, left). Notably, the two Rh–C$_{carbene}$ bonds are significantly different. The one of the IMes chelate of 2.024(2) Å is shorter than that of the monodentate IMes with 2.048(2) Å. This observation is in line with expectation as a chelate formation generally strengthens bonding to metal centers.

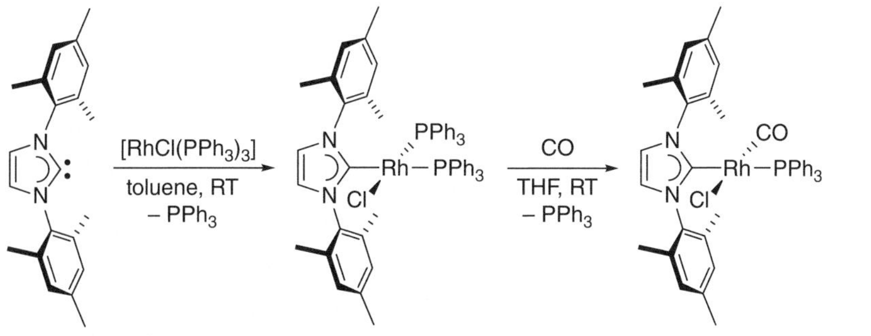

Scheme 7.20
Preparation of mixed phosphine/IMes complexes of rhodium(I).

This compound is highly reactive and exposure to dihydrogen gas affords the dihydrido complex [RhClH$_2$(IMes)$_2$]. Its formation can be explained by a heterolytic cleavage of dihydrogen at the coordination sphere giving a new hydrido ligand with protonation of the anionic methylene tether. The two hydrido ligands show a single upfield shifted resonance at –23.04 ppm, and the protonation of the methylene tether also leads to two equivalent IMes ligands on the NMR timescale. Its solid state molecular structure is depicted in Figure 7.5 (right). The loss of chelation leads to two essentially identical Rh–C$_{carbene}$ bonds of 2.018(2) and 2.022(2) Å, respectively.

Purging as a solution of [RhClH(κ^2-*C,C*-IMes)(IMes)] in tetrahydrofurane with carbon monoxide leads to reversal of the cyclometallation, and the square planar rhodium(I) complex *trans*-[RhCl(CO)(IMes)$_2$] was obtained as an off-white powder in good yields. Reductive elimination reactions of *cis*-standing mixed hydrido-alkyl complexes are common, and it is likely that such a reaction occurs here as well. The increased electron density in the resulting rhodium(I) species is efficiently backdonated into the strongly π-accepting carbonyl ligand, which efficiently stabilizes the complex. ^{1}H NMR studies shows only one set of signals for the IMes ligand as expected for a *trans* arrangement of the two carbenes. Comparison of IR wavenumber of 1935 cm^{-1} for the carbonyl stretch with the value of 1943 cm^{-1} for the phosphine analog *trans*-[RhCl(CO)(P^{i}Pr$_3$)$_2$] confirms once again the stronger donating power of NHCs compared to trialkylphosphines.

The IMes ligand has also been used to replace phosphine ligands in Wilkinson's catalyst [RhCl(PPh$_3$)$_3$]. For example, Crudden reported that a 1:1 reaction in toluene afforded the mixed carbene/phosphine complex *cis*-[RhCl(IMes)(PPh$_3$)$_2$] (Scheme 7.20) [20]. The *cis* configuration was concluded based on ^{31}P NMR spectroscopy, which showed two distinct signals at 48.7 ($^2J_{P-P} = 39$ Hz; $^1J_{P-Rh} = 210$ Hz; P *cis* to IMes) and 35.4 ppm ($^2J_{P-P} = 39$ Hz; $^1J_{P-Rh} = 119$ Hz; P *trans* to IMes) with characteristic coupling constants, respectively. In general, complexes of the type [RhCl(L)(PPh$_3$)$_2$] favor a *trans* arrangement of the two phosphine ligands. However, it is likely that the steric bulk of the IMes ligand enforces the *cis* arrangement of the phosphine donors in this given case. The complex is relatively sensitive to oxygen in solution and decomposes under dissociation of phosphine and its subsequent oxidation to phosphine-oxide.

The lability of the phosphine ligand makes the complex a suitable catalyst precursor, and it has been successfully explored in the hydroformylation of vinylarenes with good

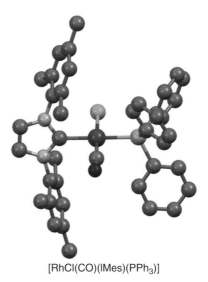

[RhCl(CO)(IMes)(PPh₃)]

Figure 7.6
Molecular structure of [RhCl(CO)(IMes)(PPh₃)].

selectivity, but marginal reactivity. The ease of phosphine dissociation was also exploited to prepare the respective carbonyl adduct as a potential intermediate in hydroformylation reaction. Exposure of a solution of *cis*-[RhCl(IMes)(PPh₃)₂] in tetrahydrofurane to a stream of carbon monoxide resulted in the selective displacement of the phosphine donor *cis* to the NHC, possibly to release steric strain. The resulting complex of the formula [RhCl(CO)(IMes)(PPh₃)] is highly stable and retains the square planar geometry. The positioning of the four different ligands relative to each other was established by ^{31}P NMR spectroscopy, which shows a single doublet at 31.7 ppm. The heteronuclear coupling constant of $^1J_{\text{P-Rh}} = 115$ Hz indicates that the phosphine is *trans* to the carbene as larger values are expected for the *cis* form. Moreover, the same arrangement of the ligands in the solid state was established by a single crystal X-ray analysis (Figure 7.6). The Rh–C$_{\text{carbene}}$ and Rh–P distances of 2.032(4) Å and 2.302(1) Å are consistent with expected values. The former are, within standard deviation, essentially the same as those observed for the rhodium(III)-hydrido complexes mentioned above.

7.2.2   *In Situ* Deprotonation of Azolium Salts

The *in situ* deprotonation of an azolium salt by a basic metal precursor—the earliest reported general method to access metal-NHC complexes—was applied to rhodium chemistry by Herrmann in 1996 [8]. The acetylacetonato-bis(carbonyl)rhodium(I) complex [Rh(acac)(CO)₂] is a useful precursor for this purpose, and its reaction with 1,3-dimethylimidazolium iodide furnishes the desired bis(carbonyl) monocarbene complex *cis*-[RhI(CO)₂(IMe)] in a good yield (Scheme 7.21). Compared to synthetic routes involving free NHCs that proceed at room temperature, this one-pot reaction requires slightly harsher conditions and occurs under reflux to efficiently activate the *in situ* deprotonation of the imidazolium salt by the acetylacetonato ligand. Although the free acetylacetonate anion is not basic enough to remove the C2–H proton of azolium salts, the presence of

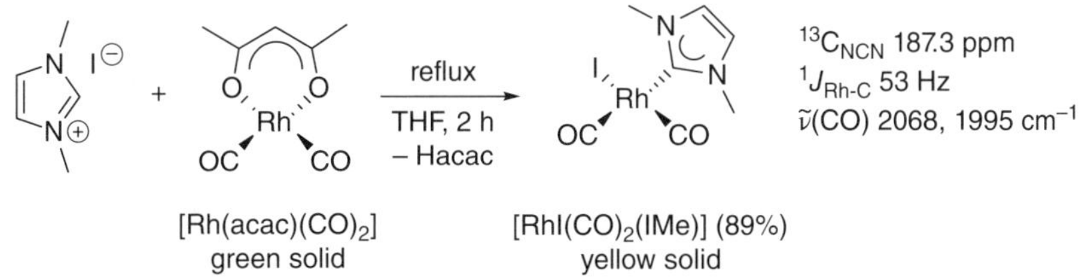

Scheme 7.21
Preparation of a rhodium-dicarbonyl complex using a basic metal precursor.

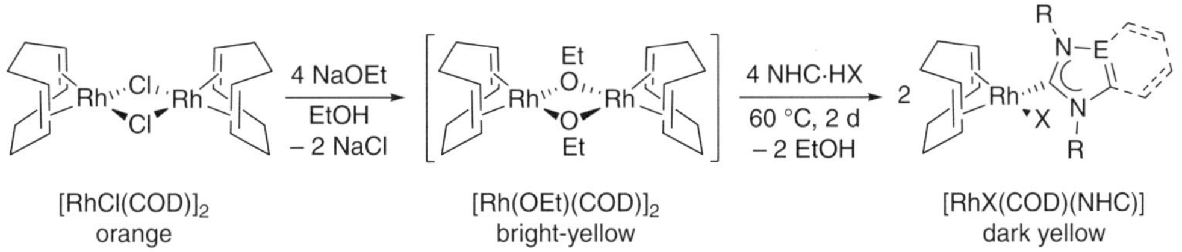

Scheme 7.22
One-pot synthesis of mixed COD/NHC complexes of rhodium(I).

suitable metal ions can drive complex formation in this case, possibly through a concerted deprotonation-coordination process. The trapping of minute amounts of free NHCs formed in such acid–base equilibria via metal coordination can also contribute to formation of the product in accordance with Le Chatelier's principle. The pentane-2,4-dione thus formed is a much poorer ligand, and easy replacement by the stronger carbene and the iodido donor takes place.

The obtained yellow iodido complex shows similar solubilities to its chlorido analog *cis*-[RhCl(CO)$_2$(IMe)]. However, comparison of spectroscopic features of the two isoelectronic and isostructural derivatives reveals an expected influence of the halido ligand. The electron richer and softer iodido ligand induces greater electron density into the complex, which is expressed in the more downfield chemical shift for the carbene carbon and enhanced backdonation to the carbonyl ligands, as evidenced by the smaller wavenumbers of their IR stretches (*cf.* $\tilde{v}(CO)_{av}$ of 2032 for I$^-$ *vs.* 2041 cm^{-1} for Cl$^-$). Since complexes of the type *cis*-[RhX(CO)$_2$(NHC)] are often used to determine donating abilities of NHCs, it is critical to avoid any possibility of halide scrambling in such complex probes. Moreover, a comparison across systems with different halido ligands is obviously not valid.

Dinuclear rhodium precursors with basic ligands have also been successfully used for this type of reaction [21]. They can be prepared *in situ* from common sources such as the bis[(μ-chlorido)dicarbonylrhodium(I)] dimer. The addition of excess sodium ethoxide to a solution of this compound in ethanol leads to a color change from orange to bright yellow, which indicates the exchange of the anionic bridging ligands, and the μ-ethoxido species [Rh(OEt)(COD)]$_2$ is eventually formed, isolation of which is not necessary. The increased basicity of the bridging alkoxido ligands can be explored for metal carbene synthesis, and on treatment with excess azolium salts concerted deprotonation and coordination takes place leading to the formation of the respective mononuclear NHC complexes of the formula [RhX(COD)(NHC)] in good yields (Scheme 7.22).

Table 7.2 Selected rhodium(I) complexes with classical NHCs and their key NMR data.

NHC			
X	Br	I	I
$\delta\,^{13}C_{carbene}$ **[ppm]**	185.3	186.1	196.4
$^{1}J_{C\text{-}Rh}$ **[Hz]**	51.5	42.6	47.8

The halides of the azolium salts complete the square planar coordination sphere as additional anionic ligands. Notably, this reaction is equally applicable to benzimidazolium, triazolium, and imidazolium salts (Table 7.2). Among these, the reaction rate was observed to decrease in the order benzimidazolium > triazolium > imidazolium in accordance with the decreasing acidities of their NCHN protons. Moreover, excess base and azolium salt did not lead to the formation of cationic bis(NHC) complexes of the type [Rh(COD)(NHC)$_2$]X, which were obtained in similar reactions involving excess free NHCs. Thus, free carbenes were ruled out as intermediates by the authors.

An analogous, but slightly modified approach was explored by Baker and coworkers [22]. Contrary to the above observation, they noted that the *in situ* use of the methoxido-bridged dimer [Rh(OMe)(COD)]$_2$ in place of its ethoxido analog with excess base and 1,3-dimethylimidazolium salts as described above does indeed lead to the formation of undesired cationic bis(carbene) complexes. This difference in reactivity between these two very similar systems is remarkable and difficult to comprehend, since the basicities of the two alkoxides are very similar. Further, it underlines the fact that small changes can often result in very different chemical reactivities. On the other hand, the application of the isolated and well-defined [Rh(OMe)(COD)]$_2$ precursor provides the exact amounts of base, and its reaction with excess carbene precursors at room temperature in aprotic solvents gave the best yields of the desired neutral, mono-NHC complexes (Scheme 7.23). Moreover, scrupulous precautions to exclude air or moisture are not required in this procedure. The bromido and iodido complexes can conveniently be prepared from their respective salts. Moreover, complexes with thiocyanato (*S*CN), selenocyanato (*Se*CN) and isocyanato (*N*CO) ligands can be prepared using a mixture of the hexafluorophosphate imidazolium salt and the potassium salt of the respective pseudohalide. An example involving 1,3-dimethylbenzimidazolium hexafluorophosphate and tetraethylammonium chloride, which led to the chlorido-benzimidazolin-2-ylidene complex (not shown) was also reported. Notably, the best route to the chlorido complex [RhCl(COD)(IMe)] required the use of the tosylate imidazolium salt in conjunction with lithium chloride. This combination offered better separation from small amounts of the bis(NHC) byproduct. Subsequent exposure of the chlorido complex to excess sodium azide furnished the azido complex [Rh(N$_3$)(COD)(IMe)] in excellent yields. With exception from the orange selenocyanato complex [Rh(*Se*CN)(COD)(IMe)], all other compounds were isolated as yellow solids.

The various co-ligands X do not show a systematic influence on the chemical shift of the carbene donor or its coupling constant $^{1}J_{C\text{-}Rh}$ with the rhodium center (Table 7.3). The lack

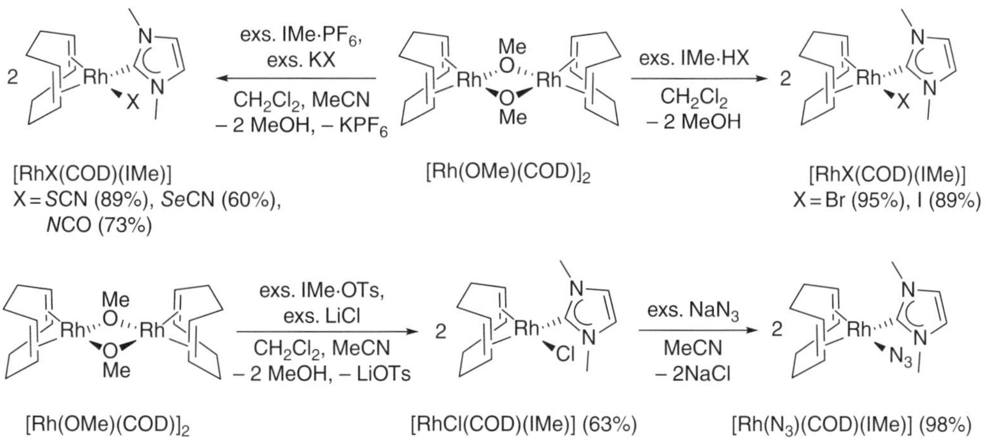

[RhX(COD)(IMe)]
X = SCN (89%), SeCN (60%),
NCO (73%)

[Rh(OMe)(COD)]₂

[RhX(COD)(IMe)]
X = Br (95%), I (89%)

[Rh(OMe)(COD)]₂

[RhCl(COD)(IMe)] (63%)

[Rh(N₃)(COD)(IMe)] (98%)

Scheme 7.23
Preparation of IMe-rhodium(I) complexes with various anionic co-ligands.

Table 7.3 Selected spectroscopic data of [RhX(COD)(IMe)] complexes.

X	$\delta\,C_{NCN}$ [ppm]	$^1J_{C\text{-}Rh}$ [Hz]	$\delta\,C=C_{trans}$ [ppm]	$^1J_{C2trans\text{-}Rh}$ [Hz]	Rh-C_{2trans} [Å][c]
Cl[a]	182.8	50.9	67.8	14.6	1.98₅
Br[b]	183.0	49.8	68.9	14.5	1.99₁
NCO[b]	182.4	51.3	69.4	13.7	1.99₄
N₃[b]	182.5	52.8	69.4	13.6	1.99₆
I[b]	182.4	48.6	71.2	14.0	2.00₆
SCN[a]	180.7	54.2	74.3	ND	2.01₃
SeCN[a]	188.5	49.6	74.5	12.5	2.01₄

[a] NMR data recorded in CDCl₃.
[b] NMR data recorded in acetone-d_6.
[c] Rhodium–olefin distances are shown relative to the centroid of the carbon–carbon bond. Where more than one molecule is present in the structure, Rh–olefin distances shown are averages.

of any simple electronic correlation is due to the fact that the two ligands are mutually *cis* to each other, and therefore interferences between electronic and steric factors cannot be avoided. This is particularly so, since the shapes of the anionic ligands and thus their orientation with respect to the IMe ligand are quite different. Figure 7.7 depicts the solid state molecular structures of complexes with selected halido and pseudohalido ligands. In particular, the orientations of three-atomic pseudohalido ligands differ greatly dependent on the number of electron pairs at the donor atoms and the overall charge distribution.

On the other hand, there appear to be some discernible electronic effects of the various anionic ligands on the *trans*-standing double bond of the COD ligand. The distance between the metal and the C=C$_{trans}$ centroid correlates well with the carbon chemical shifts of the latter (Table 7.3). This observation is in accordance with the study of Crabtree and Quirk [23] suggesting that an upfield ^{13}C NMR chemical shift reflects increasing metallocyclopropane character of the rhodium–olefin bond. This in turn would result in shorter Rh–C$_{2trans}$ distances.

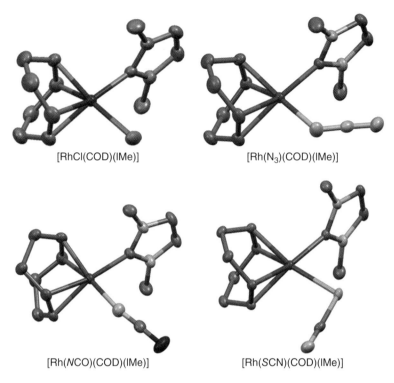

Figure 7.7
Rhodium(I)-NHC complexes with halido and pseudohalide ligands.

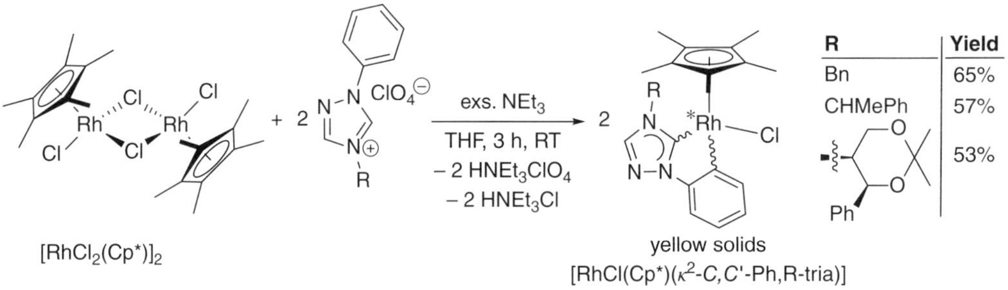

Scheme 7.24
Preparation of cyclometallated [Rh^{III}(Cp*)(NHC)] complexes via *in situ* deprotonation with external base.

The preparation of complexes via *in situ* deprotonation of azolium salts using of an external base in conjunction with a rhodium source has the advantage that the metal precursors are simpler and often cheaper as well. An early example for the syntheses of rhodium(III)-NHC complexes was reported by Enders in 1997 [12]. Stirring a mixture of the dichlorido-pentamethylcyclopentadienyl rhodium(III) dimer [RhCl$_2$(Cp*)]$_2$ with two equivalents of 1,2,4-triazolium perchlorates and excess triethylamine in tetrahydrofurane for three hours at room temperature is already sufficient to yield cyclometallated rhodium(III) carbene complexes (Scheme 7.24). In the presence of metal, which can trap any free NHCs formed, the amine base is active enough to affect *in situ* deprotonation of a triazolium salt. The latter also belongs to the more acidic ones among the classical carbene precursors. Overall, the chemistry is similar to that described for the ruthenium(II)-cymene analogs (see Section 7.1.3), but the yields and selectivities were observed to be lower. The products containing a η^5-Cp*, a chlorido and a κ^2-C,C' cyclometallated triazolin-5-ylidene ligand exhibit a half-sandwich structure and are chiral at metal. With non-chiral carbene ligands (R = Bn), a racemate of R_{Rh} and S_{Rh} complexes are obtained.

Chiral induction can be achieved with enantiopure ligands bearing a chiral wing tip group, which leads to the enrichment of one among two possible diastereomers. For the (*R*)-phenylethyl wing tip group, a moderate diastereomeric excess of *de* = 50% was achieved for the (R_{Rh},R_{NHC}) diastereomer. An improved *de* = 83% for the R_{Rh} diastereomer was obtained for the (*S,S*)-2,2-dimethyl-4-phenyl-l,3-dioxan-5-yl substituted ligand system. Nevertheless, these *de* values are significantly lower than those obtained for the ruthenium-cymene analogs, which was attributed to a slow epimerization involving halido ligand exchange that occurs favorably in coordinating solvents.

The cyclometallated complexes show very similar key ^{13}C NMR spectroscopic features. In CDCl$_3$, all carbene donors resonate in the very narrow range of 185.2–185.8 ppm with heteronuclear coupling constants of $^1J_{Rh\text{-}C}$ = 56.2–56.8 Hz. The anionic and more electron-rich σ-aryl donor atoms resonate at more shielded regions with chemical shifts of 155.5–155.6 ppm and decreased coupling constants of $^1J_{Rh\text{-}C}$ = 33.5–34.0 Hz.

For less C2–H acidic azolium salts, stronger Brønsted bases need to be used. Potassium *tert*-butoxide (KO^{t}Bu) is a popular reagent for the *in situ* deprotonation. However, its usage requires dry and aprotic solvents, such as tetrahydrofurane.

7.2.3 The Silver-Carbene Transfer Route

A particularly convenient and mild route for the preparation of rhodium(I)-NHC complexes involves the transfer of the carbene from their readily available silver(I) complexes. The latter can be easily prepared by reaction of the azolium salts with silver(I) oxide as discussed in the coinage metal chapter. An early example of this approach was reported by Crabtree and coworkers, who prepared the silver(I) complex of 1,3-di(4-tolylmethyl)imi-dazolin-2-ylidene (tmiy) [Ag(tmiy)$_2$][AgCl$_2$] and reacted it with the chlorido-cyclooctadi-ene-rhodium(I) dimer [RhCl(COD)]$_2$ in dichloromethane (Scheme 7.25) [24]. A 1:1 ratio of carbene to metal was kept in this reaction, which proceeded mildly at room temperature and was complete within one hour. The NHC ligand transfers from silver to rhodium with concurrent bridge-cleavage of the dimeric rhodium precursor driven by the favorable pre-cipitation of silver chloride. To ensure a smooth reaction, a halide source is thus required. Here, an imidazolium chloride is used, and no halide scrambling can occur, exclusively yielding the desired [RhCl(COD)(tmiy)] complex in excellent yields. The characteristic ^{13}C NMR data of the complex are comparable to isostructural analogs mentioned earlier. ^{1}H NMR spectroscopic studies reveal diastereotopy for the benzylic protons (AB pattern) and broadening of these signals was not observed in toluene-d_8 up to 105 °C, which is consistent with a Rh–C rotational barrier of >22 kcal mol^{-1}. The hampered rotation can be attributed to steric factors, as NHC complexes with smaller N-substituents show fluxionality as a result of greater rotational freedom.

As described above, the COD ligand in neutral [RhX(COD)(NHC)] type complexes can be easily displaced by carbon monoxide. Passing CO gas through a solution of [RhCl(COD)

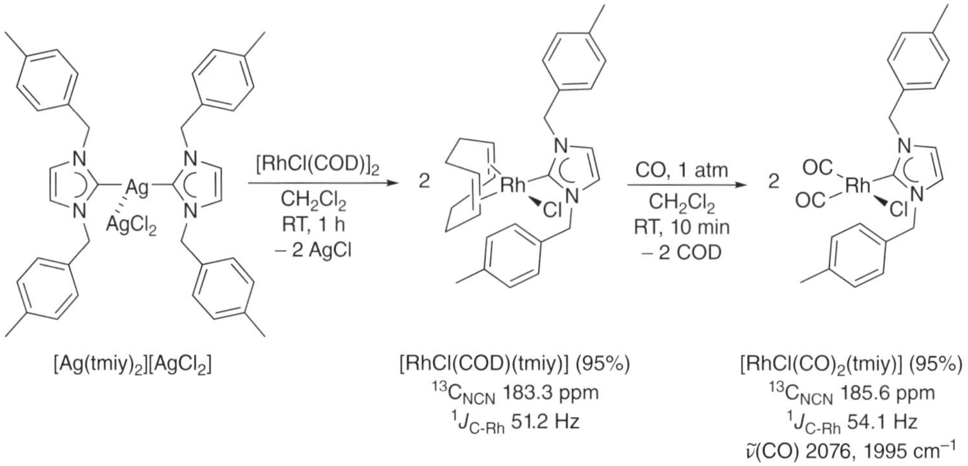

Scheme 7.25
Preparation of a rhodium(I)-NHC complex via silver-carbene transfer.

(tmiy)] for only 10 minutes readily affords the bis(carbonyl) complex *cis*-[RhCl(CO)$_2$(tmiy)] in a high yield. The spectroscopic features of this complex are very similar to that of its 1,3-dimethylimidazolin-2-ylidene counterpart *cis*-[RhCl(CO)$_2$(IMe)] (see Scheme 7.17). A detailed comparison of the average wavenumbers for the carbonyl stretches indicates that the tmiy ligand (~2036 cm^{-1}) is a slightly stronger donor than IMe (~2041 cm^{-1}).

The application of azolium halides other than chloride in this type of silver-carbene transfer reactions can result in anion scrambling leading to [RhX(COD)(NHC)] (X = halide) product mixtures that are difficult to separate due to their similarities. Nevertheless, the silver(I) ion preferably precipitates with the heavier halide, and the complex with the lighter halide is expected to be the major product.

If azolium salts with non- or weakly-coordinating anions (e.g. BF$_4^-$ or PF$_6^-$ etc.) are used in the transmetallation, it is recommendable to add an alkali chloride to prevent halido abstraction from the rhodium coordination sphere.

On the other hand, such a chlorido abstraction can be purposely exploited to prepare cationic bis(NHC) complexes of rhodium(I) as reported by Leitner and coworkers [25]. They prepared the silver(I) bis(NHC) complex [Ag(EMIM)$_2$]NTf$_2$ by reaction of 1-ethyl-3-methylimidazolium bis(trifluoromethylsulfonyl)amide (EMIM·H$^+$NTf$_2^-$) with silver(I) oxide (Scheme 7.26). Subsequent exposure of the well-defined silver complex with half an equivalent of the rhodium-COD dimer led to the concurrent transfer of both NHCs to rhodium to give the cationic bis(carbene) complex [Rh(COD)(EMIM)$_2$]NTf$_2$ in high yields. The chlorido ligand has been abstracted by the silver ion resulting in silver chloride formation as a major driving force for the double transmetallation. Due to the unsymmetrical nature of the NHC ligands, the complex was obtained as a pair of *syn*- and *anti*-rotamers, which differ in the orientation of the N-substituents with respect to each other. Scheme 7.26 depicts only the supposedly more preferred *anti*-rotamer. Notably, the same complex salt cannot be obtained using excess of the respective mono-NHC-halido silver complexes, where transmetallation stopped after a single NHC transfer. Also, previous attempts to prepare cationic [Rh(COD)(NHC)$_2$]$^+$ complexes by chlorido abstraction from [RhCl(COD)

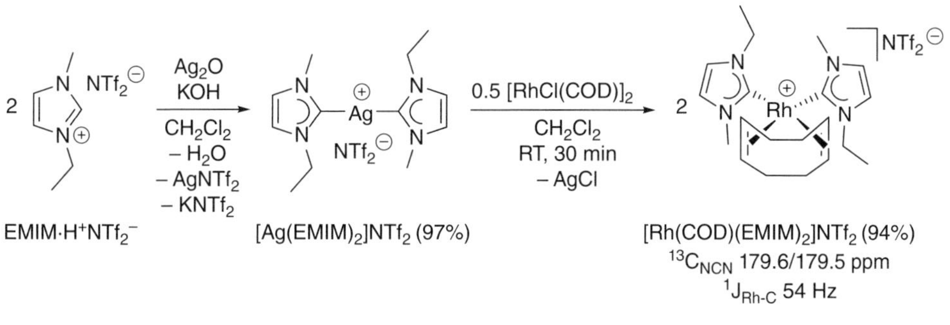

Scheme 7.26
Concurrent transmetallation of two NHCs from silver to rhodium.

(NHC)] with $AgPF_6$ and addition of $[(NHC)AgCl]_n$ (n = 1, 2) transmetallating agents were to no avail.

The $[Rh(COD)(EMIM)_2]NTf_2$ complex salt exhibits a better solubility in common organic solvents compared to its IMe analog prepared via the free NHC route (see Section 7.2.1). This can be reasoned by the unsymmetrically substituted NHC ligand, which prevents efficient packing in the solid state. Its ^{13}C NMR spectrum shows two closely spaced signals for the carbene donors indicating the presence of *syn-* and *anti*-rotamers in solution. These resonances are in the expected range and shifted upfield in comparison to those of neutral $[RhCl(COD)(NHC)]$ complexes with a single imidazolin-2-ylidene, which is in line with the cationic charge.

7.3 Iridium(I)- and Iridium(III) NHC Complexes

7.3.1 Cleavage of Enetetramines and the Free Carbene Route

Cleavage of Enetetramines

As with many transition metals, the first NHC complexes of iridium were obtained by Lappert and coworkers in 1973 via the cleavage of electron-rich enetetramines [26,27]. Depending on the stoichiometry and reaction conditions, several ligands can be displaced by the incoming carbene ligands. Thus, heating a mixture of the iridium(I)-bis(phosphine) complex *trans*-$[IrCl(CO)(PPh_3)_2]$ and sodium tetrafluoroborate in acetone at reflux led to the substitution of only the chlorido ligand by the saturated 1,3-dimethylimidazolidin-2-ylidene (SIMe) ligand, and the cationic complex *trans*-$[Ir(CO)(SIMe)(PPh_3)_2]BF_4$ is obtained as yellow-orange crystals after purification in 94% yield (Scheme 7.27, right).

The introduction of three carbenes can be achieved in a two-step reaction protocol. This is accomplished by initial heating of a suspension of the same iridium precursor and the carbene dimer in toluene under reflux for 10–15 minutes (Scheme 7.27, left). The higher reaction temperature obtained in refluxing toluene also leads to the displacement of the phosphine donors. The precipitate formed, supposedly $[Ir(CO)(SIMe)_3]Cl$, is subsequently stirred with sodium tetrafluoroborate in acetone for an additional hour, and an anion exchange occurs leading to the formation of the final tris(carbene) complex $[Ir(CO)(SIMe)_3]BF_4$ in 77% yield. Both complex salts are thermally stable to >200 °C. Based on these observations, the rate of substitution at *trans*-$[IrCl(CO)(PPh_3)_2]$ was observed to

n = 3   n = 1

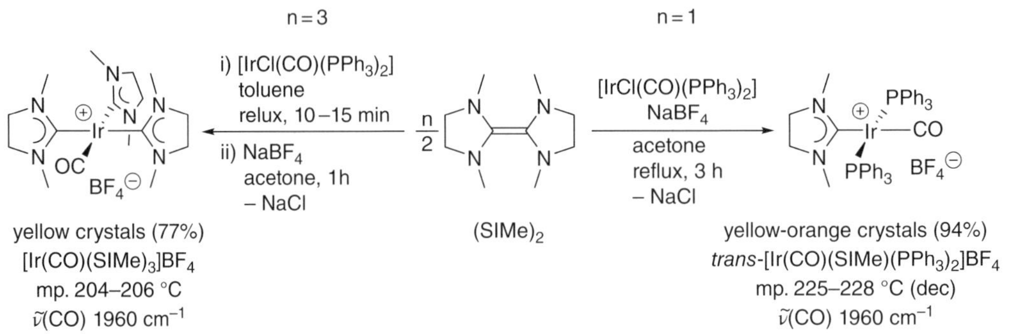

| | i) [IrCl(CO)(PPh$_3$)$_2$] toluene relux, 10–15 min | | [IrCl(CO)(PPh$_3$)$_2$] NaBF$_4$ | |
| | ii) NaBF$_4$ acetone, 1h – NaCl | $\frac{n}{2}$ | acetone reflux, 3 h – NaCl | |

yellow crystals (77%)   (SIMe)$_2$ yellow-orange crystals (94%)
[Ir(CO)(SIMe)$_3$]BF$_4$ *trans*-[Ir(CO)(SIMe)(PPh$_3$)$_2$]BF$_4$
mp. 204–206 °C   mp. 225–228 °C (dec)
$\tilde{v}$(CO) 1960 cm^{-1} $\tilde{v}$(CO) 1960 cm^{-1}

Scheme 7.27
Preparation of the first iridium-NHC complexes via olefin cleavage.

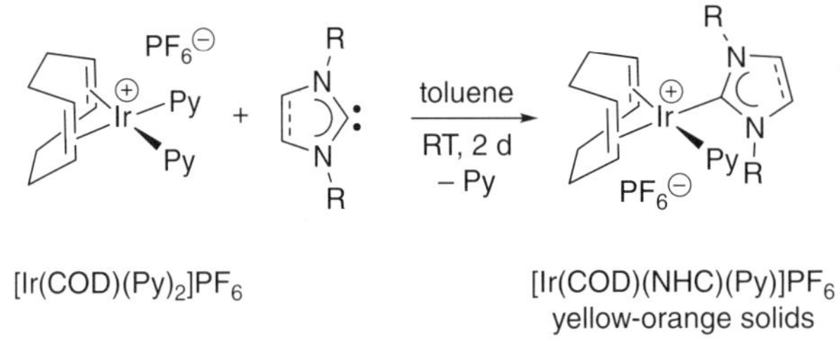

NHC	Yield	$^{13}C_{NCN}$(CD$_2$Cl$_2$)
SIMes	80%	NR
IMes	82%	174.0 ppm
IPr	82%	175.6 ppm
ICy	63%	174.9 ppm

[Ir(COD)(Py)$_2$]PF$_6$ [Ir(COD)(NHC)(Py)]PF$_6$
                                                yellow-orange solids

Scheme 7.28
Preparation of cationic iridium-NHC complexes using free carbenes.

decrease in the order Cl > PPh$_3$ > CO. However, this order cannot be generalized as phosphine replacement occurs first in the analogous rhodium complex. The CO stretch for the tris(NHC) complex is 20 wavenumbers smaller than that recorded for the mixed NHC/phosphine system. This observation undoubtedly indicates a greater amount of backbonding from iridium(I) to the carbonyl ligand in the tris(NHC) complex due to the presence of three strongly donating carbenes as opposed to only one NHC in the mixed bis(phosphine) complex.

The Free Carbene Route

Closely related to the olefin cleavage route are reactions involving free NHCs. In the simplest case, the free carbene replaces a weaker ligand from the coordination sphere of the iridium center. In an report by Nolan and coworkers, this route has been explored to prepare NHC analogs of Crabtree's catalyst [Ir(COD)(PCy$_3$)(Py)]PF$_6$ as potential catalyst precursors for hydrogenation reactions [28,29].

A series of common free NHCs were reacted with the cationic bis(pyridine)iridium(I) complex [Ir(COD)(Py)$_2$]PF$_6$ In a straightforward manner (Scheme 7.28). The less stable carbenes SIMes and ICy were freshly prepared by deprotonation of their chloride salts with KO^{t}Bu in THF, and subsequently extracted using toluene for further reactions. On the other hand, the more stable imidazolin-2-ylidenes IPr and IMes were used as pre-isolated species. Since the chosen carbenes are all bulky, the formation of bis(carbene) complexes is

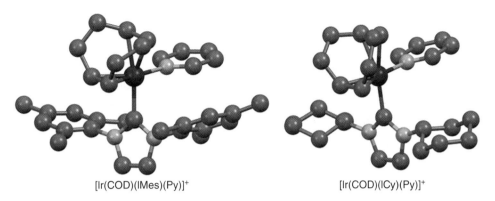

[Ir(COD)(IMes)(Py)]⁺   [Ir(COD)(ICy)(Py)]⁺

Figure 7.8
Representative molecular structures of the [Ir(COD)(IMes)(Py)]⁺ and [Ir(COD)(ICy)(Py)]⁺ complex cations.

sterically disfavored and a slight excess could be used without any problems to accelerate the substitution. Still the reactions were slow at room temperature and took about two days to complete. This can be partially attributed to the low solubility of the charged iridium(I) complex precursor in non-polar toluene, which results in a heterogeneous reaction. Moreover, the slow reaction rate also suggests that the pyridine ligands are tightly bound to the cationic and thus rather Lewis acidic iridium center in this precursor. Nevertheless, carbenes are stronger donors and can stabilize cationic complexes even better, leading to an enthalpically driven one-to-one pyridine replacement. All target complexes of the type [Ir(COD)(NHC)(Py)]PF$_6$ were obtained in moderate to good yields as yellow-orange powders after a simple filtration step and washing with hexane. They can be considered analogs of Crabtree's catalyst, in which the tricyclohexylphosphine (PCy) has been replaced with a carbene. ¹H NMR studies provide evidence for a fluxional behavior involving the rotation of the N–C$_{Ar}$ in complexes with aromatic N-wing tips. As anticipated, this rotation can be slowed down at lower temperatures. For example, two distinct *o*-CH$_3$ signals are detected for the IMes complex on cooling a NMR sample to 233 K. The ¹³C$_{carbene}$ signals of the three unsaturated imidazolin-2-ylidene complexes are detected in the narrow range of 174.0–175.6 ppm without any clear trend. NMR monitoring of samples over several weeks also show signs of decomposition due to the lability of some ligands in solution.

Single crystals of all four compounds were grown from diffusion of diethylether into saturated solutions in dichloromethane. Their analysis by X-ray diffraction confirmed the structure also found in solution, and Figure 7.8 shows those of the IMes and ICy complexes as representatives. Notably, the Ir–C$_{carbene}$ bond of 2.062(2) Å in the ICy complex was found to be the shortest, while those of all the other three complexes were essentially equal within 3σ averaging to 2.077 Å. The supposedly stronger Ir–C$_{carbene}$ bond was ascribed to a combination of the superior σ-donating strength as well as the reduced steric bulk of the ICy ligand.

Testing of the complexes for the catalytic hydrogenation of ketones and olefins, revealed that the [Ir(COD)(ICy)(Py)]PF$_6$ was the most active species. Its performance compared favorably with many previously studied ruthenium and cationic iridium complexes. Overall, the NHC complexes showed an enhanced tolerance to elevated temperatures, at which many other more active catalysts, for example Crabtree's catalyst, quickly deactivate.

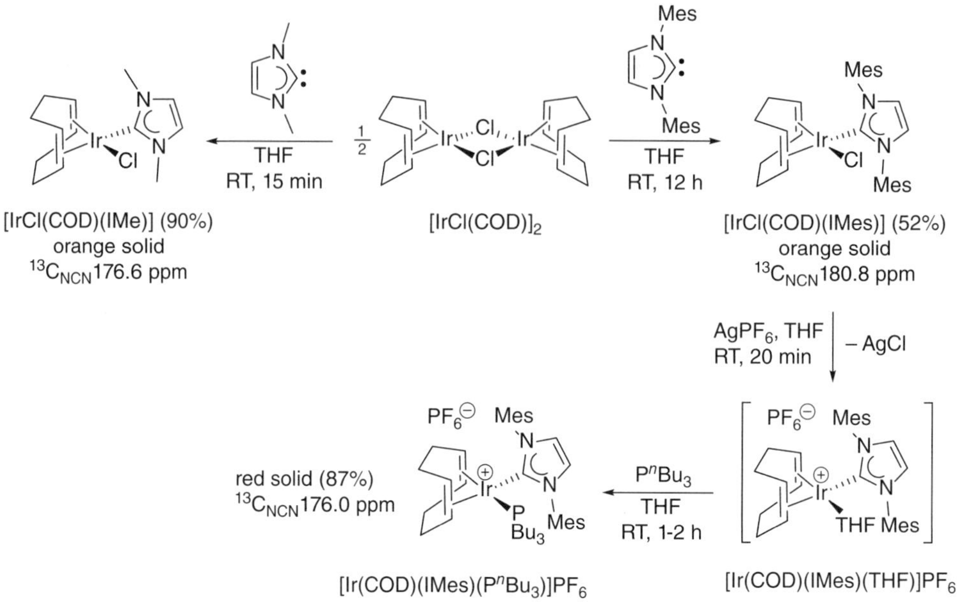

Scheme 7.29
Cleavage of a iridium(I) dimer by IMes and subsequent ligand substitutions.

In analogy to rhodium(I) NHC chemistry, it is also feasible to obtain iridium(I) analogs via bridge-cleavage of suitable halido-bridged dimeric iridium(I) precursors. For example, the dimer $[IrCl(COD)]_2$ can be easily cleaved by using two equivalents of the IMe or IMes ligand as reported by the groups of Herrmann [8] and Buriak [30] (Scheme 7.29), respectively.

These reactions proceed under mild conditions by simply stirring a mixture of the reactants in tetrahydrofurane at room temperature, and are indicated by a color change to dark orange. Notably, the reaction with the sterically unassuming IMe ligand occurred within minutes and also gave a high yield of 90%, while longer reaction times were required for the more bulky and less nucleophilic IMes ligand. Moreover, the [IrCl(COD)(IMes)] product was crystallized from a tetrahydrofurane/isopropanol solution (6:7) in a lower yield of 52% as an orange solid.

The $^{13}C_{carbene}$ signal of [IrCl(COD)(IMe)] was detected at 176.6 ppm in $CDCl_3$. The aromatic mesityl wing tip groups have a deshielding effect, and the respective signal for [IrCl(COD)(IMes)] was therefore detected downfield at 180.8 ppm in $CDCl_3$. Also, the latter resonance is as expected more downfield compared to that found for the cationic and more Lewis acidic $[Ir(COD)(IMes)(Py)]PF_6$ complex (see Scheme 7.28).

The [IrCl(COD)(IMes)] complex was subsequently used to prepare mixed carbene/ phosphine complexes as alternative analogs of Crabtree's catalyst $[Ir(COD)(PCy_3)(Py)]$ PF_6, in which the pyridine has formally been replaced by a NHC. The removal of the chlorido ligand by $AgPF_6$ in tetrahydrofurane supposedly leads to the solvate $[Ir(COD)(IMes)$ $(THF)]PF_6$ as an intermediate. The addition of tri(n-butyl)phosphine then rapidly leads to the formation of the mixed carbene/phosphine complex $[Ir(COD)(IMes)(P^nBu_3)]PF_6$, which was purified by crystallization from tetrahydrofurane/hexane (1:1) and isolated as a red solid in 87%. The $^{13}C_{carbene}$ resonance of this complex was detected upfield from its

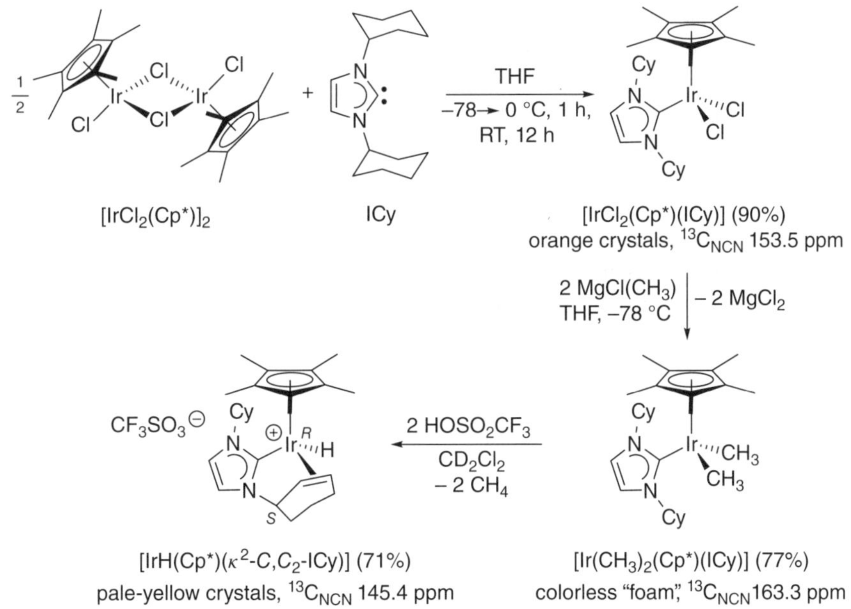

Scheme 7.30
Cleavage of a iridium(III) dimer by ICy and subsequent ligand substitutions.

neutral precursor in agreement with the built-up of positive charge by removal of an anionic ligand.

Notably, the catalytic activity of this mixed NHC/phosphine complex in the hydrogenation of alkenes was increased compared to that of the mixed NHC/pyridine complexes discussed above. For unhindered olefins, it was more active than its less bulky counterpart [Ir(COD)(IMe)($P^{n}Bu_3$)]PF_6 prepared via a different approach. However, the latter was superior in the conversion of tertiary and quaternary alkenes.

The same strategy can also be applied to dimeric iridium(III) precursors. In fact, the preparation of an iridium(III) NHC complex by this route even predates the above mentioned example. In 2000, Herrmann and coworkers reported the reaction of the iridium(III) dimer [IrCl$_2$(Cp*)]$_2$ with two equivalents of 1,3-dicyclohexylimidazolin-2-ylidene [31]. In this early example, the two reactants were carefully mixed at a very low temperature of −78 °C, which according to later reports is not strictly required. Subsequent warming to 0–°C and further stirring at room temperature eventually afforded the product [IrCl$_2$(Cp*) (ICy)], which was purified by column chromatography and obtained as air stable orange crystals in a high yield of 90% (Scheme 7.30).

The dichlorido complex was then subjected to a double ligand substitution on exposure to two equivalents of the methyl Grignard reagent MgCl(CH$_3$) at low temperature under salt metathesis conditions. The resulting dimethyl complex [Ir(CH$_3$)$_2$(Cp*)(ICy)] contains three different types of carbon donors and was isolated as a colorless foam. Further protonolysis of the methyl ligands with trifluoromethanesulfonic acid led to the immediate evolution of methane gas and formation of a coordinatively unsaturated iridium(III) intermediate. The latter triggered an interesting C–H bond activation process of one cyclohexyl substituent, and the new cationic hydrido complex [IrH(Cp*)(κ^2-C,C$_2$-ICy)] was obtained that contains an unusual cyclohexene-functionalized carbene ligand coordinating in a chelating $\kappa^2\text{-}C,C_2$ mode. The complex transformations from dichlorido to dimethyl and finally to hydrido species are accompanied by characteristic changes of the $^{13}C_{carbene}$ resonances from 153.5 ppm → 163.3 ppm → 145.4 ppm (CD_2Cl_2), which are reflective of the Lewis acidities of the metal center in each complex. The final product contains chiral centers at the iridium and at the α-carbon of the newly generated cyclohexenyl N-substituent. Overall, the four isomers Ir_RC_R, Ir_RC_S, Ir_SC_R, and Ir_SC_S are possible. However, its ^1H NMR spectrum only shows one hydrido singlet at −16.30 ppm, which indicates the presence of a single enantiomeric pair. In line with the X-ray structural data and the centrosymmetric space group, the assumption was made that only the two enantiomers Ir_SC_R and Ir_RC_S were formed.

7.3.2 *In Situ* α-Elimination

In certain cases, the treatment of azolium salts with bases does not lead to deprotonation and the formation of free carbenes. Instead, nucleophilic addition to the C2-carbon occurs, which affords neutral carbene adducts. In particular, triazolium and imidazolinium salts have been described to form such adducts, which can form free carbenes under α-elimination of small molecules.

The exploitation of this concept for the preparation of iridium-NHC complexes was described by Herrmann and coworkers in 2002 [32]. The authors generated the butanol adduct of SIMes by reacting the 1,3-dimesitylimidazolinium chloride with potassium-*tert*-butoxide in tetrahydrofurane. After formation of the adduct, which is indicated by a color change to light yellow, a solution of the dimeric iridium(I) precursor in toluene was added.

α-Elimination of *tert*-butanol supposedly occurs upon heating to 80 °C, and the free imidazolidin-2-ylidene thus formed attacks the coordinatively unsaturated iridium(I) dimer affording the four-coordinated complex [IrCl(COD)(SIMes)] as a light orange solid in a decent yield of 69% (Scheme 7.31). The saturated carbene resonates at 207.0 ppm, which is as expected more downfield compared to the carbene signal in the complex [IrCl(COD)(IMes)] containing the unsaturated NHC (see Scheme 7.29).

The same product can also be obtained by heating the iridium(I) dimer with the pentafluorophenyl adduct of the SIMes [33]. This imidazolidine can be prepared in a milder way

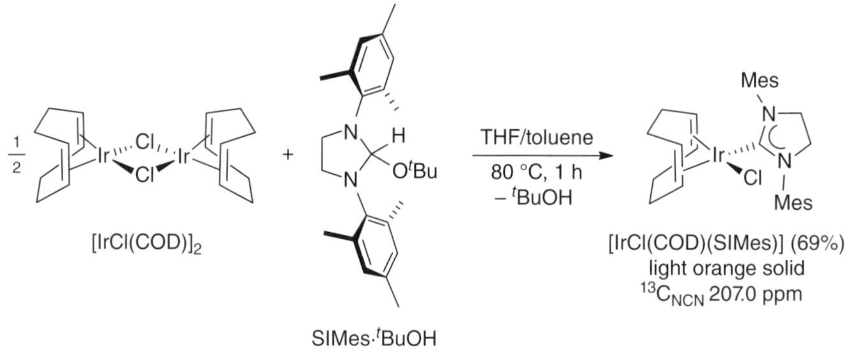

Scheme 7.31
Preparation of [IrCl(COD)(SIMes)] via α-elimination of *tert*-butanol.

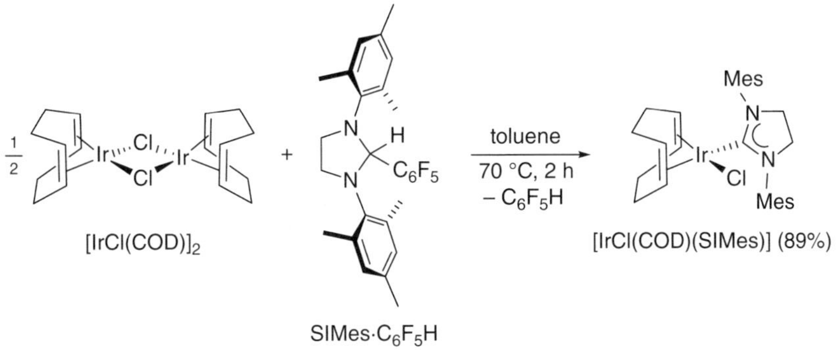

Scheme 7.32
Preparation of [IrCl(COD)(SIMes)] via α-elimination of pentafluorobenzene.

2-(trichloromethyl)
imidazolidines

2-(triethylborane)
imidazolines

2-(dimethylamino)
imidazolidines

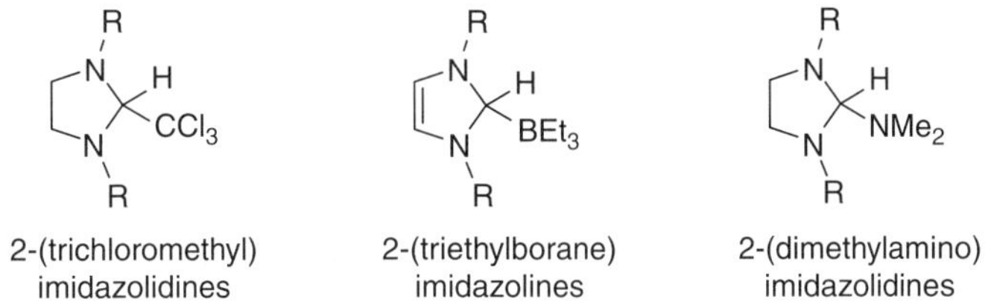

Scheme 7.33
Imidazolidines and imidazolines used for NHC generation via α-elimination.

without the use of any base by condensation of the respective substituted diamine with pentafluorobenzaldehyde. Moreover, it is much more stable and can be kept for more than nine months without any sign of decomposition. In addition, the α-elimination of pentafluorobenzene was reported to occur at milder reaction conditions. Thus the reaction was conducted at a lower temperature of 70 °C, while the yield could be improved to 89% (Scheme 7.32).

The ease of α-elimination of pentafluorobenzene from their respective imidazolidines is greatly influenced by the steric surroundings. Bulky aromatic wing tip groups with *ortho*-substituents are especially favorable. In such systems, the rotation of the pentafluorophenyl group at the C2-carbon is greatly hindered, and steric pressure reduces the activation barrier for the α-elimination.

Since the neutral imidazolidines are much more soluble than azolium salts, the α-elimination route is a useful alternative for the making of NHC complexes when other more common methods are problematic or only afford low yields. In addition to alcohol or pentafluorobenzene adducts, one could consider the use of 2-(trichloromethyl)imidazolidines [34], 2-(triethylborane)imidazolines [35] and 2-(dimethylamino)imidazolidines [36] (Scheme 7.33).

7.3.3   *In Situ* Deprotonation of Azolium Salts

Iridium complexes with sufficiently basic ligands can be directly used to react with acidic azolium salts to generate iridium-NHC complexes. This method is convenient as the handling of generally air and moisture sensitive free NHCs can be avoided. An early

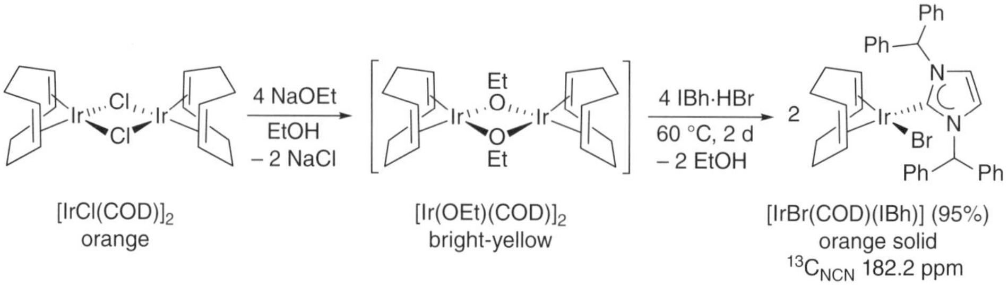

Scheme 7.34
One-pot synthesis of a mixed COD/NHC complex of iridium(I).

application of this route, which is also applicable to rhodium(I) NHC chemistry, was described by Herrmann and coworkers [21]. The suitable complex precursor [Ir(OEt) (COD)]$_2$ containing the basic ethoxide ligands was readily prepared by reaction of the chlorido-bridged dimer [IrCl(COD)]$_2$ with an excess of sodium ethoxide in ethanol (Scheme 7.34). Successful ligand exchange is indicated by a color change from orange to yellow, which occurred rapidly within five minutes. Subsequently and without isolation of the ethoxide species in substance, four equivalents of 1,3-dibenzhydrylimidazolium bromide (IBh·HBr) were introduced at once, and stirring of the mixture continued at 60 °C for two days, which gave the reaction sufficient time to complete. The formation of the target complex [IrBr(COD)(IBh)] in a high yield of 95% occurred via *in situ* deprotonation of the imidazolium salt by the bridging ethoxido ligands. In addition to the cleavage of the bridges, this process liberates one molecule of ethanol per carbene, which together with the bromide counter-anion instantaneously coordinates to the iridium center, furnishing the product. The presence of chloride anions in the reaction mixture could lead to halido scrambling, although binding of the softer bromido ligand is preferred according to the HSAB concept.

The *in situ* generation and direct reaction of the ethoxido-bridged dimer [Ir(OEt)(COD)]$_2$ using excess base may pose a problem in cases where the carbene precursors are less robust or carry non-innocent functions. Such a problematic scenario was encountered when 1,3-diallyl-benzimidazolium bromide (All$_2$-bimyH$^+$Br$^-$) was employed under the same reaction conditions with the intention to form the complex [IrBr(COD)(All$_2$-bimy)] [37]. Instead, the 1,3-di(*n*-propyl)benzimidazolin-2-ylidene complex [IrBr(COD)(Pr$_2$-bimy)] was isolated in 55% yield after stirring for two days at 60 °C (Scheme 7.35). In the course of the complex formation, additional hydrogenation of both allyl wing tip groups has occurred due to the presence of excess sodium ethoxide. In fact, base-promoted oxidation of alcohols are known to occur with ease under such conditions in the presence of catalytic amounts of iridium, rhodium, or ruthenium ions.

Here it is likely that the initially targeted complex indeed forms in the first step. However, bromido-ethoxido substitution then takes place furnishing a new alkoxido intermediate, which undergoes internal *β*-hydride abstraction to give a hydrido complex under elimination of one molecule of acetaldehyde. Subsequent olefin-insertion under formation of an alkyl complex and its protonation by the ethanol solvent completes the reduction of the olefin, while another equivalent of ethoxide is generated to close the sequence. The reaction halts after both N-allyl substituents of the NHC have been converted to N-propyl groups via the "hydrogen borrowing" mechanism, where formally the dihydrogen "borrowed" from ethanol is "returned" to the olefin.

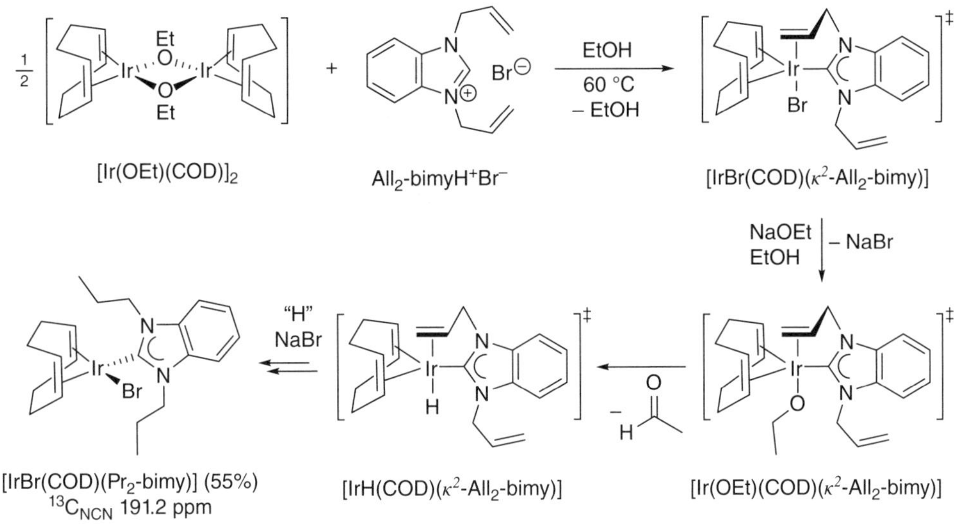

Scheme 7.35
"Self-hydrogenation" of a 1,3-diallylbenzimidazolin-2-ylidene complex.

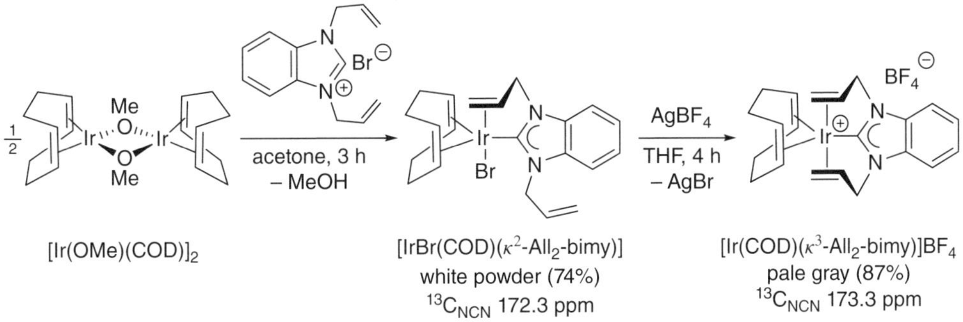

Scheme 7.36
Preparation of allyl-functionalized NHC complexes of iridium(I).

To circumvent this undesirable hydrogenation, the pre-isolated and well-defined basic dimer $[Ir(OMe)(COD)]_2$ can be used. Stirring it with two equivalents of the All_2-bimyH$^+$Br$^-$ salt leads to the clean formation of the five-coordinated complex $[IrBr(COD)(\kappa^2\text{-}All_2\text{-}bimy)]$ with intact allyl groups, one of which is η^2 coordinated to the iridium center giving overall a κ^2-chelating mode of the NHC (Scheme 7.36). This results in inequivalent allyl groups, and the NMR spectra show two distinct sets of olefinic signals. The two carbon atoms of the olefin involved in the coordination give rise to substantially upfield shifted ^{13}C NMR resonances at 51.5 and 49.9 ppm, while those of the pendant counterpart come into resonance at 133.9 and 116.3 ppm, respectively. The latter signals are comparable to those found for the benzimidazolium salt, that is, 141.3 and 113.5 ppm.

Abstraction of the bromido ligand by silver(I) tetrafluoroborate enforces the coordination of the second allyl substituent, which gives rise to the cationic iridium complex $[Ir(COD)(\kappa^3\text{-}All_2\text{-}bimy)]BF_4$ containing solely metal-carbon bonds.

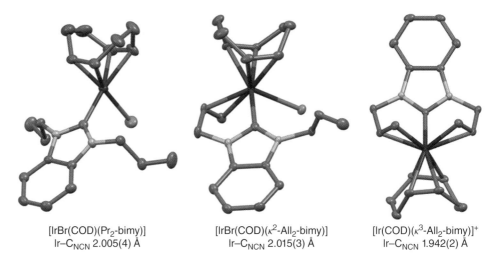

$$[IrBr(COD)(Pr_2\text{-bimy})]$$
$$Ir\text{--}C_{NCN}\ 2.005(4)\ Å$$

$$[IrBr(COD)(\kappa^2\text{-All}_2\text{-bimy})]$$
$$Ir\text{--}C_{NCN}\ 2.015(3)\ Å$$

$$[Ir(COD)(\kappa^3\text{-All}_2\text{-bimy})]^+$$
$$Ir\text{--}C_{NCN}\ 1.942(2)\ Å$$

Figure 7.9
Molecular structures of iridium-benzimidazolin-2-ylidene complexes.

The coordination number and thus the geometry of these three NHC complexes has a pronounced influence on the $^{13}C_{carbene}$ NMR signals. The four-coordinated [IrBr(COD)(Pr$_2$-bimy)] complex shows the most downfield chemical shift at 191.2 ppm for the carbene donor. Olefin binding leads to the formally five-coordinated species [IrBr(COD)(κ^2-All$_2$-bimy)] and complex [Ir(COD)(κ^3-All$_2$-bimy)]BF$_4$, for which the carbene signal was detected upfield at 172.3 and 173.3 ppm, respectively. The very similar chemical shifts are surprising given the fact that the complexes differ in their charges.

The molecular structures of the three complexes obtained from single crystal X-ray diffraction are depicted in Figure 7.9. They are fully consistent with solution data and confirm the different coordination numbers of the iridium centers. The structure of the cation [Ir(COD)(κ^3-All$_2$-bimy)]$^+$ is particularly interesting, since it resembles an armchair, where the carbene plane forms the backrest while the coordinated allyl groups represent the armrests. Notably, the iridium–carbene bond in this cationic "armchair" molecule of 1.942(2) Å is shorter than those found for the two neutral complexes of 2.015(3) and 2.005(4) Å, which are identical within 3σ despite the different coordination geometries.

The two neutral complexes [IrBr(COD)(Pr$_2$-bimy)] and [IrBr(COD)(κ^2-All$_2$-bimy)] were tested for their catalytic activities in the transfer hydrogenation of cyclohexanone to cyclohexanol, using isopropanol as hydrogen source and promoted by potassium hydroxide. With a turnover frequency (TOF) of 6000 h^{-1}, the four-coordinated species [IrBr(COD) (Pr$_2$-bimy)] turned out to be much more active compared to the five-coordinate complex [IrBr(COD)(κ^2-All$_2$-bimy)], having a TOF of only 50 h^{-1}. Apparently, additional coordination of the allyl arm blocks activation sites for incoming substrates leading to a poor conversion rate.

In addition to iridium(I) complexes, those of the +III oxidation state can also be prepared using the same strategy. In the given example by Crabtree and coworkers, the seven-coordinated pentahydrido-iridium(V) complex [IrH$_5$(PPh$_3$)$_2$] is used as a basic metal precursor. When a mixture consisting of 1-(2-pyridyl)-3-isopropylbenzimidazolium tetrafluoroborate (Py,^{i}Pr-bimy)H$^+$BF$_4$$^-$ and this compound is heated in tetrahydrofuran under reflux overnight, the well-defined six-coordinated iridium(III) complex [IrH$_2$(κ^2-*C,N*-Py,^{i}Pr-bimy)

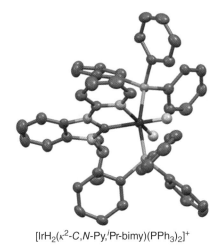

Scheme 7.37
Preparation of a pyridine-functionalized NHC complex of iridium(III).

$[IrH_2(\kappa^2\text{-}C,N\text{-}Py,^iPr\text{-}bimy)(PPh_3)_2]^+$

Figure 7.10
Molecular structure of the $[IrH_2(\kappa^2\text{-}C,N\text{-}Py,^iPr\text{-}bimy)(PPh_3)_2]^+$ cation.

$(PPh_3)_2]BF_4$ is obtained in 80% yield [38]. Notably, the reaction can be carried out in air (Scheme 7.37). The deprotonation of the benzimidazolium precursor occurs via one of the bound hydrido ligands releasing one molecule of dihydrogen. Moreover, the carbene complex formation is accompanied by a two-electron reduction of the iridium center changing it formal oxidation state from +V to +III. It is reasonable to assume that the chelation of the bulky bidentate NHC ligand to the already crowded iridium moiety would formally trigger reductive elimination of two additional hydrido ligands liberating a second molecule of dihydrogen.

The product can be isolated as a yellow solid, and the $^{13}C_{carbene}$ signal was detected as a triplet at 192.5 ppm in $CDCl_3$ due to heteronuclear coupling of $^2J_{PC} = 6.3$ Hz with two equivalent phosphine ligands, which therefore must adopt a *trans* arrangement. This chemical shift is rather downfield shifted due to the presence of four additional strongly σ-donating co-ligands. A single crystal X-ray diffraction study confirmed the distorted octahedral geometry of the dihydrido-complex cation with a $\kappa^2\text{-}C,N$ chelating mode of the functionalized NHC and a *trans* orientation of the two triphenylphosphine ligands with respect to each other (Figure 7.10).

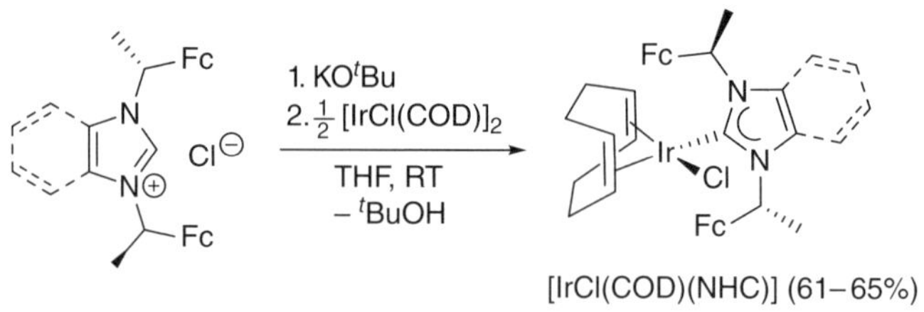

[IrCl(COD)(NHC)] (61–65%)

Scheme 7.38
Preparation of iridium(I) NHC complexes via *in situ* deprotonation with external base.

Similar reactions involving pyridine-functionalized imidazolium salts led to the preferred formation of *mesoionic* carbene complexes of iridium(III), which contain C4-bound imidazolin-4-ylidenes as opposed to the normal coordination mode via the C2-carbon. It is likely that the competing deprotonation at the C4-carbon is due to steric factors, since it is sterically less shielded than the more acidic C2-carbon atom.

In situ deprotonation can also occur with an external base, which is not bound to the metal center. This one-pot variation is often even more straightforward, because simpler metal sources can be used in conjunction with the base and azolium salt without any prior isolation of any intermediates.

An example involving the use of chiral 1,3-diferrocenylethyl-substituted azolium chlorides is shown in Scheme 7.38. The imidazolium or benzimidazolium carbene precursor is simply stirred with the *tert*-butoxide base and half an equivalent of the iridium(I) dimer in tetrahydrofuran at room temperature overnight [39]. Supposedly, the strong base deprotonates the azolium salt generating on equivalent of *tert*-butanol and the NHC ligand *in situ*. The latter is immediately trapped via coordination to iridium under bridge-cleavage conditions. The use of the strong base requires dry solvents, although the work up can be conveniently conducted in air affording the desired complexes of the type [IrCl(COD)(NHC)] in moderate yields ranging from 61–65%.

The spectroscopic features of the chiral complexes are unexceptional and in agreement with their non-chiral analogs. Their application as catalysts in asymmetric transfer hydrogenation of ketones and alkenes using isopropanol gave generally low enantioselectivities. Nevertheless, it was noted that the benzimidazolin-2-ylidene complex was significantly superior compared to its imidazolin-2-ylidene analog.

7.3.4 The Silver-Carbene Transfer Route

The transfer of carbenes from readily available silver-NHC complexes to iridium has been explored for the preparation of complexes in the +I and +III oxidation states. In most cases, the transfer occurs readily, and precautions to exclude moisture and air are usually not required. Due to this convenience, the silver-carbene transfer route has become very popular not only in iridium chemistry, but in transition metal chemistry in general. An early example employing iridium(I) that has also been applied to prepare analogous rhodium(I) complexes was detailed by Crabtree and coworkers in 2003 [24]. The procedure is initiated by exposure of simple imidazolium chlorides to silver(I) oxide in dichloromethane, which furnishes the desired silver-carbene transfer agents in good yields. Treatment of these intermediate complexes with the common chlorido-bridged iridium(I) dimer [IrCl(COD)]$_2$

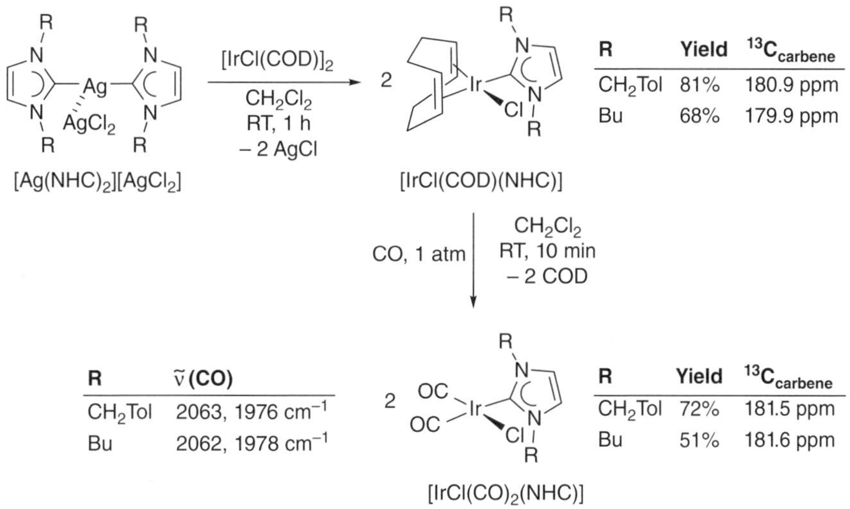

Scheme 7.39
Silver-carbene transfer to iridium(I) and subsequent carbonylation.

leads to the immediate transfer of the NHC from silver(I) to iridium(I), while the precipitation of silver chloride provides a strong driving force (Scheme 7.39). The targeted mononuclear iridium(I) NHC complexes of the general formula [IrCl(COD)(NHC)] that are formally obtained via bridge-cleavage reactions are often well soluble in common organic solvents and can be separated from the silver halide by simple filtration. Subsequent removal of the solvent under vacuum provides the desired products as yellow powders in moderate to good yields.

^{1}H NMR spectroscopy on the complexes indicates diastereotopy of the N–CH$_2$ groups as a result of restricted rotation around the iridium–carbon bond due to steric reasons. In particular, a well-defined AB pattern was observed for the two benzylic hydrogen atoms in the complex with the more bulky 4-tolylmethyl (–CH$_2$Tol) substituent with a geminal coupling constant of $^2J_{H-H} = 14.7$ Hz. Broadening of these signals were not observed upon heating a sample in toluene-d_8 up to 105 °C, which implies an energy barrier of >22 kcal mol^{-1} (92 kJ mol^{-1}) for the iridium–carbon bond rotation. Their ^{13}C NMR spectra recorded in CDCl$_3$ reveal very similar carbene resonances at 180.9 and 179.9 ppm, respectively.

Attempts to prepare bis(carbene) complexes bearing two NHC ligands via direct reaction of the parent iridium(I) dimer with excess silver-carbene or by abstraction of the chlorido ligand and addition of extra silver-carbene transfer agent to the monocarbene complexes were met with limited success. Notably, these attempts were likely conducted using silver(I) complexes of the type [Ag(NHC)$_2$][AgCl$_2$] or [AgCl(NHC)] that contain halido ligands.

On the other hand, the substitution of the COD ligand occurs cleanly and within minutes on bubbling carbon monoxide gas through solutions of the complexes in dichloromethane. The formation of the carbonyl complexes of the type *cis*-[IrCl(CO)$_2$(NHC)] occurs with a color change of the solutions from bright to pale yellow. Diastereotopy of the N-methylene hydrogen atoms are retained, which is also an indicator for the *cis* geometry of the two new carbonyl ligands. In comparison to the precursor COD complexes, fluxionality with respect to the iridium–carbon bond rotation was observed as indicated by peak broadening of the

Table 7.4 Preparation of hetero-bis(carbene) complexes of iridium(I) via carbene transfer.

R	E	T	t[h]	Yield [%]	$^{13}C_{NCN}$ [ppm]	$^{13}C_{NCN'}$ [ppm]
nBu	C	reflux	36	48	179.0	176.0
Me	C	RT	72	75	179.2	177.9
Me	N	RT	54	88	177.7	181.1

initially observed AB pattern, followed by coalescence into a singlet at ~55 °C. It is apparent that the replacement of the COD chelate with two smaller carbonyl ligands leads to a less congested environment around the iridium center to allow for a less hampered bond rotation.

The introduction of the carbonyl ligands only induces a small downfield shift of <1 ppm for the $^{13}C_{carbene}$ resonances, which are thus less diagnostic. Nevertheless, two new and distinct carbonyl signals are observed at ~173 and ~168 ppm in the ^{13}C NMR spectrum providing evidence for the formed complexes and also pointing to a *cis* arrangement. Moreover, two carbonyl stretches for each complex were recorded by solution IR spectroscopy in dichloromethane. The comparison of the averaged wavenumbers in members of the *cis*-[IrCl(CO)$_2$(NHC)] family prepared using this route is currently often used to compare the donating strength of the respective NHCs.

While the preparation of bis(carbene) iridium(I) complexes failed with silver-carbenes containing halido ligands, the use of cationic [Ag(NHC)$_2$]$^+$ type complexes with non-coordinating anions was more rewarding. For example, the reaction of [IrCl(COD)(NHC)] (NHC = 1,3-di-*p*-tolylmethylimidazolin-2-ylidene) with various [Ag(NHC')$_2$]PF$_6$ transmetallating agents in excess led to the formation of hetero-bis(carbene) complexes of iridium(I) which contain two different NHC ligands (Table 7.4) [40]. The reaction is rather slow and even under reflux conditions requires prolonged reaction times to complete. Moreover, the reactions had to be carried out in the dark to prevent decomposition of the silver species. After column chromatography, the cationic bis(NHC) complexes were isolated as bright orange solids.

In addition to the already mentioned diastereotopy of the N-methylene units, NMR spectroscopy also as expected reveals the presence of two different NHC ligands. Their respective chemical shifts are summarized in Table 7.4.

Even more straightforward is the preparation of homo-bis(carbene) complexes. Treating the parent iridium(I) dimer [IrCl(COD)]$_2$ directly with the cationic bis(carbene) species [Ag(EMIM)$_2$]NTf$_2$ affords the bis(NHC) complex [Ir(COD)(EMIM)$_2$]NTf$_2$ in 48% yield already after 1 h (Scheme 7.40) [25]. Notably, this reaction did not require excess of the transmetallating agent and could proceed at room temperature.

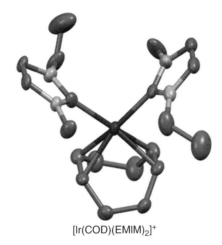

Scheme 7.40
Preparation of a homo-bis(carbene) iridium(I) complex via carbene transfer.

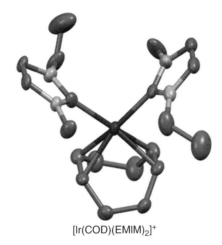

[Ir(COD)(EMIM)$_2$]$^+$

Figure 7.11
Molecular structure of the anti-[Ir(COD)(EMIM)$_2$]$^+$ complex cation.

Due to the unsymmetrically substituted NHC ligands, a mixture of *anti* and *syn* rotamers was obtained. In accordance, two closely spaced $^{13}C_{carbene}$ signals were detected in the ^{13}C NMR spectrum. The same observation was made for the analogous rhodium(I) species. The solid state molecular structure of the complex cation obtained by single crystal X-ray diffraction is shown in Figure 7.11. Notably, the *anti* arrangement of the two carbene ligands was found, whereby equivalent N-substituents point in different directions away from each other. This orientation is supposedly more preferred on steric grounds.

Overall, these examples highlight the substantial reactivity differences observed for the seemingly similar silver-transfer agents.

The preparation of iridium(III) NHC complexes via silver-carbene transfer is also feasible. For example, Thompson and Forrest reported that iridium(III) phosphors containing cyclometallated NHC ligands can be prepared in a one-pot procedure by simply mixing and heating the respective imidazolium or benzimidazolium iodides with half-equivalent of silver(I) oxide and one equivalent of the simple hydrated iridium(III) trichloride salt in ethoxyethanol for an extended period of time (Scheme 7.41) [41].

The silver(I) oxide serves several purposes. It acts as a base to deprotonate the azolium salts, and at the same time provides a Lewis acid to stabilize the NHC formed *in situ*. The

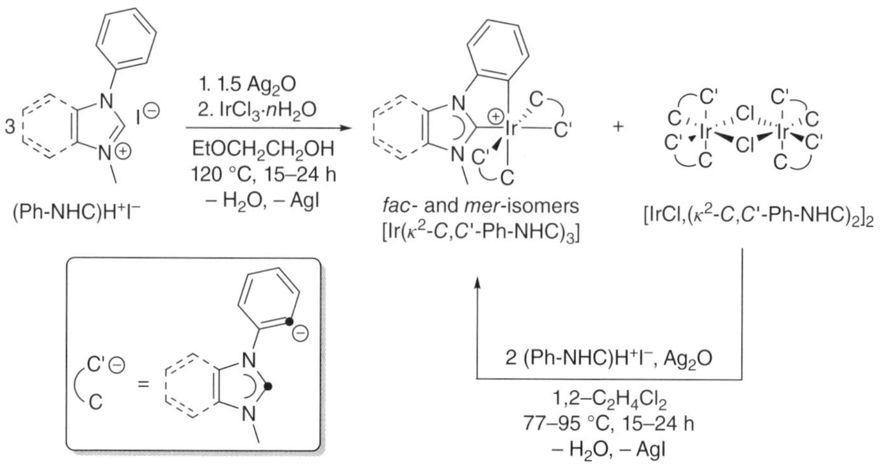

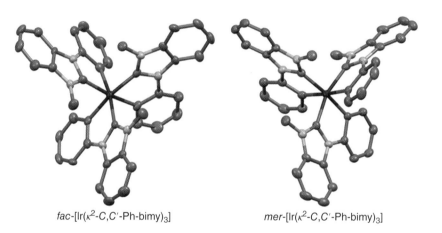

Scheme 7.41
Preparation of cyclometallated iridium(III) NHC complexes using Ag$_2$O.

Figure 7.12
Molecular structures of *fac*- and *mer*-isomers of [Ir(κ^2-*C*,*C*'-Ph-bimy)$_3$].

resulting silver(I)-NHC complex subsequently acts as the key reagent to transfer the carbene to iridium, while precipitation of silver(I) iodide provides a strong driving force. In this particular case, C–H activation of the phenyl N-substituent occurs concurrently, which is likely base-assisted and facilitated by excess silver(I) oxide as well. Overall, a mixture of NHC compounds was obtained in low yields primarily containing the bis(chelate) dimer [IrCl(κ^2-*C*,*C*'-Ph-NHC)$_2$]$_2$ as well as smaller amounts of the facial- (*fac*) and meridional (*mer*) isomers of the desired tris(chelate) complex [Ir(κ^2-*C*,*C*'-Ph-NHC)$_3$]. Separation of the complex mixture is difficult, but can be accomplished by washing with appropriate solvents, recrystallization, and column chromatography. Moreover, the dinuclear bis(chelate) complexes could be converted into isomers of the mononuclear tris(chelate) complexes by bridge-cleavage reactions with additional silver(I) oxide and carbene precursor in 1,2-dichloroethane at elevated temperatures. The *fac*-isomers were noted to be less soluble compared to the *mer*-configured counterparts.

Single crystals of both *fac*- and *mer*-isomers were obtained for the benzimidazolin-2-ylidene complex. Their molecular structures obtained by X-ray diffraction are depicted in Figure 7.12. Both complexes have pseudo-octahedral geometries with the greater distortion observed for the meridional species. This is best exemplified by the small C–Ir–C angle of only ~162° between two *trans*-standing carbene donors and the metal center, which deviates significantly from linearity.

The two geometrical isomers also show distinct differences in their NMR spectra recorded in $CDCl_3$. The more symmetrical *fac*-isomer gives rise to one carbene signal at 189.6 ppm in $CDCl_3$ indicating the equivalence of the three chelating ligands. In this complex all carbene donors are *trans* to a strongly σ-donating anionic aryl moiety. In the *mer*-arrangement, symmetry is significantly reduced, and three resonances are observed at 188.2, 186.0 and 184.9 ppm for the three carbene donors, respectively. By comparison with the *fac*-isomer, the most downfield signal at 188.2 ppm can be assigned to the NHC moiety *trans* to the aryl donor, while the two more similar upfield chemical shifts at 186.0 and 184.9 ppm can be attributed to the two inequivalent carbene carbon atoms *trans* to each other.

The *mer*- and *fac*-isomers do not interconvert by means of photochemical or thermal activation, but instead undergo decomposition on heating or irradiation. In contrast *mer* to *fac* isomerization occurs easily for pyrazole analogs. The observed difficulty for such transformations in the carbene complexes is attributed to the strong carbene–iridium bond. The replacement of the classical pyrazole N donors in this type of cyclometallated Ir-based phosphors by NHC donors leads to higher triplet energies, and the resulting complexes display near-UV luminescence at room temperature.

References

[1] P. B. Hitchcock, M. F. Lappert, P. L. Pye, *J. Chem. Soc., Chem. Comm.* **1976**, 644.
[2] P. B. Hitchcock, M. F. Lappert, P. L. Pye, *J. Chem. Soc., Dalton Trans.* **1978**, 826.
[3] T. Weskamp, W. C. Schattenmann, M. Spiegler, W. A. Herrmann, *Angew. Chem., Int. Ed.* **1998**, *37*, 2490.
[4] J. Huang, E. D. Stevens, S. P. Nolan, J. L. Petersen, *J. Am. Chem. Soc.* **1999**, *121*, 2674.
[5] M. Scholl, T. N. Trnka, J. P. Morgan, R. H. Grubbs, *Tetrahedron Lett.* **1999**, *40*, 2247.
[6] A. Fürstner, L. Ackermann, B. Gabor, R. Goddard, C. W. Lehmann, R. Mynott, F. Stelzer, O. R. Thiel, *Chem. Eur. J.* **2001**, *7*, 3236.
[7] T. M. Trnka, J. P. Morgan, M. S. Sanford, T. E. Wilhelm, M. Scholl, T.-L. Choi, S. Ding, M. W. Day, R. H. Grubbs, *J. Am. Chem. Soc.* **2003**, *125*, 2546.
[8] W. A. Herrmann, M. Elison, J. Fischer, C. Köcher, G. R. J. Artus, *Chem. Eur. J.* **1996**, *2*, 772.
[9] J. Huang, H.-J. Schanz, E. D. Stevens, S. P. Nolan, *Organometallics* **1999**, *18*, 2370.
[10] W. Baratta, W. A. Herrmann, P. Rigo, J. Schwarz, *J. Organomet. Chem.* **2000**, *593–594*, 489.
[11] M. Scholl, S. Ding, C. W. Lee, R. H. Grubbs, *Org. Lett.* **1999**, *1*, 953.
[12] D. Enders, H. Gielen, G. Raabe, J. Runsink, J. Henrique Teles, *Chem. Ber./Recueil* **1997**, *130*, 1253.
[13] P. Csabai, F. Joó, *Organometallics* **2004**, *23*, 5640.
[14] S. Pei Shan, X. Xiaoke, B. Gnanaprakasam, T. T. Dang, B. Ramalingam, H. V. Huynh, A. M. Seayad, *RSC Adv.* **2015**, *5*, 4434.
[15] D. J. Cardin, M. J. Doyle, M. F. Lappert, *J. Chem. Soc., Chem. Commun.* **1972**, 927.
[16] (a) D. M. Lemal, R. A. Lovald, K. I. Kawano, *J. Am. Chem. Soc.* **1964**, *86*, 2518. (b) H. E. Winberg, J. E. Carnahan, D. D. Coffman, M. Brown, *J. Am. Chem. Soc.* **1965**, *87*, 2055.
[17] E. Çetinkaya, P. B. Hitchcock, H. Küçükbay, M. F. Lappert, S. Al-Juaid, *J. Organomet. Chem.* **1994**, *481*, 89.
[18] W. A. Herrmann, L. J. Goossen, G. R. J. Artus, C. Köcher, *Organometallics* **1997**, *16*, 2472.

[19] J. Huang, E. D. Stevens, S. P. Nolan, *Organometallics* **2000**, *19*, 1194.

[20] A. C. Chen, L. Ren, A. Decken, C. M. Crudden, *Organometallics* **2000**, *19*, 3459.

[21] C. Köcher, W. A. Herrmann, *J. Organomet. Chem.* **1997**, *532*, 261.

[22] M. V. Baker, S. K. Brayshaw, B. W. Skelton, A. H. White, *Inorg. Chim. Acta* **2004**, *357*, 2841.

[23] R. H. Crabtree, J. M. Quirk, *J. Organomet. Chem.* **1980**, *199*, 99.

[24] A. R. Chianese, X. Li, M. C. Janzen, J. W. Faller, R. H. Crabtree, *Organometallics* **2003**, *22*, 1663.

[25] U. Hintermair, U. Englert, W. Leitner, *Organometallics* **2011**, *30*, 3726.

[26] B. Çetinkaya, P. Dixneuf, M. F. Lappert, *J. Chem. Soc., Chem. Commun.* **1973**, 206.

[27] B. Çetinkaya, P. Dixneuf, M. F. Lappert, *J. Chem. Soc., Dalton Trans.* **1974**, 1827.

[28] H. M. Lee, T. Jiang, E. D. Stevens, S. P. Nolan, *Organometallics* **2001**, *20*, 1255.

[29] A. C. Hillier, H. M. Lee, E. D. Stevens, S. P. Nolan, *Organometallics* **2001**, *20*, 4246.

[30] L. D. Vázquez-Serrano, B. T. Owens, J. M. Buriak, *Chem Commun.* **2002**, 2518.

[31] M. Prinz, M. Grosche, E. Herdtweck, W. A. Herrmann, *Organometallics* **2000**, *19*, 1692.

[32] K. Denk, P. Sirsch, W. A. Herrmann, *J. Organomet. Chem.* **2002**, *649*, 219.

[33] A. P. Blum, T. Ritter, R. H. Grubbs, *Organometallics* **2007**, *26*, 2122.

[34] T. M. Trnka, J. P. Morgan, M. S. Sanford, T. E. Wilhelm, M. Scholl, T.-L. Choi, S. Ding, M. W. Day, R. H. Grubbs, *J. Am. Chem. Soc.* **2003**, *125*, 2546.

[35] Y. Yamaguchi, T. Kashiwabara, K. Ogata, Y. Miura, Y. Nakamura, K. Kobayashi, T. Ito, *Chem. Commun.* **2004**, *19*, 2160.

[36] J. A. Chamizo, J. Morgado, *Transition Met. Chem.* **2000**, *25*, 161.

[37] F. E. Hahn, C. Holtgrewe, T. Pape, M. Martin, E. Sola, L. A. Oro, *Organometallics* **2005**, *24*, 2203.

[38] S. Gründemann, A. Kovacevic, M. Albrecht, J. W. Faller, R. H. Crabtree, *J. Am. Chem. Soc.* **2002**, *124*, 10473.

[39] H. Seo, B. Y. Kim, J. H. Lee, H.-J. Park, S. U. Son, Y. K. Chung, *Organometallics* **2003**, *22*, 4783.

[40] L. N. Appelhans, C. D. Incarvito, R. H. Crabtree, *J. Organomet. Chem.* **2008**, *693*, 2761.

[41] T. Sajoto, P. I. Djurovich, A. Tamayo, M. Yousufuddin, R. Bau, M. E. Thompson, R. J. Holmes, S. R. Forrest, *Inorg. Chem.* **2005**, *44*, 7992.

8 Beyond Classical N-heterocyclic Carbenes I

The organometallic chemistry of N-heterocyclic carbenes has developed rapidly driven by the unique properties of these special ligands, their ease of synthesis, and the numerous applications they have found. In addition to the classical NHCs based on the imidazole, triazole, benzimidazole, and imidazoline backbone (Chart 8.1), which are the primary focus of this book, many "non-classical" variations have emerged in the scientific literature. Their development is mostly a result of curiosity-driven fundamental research primarily linked to the search for stronger and more versatile donors.

Compared to the classical NHCs, these are still less explored, although some types are becoming increasingly popular. Many statements made about the classical types in the previous chapters are also true for their non-classical counterparts. Moreover, they often share the described general synthetic routes for their preparation and metallation. This chapter provides an overview on some aspects of the chemistry of various selected non-classical NHCs. In order to qualify as a non-classical NHC, the heterocyclic structure of the species must differ from those depicted in Chart 8.1. Moreover, the "carbene" must be cyclic and contain at least one nitrogen atom. Thus, other heterocyclic carbenes without nitrogen atoms or acyclic carbenes are not covered in this chapter.

8.1 N,S-Heterocyclic Carbenes (NSHCs)

Early common variations of the classical NHCs involve the simple replacement of one nitrogen atom in the five-membered ring with another heteroatom. Generally, the formal substitution of an N-R group in the four classical NHCs **I–IV** (Chart 8.1) with a sulfur atom could result in the five new *N,S*-heterocyclic carbenes **Ia–IVa** and **IVa′**, which are depicted in Chart 8.2. Among these, only thiazolin-2-ylidene **IIa** has been isolated as a stable free carbene by Arduengo and coworkers by deprotonation of the respective thiazolium chloride with potassium hydride in tetrahydrofurane (Scheme 8.1) [1]. In addition, potassium chloride and dihydrogen gas are formed as stable byproducts. The free and almost colorless thiazolin-2-ylidene with a bulky Dipp (2,6-diisopropylphenyl) wing tip slowly undergoes proton-catalyzed reversible dimerization to its electron-rich and red-colored olefin, while dimerization in the absence of any acid catalyst was not observed for weeks (Scheme 8.1). Notably, this behavior is different from that observed for unsaturated imidazolin-2-ylidenes (**II**), but resembles more that of saturated imidazolidin-2-ylidenes (**I**) and benzimidazolin-2-ylidenes (**III**) instead.

The Organometallic Chemistry of N-heterocyclic Carbenes, First Edition. Han Vinh Huynh.
© 2017 John Wiley & Sons Ltd. Published 2017 by John Wiley & Sons Ltd.

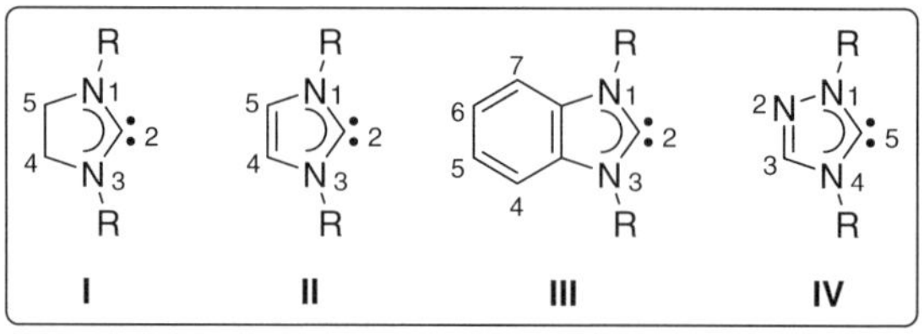

Chart 8.1
Four types of classical N-heterocyclic carbenes.

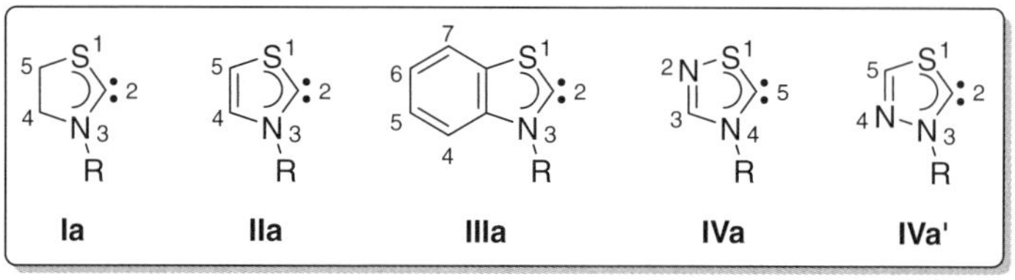

Chart 8.2
Formally possible species of *N*,*S*-heterocyclic carbenes (NSHCs).

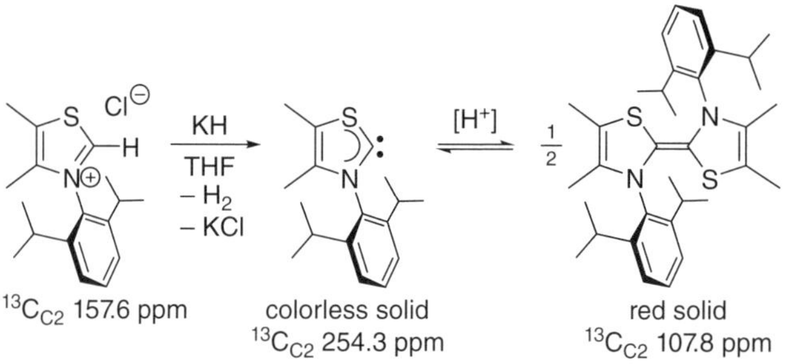

Scheme 8.1
Preparation of a stable thiazolin-2-ylidene and its subsequent dimerization.

As generally observed for NHCs, a stark downfield shift is observed for the carbene carbon of this free NSHC (i.e. 254.3 ppm in THF-d_8) upon formation from the salt (i.e. 157.6 ppm in DMSO-d_6). The substitution of the N-aryl or N-alkyl group in imidazolin-2-ylidenes with a larger sulfur atom leads to a reduced p_π–p_π resonance stabilization. In addition, the carbene center cannot be efficiently shielded by sterics anymore due to the lack of a second wing tip group. Thus, the free thiazolin-2-ylidene is more reactive and difficult to handle compared to its classical NHC counterparts. In air, it decomposes rapidly under color change to brown. Steric protection of the monomeric carbene has become much more important, and a very bulky N-substituent is required for the isolation of free thiazolin-2-ylidenes. For example, the free N-mesitylthiazolin-2-ylidene ($^{13}C_{C2}$ 251.6 ppm) could only be observed by solution NMR spectroscopy below 0 °C. At higher temperatures, only the dimeric form is present. The elusive free NSHC with N-methyl substituent could not even be detected at low temperatures.

The molecular structures of the precursor salt, monomeric, and dimeric forms of 3-(2,6-diisopropylphenyl)thiazolin-2-ylidene could be established by single crystal X-ray diffraction

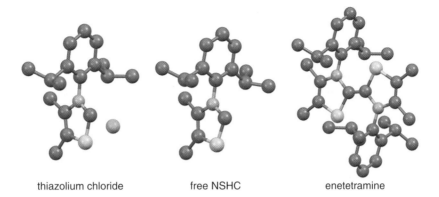

Figure 8.1
Molecular structures of precursor salt, free NHC, and the dimeric enetetramine of 3-(2,6-diisopropylphenyl)thiazolin-2-ylidene. Hydrogen atoms have been omitted for clarity.

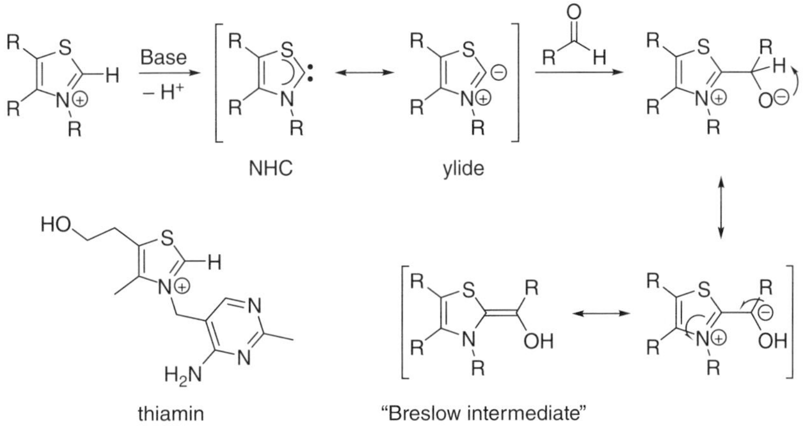

Scheme 8.2
Thiamin and key intermediates in the activation of carbonyl compounds by an NSHC.

analyses and are depicted in Figure 8.1. Notably, this carbene is the only one for which molecular structures of both the monomer and its dimer could be determined. The plane of the Dipp substituent is in all three cases almost perpendicular to the heterocyclic plane. The angle at the carbene carbon amounts to ~104°, which is as commonly observed smaller than that for the respective precursor salt (112°). The dimer is arranged in the sterically more favorable *anti*-form with respect to the C=C double bond. Moreover, the angles around the equivalent central carbon atoms are slightly increased to ~107°.

Free thiazolin-2-ylidenes have long been proposed to be reactive intermediates in the thiamin (vitamin B1) catalyzed C–C coupling reactions and related transformations (Scheme 8.2). In these reactions, thiamin needs to be activated by a base to form the free carbene, which is supposedly in equilibrium with its resonance ylidic form. Both NSHC and ylide are strong nucleophiles that can initiate the catalytic cycle by attacking the carbonyl carbon of an aldehyde. Formal (base-assisted) proton shift leads to "umpolung," and the initially electrophilic carbonyl carbon is transformed into a nucleophilic center with

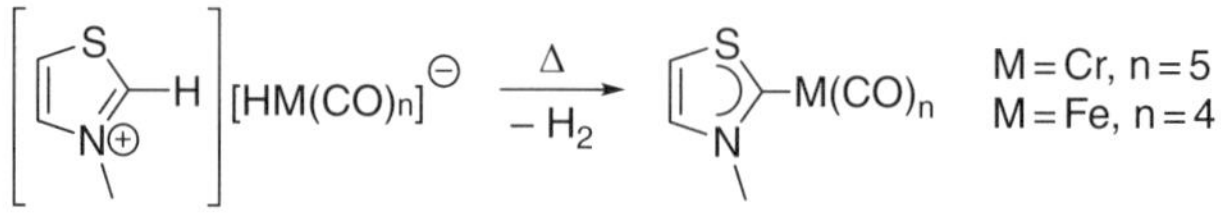

Scheme 8.3
First preparation of thiazolin-2-ylidene complexes.

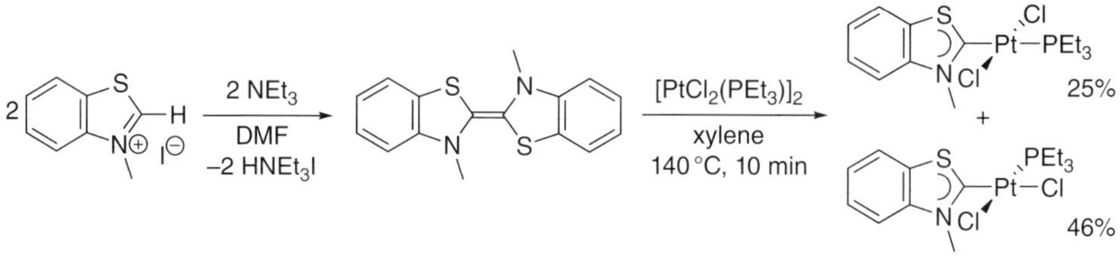

Scheme 8.4
Preparation of benzimidazolin-2-ylidene complexes using carbene dimers.

the formation of the so-called "Breslow intermediate." The latter is the key intermediate in many NHC organocatalyzed reactions and can undergo various subsequent transformations with other substrates leading to a wide range of coupling reactions.

Complexes of thiazolin-2-ylidenes were also described a long time before the isolation of the free ligand. Karl Öfele reported as early as 1969 that such complexes could be prepared by heating thiazolium salts containing hydridic chromium or iron carbonyl anions (Scheme 8.3) [2]. By analogy to the preparation of the imidazolin-2-ylidene analog, the neutral mixed carbene/carbonyl complexes are formed via *in situ* deprotonation of the 3-methylthiazolium cation under elimination of dihydrogen gas.

The resulting complexes were reported to be air and moisture stable solids. However, other data were not provided in the original work.

The chemistry of benzothiazolium salts is very similar to that of their non-annulated derivatives. Upon C2–H deprotonation, the carbene dimers are formed directly [3], but free benzothiazolin-2-ylidenes **IIIa** (Chart 8.2) are unknown to date.

The easily available carbene dimers can be used to prepare benzimidazolin-2-ylidene complexes as reported by Lappert and coworkers [4]. For example, this can be achieved by heating the electron-rich olefin with the dimeric platinum(II) precursor [PtCl$_2$(PEt$_3$)]$_2$ in xylene to 140 °C for only 10 minutes (Scheme 8.4). This reaction occurs via cleavage of the chlorido bridges of the metal precursor as well as the C=C double bond of the carbene dimer. The reaction product was obtained as a mixture of *cis* (white) and *trans* isomers (yellow). The former was noted to be thermodynamically favorable due to the similar *trans* influences of carbene and phosphine, which are greater than that of the chlorido ligand.

The preparation of the carbene dimers is not necessary. Complexes of benzothiazolin-2-ylidene ligands can be more conveniently prepared via the direct and *in situ* deprotonation of benzimidazolium salts by basic metal precursors. Caló and coworkers reported the straightforward preparation of the diiodido-bis(3-methylbenzothiazolin-2-ylidene) palladium(II) complex by heating a mixture of two equivalents of the respective 3-methylbenzothiazolium iodide salt with palladium(II) acetate in tetrahydrofurane to reflux (Scheme 8.5).

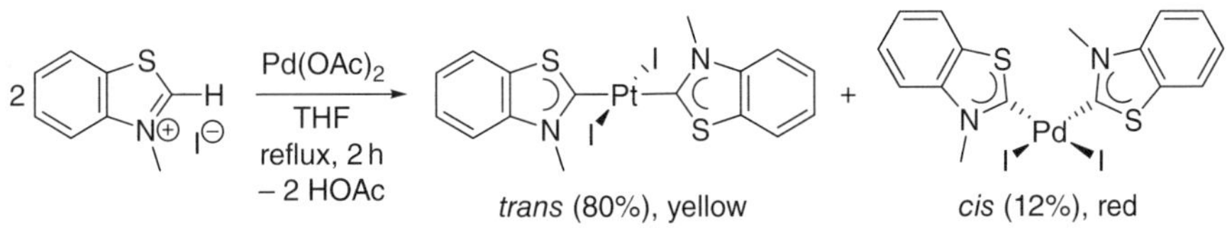

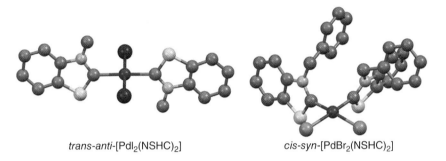

Scheme 8.5
cis- and *trans*-Bis(benzothiazolin-2-ylidene)palladium(II) complexes via *in situ* deprotonation with Pd(OAc)$_2$.

trans-anti-[PdI$_2$(NSHC)$_2$]   *cis-syn*-[PdBr$_2$(NSHC)$_2$]

Figure 8.2
Molecular structures of *trans-* and *cis*-configured bis(benzothiazolin-2-ylidene) palladium(II) complexes. Hydrogen atoms have been omitted for clarity.

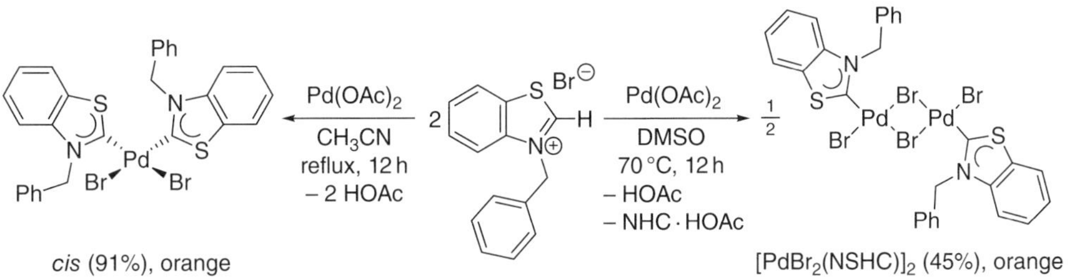

cis (91%), orange   [PdBr$_2$(NSHC)]$_2$ (45%), orange

Scheme 8.6
Mono- and bis(benzothiazolin-2-ylidene)palladium(II) complexes via *in situ* deprotonation with Pd(OAc)$_2$.

The complex was obtained as a mixture of *cis* and *trans* isomers of different colors that could be separated by column chromatography. The unsymmetrical nature of this ligand type would in general also generate rotamers that differ in their orientation with respect to the metal-carbon bond. ^{13}C NMR data were only reported for the major *trans* isomer revealing a very downfield shift for the carbene carbon at 210.5 ppm in CDCl$_3$. The molecular structure of the *trans* isomer could also be established, which shows that the complex exists in the more favorable *anti*-configuration (Figure 8.2, left).

Huynh and Hor reported the same approach using 3-benzylbenzothiazolium bromide. However, the use of more polar and higher boiling solvents and longer reaction times led to a different outcome [5]. The complex *cis*-dibromido-bis(3-benzylbenzothiazolin-2-ylidene)palladium(II) was obtained in 91% yield when two equivalents of the salt and palladium(II) acetate were heated in acetonitrile under reflux overnight (Scheme 8.6). The predominant formation of the *cis* isomer is probably due to the higher polarity of acetonitrile compared to tetrahydrofurane. The $^{13}C_{carbene}$ NMR signal recorded in DMSO-d_6 is at 203.8 ppm, and the molecular structure determined by X-ray diffraction also confirms the *cis-syn*-arrangement (Figure 8.2, right).

Heating the same ratio of reactants in dimethylsulfoxide at 70 °C led primarily to the isolation of the dimeric monocarbene complex of the formula [PdBr$_2$(NSHC)]$_2$ instead. At the moderate temperature, only one carbene ligand was formed giving rise to a monocarbene complex, while the second equivalent of precursor salt merely served as a bromide source. Initially, a monomeric DMSO solvate is formed, which can be isolated as the solvent-free, bromido-bridged dimeric complex as evidenced by its solid state molecular structure depicted in Figure 8.3.

An elegant alternative two-step route to benzothiazolin-2-ylidene complexes via oxidative addition of the respective 2-chlorobenzothiazole to the electron-rich Vaska-type complex *trans*-[IrCl(CO)(PMe$_2$Ph)] was already reported by Stone and coworkers in 1973 [6]. After oxidative addition, an iridium(III) species containing a anionic benzothiazolyl ligand is obtained (Scheme 8.7). The latter is subsequently protonated by tetrafluoroboric acid to furnish a neutral NH-benzothiazolin-2-ylidene ligand, while N-alkylation using trialkyloxonium salts failed supposedly due to steric reasons.

Notably, the methodology can also be applied to prepare complexes of thiazolin-2-ylidenes and benzoxazolin-2-ylidenes, where the sulfur is replaced by an oxygen atom.

In addition, the preparation of complexes with N-alkyl carbenes is also possible in a single step starting from the N-alkyl thiazolium salts as exemplified in Scheme 8.8 for the preparation of a cationic platinum(II) complex [7]. The approach has subsequently also

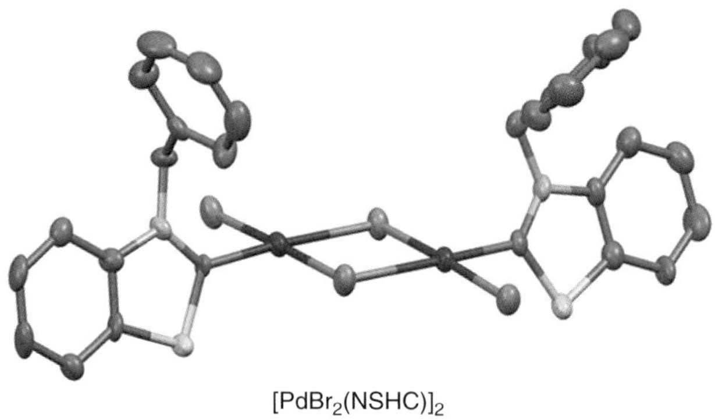

[PdBr$_2$(NSHC)]$_2$

Figure 8.3
Molecular structure of a dimeric mono-benzothiazolin-2-ylidene complex of palladium(II). Hydrogen atoms have been omitted for clarity.

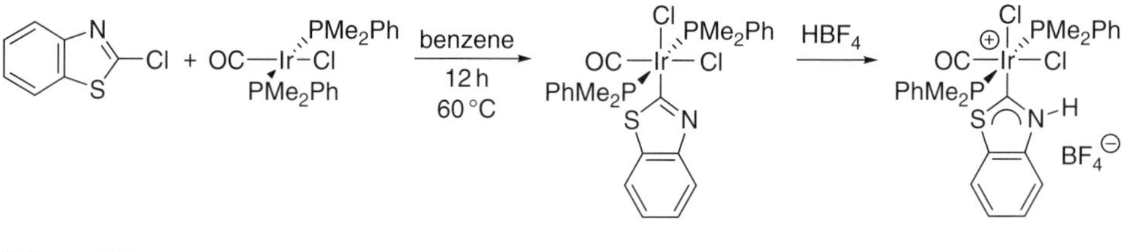

Scheme 8.7
Preparation of an iridium(III)-NH-benzothiazolin-2-ylidene complex via oxidative addition and protonation.

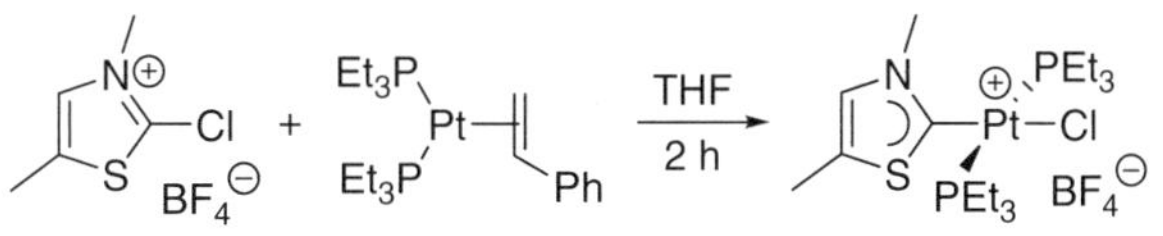

Scheme 8.8
Preparation of a platinum(II)-thiazolin-2-ylidene complex via oxidative addition of a thiazolium salt.

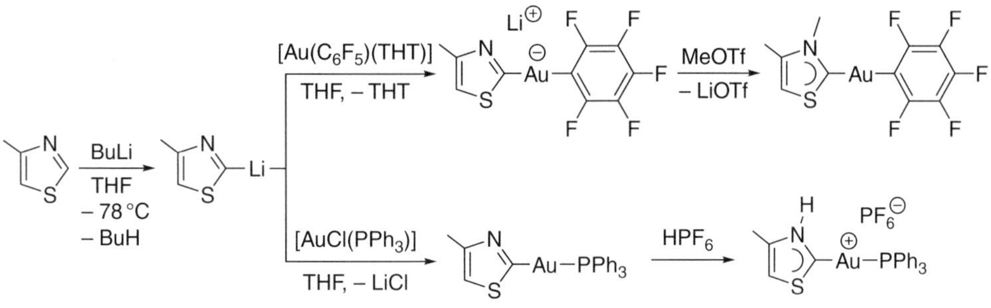

Scheme 8.9
Preparation of neutral and cationic gold(I) thiazolin-2-ylidenes using 2-lithium-thiazolate.

been applied to prepare complexes of nickel, palladium, rhodium, manganese, chromium, iron, and osmium using various low-valent metal precursors [7,8].

Another widely applicable approach to thiazolin-2-ylidene complexes was described by Raubenheimer in 1990 [9]. In this procedure, carbene complex formation is achieved in three steps including (i) C2-lithiation of thiazole, (ii) complexation to transition metals followed by (iii) N-protonation or N-alkylation of the anionic thiazolyl ligand to form the respective carbene complex. The lithiation of thiazole is site-selective and occurs at the C2 carbon. The C2–H proton is the most acidic in the molecule due to the polarization from two adjacent electronegative heteroatoms. Subsequently, the thiazolate can be transmetallated to a wide range of metal centers. For example, reaction of the lithio-heterocycle with [Au(C$_6$F$_5$)(THT)] leads to the displacement of the weak tetrahydrothiophene (THT) ligand giving rise to a new monoanionic aurate(I) complex (Scheme 8.9, top). Final N-methylation occurs via reaction with the strong electrophile methyltriflate, which affords the neutral gold(I) carbene complex.

The reaction of the lithium-thiazolate with chlorido-triphenylphosphinegold(I) on the other hand eliminates lithium chloride, and a neutral mixed phosphine/thiazolato complex is obtained (Scheme 8.9, bottom). Its N-protonation with HPF$_6$ then yields the cationic mixed NH-thiazolin-2-ylidene/phosphine complex, the charge of which is balanced by the large PF$_6^-$ counter-anion. Thiazolin-2-ylidene complexes of gold, copper, chromium, molybdenum, manganese, and iron were prepared using this methodology [10].

Complexes of saturated thiazolidin-2-ylidenes (**Ia**) are much rarer compared to those of the unsaturated (**IIa**) and benzannulated types (**IIIa**) described thus far. In an early report, Öfele and coworkers mentioned that these can be obtained by reductive hydrogenation of the C$_4$=C$_5$ double bond in thiazolin-2-ylidene complexes without giving further preparative details [11]. Based on mass spectrometric analyses of the [Fe(CO)$_4$(NHC)] complex (NHC=3-methylthiazolidin-2-ylidene), the authors concluded that the loss in aromaticity in the ligand also leads to an overall destabilization of the complex.

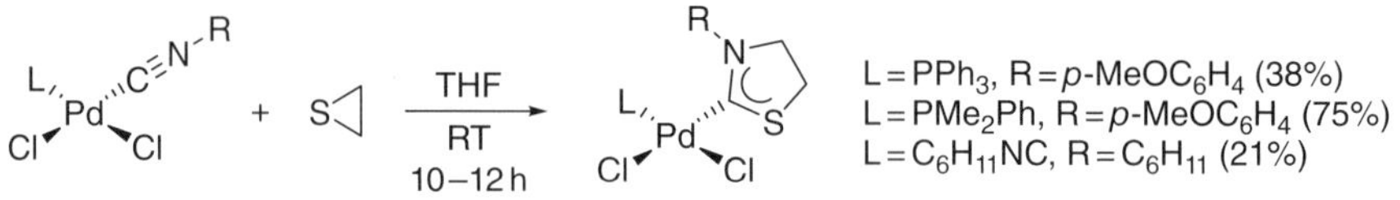

Scheme 8.10
Palladium(II) complexes of saturated thiazolidin-2-ylidene ligands via nucleophilic attack on coordinated isocyanides.

A more common synthetic route involves the nucleophilic attack of a sulfur nucleophile on an isocyanide ligand. The use of thiirane has the additional advantage that the release of ring strain aids in the formation of the NHC (Scheme 8.10). The outcome of these template-directed syntheses, however, is heavily influenced by the stereoelectronic properties of the co-ligands in the complex precursor and the nature of the isocyanide ligand used. For example, Michelin and coworkers observed that the replacement of the triphenylphosphine with a more donating, but less bulky dimethylphenylphosphine ligand would lead to an enhancement of the yield for the respective thiazolidin-2-ylidene palladium(II) complex from 38 to 75%. In this case, an aromatic isocyanide ligand was applied [12].

Exposure of a bis(cyclohexylisocyanide) complex to excess thiirane led to the conversion of only one isonitrile into the thiazolidin-2-ylidene ligand, and even so, the mixed-ligand complex was obtained in a rather low yield of only 21%.

Sulfur containing NHCs and their complexes of type **IVa** and **IVa′** are either very rare or unknown. Future fundamental investigations into such species will be certainly challenging, but also academically and intellectually stimulating.

8.2 N,O-Heterocyclic Carbenes (NOHCs)

Formal replacement of nitrogen with an oxygen atom in the stereotypical NHCs would result in the oxygen-containing NOHCs depicted in Chart 8.3. In comparison to their direct sulfur analogs, weaker donor strengths are expected for these species due the stronger electronegativity of the oxygen. Free oxygen-containing NHCs have not been characterized in substance, but their complexes are readily available most commonly via template-directed approaches involving isocyanide transition metal complexes.

An early example was described by Fehlhammer and coworkers in 1982, who studied the reactions of tosylmethylisocyanide complexes with nucleophiles and 1,3-dipolarophiles [13]. This isocyanide ligand contains α-acidic methylene protons, and coordination to Lewis acids further enhances their α-acidity. Thus, deprotonation can already be achieved with the moderately basic triethylamine as exemplified in Scheme 8.11 with a cationic bis(phosphine)platinum(II) complex. The resulting active 1,3-dipolar platinum-nitrilylide undergoes subsequent cycloaddition with various dipolarophiles. For example, reaction with methylisocyanate affords a neutral oxazolato complex in a low yield of 17%, which can be alkylated with triethyloxonium tetrafluoroborate to the cationic 3,4,5-trisubstituted oxazolin-2-ylidene complex (type **IIb**).

The reaction of the metal-ylide with benzaldehyde, on the other hand, leads to an oxazolinato complex in 71% yield, which can be reversibly N-protonated to give a saturated oxazolidin-2-ylidene complex (type **IIa**). NMR spectroscopic data were not reported in this work.

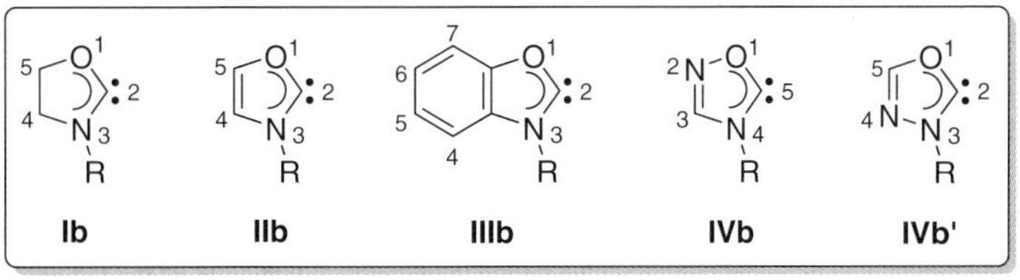

Chart 8.3
Formally possible species of oxygen-containing NOHCs.

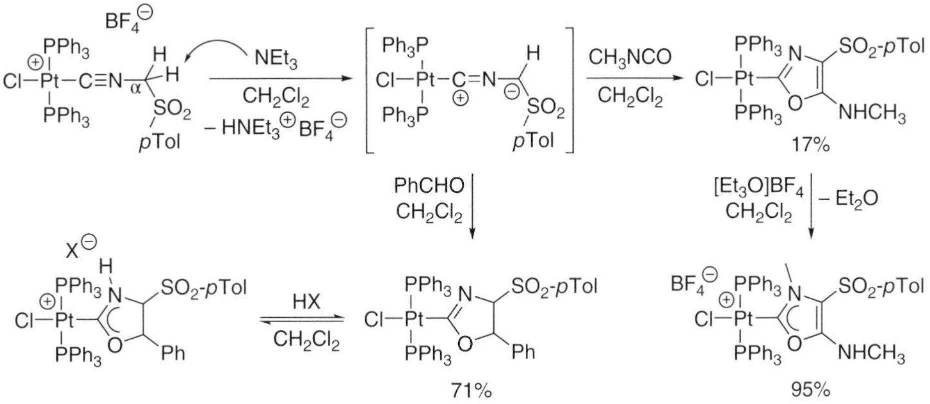

Scheme 8.11
Early preparation of oxazolin- and oxazolidin-2-ylidene platinum(II) complexes.

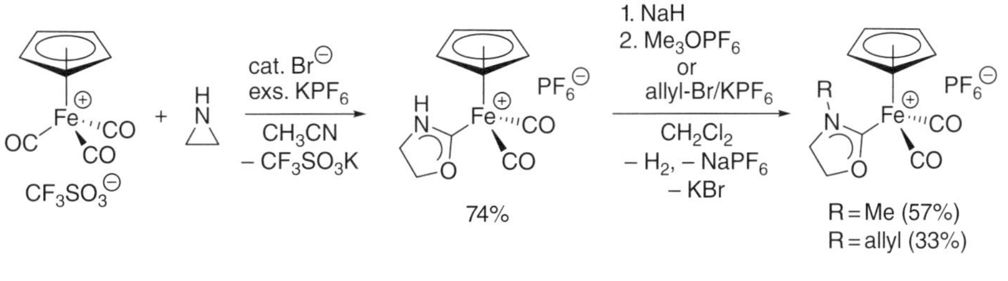

Scheme 8.12
Oxazolidin-2-ylidene complexes via aziridine attack on carbonyl ligands.

Another template-directed approach to saturated oxazolidin-2-ylidene complexes that makes use of cationic metal carbonyl complexes was described by Angelici and coworkers shortly after in 1984 [14]. As shown in Scheme 8.12, reaction of the cyclopentadienyl-tricarbonyl-iron cation with aziridine in the presence of catalytic amounts of bromide and excess potassium hexafluorophosphate produces the respective oxazolidin-2-ylidene complex in good yields. The formation of the oxygen-containing NOHC ligand occurs via attack of aziridine on a bound carbonyl ligand, which must be catalyzed by a bromide anion. In the given example, 5 mol% 2-bromoethylamine hydrobromide was used as the catalyst. With stoichiometric amounts of the ammonium bromide, the reaction is quantitative, while its replacement with triethylammonium bromide leads to a drop in yield to 64%.

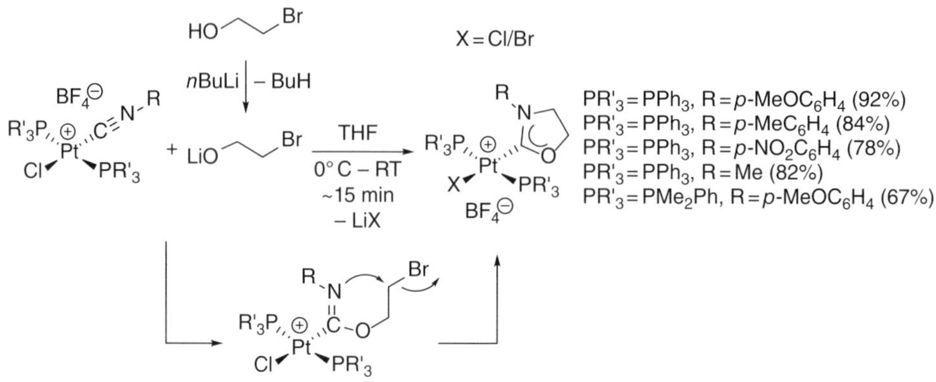

Scheme 8.13
Oxazolidin-2-ylidene complexes via attack of 2-bromoethoxide on platinum-isocyanide complexes.

The carbene carbon in this complex was reported to resonate at 220.2 ppm in CD_3CN. Notably, this value is significantly more downfield than those of reported for the stereotypical NHCs containing two α-nitrogen atoms.

The N–H moiety of the carbene ligand can be easily deprotonated using sodium hydride to give an intermediate imidoyl complex. Subsequent reactions with electrophiles such as oxonium salts or alkyl halides lead to the formation of complexes bearing N-alkyl oxazolidin-2-ylidene ligands [15].

Due to the similarity of isocyanide and carbonyl ligands, it is not surprising that carbene complexes can also be obtained from nucleophilic attack on metal-bound isocyanides. Accordingly, a series of cationic heteroleptic bis(phosphine)/isocyanide complexes were converted into the respective oxazolidin-2-ylidene complexes (Scheme 8.13) [16]. The key to these reactions is the use of the bifunctional 2-bromoethanol. After deprotonation of the alcohol using *n*-butyllithium, a stronger nucleophilic alkoxide is generated, which can attack the electrophilic isocyanide carbon donor to generate an imidoyl intermediate. The nitrogen atom of this species is sufficiently nucleophilic to affect cyclization under liberation of a bromide ion. The latter in turn can coordinate to the platinum center leading to a mixture of halido-scrambled carbene complexes.

The reaction is sensitive to a number of different factors. The use of stronger donating phosphines as co-ligands reduces to electrophilicity of the isocyanide carbon leading to a lower yield. On the other hand, electron richer and less bulky isocyanide ligands are favorable and give higher conversions.

In general, successful nucleophilic attack on isocyanide ligands appears to occur predominantly when the difference of the wavenumbers between free and coordinated isocyanide is large, that is, $\Delta\tilde{v} = \tilde{v}(C \equiv N)_{bound} - (C \equiv N)_{free} > 40 cm^{-1}$. However, this is just a prerequisite and not a sufficient condition. For example, the *tert*-butyl isocyanide complex ($\Delta\tilde{v} = 81 cm^{-1}$) does not react at all, due to steric reasons. Furthermore, it was observed that palladium(II) analogs did not give the respective carbene complexes under the same reaction conditions. Instead, decomposition to palladium black was observed.

Complexes of oxazolidin-2-ylidenes can also be directly obtained by the reaction of 2-hydroxyethylisocyanides with electron poor metal centers. As exemplified in Scheme 8.14, addition of four equivalents of the 2-hydroxy isocyanide to palladium(II) halides quantitatively furnishes homoleptic tetrakis(oxazolidin-2-ylidene) complexes of palladium(II) [17].

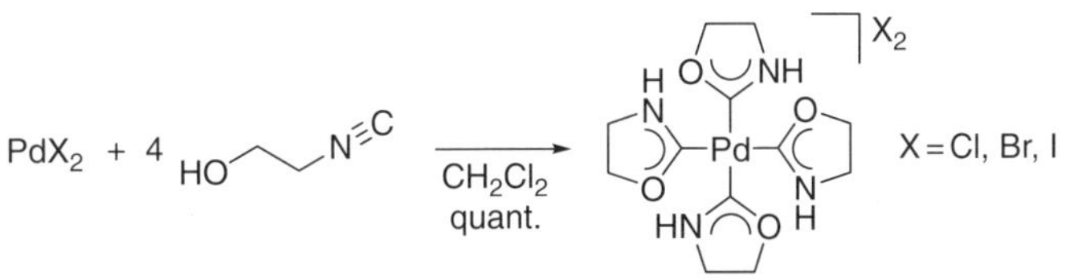

Scheme 8.14
Syntheses of tetrakis(oxazolidin-2-ylidene)palladium(II) complexes.

The complexes are dicationic, and their charge is balanced by the respective halides, which can undergo subsequent anion exchange.

Coordination of the isocyanide to the Lewis acidic metal center increases the electrophilicity at the carbon donor, which in turn triggers nucleophilic attack of the hydroxo group. Cyclization gives a favorable five-membered heterocyclic system, and concurrent proton migration to the nitrogen atom completes the reaction sequence and the formation of the carbene. Similar reactions have also been carried out with other transition metals such as nickel(II) and platinum(II).

The template-assisted cyclization of functionalized isocyanide ligands often affords saturated oxazolidin-2-ylidene complexes, and compounds containing the unsaturated oxazolin-2-ylidene are less commonly prepared using this route. An exception is represented by the metal-mediated coupling of propargylic alcohol and isocyanides reported by Ruiz and coworkers [18].

In this reaction, a cationic phenylisocyanide complex of manganese, which also contained three carbonyl and one bidentate 2,2′-bipyridine co-ligands, was used as a template. Exposure of this complex to sodium propargylic alkoxide in the respective alcohol as a solvent led to nucleophilic attack of the alkoxide at the bound isocyanide. The intermediate imidoyl species cyclizes via attack at the internal methine carbon atom, and subsequent protonation by the alcohol solvent affords an oxazolidin-2-ylidene ligand with an exocyclic double bond. This intermediate slowly converts to the more stable oxazolin-2-ylidene complex under hydrogen-shift, which generates the 4-methyl substituent of the NOHC ligand (Scheme 8.15). The transformation of a saturated into an unsaturated carbene is accompanied by a significant upfield shift of the carbon donor resonance from 241.0 ppm to 215.9 ppm.

The more "obvious" route by deprotonation of oxazolium salts was only reported by Biffis and Cavell and coworkers in 2005 [19]. The required oxazolium salts were prepared by reaction of the parent oxazole with alkyl halides, but only stronger electrophiles would alkylate the nitrogen atom, giving low, yet still acceptable yields (Scheme 8.16). This is due to the general low nucleophilicity of the oxazole nitrogen due to the presence of an electronegative oxygen atom in the ring. The reaction of oxazole with picolyl bromide (2-PyCH$_2$Br) also gave the respective salt in a low yield. However, it was noted that this N-donor-functionalized salt slowly decomposes in solution, limiting its use as a carbene precursor. Due to the polarizing effect of the oxygen atom, the C2–H proton of the oxazolium salt may become sufficiently acidic to allow for self-deprotonation by the basic pyridine moiety of the picolyl substituent.

The simple N-alkyloxazolium halides can be routinely treated with palladium(II) acetate in dimethylsulfoxide at elevated temperatures, which affords the dihalido-bis(carbene) complexes as stable compounds in moderate to good yields. Concurrent *in situ* NOHC generation and palladation occurs with the use of the metal precursor bearing basic ligands.

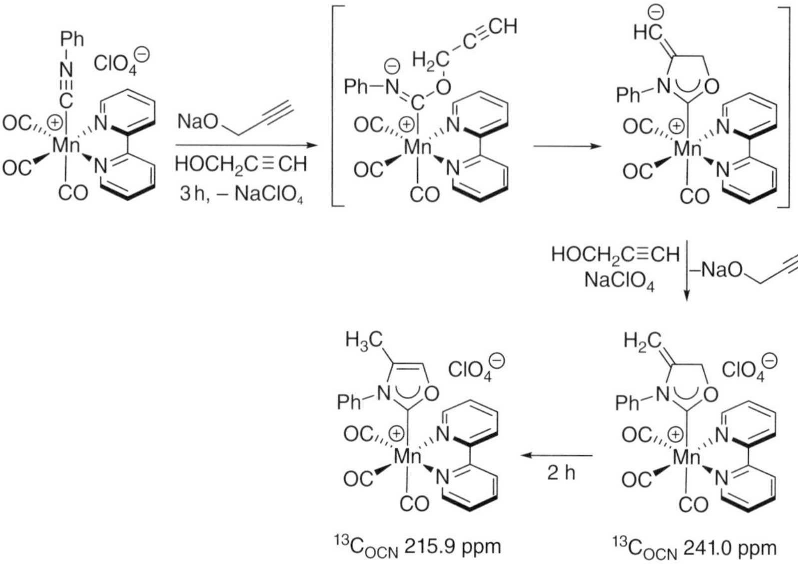
NaO
HOCH2C≡CH
3h, – NaClO4
HOCH2C≡CH
NaClO4
–NaO
2 h
13COCN 215.9 ppm
13COCN 241.0 ppm

Scheme 8.15
Mechanistic pathway to an oxazolin-2-ylidene via an intermediate oxazolidin-2-ylidene complex by coupling of an isocyanide ligand with propargylic alcohol.

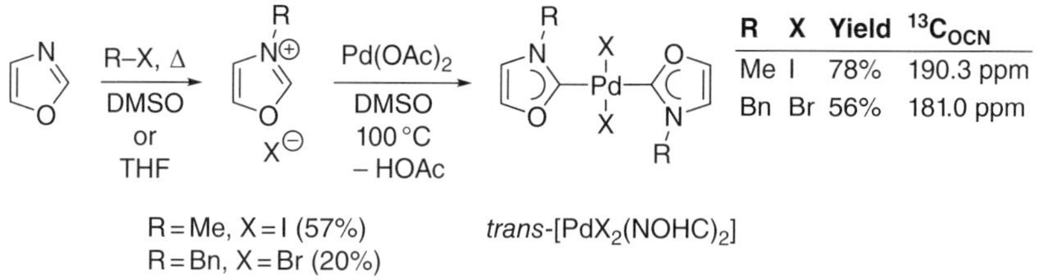
R–X, Δ
DMSO
or
THF
Pd(OAc)2
DMSO
100 °C
– HOAc
R X Yield 13COCN
Me I 78% 190.3 ppm
Bn Br 56% 181.0 ppm
R = Me, X = I (57%)
R = Bn, X = Br (20%)
trans-[PdX2(NOHC)2]

Scheme 8.16
Palladium(II) oxazolin-2-ylidene complexes via *in situ* deprotonation of oxazolium salts with $Pd(OAc)_2$.

Interestingly, only *trans* configured bis(NOHC) isomers were observed in these reactions. Compared to the direct imidazolin-2-ylidene analogs, the ^{13}C NMR signals for the carbene donors appear at lower field, which is attributed to the deshielding effect of the electron-withdrawing oxygen atom. Catalytic studies revealed high activity of the complexes in the Mizoroki-Heck coupling of activated aryl bromides with *n*-butylacrylate.

The *in situ* generation of free NOHCs can also be achieved with an external base in analogy to common NHCs. Thus, three equivalents of N-methyloxazolium tetrafluoroborate can be deprotonated using excess potassium acetate in the presence of a platinum(IV) complex to give the cationic trimethyltris(carbene)platinum(IV) complex in a simple ligand substitution reaction, as reported by Steinborn and coworkers [20]. The N,O-heterocyclic carbenes generated *in situ* easily replace the much weaker acetone donors from the platinum(IV) center (Scheme 8.17, right).

Another example, reported by Cabeza and coworkers, involves the use of potassium bis(trimethylsilyl)amide as a much stronger base [21]. The oxazolin-2-ylidene generated in

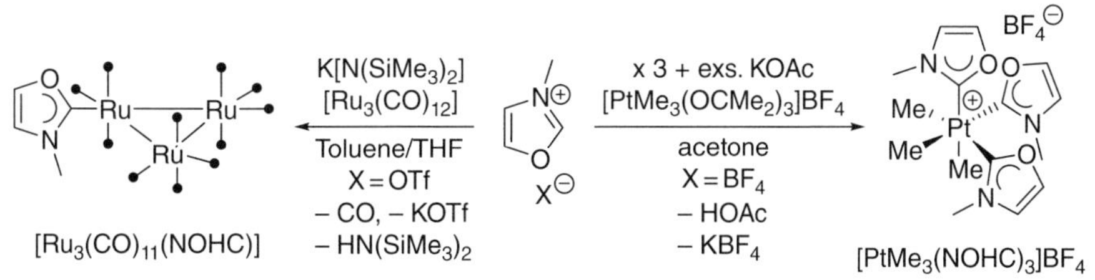

Scheme 8.17
Oxazolin-2-ylidene complexes via *in situ* deprotonation of oxazolium salts with an external base.

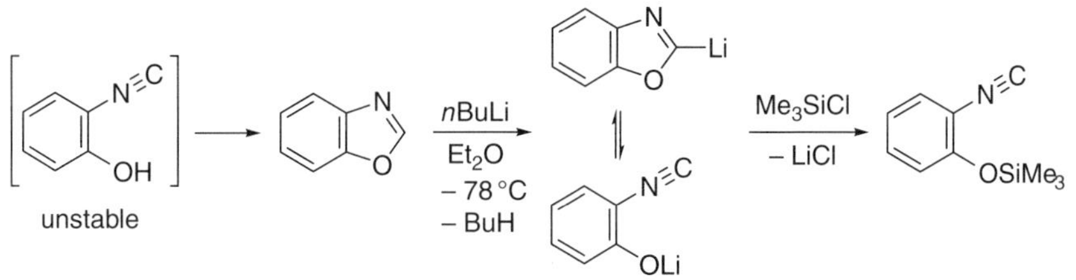

Scheme 8.18
Elusive free 2-hydroxyphenyl isocyanide and preparation of its silyl protected form.

the presence of the triruthenium-carbonyl cluster [$Ru_3(CO)_{12}$] displace one carbonyl ligand while the triruthenium core is maintained (Scheme 8.17, left). Unfortunately, only a limited number of oxazolium salts can be efficiently prepared, which limits the wider application of *in situ* deprotonation methods.

Benzoxazolin-2-ylidene (type **IIIb**) complexes can be elegantly obtained from 2-trimethyl-siloxyphenyl isocyanide, which can be produced by lithiation of benzoxazole followed by treatment with trimethylsilyl chloride (Scheme 8.18). The molecule can be considered a synthetic equivalent for the elusive 2-hydroxyphenyl isocyanide, which does not exist in its free form, but spontaneously cyclizes to benzoxazole.

Hahn and Tamm reported its complexation to an iron carbonyl complex fragment by heating it with pentacarbonyliron(0) in an evacuated autoclave to 80 °C (Scheme 8.19) [22]. The isocyanide displaces one carbonyl ligand giving the stable isocyanide complex in a good yield. Subsequently, the Si–O bond had to be cleaved, which occurs with catalytic amounts of fluoride in methanol, leading to the immediate formation of an N-protonated benzoxazolin-2-ylidene complex via intramolecular cyclization. Despite the use of a low-valent and electron-rich metal center, which should increase backdo-nation to the isocyanide function to prevent cyclization, the expected tetracarbonyl-2-hydroxyphenylisocyanide-iron(0) complex could not be observed as an intermediate. The rapid formation of the carbene was attributed to the aromaticity of the resulting cyclic system. Notably, the N–H carbene complex was found to be highly stable. Thus deprotonation with sodium hydride and subsequent N-methylation occurred smoothly giving the respective 3-methylbenzoxazolin-2-ylidene iron(0) complex as the final product.

Comparison of the chemical shifts for carbon donors in the three complexes revealed distinct characteristics. Although measured in different solvents, it is obvious that the

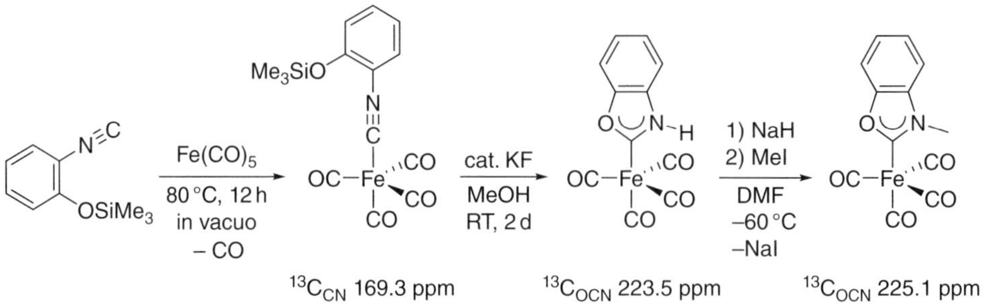

Scheme 8.19
Template-directed synthesis of a benzoxazolin-2-ylidene iron(0) complex.

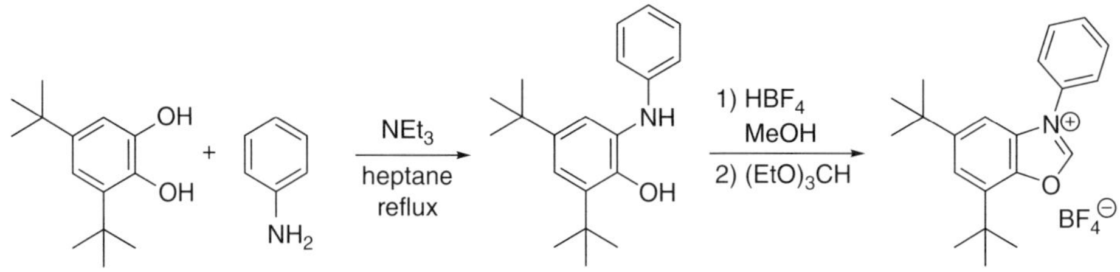

Scheme 8.20
Preparation of an N-phenyl-substituted benzoxazolium salt.

benzoxazolin-2-ylidene complexes show very downfield resonances of >200 ppm, which are in line with the formation of carbene complexes. Using this strategy, a range of benzoxazolin-2-ylidene complexes has been prepared [23].

Derivatives with N-aryl substituents cannot be obtained using the above mentioned strategies. A different approach that can yield such compounds was devised by Bellemin-Laponnaz using an appropriately substituted benzoxazolium salt [24]. The respective carbene precursor salt was prepared via condensation of 3,5-di(*tert*-butyl) catechol with aniline, and subsequent cyclization with triethylorthoformate in the presence of tetrafluoroboric acid (Scheme 8.20). Rhodium and copper complexes can then be prepared in good yields by mixing the benzoxazolium salt with the suitable metal precursors followed by *in situ* deprotonation with alkali *tert*-butoxide (Scheme 8.21). The carbene ligand thus generated is immediately stabilized by coordination to the metal. Notably, attempts to apply the common silver-carbene transfer method were unsuccessful, since action of silver(I) oxide on the benzoxazolium salt only yielded an intractable mixture of decomposed materials. Exposure of the rhodium(I)-COD complex to carbon monoxide gas with the intention to prepare the dicarbonyl-derivative for donor strength determination also led to substantial decomposition, which highlights a different property of N,O-heterocyclic carbenes compared to their N,N-substituted counterparts.

The complexes show, as expected, downfield ^{13}C NMR signals for the carbene carbon atoms at 214.4 and 200.0 ppm, respectively. The solid state molecular structure could be established for the rhodium(I)-NHC complex, which confirmed its identity as an N-phenylbenzoxazolin-2-ylidene complex (Figure 8.4).

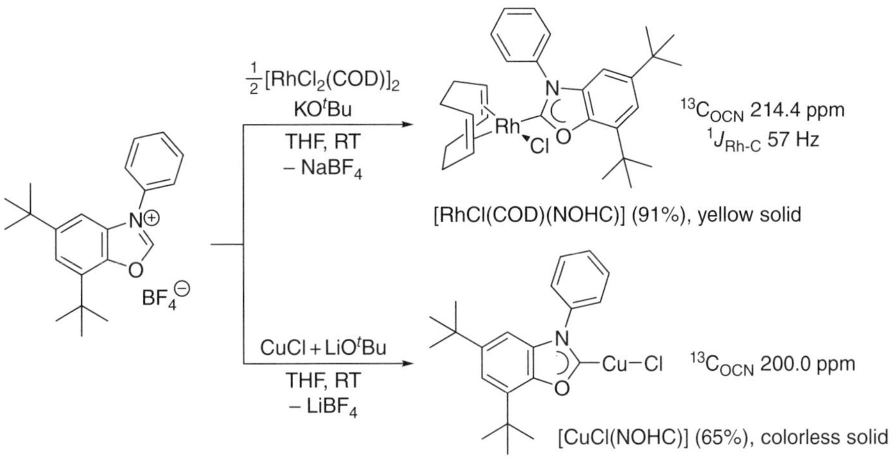

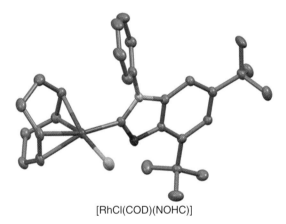

Scheme 8.21
Preparation of N-phenyl-substituted benzoxazolin-2-ylidene complexes.

[RhCl(COD)(NOHC)]

Figure 8.4
Molecular structure of a rhodium(I)-benzoxazolin-2-ylidene complex. Hydrogen atoms have been omitted for clarity.

8.3 Expanded Six-Membered NHCs

Another simple way to modify the stereotypical five-membered N-heterocyclic carbenes is to extend the ring size. In doing so, cyclic diaminocarbenes with six-, seven- and even eight-membered rings can be obtained as representatives of the so-called expanded-ring NHCs. In general, increasing the ring size, but maintaining the carbene donor adjacent to two nitrogen atoms will lead to an increase of the endocyclic NCN angle (Figure 8.5). Nevertheless, the changes are insufficient to propose any change of hybridization at the carbene atom. Concurrently, the exocyclic RNC (α) angles are expected to decrease, which brings the substituent R closer to any coordinated metal center, thus increasing its steric impact.

For simpler systems containing an unsubstituted and saturated backbone, one would expect a slight increase of the electron donating ability with increasing ring size, since the

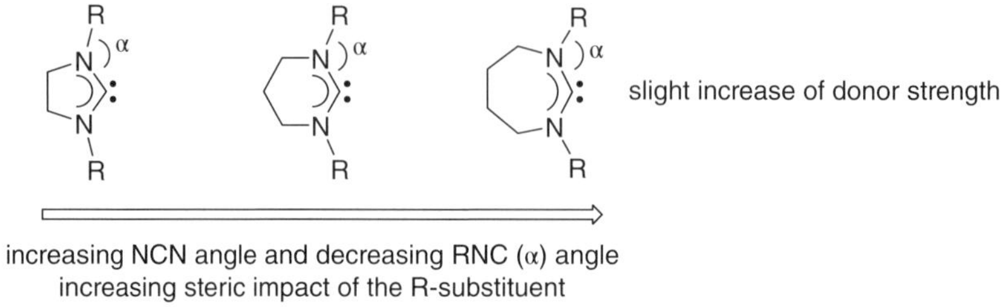
R
N
N
R
α
slight increase of donor strength
increasing NCN angle and decreasing RNC (α) angle
increasing steric impact of the R-substituent

Figure 8.5
Structural and electronic trends of expanded-ring NHCs.

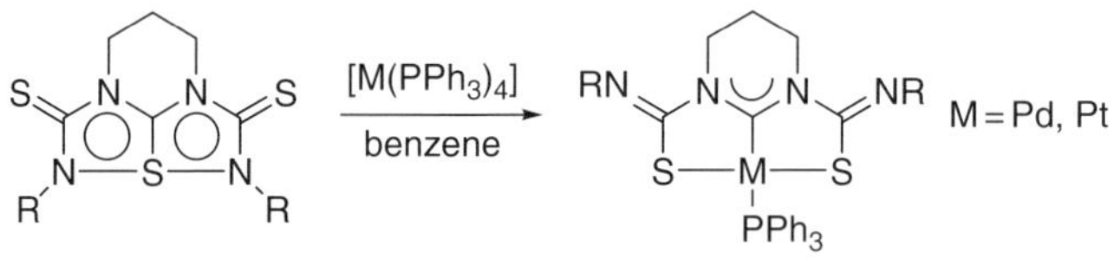
[M(PPh3)4]
benzene
M = Pd, Pt
PPh3

Scheme 8.22
Preparation of palladium(II) and platinum(II) *S,C,S*-pincer-type complexes of six-membered NHCs.

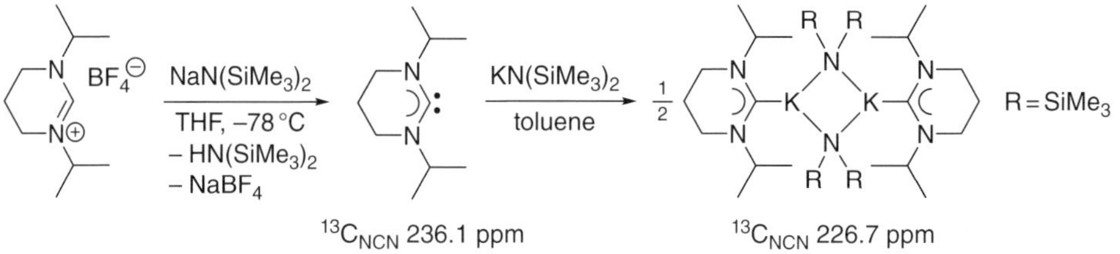
NaN(SiMe3)2
THF, –78 °C
– HN(SiMe3)2
– NaBF4
KN(SiMe3)2
toluene
R = SiMe3
13CNCN 236.1 ppm
13CNCN 226.7 ppm

Scheme 8.23
Preparation the first 6-membered NHC and its potassium complex.

stepwise incorporation of additional CH_2 groups should increase the positive inductive effect of the backbone.

Complexes of six-membered NHCs were reported by Iwasaki and coworkers in 1996 [25]. In this work, a complicated poly-heterocyclic ring system was exposed to zero-valent, homoleptic palladium or platinum-phosphine complexes in benzene. In this reaction, the C–S bond is oxidatively activated by the zero-valent late transition metals, and SCS-type pincer complexes are obtained in a single step (Scheme 8.22). The pincer ligand contains a central six-membered saturated NHC moiety, which is flanked by two thio-amido wing tip groups. The same methodology was also applied for the preparation of rhodium(III) pincer complexes [26].

The first free NHC with a six-membered ring was reported by Alder and coworkers three years later [27]. The carbene precursor was conveniently prepared by heating an equimolar mixture of 1,3-diisopropylaminopropane and ammonium tetrafluoroborate in triethyl-orthoformate under reflux. Subsequent deprotonation with sodium-bis(trimethylsilyl) amide in tetrahydrofurane at –78 °C afforded the free carbene in 40% yield (Scheme 8.23). The free six-membered NHC melts close to ambient temperature and can be purified by distillation.

Interestingly, the free species is thermodynamically stable toward dimerization, which may hint at a greater steric impact of the N-isopropyl groups in the six-membered ring. Its direct five-membered analog 1,3-diisopropylimidazolidin-2-ylidene, on the other hand, can only be isolated as the dimeric enetetramine. The carbene carbon of the free NHC resonates in the typical downfield region with a chemical shift of 236.1 ppm recorded in toluene-d_8. In the presence of alkali metals, the carbene readily forms alkali metal adducts. Thus, its treatment with alkali bis(trimethylsilyl)amides leads to the first examples of group I carbene complexes. On coordination to the potassium amide, a slight upfield shift of the carbene donor to 226.7 ppm was observed. Moreover, single crystals of this complex could be obtained, which analysis revealed a dimeric structure with two bridging amido ligands (Figure 8.6).

The use of the same carbene precursor salt and its N-mesityl substituted analog to prepare transition metal complexes were subsequently reported by Buchmeiser and coworkers in 2004 [28]. The heterocyclic salts were prepared in a three-step protocol initiated by condensation of the respective N,N-disubstituted 1,3-diaminopropanes with formaldehyde in methanol with moderate heating (Scheme 8.24). The resulting cyclic diamines were then oxidized with N-bromosuccinimide to the respective quaternary ammonium bromides, which were subjected to anion exchange with silver tetrafluoroborate to afford the desired tetrahydropyrimidinium salts.

Metallation to rhodium(I) was achieved by *in situ* deprotonation of the salts with lithium *tert*-butoxide in the presence of the chlorido-bridged rhodium-COD dimer, which afforded

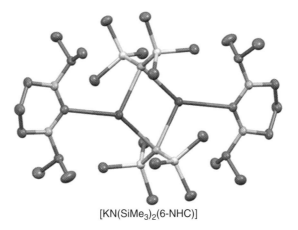

[KN(SiMe$_3$)$_2$(6-NHC)]

Figure 8.6
Molecular structure of a dimeric potassium six-membered NHC complex. Hydrogen atoms have been omitted for clarity.

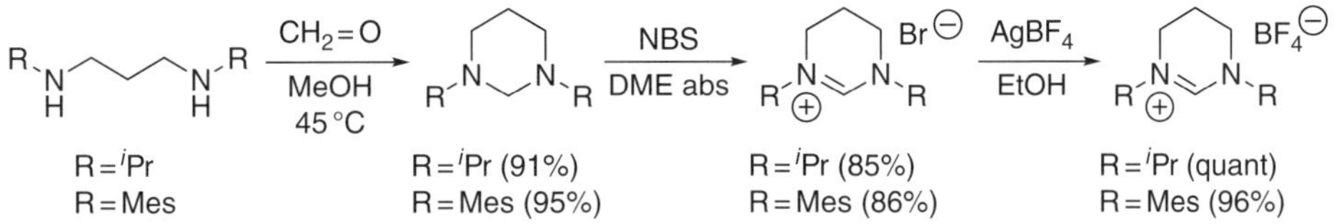

Scheme 8.24
Preparation of tetrahydropyrimidinium salts as carbene precursors.

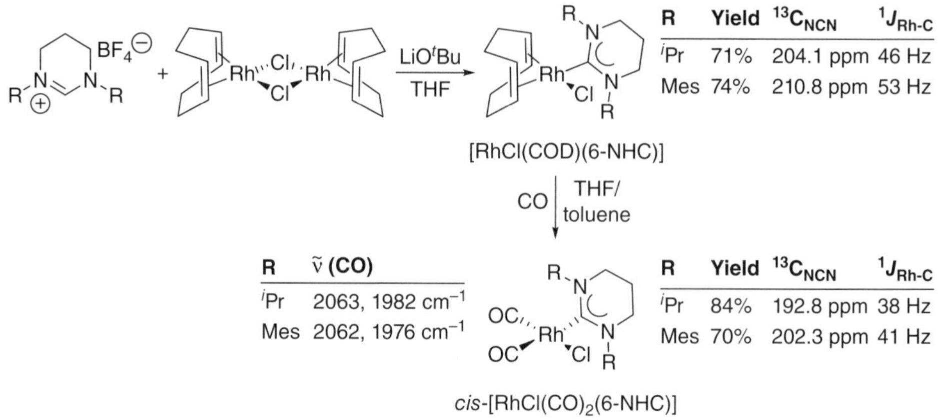

R	Yield	$^{13}C_{NCN}$	$^{1}J_{Rh-C}$
^{i}Pr	71%	204.1 ppm	46 Hz
Mes	74%	210.8 ppm	53 Hz

[RhCl(COD)(6-NHC)]

R	$\tilde{v}$ (CO)
^{i}Pr	2063, 1982 cm^{-1}
Mes	2062, 1976 cm^{-1}

R	Yield	$^{13}C_{NCN}$	$^{1}J_{Rh-C}$
^{i}Pr	84%	192.8 ppm	38 Hz
Mes	70%	202.3 ppm	41 Hz

cis-[RhCl(CO)$_2$(6-NHC)]

Scheme 8.25
Preparation of rhodium(I) complexes bearing 6-membered NHCs.

the yellow [RhCl(COD)(6-NHC)]-type complexes in good yields of >70% (Scheme 8.25). The carbene resonances of these two complexes are detected at downfield regions of >200 ppm in CDCl$_3$. As expected the aromatic mesityl wing tips exert a deshielding effect in comparison to the aliphatic isopropyl substituents.

Bubbling CO gas through solutions of the COD complexes leads to displacement of the diolefin with two carbonyl ligands, affording *cis*-[RhCl(CO)$_2$(6-NHC)]-type compounds in decent yields. Interestingly, this ligand substitution results in a slight upfield shift of the carbene resonances and a decrease of the heteronuclear coupling constant. Evaluation of the carbonyl stretching frequencies led to the conclusion that the saturated six-membered NHCs are stronger donors than their classical and saturated five-membered analogs, that is, imidazolidin-2-ylidenes (SIMes: 2081, 1996 cm^{-1}). Notably, comparison of the wavenumbers obtained for the N-isopropyl and N-mesityl analogs may lead to the hypothesis that the former is a weaker donor, which is not intuitive and difficult to comprehend. This highlights once more that carbonyl based methods may not be suitable to discern smaller changes within a given ligand system.

One year earlier, Richeson and coworkers reported the isolation of the first six-membered NHC with an aromatic backbone [29]. The synthesis of this free 6-NHC begins with a N,N'-diisopropyl substituted diaminonaphthalene, which was subjected to ring-closure with triethylorthoformate and *in situ* protonation with hydrochloric acid to yield the perimidinium chloride salt as a suitable precursor to a benzannulated 6-NHC. Its deprotonation to the free carbene was only successful with the bulky and non-nucleophilic LiHMDS base, while the use of *tert*-butoxide bases led to the formation of neutral alcohol adducts (Scheme 8.26). The free carbene embedded in a tricyclic system was isolated as a monomeric species, which is probably due to the steric impact of the two bulky isopropyl substituents. Its carbene resonance detected at 241.7 ppm in C$_6$D$_6$ is deshielded compared to Alder's fully saturated 6-NHC bearing the same wing tips, which can be explained by the benzannulation.

Exposure of the free 6-NHC to the chlorido-bridged rhodium(I)-COD dimer afforded the stable rhodium(I)-6-NHC complex [RhCl(COD)(6-NHC)], which shows as expected a significant upfield shift of the carbene donor to 213.3 ppm upon coordination. In an attempt to probe the donating ability of the ligand, conversion to the dicarbonyl complex

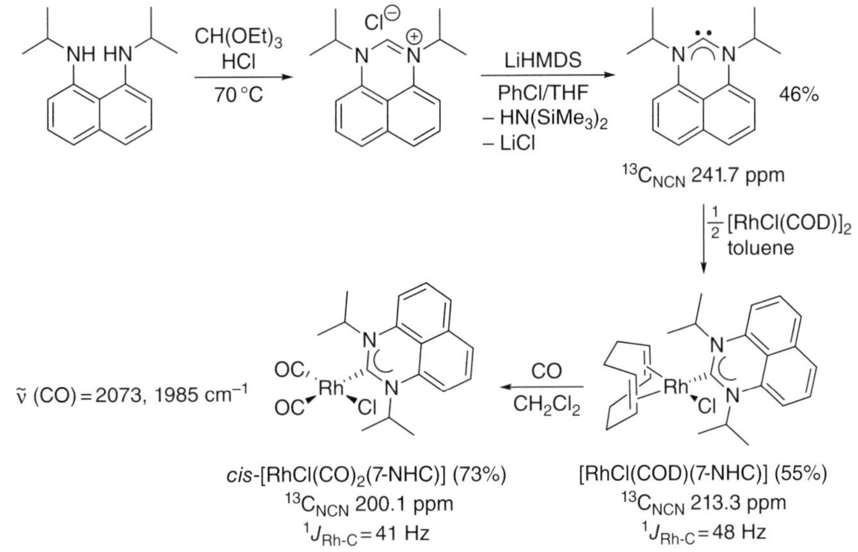

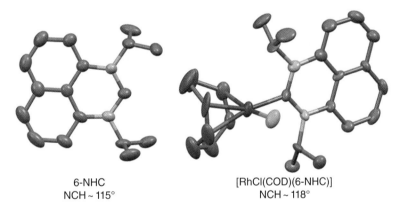

Scheme 8.26
Preparation of a benzannulated 6-NHC and its rhodium(I) complexes.

6-NHC
NCH ~ 115°

[RhCl(COD)(6-NHC)]
NCH ~ 118°

Figure 8.7
Molecular structures of a benzannulated 6-NHC and its rhodium(I) complex. Hydrogen atoms have been omitted for clarity.

cis-[RhCl(CO)$_2$(6-NHC)] was carried out by bubbling carbon monoxide gas through a solution of the COD complex. An upfield shift the carbon donor to 200.1 ppm indicates a reduced electron density. More importantly, two carbonyl stretches were found at 2073 and 1985 cm^{-1}, which compared to those detected for Alder's analog reveals a weaker electron donating ability as a result of benzannulation. Nevertheless, this 6-NHC was proposed to be a stronger donor than the stereotypical five-membered NHCs such as SIMes by comparison of the carbonyl stretches.

X-ray analysis on single crystals obtained for the free 6-NHC and the [RhCl(COD)(6-NHC)] complex confirmed the identity of the compounds, and their molecular structures are depicted in Figure 8.7. The NCN angle of the free 6-NHC was determined to be ~115°, which is, as expected, enlarged compared to those of typical 5-NHCs ranging from 100–110°. There is a slight increase of the NCN angle upon complex formation. The metal-carbene bond distance is in the typical range also observed for analogs bearing 5-NHC complexes.

8.4 Expanded Seven- and Eight-Membered NHCs

The first complexes with seven-membered NHCs were reported by the Stahl group in 2005 [30]. The ligand precursor was prepared via cyclization of 2,2′-di(2-adamantylamino) biphenyl using triethylorthoformate in the presence of ammonium tetrafluoroborate, while attempts to cyclize the dimesityl derivative were futile due to its reduced nucleophilicity (Scheme 8.27). Palladation was effected by *in situ* deprotonation of the acidic C2–H proton in the heterocyclic salt with potassium *tert*-butoxide in the presence of the η^3-allylchloridopalladium(II) dimer using dry tetrahydrofurane as the solvent, which afforded the yellow [PdCl(allyl)(7-NHC)] complex as a diastereomeric mixture in high yield. The complex is air stable and can be purified by column chromatography on silica gel. Protonolysis of the allyl ligand to propene with hydrochloric acid solution in diethyl ether leads to dissociation of the latter, and subsequent dimerization of the coordinatively unsaturated monocarbene complex fragment cleanly affords the chlorido-bridged dimer [PdCl$_2$(7-NHC)]$_2$ in essentially quantitative yield. The carbene signal of this dinuclear complex was detected at ~201.4 ppm in CDCl$_3$. Single crystals obtained for both complexes were analyzed by X-ray diffraction, and their molecular structures depicted in Figure 8.8 confirm the non-planar nature of the seven-membered carbene ligand. In particular, the 7-NHC ligand exhibits a tortional twist, which gives rise to C_2 symmetry.

The first seven-membered NHCs with a fully saturated backbone were prepared by Cavell, Dervisi, Fallis, and coworkers [31]. The precursors to the 7-NHCs were prepared

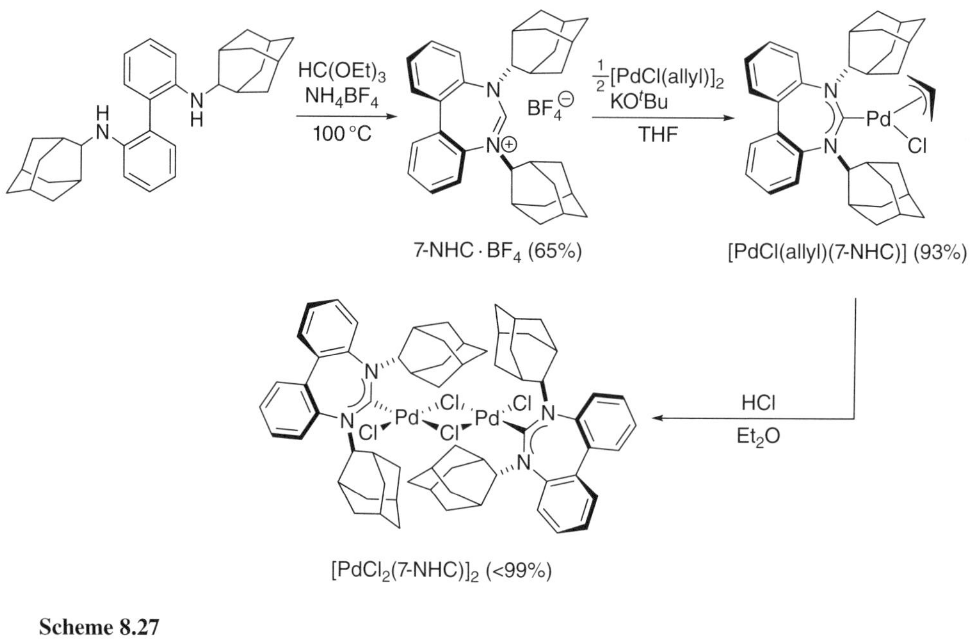

Scheme 8.27
Preparation of the first seven-membered NHC complexes.

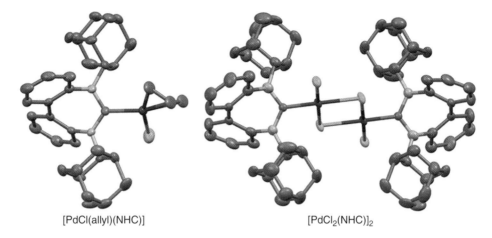

Figure 8.8
Molecular structures of palladium(II) complexes with seven-membered NHCs. Hydrogen atoms have been omitted for clarity.

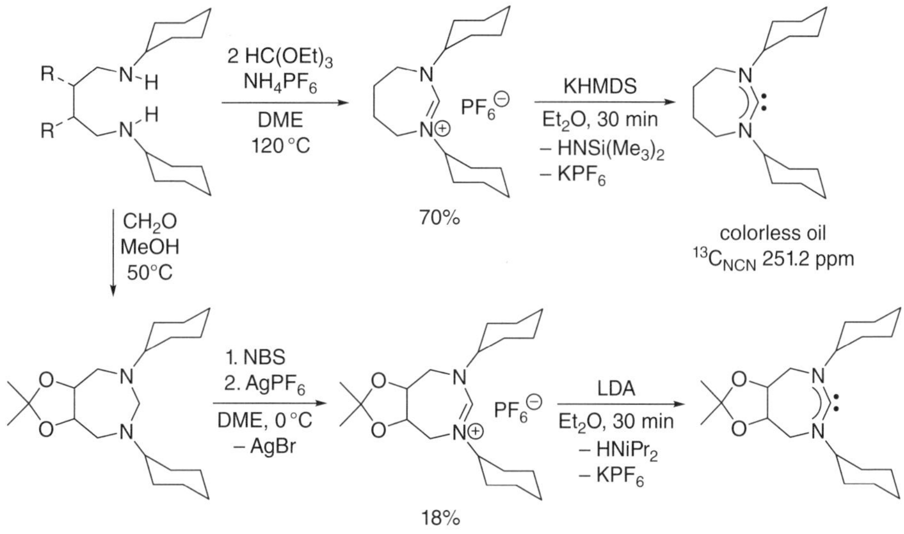

Scheme 8.28
Preparation of the first seven-membered NHCs with a saturated backbone.

via routine condensation of the respective substituted diamines with either orthoformate or formaldehyde. In the first case, addition of ammonium hexafluorophosphate led directly to the formation of the desired heterocyclic salt in 70% yield, while in second case a saturated alicyclic diamine was obtained first, that had to be oxidized using NBS followed by anion exchange with silver hexafluorophosphate giving the carbene precursor in only 18% yield (Scheme 8.28). It was also noted that the use of neat orthoformate led to a dramatic drop in the yield to 10–20%.

This low yielding and often poorly reproducible methodology has since been replaced by the authors with a more elegant and modified route originally reported by Bertrand

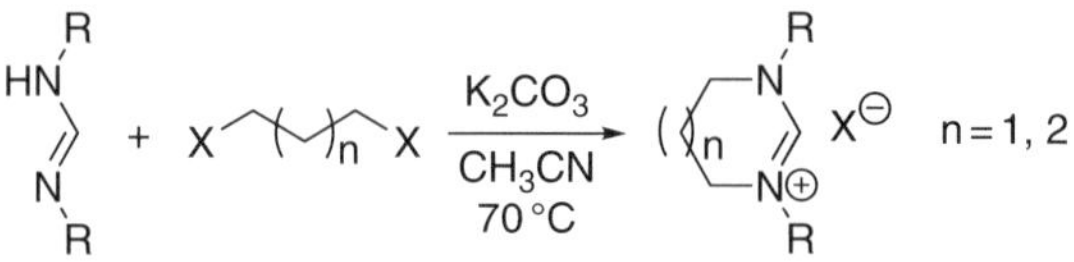

Scheme 8.29
Improved route to six- and seven-membered amidinium salts.

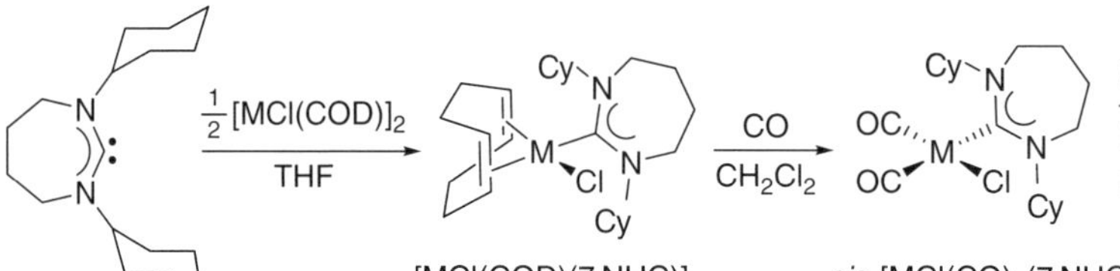

M	$\tilde{v}$ **(CO)**
Rh	2071, 1990 cm^{-1}
Ir	2058, 1973 cm^{-1}

[MCl(COD)(7-NHC)]

M	Yield	$^{13}C_{NCN}$	$^{1}J_{Rh-C}$
Rh	66%	215.3 ppm	44 Hz
Ir	62%	208.3 ppm	-

cis-[MCl(CO)$_2$(7-NHC)]

M	Yield	$^{13}C_{NCN}$	$^{1}J_{Rh-C}$
Rh	84%	206.2 ppm	38 Hz
Ir	82%	200.5 ppm	-

Scheme 8.30
Preparation of seven-membered NHC complexes with a saturated backbone.

and coworkers [32]. In this synthetic method, ring-closure to the cyclic amidinium salt is achieved from the reaction of a dielectrophile with an appropriately substituted formamidine at elevated temperatures in acetonitrile (Scheme 8.29) [33]. Both alkyl- and aryl-substituted species can be obtained in this way.

Deprotonation to the free carbene proved difficult and could only be achieved using specific potassium or lithium amide bases (Scheme 8.28). The carbene signal for the simpler 7-NHC could be detected at 251.2 ppm in benzene, which is downfield shifted compared to a similar free 6-NHC (Scheme 8.24).

Attempts to prepare silver-7-NHC complexes from the amidinium salts and silver(I) oxide for transmetallation failed, and instead acyclic hydrolysis products were obtained. On the other hand, the reaction of *in situ* generated 7-NHCs with dimeric iridium(I) or rhodium(I) precursors was successful and afforded the respective complexes in decent yields, as exemplified by the cleavage reactions of the chlorido-bridged COD complexes of rhodium(I) and iridium(I) (Scheme 8.30). The properties of the yellow air stable complexes are similar to those containing classical five-membered NHCs. However, the carbene signals for the expanded-ring NHC donors resonate more downfield with values of >200 ppm.

Displacement of the diolefin ligands occurs smoothly by passing carbon monoxide through a solution in dichloromethane, and the *cis*-dicarbonyl complexes are obtained in good yields. The introduction of two good π-acceptors reduces the electron density at metal and upfield shifts of ~9 ppm were observed for the ^{13}C NMR signals of the carbene donors. More importantly, the carbonyl stretches of these complexes could be used to evaluate the donating ability of this 1,3-dicyclohexyl-substituted 7-NHC ligand. Comparing the average wavenumber for the two stretches observed in the rhodium complex with those obtained for some 5- and 6-NHC analogs may lead to the conclusion that the 7-NHC is a stronger donor than a SIMes, but also a weaker donor compared 6-NHC with either aromatic or aliphatic wing tips (Chart 8.4). Nevertheless, such a conclusion has to be taken cautiously, since the NHCs differ both in backbone and in N-substituents.

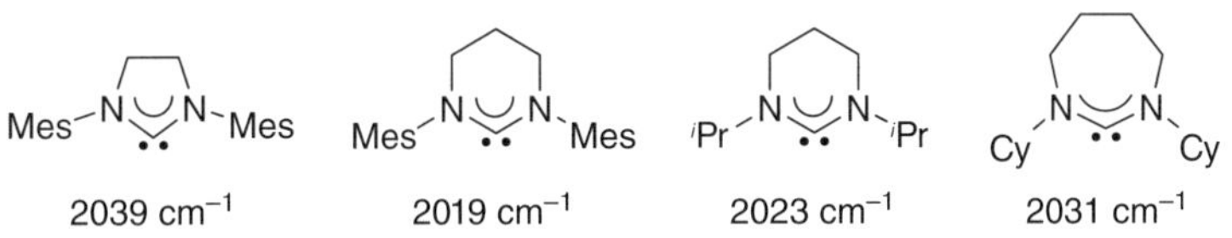

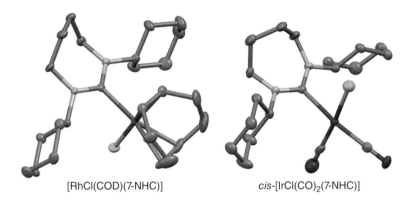

Mes–N⌣N–Mes
2039 cm⁻¹

Mes–N⌣N–Mes
2019 cm⁻¹

ᴵPr–N⌣N–ᴵPr
2023 cm⁻¹

Cy–N⌣N–Cy
2031 cm⁻¹

Chart 8.4
Averaged stretching frequencies for some *cis*-[RhCl(CO)(NHC)] complexes.

[RhCl(COD)(7-NHC)]   *cis*-[IrCl(CO)₂(7-NHC)]

Figure 8.9
Molecular structures for [RhCl(COD)(7-NHC)] and *cis*-[IrCl(CO)₂(7-NHC)]. Hydrogen atoms have been omitted for clarity.

Solid state molecular structures were obtained for the [RhCl(COD)(7-NHC)] and *cis*-[IrCl(CO)₂(7-NHC) complexes, which allowed for the evaluation of NCN angle (Figure 8.9). As anticipated, the NCN angle is enlarged compared to smaller ring systems. An increase was noted going from 5-NHC (103–107°) to 6-NHC (115–118°) to 7-NHC (120–123°). With increasing NCN angles, the wing tips are forced out toward the metal, making it sterically more congested. Moreover, the rhodium–carbon distance of 2.056(2) Å was found to be similar to that formed with a 6-NHC of 2.047(3) Å, but both are notably longer than those found for 5-NHCs, being in a typical range of 1.994–2.032 Å. In this case, the elongation can reasonably be attributed to a combination of steric and electronic factors.

Expanded-ring NHCs with the largest ring size so far have been reported by Cavell and coworkers in 2011. They applied the "amidine route" to build up N-heterocyclic salts with an eight-membered ring. The reaction of the aromatic formamidines with 1,5-dibromopentane is very slow and takes about 10–14 days even under reflux conditions (Scheme 8.31), which highlights the difficulty in building larger ring systems. Nevertheless, the eight-membered heterocyclic salts can be obtained in generally acceptable yields of ≥ 75%. The chemical shifts for the C2–H proton range from 7.31–7.61 ppm, which in comparison to typical 5-NHC precursors is rather upfield shifted indicating a reduced acidity as a consequence of a longer alkyl chain in the backbone. Thus, more basic free 8-NHCs are expected to form upon deprotonation of the salts with the strong KHMDS base in tetrahydrofuran. As observed for other expanded-ring carbenes, the ylidene resonances for the 8-NHCs appear in the more downfield region of 250–260 ppm. The xylyl-substituted 8-NHC could also be characterized by single crystal X-ray diffraction, which reveals a "boat" conformation of the eight-membered ring (Figure 8.10). In addition, a large NCN angle of ~120° was

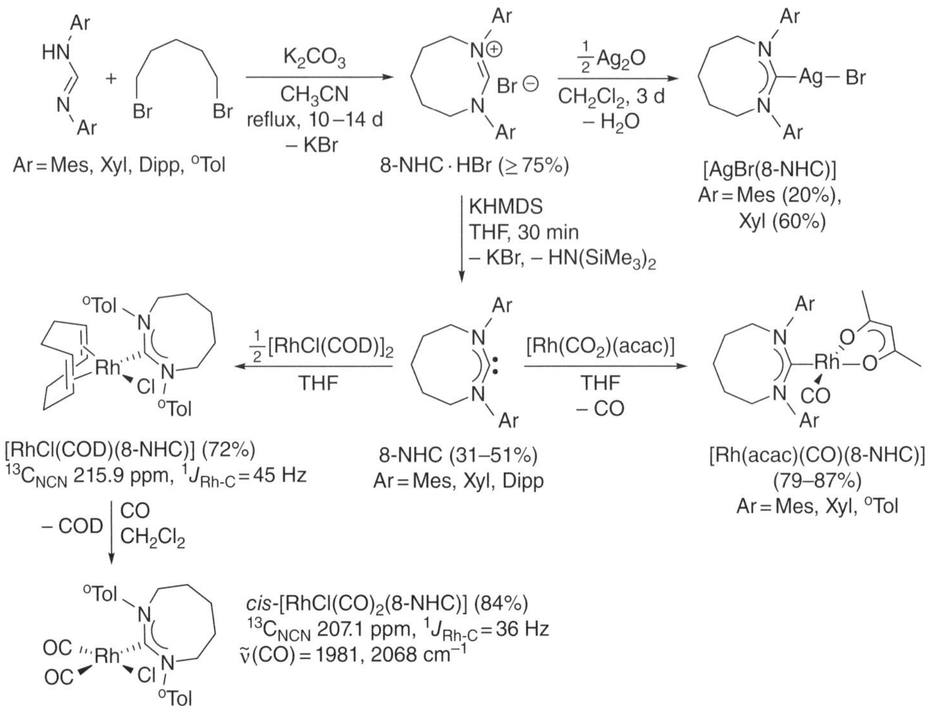

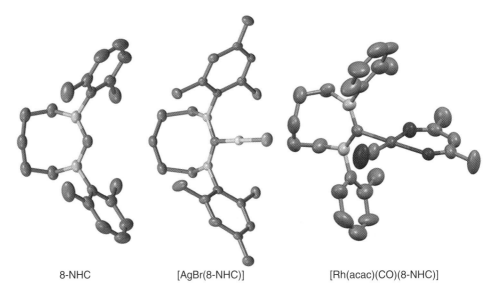

Scheme 8.31
Preparation of 8-NHCs and their coordination chemistry with silver(I) and rhodium(I).

Figure 8.10
Molecular structures of a 8-NHC and its silver(I) and rhodium(I) complex. Hydrogen atoms have been omitted for clarity.

observed, as expected, while the exocyclic CNC_{xyl} angle of ~114° is very small, pushing the xylyl wing tip closer to the carbene carbon.

The direct reaction of the salts with silver(I) oxide produces the bromido-silver(I) monocarbene complexes. These reactions are also very slow due to the reduced acidity of the C2–H proton. Moreover, the feasibility of the reaction largely depends on the steric bulk of the respective carbenes, whereby lower yields are obtained for more bulky systems. For the most bulky 8-NHC bearing Dipp substituents, no formation of silver-carbene was observed due to steric reasons. However, once formed the silver complexes show little sign of ligand dissociation and fluxionality. The enhanced steric bulk makes the formation of bis(NHC) complexes impossible and renders the monocarbene complexes very stable, which even allows for the determination of the carbene chemical shifts and the heteronuclear $^1J_{Ag-C}$ coupling constants (Ar = Mes: $^{13}C_{NCN}$ 217.2 ppm, $^1J^{107}_{Ag-C}$ = 224 Hz, $^1J^{109}_{Ag-C}$ = 256 Hz; Ar = Xyl: $^{13}C_{NCN}$ 217.4 ppm, $^1J^{107}_{Ag-C}$ = 223 Hz, $^1J^{109}_{Ag-C}$ = 255 Hz). On the other hand, the increased stability does not allow for their application as carbene transfer reagents. Single crystals for X-ray analyses were obtained for complexes of both 8-NHCs.

The extreme steric congestion around the carbene donor also makes coordination to larger complex fragments difficult. The routine reaction of free carbenes with the chlorido-bridged rhodium-COD dimer failed for most of the 8-NHCs, which was attributed to the steric repulsion between the bulky 8-NHCs and the COD ligand. Only the smallest *ortho*-tolyl-substituted one could be successfully converted into the desired [RhCl(COD)(8-NHC)] complex and subsequently into the bis(carbonyl) complex *cis*-[RhCl(CO)$_2$(8-NHC)] for the evaluation of its electronic properties. Complex formation with smaller complex fragments is easier, and exposure of the 8-NHCs to [Rh(acac)(CO)$_2$] leads to carbonyl substitution affording complexes of the type [Rh(acac)(CO)(8-NHC)].

The solid state molecular structures depicted in Figure 8.10 confirm the large NCN angle and the boat conformation of the eight-membered heterocycle, which is maintained upon complex formation.

8.5 Expanded Diamidocarbenes (DAC)

A special type of expanded-ring NHCs are the so-called diamidocarbenes (DAC). The electron-withdrawing nature of amide groups in these carbenes increases the electrophilicity and reduces the nucleophilicity of the donor atom. In 2009, Bielawski and coworkers reported the preparation of the N-heterocyclic salt 6-DAC·H$^+$OTf$^-$ by reaction of N,N'-bis(2,6-diisopropylphenyl)-N-trimethylsilylformamidine with dimethylmalonyl dichloride in toluene and trimethylsilyl triflate (TMS·OTf) in nearly quantitative yield as a suitable precursor to a 6-DAC (Scheme 8.32, R = SiMe$_3$) [34]. Almost simultaneously, César and Lavigne disclosed a more elegant route to related, but neutral precursors 6-DAC·HCl by mixing N,N'-dimesitylformamidine with dimethylmalonyl dichloride in dichloromethane and an excess of triethylamine (Scheme 8.32, R = H) [35]. The same reaction can also be conducted with the formamidine with Dipp substituents.

All three precursors furnished the free 6-DAC upon deprotonation with NaHMDS in benzene [34,36], while the use of more polar tetrahydrofurane proofed to be less suitable for their isolation [35]. The carbene signals of the two free DACs were detected at 277.7 and 278.4 ppm in C$_6$D$_6$, respectively, which are markedly more downfield compared to those of 6-NHC analogs due to the electron-withdrawing carbonyl groups. Single crystals

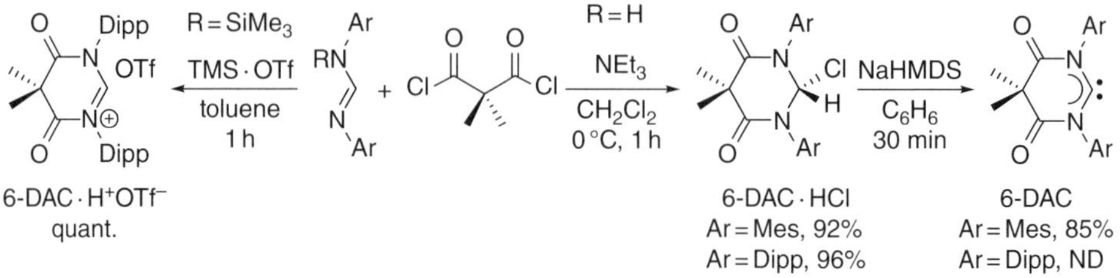

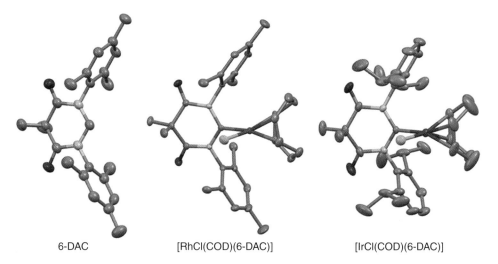

Scheme 8.32
Preparation of 6-DAC precursors and their deprotonation to free 6-DACs.

6-DAC   [RhCl(COD)(6-DAC)]   [IrCl(COD)(6-DAC)]

Figure 8.11
Molecular structures of a 6-DAC and a rhodium(I) and iridium(I) complex. Hydrogen atoms have been omitted for clarity.

for the DAC bearing mesityl wing tips could be obtained for the determination of its solid state molecular structure (Figure 8.11). On the other hand, the Dipp-substituted carbene slowly decomposes in solution under C–H activation of the isopropyl methine group, which highlights a different reactivity of 6-DACs compared to their normal NHC cousins (Scheme 8.33). This was further demonstrated for both 6-DACs by the remarkable activation of several small molecules such as carbon monoxide, isonitrile, and even ammonia. The desirable and challenging activation of dihydrogen, however, could not be achieved. Nevertheless, reactivities typical for NHCs were also observed as they also react with elemental sulfur to form thiourea derivatives.

Despite their increased electrophilicity, the 6-DACs are still sufficiently nucleophilic to form complexes with transition metals, which has been explored with rhodium(I) [35] and iridium(I) [34] for an estimation of their electronic properties. Thus, the freshly prepared free DACs were routinely treated with rhodium and iridium-COD dimers to give the respective [MCl(COD)(6-DAC)] complexes in moderate yields (Scheme 8.34). Successful attempts were also made to obtain single crystals for structural elucidation of these complexes (Figure 8.11). Compared to their classical counterparts, the carbon donors of the DACs resonate at significantly more downfield regions due to the deshielding effect of the

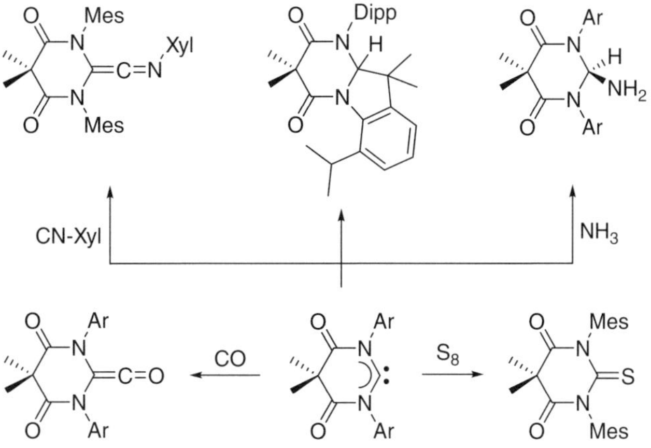

Scheme 8.33
Reactivities of 6-DACs with some small molecules.

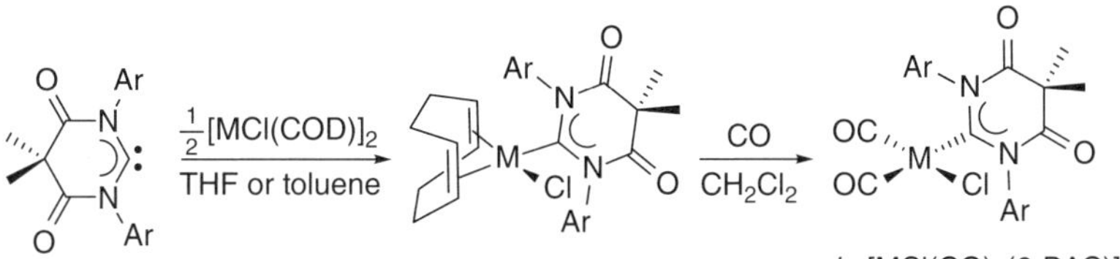

M	$\tilde{v}$(CO) [cm^{-1}]
Rh	2005, 2086
Ir	1991, 2074

6-DAC
Ar = Mes, Dipp

[MCl(COD)(6-DAC)]

M	Ar	Yield	$^{13}C_{NCN}$	$^{1}J_{Rh-C}$
Rh	Mes	68%	245.2 ppm	49 Hz
Ir	Dipp	55%	231.3 ppm	-

cis-[MCl(CO)$_2$(6-DAC)]

M	Ar	Yield	$^{13}C_{NCN}$	$^{1}J_{Rh-C}$
Rh	Mes	98%	230.8 ppm	43 Hz
Ir	Dipp	91%	224.1 ppm	-

Scheme 8.34
Preparation of iridium(I) and rhodium(I) complexes with DACs.

electron-withdrawing amide functions. Diolefin substitution occurred cleanly by exposing the two complexes to carbon monoxide, giving the desired *cis*-[MCl(CO)$_2$(6-DAC)], with slightly upfield shifted carbene signals, in high yields. Comparison of the values found for their carbonyl stretches revealed a reduced donating ability relative to 6-NHCs, which compares well to that of tricyclohexylphosphine.

The molecular structures of the free mesityl substituted DAC and one representative each for a rhodium and an iridium complex are depicted in Figure 8.11. The structural features of the NCO groups in the DAC ring system can be characterized as hybrids of canonical amides and ketones, and the presence of a sp^3-hybridized C3 atom in the backbone results in a non-planar heterocycle. In the complexes, one of the methyl groups attached to this carbon atom points in the direction of the chlorido ligand. Moreover, the metal-carbon bond distances were found to be smaller than those of analogous 6-NHC complexes. The supposedly stronger metal-carbene interaction could be a result of either (i) enhanced backdonation into a formally vacant p_π orbital at the carbene carbon, or (ii) a more Lewis acidic metal center because of weaker carbene donation, which draws the ligands closer. A combination of both factors is probably most likely.

Using essentially the same approach, Bielawski and coworkers also prepared a seven-membered diamidocarbene (7-DAC) and studied its coordination chemistry [37]. In this case, phthaloyl chloride was used for the cyclization with N,N′-dimesitylformamidine, which, by analogy to the reaction above, gave the neutral adduct 7-DAC·HCl as a suitable carbene precursor in a high yield (Scheme 8.35). Upon deprotonation, the free 7-DAC was obtained, which was fully characterized including single crystal X-ray diffraction (Figure 8.12). In comparison to the 6-DAC analogs, the 7-DAC exhibits a more highfield chemical shift of 268.4 ppm for the carbene carbon, providing proof that installation of an additional carbon atom in the heterocycle reduces the deshielding effect of the two amide groups overall. As expected, the increased ring size also enlarges the NCN angle from ~115° observed in the 6-DAC analog to ~123° in this 7-DAC, which is among the widest angles observed for seven-membered NHCs. The reduced electrophilicity of the 7-DAC also leads to decreased reactivity with small molecules such as CO and NH_3. Although attempted reactions of the 7-DAC with these molecules gave rise to some spectroscopic hints for activation, defined reaction products could not be obtained.

The coordination ability of the ligand was studied with rhodium(I), iridium(I) and gold(I) centers (Scheme 8.36). The chemistry with the first two metal centers closely resembles that of the 6-DACs discussed earlier, and straightforwardly gives the complexes [RhCl(COD)(7-DAC)] and *cis*-[RhCl(CO)$_2$(7-DAC)] as stable compounds. The stretching frequencies

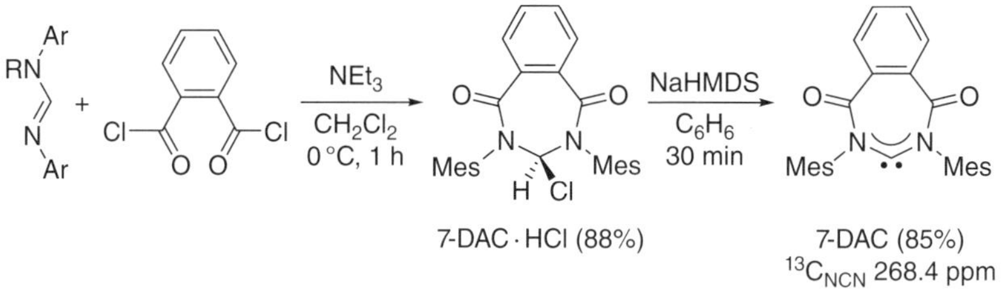

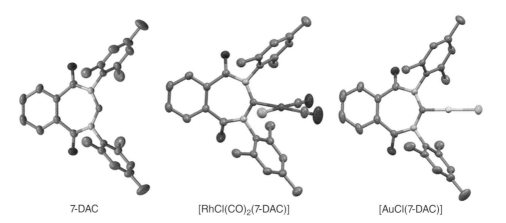

Scheme 8.35
Preparation of a 7-DAC.

7-DAC   [RhCl(CO)$_2$(7-DAC)]   [AuCl(7-DAC)]

Figure 8.12
Molecular structures of a 7-DAC and its rhodium(I) and gold(I) complex. Hydrogen atoms have been omitted for clarity.

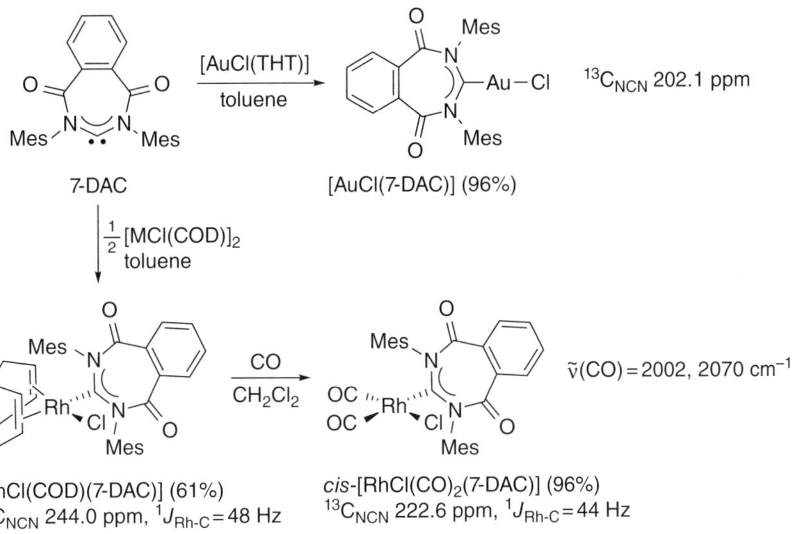

Scheme 8.36
Preparation of some 7-DAC complexes of rhodium(I) and gold(I).

of the carbonyl ligands observed for the latter (averaging to $2036\,cm^{-1}$) indicate a slightly increased donor strength of 7-DAC in comparison to the 6-DACs, which is in good agreement with the slightly more shielded carbon donor observed by ^{13}C NMR spectroscopy of the free 7-DAC. Treatment of the 7-DAC with chlorido-tetrahydrothiophenegold(I) leads to the simple displacement of the labile thioether ligand and affords the first gold(I)-DAC complex. The carbene signal in this [AuCl(7-DAC)] complex was detected at 202.1 ppm, while other spectroscopic properties are unexceptional.

The molecular structures determined for the free 7-DAC, *cis*-[RhCl(CO)$_2$(7-DAC)] and [AuCl(7-DAC)] shown in Figure 8.12 reveal that the bicyclic carbene is less bent, but instead approaching planarity, compared to the 6-DACs, which still contain a sp^3-hybridized carbon atom in the backbone. The coordination geometry of the d^{10} gold(I) complex is linear, as expected. Notably, the increased NCN angle brings the mesityl substituents closer to the metal centers.

References

[1] A. J. Arduengo III, J. R. Goerlich, W. J. Marshall, *Liebig's Ann.* **1997**, 365.

[2] K. Öfele, *Angew. Chem. Int. Ed. Engl.* **1969**, *8*, 916.

[3] H.-W. Wanzlick, H.-J. Kleiner, I. Lasch, H. U. Füldner, H. Steinmaus, *Liebigs Ann. Chem.* **1967**, *708*, 155.

[4] D. J. Cardin, B. Çetinkaya, E. Çetinkaya, M. F. Lappert, Lj. Manojlovic-Muir, K. W. Muir, *J. Organomet. Chem.* **1972**, *44*, C59.

[5] S. K. Yen, L. L. Koh, F. E. Hahn, H. V. Huynh, T. S. A. Hor, *Organometallics* **2006**, *25*, 5105.

[6] P. J. Fraser, W. R. Roper, F. G. A. Stone, *J. Organomet. Chem.* **1973**, *50*, C54.

[7] P. J. Fraser, W. R. Roper, F. G. A. Stone, *J. Chem. Soc., Dalton Trans.* **1974**, 102.

[8] (a) P. J. Fraser, W. R. Roper, F. G. A. Stone, *J. Chem. Soc., Dalton Trans.* **1974**, 760. (b) M. Green, F. G. A. Stone, M. Underhill, *J. Chem. Soc., Dalton Trans.* **1975**, 939. (c) C. H. Game, M. Green, F. G. A. Stone, *J. Chem. Soc., Dalton Trans.* **1975**, 2280.

[9] (a) H. G. Raubenheimer, F. Scott, M. Roos, R. Otte, *J. Chem. Soc., Chem. Commun.* **1990**, 1722. (b) H. G. Raubenheimer, F. Scott, G. J. Kruger, J. G. Toerien, R. Otte, W. van Zyl, I. Taljaard, P. Olivier, L. Linford, *J. Chem. Soc., Dalton Trans.* **1994**, 2091.

[10] H. G. Raubenheimer, S. Cronje, *J. Organomet. Chem.* **2001**, 617–618, 170 and references therein.

[11] J. Müller, K. Öfele, G. Krebs, *J. Organomet. Chem.* **1974**, *82*, 383.

[12] R. Bertani, M. Mozzon, R. A. Michelin, *Inorg. Chem.* **1988**, *27*, 2809.

[13] W. P. Fehlhammer, K. Bartel, A. Völkl, D. Achatz, *Z. Naturforsch.* **1982**, *37b*, 1044.

[14] M. M. Singh, R. J. Angelici, *Inorg. Chem.* **1984**, *23*, 2691.

[15] L. K. Johnson, R. Angelici, *Inorg. Chem.* **1987**, *26*, 973.

[16] R. A. Michelin, L. Zanotto, D. Braga, P. Sabatino, R. J. Angelici, *Inorg. Chem.* **1988**, *27*, 85.

[17] W. P. Fehlhammer, K. Bartel, U. Plaia, A. Völkl, A. T. Liu, *Chem. Ber.* **1985**, *118*, 2235.

[18] J. Ruiz, B. F. Perandones, G. García, M. E. G. Mosquera, *Organometallics* **2007**, *26*, 5687.

[19] C. Tubaro, A. Biffis, M. Basato, F. Benetollo, K. J. Cavell, L.-L. Ooi, *Organometallics* **2005**, *24*, 4153.

[20] R. Lindner, C. Wagner, D. Steinborn, *J. Am. Chem. Soc.* **2009**, *131*, 8861.

[21] J. A. Cabeza, I. del Río, D. Miguel, E. Pérez-Carreño, M. G. Sánchez-Vega, *Organometallics* **2008**, *27*, 211.

[22] F. E. Hahn, M. Tamm, *J. Chem. Soc., Chem. Commun.* **1993**, 842.

[23] M. Tamm, F. E. Hahn, *Coord. Chem. Rev.* **1999**, *182*, 175.

[24] S. Bellemin-Laponnaz, *Polyhedron* **2010**, *29*, 30.

[25] M. Yasui, S. Yoshida, S. Kakuma, S. Shimamoto, N. Matsumura, F. Iwasaki, *Bull. Chem. Soc. Jpn.* **1996**, *69*, 2739.

[26] F. Iwasaki, M. Yasui, S. Yoshida, H. Nishiyama, S. Shimamoto, N. Matsumura, *Bull. Chem. Soc. Jpn.* **1996**, *69*, 2759.

[27] R. W. Alder, M. E. Blake, C. Bortolotti, S. Bufali, C. P. Butts, E. Linehan, J. M. Oliva, A. G. Orpen, M. J. Quayle, *Chem. Commun.* **1999**, 241.

[28] M. Mayr, K. Wurst, K.-H. Ongania, M. R. Buchmeiser, *Chem. Eur. J.* **2004**, *10*, 1256.

[29] P. Bazinet, G. P. A. Yap, D. S. Richeson, *J. Am. Chem. Soc.* **2003**, *125*, 13314.

[30] C. C. Scarborough, M. J. W. Grady, I. A. Guzei, B. A. Gandhi, E. E. Bunel, S. S. Stahl, *Angew. Chem. Int. Ed.* **2005**, *44*, 5269.

[31] M. Iglesias, D. J. Beetstra, A. Stasch, P. N. Horton, M. B. Hursthouse, S. J. Coles, K. J. Cavell, A. Dervisi, I. A. Fallis, *Organometallics* **2007**, *26*, 4800.

[32] R. Jazzar, H. Liang, B. Donnadieu, G. Bertrand, *J. Organomet. Chem.* **2006**, *691*, 3201.

[33] (a) M. Iglesias, D. J. Beetstra, J. C. Knight, L.-L. Ooi, A. Stasch, S. Coles, L. Male, M. B. Hursthouse, K. J. Cavell, A. Dervisi, I. A. Fallis, *Organometallics* **2008**, *27*, 3279. (b) J. J. Dunsford, D. S. Tromp, K. J. Cavell, C. J. Elsevier, B. M. Kariuki, *Dalton Trans.* **2013**, *42*, 7318.

[34] T. W. Hudnall, C. W. Bielawski, *J. Am. Chem. Soc.* **2009**, *131*, 16039.

[35] V. César, N. Lugan, G. Lavigne, *Eur. J. Inorg. Chem.* **2010**, 361.

[36] T. W. Hudnall, J. P. Moerdyk, C. W. Bielawski, *Chem. Commun.* **2010**, *46*, 4288.

[37] T. W. Hudnall, A. G. Tennyson, C. W. Bielawski, *Organometallics* **2010**, *29*, 4569.

9 Beyond Classical N-heterocyclic Carbenes II

The carbene carbon in stereotypical N-heterocyclic carbenes is flanked by two nitrogen atoms, which stabilize it via the so-called *push-pull* effect. Removal of one or even both α-nitrogen atoms from this formally divalent carbon would decrease or diminish this heteroatom stabilization and lead to species with significantly altered structures and interesting properties [1,2]. For many of such species, it is even debatable if one should still consider them carbenes or if they are completely different species. This difficulty in understanding their nature has led to the appearance of various descriptive terms in the original literature, which however can confuse scientists and students alike. For example, terms such as *"wrong way," "unusual," "non-classical," "abnormal," "mesoionic"* and lastly *"remote"* have been coined for their description. Here we will use the term *"non-classical"* to refer to all N-heterocyclic carbenes which do not belong to the four ubiquitous, dominant, and classical four types **I–IV**, that is, imidazolin-, triazolin-, benzimidazolin-, and imidazolidin-2-ylidenes.

Among the species of focus in this chapter, the last three terms are prevalent and often appear in original works in conjunction with their respective acronyms *a*NHCs (for *abnormal* NHCs), MICs (for *mesoionic* carbenes) and *r*NHCs (for *remote* NHCs) to set them apart from their *"normal"* counterparts, that is, NHCs or *n*NHCs. To shed some light into their appropriate usage, a brief description is given in the following paragraph.

The term *"abnormal"* was originally used to describe imidazolin-4/5-ylidenes, where the carbene center is the C4 or C5 atom depending on the priority of the N-substituents in the five-membered heterocycle. The carbene atom in this overall electrically neutral compound is adjacent to a single nitrogen atom, but more importantly, no neutral and closed-shell Lewis (carbene or metal carbene) resonance structure can be drawn for such species as detailed in Scheme 9.1. Since all resonance forms require charge separation, the defining charge neutrality for carbenes cannot be fulfilled, which led to its description as *"abnormal."* This is in stark contrast to its direct and *normal* imidazolin-2-ylidene isomer, for which a neutral resonance structure can be drawn without difficulty. It must be noted that many charge-separated representations (not shown) are also possible for the latter. In NHC chemistry, the usage of *"abnormal"* has been extended to all carbene-like compounds that cannot be represented by any neutral resonance structure. More recently, it has increasingly been replaced by the IUPAC term *"mesoionic,"* [3] which is used to describe dipolar five- or six-membered heterocycles with charge delocalization, but for which a totally covalent (neutral) structure cannot be drawn. In these compounds, the formal positive charge is associated with the ring atoms, while the formal negative charge can be part of the

The Organometallic Chemistry of N-heterocyclic Carbenes, First Edition. Han Vinh Huynh.
© 2017 John Wiley & Sons Ltd. Published 2017 by John Wiley & Sons Ltd.

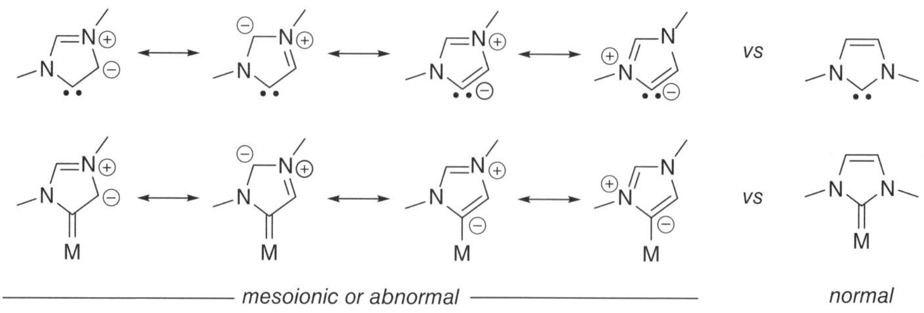

Scheme 9.1
Lewis structures highlighting the difference between *normal* and *mesoionic* (*abnormal*) imidazolinylidenes.

ring on an exocyclic nitrogen or chalcogen atom. The terms "*abnormal*" and "*mesoionic*" are in this context equivalent and could be used interchangeably. However, in terms of proper nomenclature, the use of the IUPAC term "*mesoionic*" is encouraged. Nevertheless, the term "*abnormal*" or *a*NHC is used for the *mesoionic* imidazolin-4/5-ylidenes in this chapter for historical reasons.

In this section, the most common types of NHCs with reduced heteroatom stabilization will be described. Due to their resemblance to NHCs in synthetic and various other aspects, the "carbene" nomenclature (i.e. ylidene) will be applied as well.

9.1 *Abnormal* Imidazolin-4/5-ylidenes (*a*NHC)

The metallation of imidazolium salts usually occurs at the C2 position, which carries the most acidic proton of the heterocycle, and gives rise to imidazolin-2-ylidene complexes as the most common type of NHC complexes. In certain cases, however, metallation can also occur at the supposedly less reactive C4 or C5 positions. Then complexes of imidazolin-4/5-ylidenes are obtained that have commonly been referred to as *abnormal* N-heterocyclic carbene (*a*NHC) complexes. The first examples of such complexes were reported by Crabtree and coworkers, who studied the reaction of pyridine-functionalized salts with the hydrido-iridium complex [IrH$_5$(PPh$_3$)$_2$] [4]. The two reactants were heated in tetrahydrofurane under reflux for two hours in an attempt to prepare pyridine-functionalized NHC complexes of iridium(III). Unexpectedly, the main products turned out to be dihydrido complexes, in which metallation occurred at the C5 and not at the C2 carbon atom (Scheme 9.2).

The ^{1}H NMR spectra of the complexes are in line with the proposed *abnormal* coordination of the carbene moiety and still show rather downfield signals for the C2–H proton at >8 ppm and another more upfield resonance at ~5 ppm of equal intensity for the remaining C4 (or C5 when R = Mes) hydrogen of the imidazole-derived heterocycle. Moreover, the ^{13}C NMR resonance for the carbene at ~140 ppm is substantially more downfield ($\Delta\delta$ ~20 ppm) compared to the equivalent signal in the salt precursor. Nevertheless, these signals are still significantly upfield compared to those in *normal* imidazolin-2-ylidene complexes, since the carbon donor in the *a*NHC complex is flanked by only one electronegative nitrogen atom.

Clear evidence for this *abnormal* coordination was also obtained by single crystal X-ray diffraction. Portions of two representative molecular structures are shown in Figure 9.1,

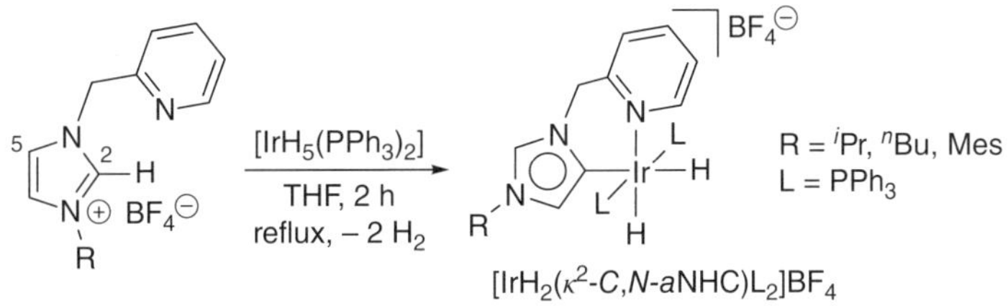

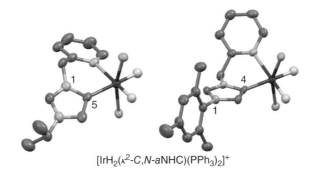

Scheme 9.2
Unexpected formation of the first *a*NHC complexes.

$[IrH_2(\kappa^2\text{-}C,N\text{-}a\text{NHC})(PPh_3)_2]^+$

Figure 9.1
Molecular structures of the first *a*NHC complexes. Only the complex cations with the chelating carbene, phosphorous, and hydrido donor atoms are depicted for clarity.

and the attachment of the C4 or C5 carbon donors to the iridium center in the *trans*-bis(phosphine) complexes can be clearly seen. The metal–carbon distances of ~2.1 Å are essentially in the range of single bonds and not different from those detected for their normal NHC counterparts.

The C4/5 metallation with iridium hydrido complexes occurs via an oxidative addition pathway. Indeed, subsequent studies indicate that C4/5–H activation is more preferred in oxidative additions, while C2–H activation is more common in deprotonation processes. Notably, these two distinctively different routes can be influenced by the counter-anion of the imidazolium salt to a certain extent. For example, halides are better hydrogen bond acceptors and can increase the acidity of the C2–H proton, leading to the preference for acid-base reactions and thus C2-metallation products. Large anions such as BF_4^- engage less in hydrogen bonding, and thus oxidative addition pathways are more prevalent with their imidazolium salts leading to C4/5–bound complexes provided the metal center is sufficiently electron rich.

Another important factor which can affect C4/5 over C2 metallation is steric bulk. Among the three ring-hydrogen atoms of a typical imidazolium salt, the C2–H proton is apparently the one most affected by steric variations of the two N-substituents. Although it is more acidic than the other two hydrogen atoms, its removal becomes increasingly difficult with increasing bulk of the wing tip groups. The hydrogen atoms at C4/5 are less affected, since each of them is only adjacent to one N-substituent. The influence of steric factors also plays a significant role for the reactions summarized in Scheme 9.2, where C4/5 metallation occurs almost exclusively.

Reducing the size of the wing tip group should therefore lead to the formation of normal NHC complexes as additional products. Indeed this is the case when the N-methyl-substituted derivative is used, where an almost 1:1 isomeric ratio of complexes is obtained (Scheme 9.3). The ratio was estimated by NMR spectroscopy. However, due to similar solubilities their separation was not possible.

Additives can also have a pronounced influence on the formation of *normal* versus *abnormal* NHC complexes. In 2004, Nolan and coworkers studied the palladation of 1,3-dimesitylimidazolium chloride [5]. In an attempt to prepare bis(NHC) complexes, two equivalents of the azolium salt were heated with Pd(OAc)$_2$ in dioxane. Such conditions are routinely applied to prepare palladium(II) NHC complexes. Surprisingly, the authors obtained almost exclusively an unusual compound of the formula *trans*-[PdCl$_2$(*a*IMes) (IMes)] that contains one IMes and its *abnormal* isomer as carbon donors in a good yield of 74% (Scheme 9.4, route a). NMR analyses corroborate the existence of two independent ligands by two sets of signals, whereby the carbene resonances appear at 175.9 (IMes) and 150.7 ppm (*a*IMes), respectively. Notably, the addition of excess cesium carbonate under otherwise similar conditions led to the predominant formation of the *normal* complex *trans*-[PdCl$_2$(IMes)$_2$], and no *abnormal* complexes were observed. The preparation of the normal bis(IMes) complex was subsequently optimized by using palladium(II) chloride instead (Scheme 9.4, route b). It is likely that excess of the additional base facilitates deprotonation of the C2-H proton leading to the normal bis(IMes) complex. Without addition of an external base, deprotonation must occur at the metal by the acetato ligands probably

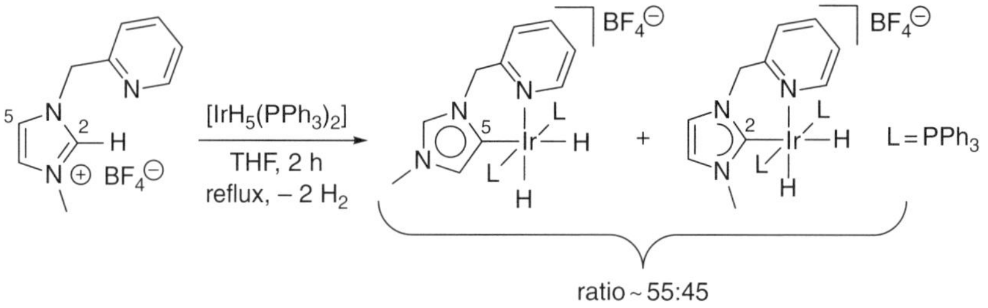

Scheme 9.3
Abnormal and *normal* NHC complexes in oxidative additions of smaller salts.

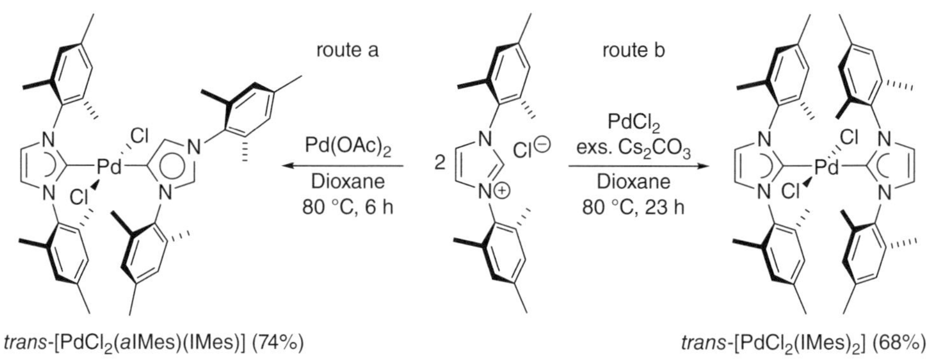

trans-[PdCl$_2$(aIMes)(IMes)] (74%)   *trans*-[PdCl$_2$(IMes)$_2$] (68%)

Scheme 9.4
Abnormal versus *normal* palladation controlled by base additives.

following a concerted metallation-deprotonation pathway. Intuitively, this should occur with relative ease at the C2 carbon bearing the most acidic proton for the first equivalent of azolium chloride. This would furnish the bulky and electron-rich intermediate [PdCl(κ^2-OAc)(IMes)]. Activation of the second carbene precursor at the C2 carbon involving this complex intermediate is sterically disfavored, and thus happens at the more accessible C5 carbon. An alternative oxidative addition pathway cannot be ruled out, since coordination of one IMes already increases the electron density at the palladium center.

The molecular structures for both the *abnormal* and *normal* isomers of the bis(carbene) palladium(II) complex could be established by single crystal X-ray diffraction (Figure 9.2). The two complexes are sterically very different, yet analysis of the two types of palladium–carbon bonds revealed that their lengths of 2.019(13) and 2.021(11)Å for the *normal* and *abnormal* carbene, respectively, are identical within experimental error, perhaps indicating a similar bonding interaction.

Another example where the steric bulk obviously dictates the formation of *normal* versus *abnormal* complexes comes from works of Esteruelas and Oliván [6]. They treated 1-mesityl,3-methylimidazolium or 1-benzyl,3-methylimidazolium tetraphenylborate with the basic hydrido-osmium species [OsH$_6$(P^{i}Pr$_3$)$_2$] (Scheme 9.5). The *in situ* deprotonation for the bulky N-mesityl derivative takes place at the least hindered C4 position, while that

trans-[PdCl$_2$(alMes)(IMes)]   trans-[PdCl$_2$(IMes)$_2$]

Figure 9.2
Molecular structures of complexes containing *abnormal* and *normal* IMes. Hydrogen atoms have been omitted for clarity.

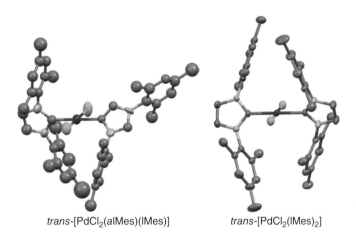

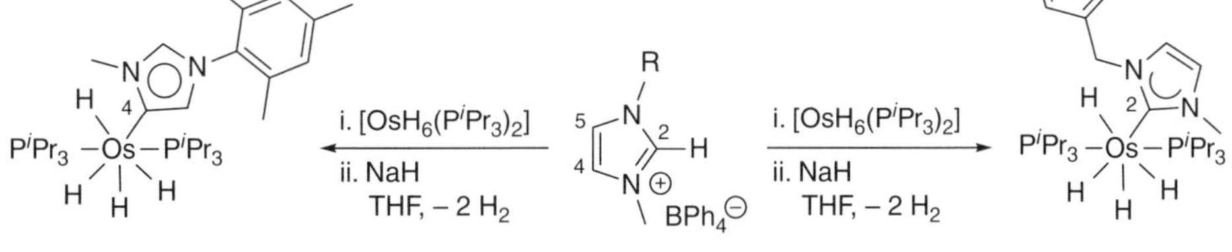

Scheme 9.5
Influence of steric bulk on C2 versus C4 metallation.

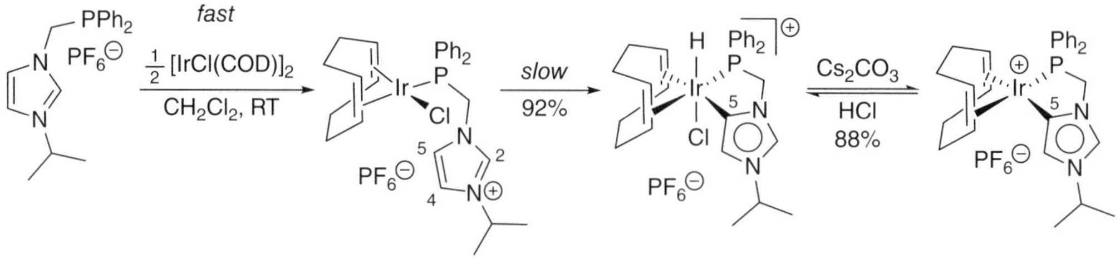

Scheme 9.6
Selective *abnormal* NHC complex formation via pre-coordination.

for the more flexible N-benzyl species occurs at the C2 carbon. Repulsion between the incoming salt and the two bulky triisopropylphosphine ligands at osmium certainly amplifies the discrimination between the different hydrogen atoms in this case. This highlights that the steric hindrance at the metal center imposed by large co-ligands is equally as important as that of the carbene precursor.

The increase of steric bulk at metal to trigger C4/5–H activation can also occur via pre-coordination of suitably donor-functionalized NHC precursors. In agreement with this notion, Li and coworkers observed the exclusive formation of *abnormal* NHC iridium(III) complexes in the reaction of phosphine-tethered imidazolium salts with the chlorido-bridged iridium-COD dimer [IrCl(COD)]$_2$ [7]. Upon mixing the two reactants, the authors observed the fast cleavage of the iridium(I) complex dimer with the formation of the phosphine adduct (Scheme 9.6). This pre-coordination results in an electron-rich and crowded iridium(I) center, which is sterically shielded by a diolefin chelate and a bulky phosphine donor. As such, C–H activation can only occur at the C5 carbon, while the reaction at the C2 carbon is sterically hindered. Moreover, the coordination of the phosphorus donor also makes the oxidative addition across the C4–H bond geometrically impossible.

Even so, the oxidative addition occurs rather slowly, affording the iridium(III) hydrido complex [IrClH(COD)(κ^2-*C,P-a*NHC)]PF$_6$. Key NMR spectroscopic features include a hydrido doublet at –15.81 ppm, which couples to the phosphine donor with a constant of $^2J_{H-P} = 7.5$ Hz suggesting their *cis* arrangement. Also, a downfield signal at 8.57 ppm was detected for the acidic C2–H proton, and the *abnormal* carbene donor resonates as a doublet at 135.4 ppm ($^2J_{C-P} = 5.7$ Hz).

The hexa-coordinated iridium(III) complex can undergo reversible base-assisted reductive elimination of HCl to form the iridium(I) complex [Ir(COD)(κ^2-*C,P-a*NHC)]PF$_6$ in 88% yield. The square planar complex exhibits resonances at 8.51 and 130 ppm (d, $^2J_{C-P} = 11.7$ Hz) for the C2–H proton and the carbene donor in the ^{1}H and ^{13}C NMR spectra, respectively.

The proposed pathway of C5 metallation does not involve the participation of the counter-anion. In agreement with the proposal, it was observed that C–H activation consistently occurred at the C5 position regardless of the anions (Cl$^-$, PF$_6^-$, or SbF$_6^-$) used, which emphasizes again that sterics can overrule electronic preferences.

Interestingly, the thermodynamics of the reaction can be influenced by the tether length. Thus, changing the methylene (C$_1$) to an ethylene (C$_2$) spacer between the heterocycle and phosphine function leads to a reversible C–H activation process. This observation is rooted in the generally observed preference for five-membered metallacycles over six-membered ones. Extension to a propylene (C$_3$) spacer affects the reaction most dramatically, and only intractable mixtures are obtained instead.

Table 9.1 Factors affecting the preference for C2 versus C4/5 metallation of imidazoliums.

Factors	*normal* C2 metallation	*abnormal* C4/5 metallation
Pathway	Deprotonation	Oxidative addition
Counter-anion	Small anions prone to hydrogen bonding, e.g. halides	Large and weakly coordinating anions, e.g. BF_4^-, PF_6^-
Wing tip groups	Small N-substituents	Large and bulky N-substituents
Metal precursor	Higher valent metal centers with small co-ligands	Low valent metal centers with bulky co-ligands

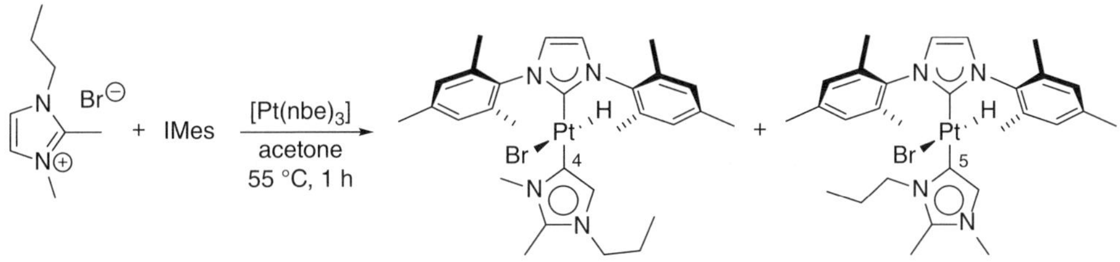

Scheme 9.7
Oxidative addition of a C2-blocked imidazolium bromide to Pt^0.

Overall, the studies suggest that the competition between C2 versus C4/5 metallation can be influenced by several factors to increase selectivity. Table 9.1 gives a summary of these factors and can be referred to in future synthetic works.

An obvious way to increase selectivity for C4/5 metallation is to completely remove competition by selectively blocking the C2 carbon with a suitable function. Most commonly, alkyl and aryl substituents have been explored for this purpose. An example reported by Cavell and coworkers is shown in Scheme 9.7 [8]. Oxidative addition of 1-*n*-propyl-2,3-dimethylimidazolium bromide to the tris(norbornene)platinum(0) complex in the presence of free IMes gave a mixture of hetero-bis(carbene)platinum(II) complexes bearing a *normal* IMes ligand and a second C4 or C5-activated *abnormal* NHC in a 3:1 ratio. Notably, the C4 complex was found to be the major isomer, where platination took place adjacent to the smaller N-methyl substituent because of steric discrimination between the C4 and C5 carbon atoms, respectively. As expected, the complexes showed very similar NMR spectra with the *abnormal* carbenes resonating at 151.6 and 152.4 ppm, respectively.

The complexes are stable in the solid state and in acetone solutions at room temperature for several days. However, they underwent reductive elimination in the presence of alkenes under formation of [Pt(η^2-alkene)$_2$(IMes)] complexes. Interestingly, no alkene insertion into the Pt–H bond was noted, and only the *abnormal* NHC was eliminated, while the *normal* IMes was retained at the resulting platinum(0) center.

Activation of C2-blocked imidazolium salts by deprotonation is also rationally a very attractive method to prepare *abnormal* NHC complexes. Research toward this direction has shown that the nature of the C2 "protecting" group has an implication on the suitability of certain bases. For example, silver(I) oxide is a very common basic metal precursor used to prepare silver-NHC complexes, which in turn serve as carbene transfer agents for the syntheses of many other transition metal NHC complexes. However, as silver(I) ions are inherently involved in this reaction, potential oxidation reactions of the organic reactants could also occur (*cf.* $Ag_{(aq)}^+ + e^- \rightarrow Ag_{(s)}$; $E^0 = +0.7994\,V$). Crabtree and coworkers have observed that 2-methyl- and 2-benzyl imidazolium salts in particular suffer from oxidative C–C bond cleavage on exposure to silver(I) oxide. The primary oxidation products are 2-acylimidazolium salts (i.e. 2-formyl and 2-benzoyl, respectively) and elemental silver is formed (silver mirror). In the presence of additional silver(I) oxide and water, the 2-acyl compounds are readily hydrolyzed to give the C2 metallated *normal* silver(I) NHC complexes and formic or benzoic acids, respectively (Scheme 9.8) [9].

It was also found that the 2-ethylimidazolium analog was oxidized significantly more slowly, while 2-isopropyl and 2-phenyl derivatives resisted oxidation. Thus, one can conclude that branched and aromatic substituents incapable of acyl formation are the best blocking groups when silver(I) oxide is used as the basic precursor.

Accordingly, the 2,4-diphenylimidazolium chloride is cleanly deprotonated by silver(I) oxide, supposedly forming a silver-*a*NHC species, which undergoes subsequent transmetallation to give the iridium(I) complex [IrCl(COD)(*a*NHC)] in 43% yield (Scheme 9.9) [10]. The diolefin complex can be easily converted into the dicarbonyl complex *cis*-[IrCl(CO)$_2$(*a*NHC)] in 91% yield by passing carbon monoxide gas through a solution in dichloromethane. An expected highfield shift of the *abnormal* carbene donor from 162.0 (acetone-d_6) to 153.2 ppm (CDCl$_3$) was observed upon ligand substitution, which cannot be accounted for by simple solvent effects. The IR stretches obtained for the carbonyl ligands in the latter compound allow for an estimation of the electron donor power of the respective *a*NHC ligand. The averaged value of 2039 cm^{-1} (i.e. TEP of ~2039 cm^{-1}) indicates a donating strength superior to classical NHCs and the most donating trialkylphosphines, which can be rationalized by the presence of only one adjacent inductively electron-withdrawing nitrogen atom.

Notably, the described transfer of the *a*NHC from silver(I) to iridium(I)-COD dimer requires the protection of both C2 and C4 positions, while the use of *a*NHCs without C4 substituents led to decomposition. On the other hand, a related approach involving *a*NHC

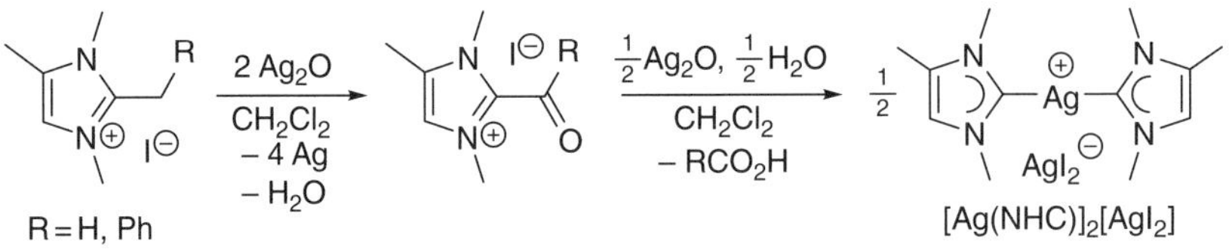

Scheme 9.8
Oxidative C–C bond cleavage of 2-methyl and 2-benzyl imidazolium salts.

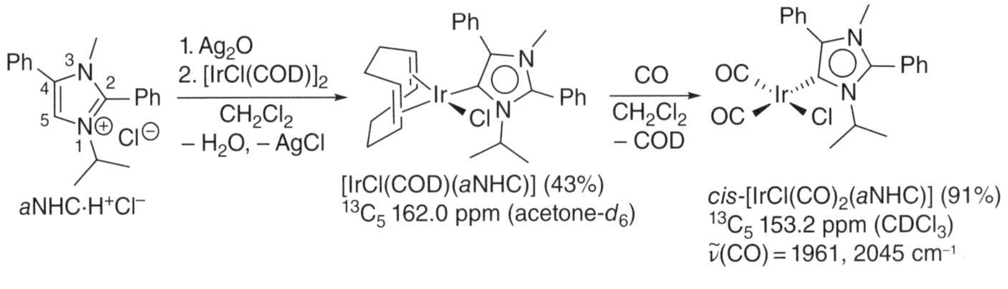

Scheme 9.9
Preparation of iridium(I) complexes of *a*NHCs by silver-carbene transfer.

transfer from silver(I) to palladium(II) does not require a fourth substituent for steric protection. Here, one-pot mixing of a 1,2,3-trisubstitued imidazolium salt with the dimeric monocarbene-palladium complex [PdBr$_2$(^{i}Pr$_2$-bimy)]$_2$ and silver(I) oxide in dichloromethane cleanly affords the mixed *normal/abnormal* NHC complex *trans*-[PdBr$_2$(^{i}Pr$_2$-bimy)(*a*NHC)] as a yellow air and moisture stable solid in 95% yield (Scheme 9.10) [11]. The carbon donor of the *a*NHC was observed at 150.4 ppm (CDCl$_3$), while that for the classical benzimidazolin-2-ylidene occurs more downfield at 181.9 ppm. Comparing the latter value with those obtained for *trans*-[PdBr$_2$(^{i}Pr$_2$-bimy)(NHC)] complexes bearing classical NHCs also indicates that imidazolin-4/5-ylidenes are generally much stronger donors than classical NHCs.

In contrast to the use of silver(I) oxide, metallation with palladium(II) acetate proceeds more straightforwardly, which allows for a greater flexibility in the design of the imidazolium precursors. Here, 2-alkyl substituents are tolerated, and no C–C bond cleavages are observed. For example, Albrecht and coworkers demonstrated that the palladation of various C2-protected diimidazolium diiodides proceeds with ease under standard conditions. Heating the salts with palladium(II) acetate in DMSO at elevated temperatures afforded the desired di*a*NHC palladium(II) complexes in generally high yields, and protection of the second C4/5 carbon is also not required (Scheme 9.11) [12].

The carbene donors of the di*a*NHC palladium(II) complexes resonate at much higher fields compared to similar C2-bound analogs. Moreover, the complexes are much more electron rich, since imidazolin-4/5-ylidenes are pronouncedly more basic than their normal C2-counterparts. This and the reduced steric protection of the metal centers lead to their lower stability in acidic media, where they decompose under protonolysis. On the other hand, the increased electron density was efficiently exploited in catalysis. Upon iodido-acetonitrile ligand exchange, highly efficient catalyst precursors were obtained for the catalytic hydrogenation of cyclooctene with only one atmosphere of dihydrogen at room temperature.

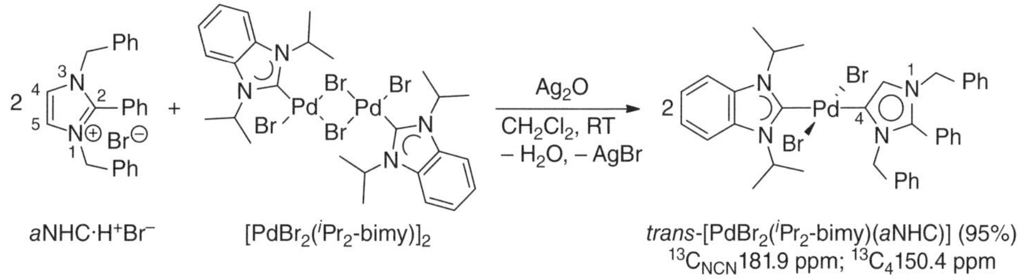

*a*NHC·H$^+$Br$^-$ [PdBr$_2$(^{i}Pr$_2$-bimy)]$_2$ *trans*-[PdBr$_2$(^{i}Pr$_2$-bimy)(*a*NHC)] (95%)
^{13}C$_{NCN}$ 181.9 ppm; ^{13}C$_4$ 150.4 ppm

Scheme 9.10
Preparation of an *a*NHC palladium(II) complex by silver-carbene transfer.

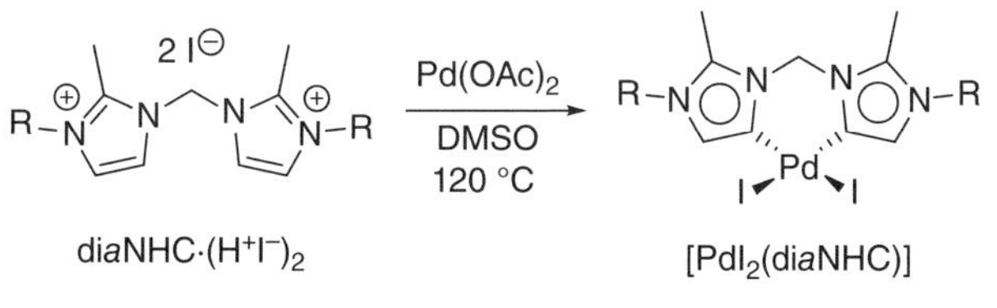

diaNHC·(H$^+$I$^-$)$_2$ [PdI$_2$(diaNHC)]

R	Yield [%]	^{13}C$_{4/5}$ [ppm]
Me	93	123.1
iPr	95	121.9
Mes	72	123.5

Scheme 9.11
Preparation of di*a*NHC complexes with Pd(OAc)$_2$.

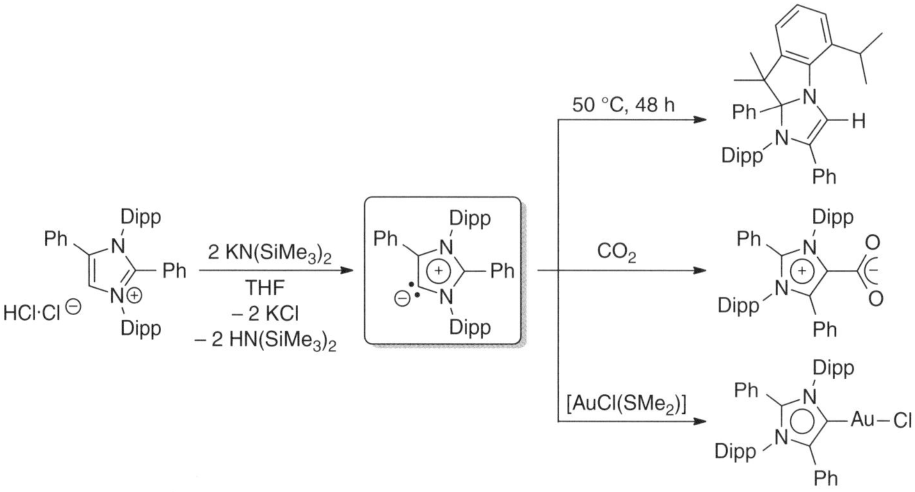

Scheme 9.12
Preparation of a free *a*NHC and its reactivity.

Since the C2–H proton is the most acidic ring position in imidazolium salts, any rational attempt to prepare free *abnormal* NHCs by deprotonation must involve C2-protected species. Bertrand and coworkers succeeded in the preparation and isolation of the only free *a*NHC [13]. They based their attempts on a tetrasubstituted, bulky imidazolium salt. C5–H deprotonation only occurred when the carbene precursor contains small halides as counteranions, which can polarize and increase the acidity of the ring-proton via hydrogen bonding. In the present case a hydrochloric acid adduct of an imidazolium chloride was used. Initial attempts using strong lithium bases (LDA and *n*BuLi) led to the formation of lithium adducts instead of free carbenes. Metal adduct formation could be avoided by using the strong KHMDS base in tetrahydrofurane, which resulted in the clean formation of the first free *abnormal* imidazolin-5-ylidene (Scheme 9.12). The *a*NHC was isolated as a green solid and proves stable in the solid state and in solution for a few days at room temperature. Thus, single crystals could be obtained for confirmation of its identity by X-ray diffraction. Heating a sample in toluene resulted in quantitative rearrangement with involvement of the isopropyl groups of one Dipp substituent, and exposure to carbon dioxide furnished a zwitterionic compound. The coordination behavior of the free *a*NHC was demonstrated by smooth formation of the gold(I) complex [AuCl(*a*NHC)] on mixing with [AuCl(SMe$_2$)] under thioether displacement.

Another approach to selective C4/5 metallation involves oxidative addition of imidazolium salts containing C4/5-halo functions reminiscent of Stone's work described earlier. The group of Albrecht reported an example involving the activation of the C4–I bond of a pyridine-functionalized imidazolium bromide with low-valent palladium(0) (Scheme 9.13) [14]. This reaction gave the desired *a*NHC chelate as the only carbene complex even though the C2 position was left unprotected. In the ^{1}H NMR spectrum recorded in DMSO-d_6, the signal for the acidic C2–H proton was still found at 9.25 ppm.

Interestingly, a single crystal X-ray analysis (Scheme 9.13, right) showed that the complex contains both iodido and bromido ligands, and halide scrambling was not described. The two ligands of the greatest *trans* influences, that is, the *a*NHC and iodido ligands, are found *cis* to each other.

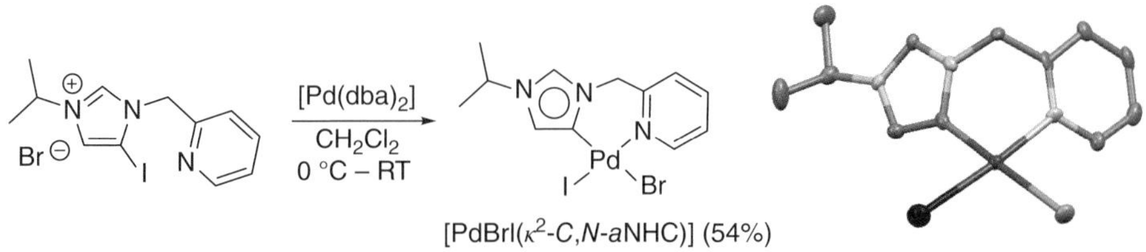

Scheme 9.13
Preparation of a κ^2-*C,N* chelating *a*NHC complex via oxidative addition (left). Molecular structure of [PdBrI(κ^2-*C,N*-*a*NHC)] (right). Hydrogen atoms have been omitted for clarity.

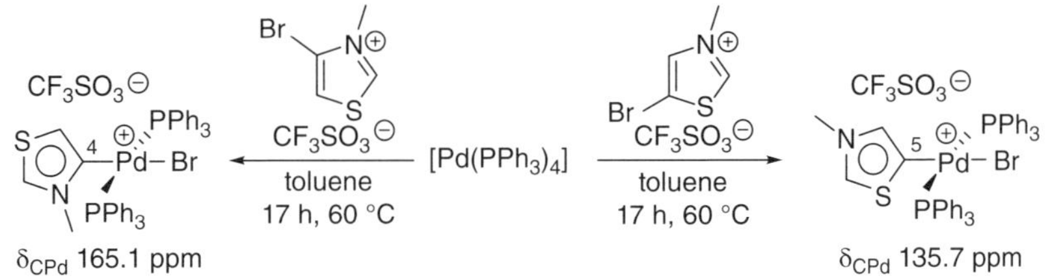

Scheme 9.14
Preparation of *mesoionic* thiazolinylidene complexes of palladium(II).

More recently, oxidative additions of isomeric 2-, 4- and 5-bromo-3-methyl thiazolium triflate salts to palladium(0) and nickel(0) precursors were explored [15]. While the reaction with 2-halo-thiazolium salts leads to complexes clearly containing a *C*2-bound NSHC ligand (thiazolin-2-ylidene), the case is more complicated for the 4- and 5-halo derivatives. Complexes obtained from the latter contain "thiazolinylidene" ligands bound via *C*4 and *C*5, for which no neutral and closed-shell resonance structure can be drawn to meet the general charge neutrality requirement for a carbene (Scheme 9.14). They are thus being described as "*abnormal*" or more precisely "*mesoionic*" cyclic carbenes in analogy to their imidazolin-4/5-ylidenes counterparts.

Although a significant downfield shift is observed for the carbon donor upon metallation in all three cases, there are great differences in their respective values. The most deshielded signal at 197.8 ppm was observed for the *normal* thiazolin-2-ylidene complex (see Section 8.1) followed by the 4-ylidene (165.1 ppm) and the 5-ylidene (135.7 ppm) counterparts. For the former two, downfield shifts of ~40 ppm were observed, while a much smaller change of only ~20 ppm was noted for the thiazolin-5-ylidene complex.

All the different approaches together highlight the wealth of organometallic chemistry involving imidazolin-4/5-ylidenes and related species. A better understanding of factors affecting their formation will lead to further advances in this area, which is driven by the search for more active complexes.

9.2 *Mesoionic* 1,2,3-Triazolin-5-ylidenes (MIC)

The application of 1,2,4-triazoles as precursors to classical and *normal* NHC ligands has been well known since the early beginnings of NHC chemistry. In contrast, the study of carbenes based on the isomeric 1,2,3-triazole system is a relatively recent development [16].

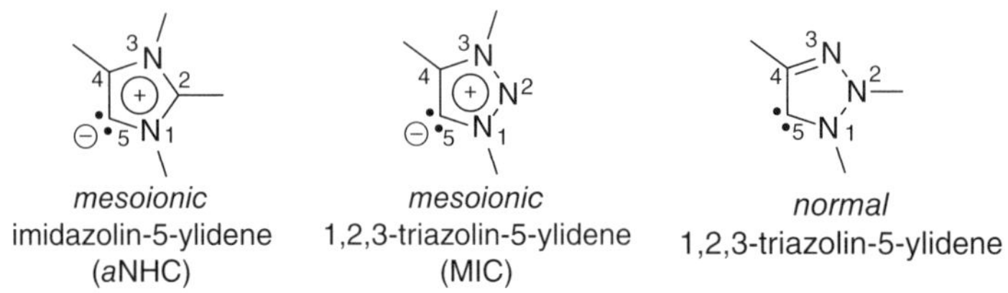

mesoionic
imidazolin-5-ylidene
(aNHC)

mesoionic
1,2,3-triazolin-5-ylidene
(MIC)

normal
1,2,3-triazolin-5-ylidene

Chart 9.1
Similarities between imidazolin-5-ylidenes with triazolin-5-ylidenes.

Formally, such ligands can be derived by replacement of a C2–R unit of *abnormal/mesoionic* imidazolin-5-ylidenes with a nitrogen atom (Chart 9.1). The vast majority of triazolinylidenes contain substituted 1,3-nitrogen atoms, and in these the *mesoionic* character is retained. However, a few examples containing 1,2-substituted nitrogen atoms have also been reported. For these analogs, a neutral carbene resonance form can be drawn, which makes them *normal* 1,2,3-triazole based NHCs.

The first complexes of 1,2,3-triazolin-5-ylidenes were only reported in 2008 by Albrecht and coworkers, but these carbenes have rapidly become widespread due to their high modularity and ease of preparation [17]. The respective 1,2,3-triazole parent compounds are easily accessible via the so-called copper(I)-catalyzed alkyne-azide cycloaddition (CuAAc), which is the catalyzed version of the well-established Huisgen 1,3-dipolar cycloaddition between an azide and an alkyne. This reaction is also the most popular form of a "click" reaction as a result of its simplicity and the often high yields of 1,4-disubstiuted 1,2,3-triazoles it can provide.

Methylation of the 1,4-disubstituted 1,2,3-triazoles thus obtained with iodomethane occurs at the most nucleophilic N3 nitrogen atom, which gives rise to triazolium iodides as direct carbene precursors. These salts contain only one ring-proton at the C5 carbon atom, which is also sufficiently acidic to undergo routine metallation reactions typically used for the making of classical NHC complexes. Their versatile organometallic chemistry is highlighted in Scheme 9.15. *In situ* palladation using the basic palladium(II) acetate in a 1:1 ratio easily occurs under standard conditions and gives the iodido-bridged dimeric mono-MIC complex [PdI$_2$(MIC)]$_2$. Notably, one equivalent of the triazolium salt in this reaction merely acts as the iodide source leading to the lower yields (Table 9.2). Similarly, reaction with half an equivalent of silver(I) oxide furnishes silver(I)-MIC ion pairs of the type [Ag(MIC)$_2$][AgI$_2$], which can transfer the MIC to a wide range of other transition metals with precipitation of silver(I) iodide as a crucial driving force. Thus, the metal complexes [RuCl$_2$(cym)(MIC)], [RhCl(COD)(MIC)], and [IrCl(COD)(MIC)] are obtained in bridge-cleavage reactions on mixing with the respective dimeric metal precursors. The last two routinely react with carbon monoxide to form the *cis*-dicarbonyl adducts, in which carbonyl stretches allow for a first estimation of the donating abilities of MICs.

The yields for these reactions are summarized in Table 9.2 along with selected key spectroscopic parameters. The ^{13}C NMR signals for the carbene donor are slightly more deshielded compared those in analogous complexes bearing imidazolin-4/5-ylidenes (aNHCs). This is within expectation due to the overall increased negative inductive effect of the additional third nitrogen atom in the heterocycle. However compared to classical NHCs, the chemical shifts are slightly more highfield, since the carbene donor is directly flanked by only one nitrogen atom. Comparison of the averaged carbonyl wavenumbers of $\tilde{v}(CO)_{av} = 2036$ and $2021\,cm^{-1}$ for the rhodium(I) and iridium(I) dicarbonyl complexes with

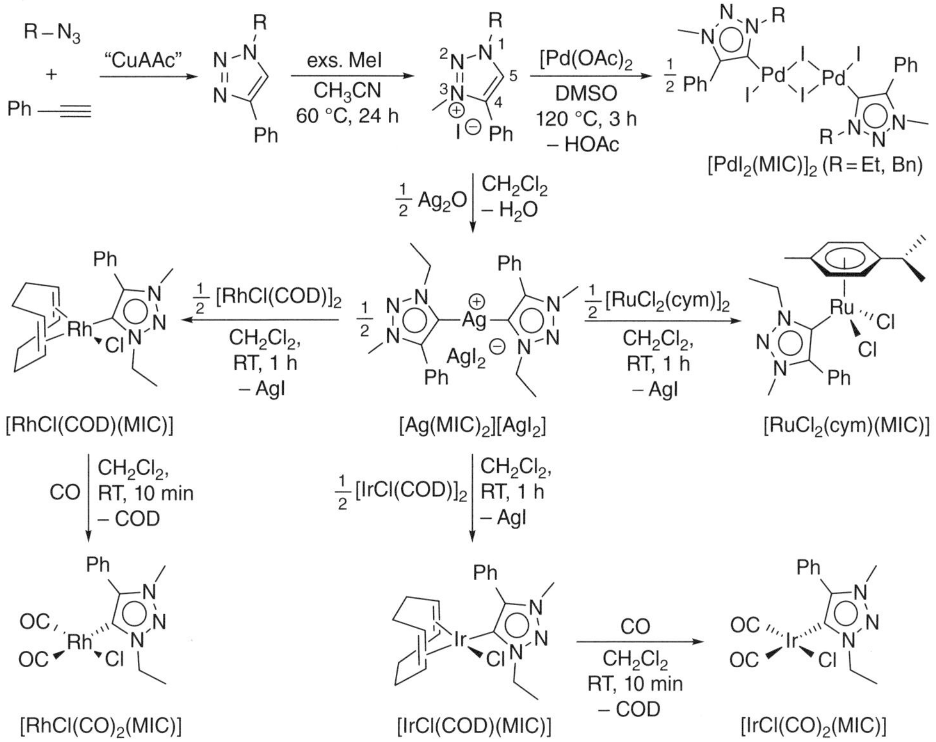

Scheme 9.15
Preparation of Pd^{II}, Ag^{I}, Ru^{II}, Rh^{I} and Ir^{I} complexes of 1,2,3-triazolin-5-ylidenes.

Table 9.2 Yields and selected spectroscopic data for MIC complexes.

Formula	Yield [%]	$^{13}C_{C5}$ [ppm]	$^{1}J_{Rh\text{-}C}$ [Hz]	$\tilde{v}$ (CO) [cm^{-1}]
[PdI$_2$(MIC)]$_2$ (R = Et)	42	159.3[a]	-	-
[PdI$_2$(MIC)]$_2$ (R = Bn)	39	159.6[a]	-	-
[Ag(MIC)$_2$][AgI$_2$]	83	166.3[b]	-	-
[RuCl$_2$(cym)(MIC)]	85	160.9[c]	-	-
[RhCl(COD)(MIC)]	94	170.6[b]	-	-
cis-[RhCl(CO)$_2$(MIC)]	92	161.2[b]	46.6	1996, 2075
[IrCl(COD)(MIC)]	99	169.5[b]	39.4	-
cis-[IrCl(CO)$_2$(MIC)]	94	162.0[b]	-	1979, 2062

[a] DMSO-d_6;
[b] CDCl$_3$;
[c] CD$_2$Cl$_2$.

those obtained for other NHC ligands led to the conclusion that the MICs are stronger donors than classical NHCs, but weaker than *abnormal* imidazolin-4/5-ylidenes. In particular, a TEP value of 2047 cm^{-1} was calculated for the MIC ligand using the iridium(I) system, which is lower than those for the most basic classical NHCs.

Single crystals of the first MIC complexes of palladium(II), rhodium(I), and iridium(I) could be obtained for their structural elucidation in the solid state. Figure 9.3 depicts

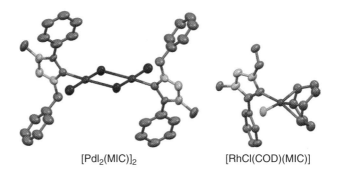

[PdI$_2$(MIC)]$_2$ [RhCl(COD)(MIC)]

Figure 9.3
Molecular structures of the first palladium(II) and rhodium(I) MIC complexes. Hydrogen atoms have been omitted for clarity.

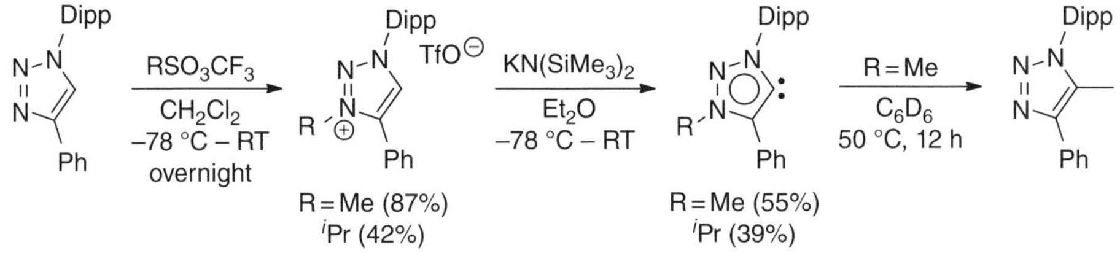

Scheme 9.16
Preparation of the first free 1,2,3-triazolin-5-ylidenes.

two molecular structures of these as representatives, which show as proposed the attachment of the ligand to the metal at the C5 position. Despite their stronger donating power, the structural features of MIC complexes are generally similar to their classical NHC counterparts.

Shortly after the first report on complexes of 1,2,3-triazolin-5-ylidene, Bertrand and coworkers succeeded in the isolation of the first free ligands [18]. Theoretical calculations predicted a markedly large singlet–triplet gap of 56 kcal mol^{-1}, which indicates carbene stability. In addition, a large Dipp substituent was chosen to provide kinetic stabilization to the free MIC ligand via steric protection. Thus, two Dipp-substituted 1,2,3-triazolium salts were prepared using standard protocols (Scheme 9.16). Subsequent C5–H deprotonation using a strong potassium base in diethyl ether indeed afforded the first examples of free mesoionic 1,2,3-triazolin-5-ylidenes, which exhibit ^{13}C NMR resonances at 202.1 (R = Me) and 198.3 ppm (R = ^{i}Pr) for the C5 carbon atoms in C$_6$D$_6$, respectively. The chemical shifts for the ylidene atom in classical NHCs are comparatively more downfield due to the deshielding effects of two adjacent nitrogen atoms.

Interestingly, the N3 substituent turned out to have substantial influence on the stability of the mesoionic species under inert conditions. For example, the N3-methyl compound was found to be stable in the solid state and in solution at –30 °C for several days, while it decomposed in the solid state at room temperature after three days. Heating a sample in C$_6$D$_6$ to 50 °C for 12 hours also led to decomposition. One of the decomposition products could be identified as the rearranged 1-(2,6-diisopropylphenyl)-4-phenyl-5-methyl-1,2,3-triazole, where the methyl group has formally migrated from the N3 to the C5 position. The N3-isopropyl species was found to be substantially more stable, and

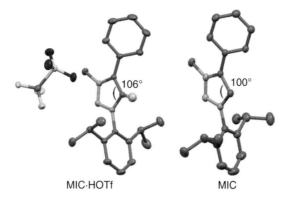

Figure 9.4
Molecular structures of a free mesoionic 1,2,3-triazolin-2-ylidene (right) and its precursor salt (left). Hydrogen atoms have been omitted for clarity with the exception of the C5–H in the salt.

no decomposition products were detected in the solid state after three days. The enhanced stability was attributed to the lower electrophilicity of the N-isopropyl group, which resists rearrangement.

Despite its lower stability, single crystals of the free N3-methyl MIC were obtained, which allowed for a structural elucidation in the solid state. Its molecular structure is depicted in Figure 9.4 along with that of its direct precursor salt. Upon C5–H deprotonation, the endocyclic angle around the carbenoid atom slightly shrinks by ~6°. This behavior resembles that of classical NHCs. Moreover, the heterocycle remains planar indicating electronic delocalization in the free MIC.

In agreement with the above observation, one could generally anticipate that the electronic properties of 1,2,3-triazolin-5-ylidenes are also modified via changes in the substituents. Nevertheless, later studies showed that carbonyl based methodologies, that is, using *cis*-[RhCl(CO)$_2$(MIC)] [19] or *cis*-[IrCl(CO)$_2$(MIC)] [20] complexes, are not capable of discerning such electronic substituent effects in 1,2,3-triazolin-5-ylidenes.

On the other hand, such effects can be detected using more sensitive ^{13}C NMR spectroscopy. For this purpose, Yuan and Huynh treated differently substituted 1,2,3-triazolium salts with silver(I) oxide in the presence of the dimeric palladium-benzimidazolin-2-ylidene complex [PdBr$_2$(^{i}Pr$_2$-bimy)]$_2$ (Scheme 9.17) [21]. In this one-pot procedure, intermediate formation of silver(I)-MIC species is proposed, which immediately transfer the MIC ligands to palladium(II) under cleavage of the bromido-bridges enthalpically driven by the formation of new palladium-carbene bonds. The air and moisture stable hetero-bis(carbene) complexes were obtained as yellow powders in moderate to good yields.

Of particular interest was the influence of the differently substituted MIC ligands on the ^{13}C$_{carbene}$ NMR signal of the ^{i}Pr$_2$-bimy reporter ligand. This study revealed that the N1-substituent can indeed fine-tune the electronic properties of a MIC ligand in accordance with its inductive effect. Changing the N1 wing tip group from phenyl to benzyl to isopropyl leads to a successive downfield shift of the ^{i}Pr$_2$-bimy carbene donor in the order 180.3 ppm (Ph) < 180.8 ppm (Bn) < 181.2 ppm (^{i}Pr), reflecting a stepwise increase of the MIC ligand's donor ability (Figure 9.5). On the other hand, a more remote and subtle change of the N3 substituent from benzyl to methyl, which is six bonds away from the reporter nuclei, can no longer be differentiated even by this method.

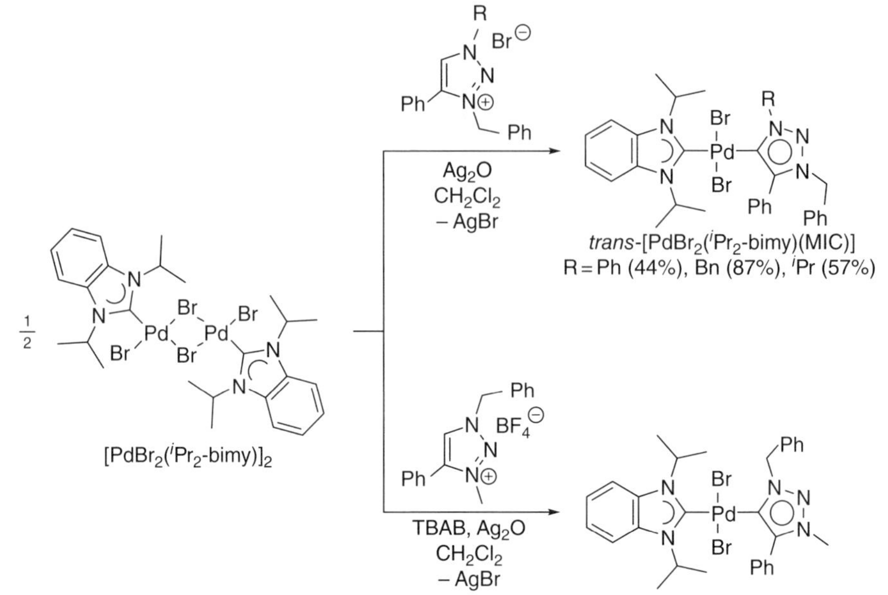

Scheme 9.17
Preparation of hetero-bis(carbene) palladium complexes for the evaluation of substituent effects in MIC ligands.

Moreover, and in agreement with carbonyl based evaluations, all MICs studied in this work have been detected to be better donors compared to saturated imidazolidin-2-ylidene, which are the strongest among classical NHCs. The additional nitrogen atom, however, makes them inferior to the *a*NHCs in terms of donating power, as already discussed earlier (Figure 9.5).

The vast majority of 1,2,3-triazolin-5-ylidenes contain nitrogen substituents at N1 and N3 positions and are thus mesoionic compounds, for which no neutral (closed-shell) resonance structure can be drawn. More recently, Herrmann and Kühn reported the first examples of 1,2,3-triazole-based carbenes with a 1,2-nitrogen substitution pattern [22]. These ligands can be represented by a neutral carbene resonance form, and therefore they are classified as normal 1,2,3-triazole-derived NHCs (trz). The preparation of the respective triazolium salts occurs via a ring-closing procedure using hydrazonoyl chlorides and isocyanides [23]. Attempted isolation of the free carbenes via deprotonation using conventional methods failed, but the reaction with sodium hydride in ammonia/tetrahydrofurane gave rise to the formation of an ammonia adduct (Scheme 9.18). DFT calculations suggested an energetically viable pathway via the generation of a free carbene, which then inserts into an N–H bond of ammonia furnishing the neutral adduct. Heating the latter in benzene-d_6 only led to decomposition, while free carbenes formed *in situ* by α-elimination of ammonia could not be detected. In the presence of a suitable metal source, however, the supposed free NHCs could be trapped via metal coordination. Thus, bridge-cleavage reactions of dimeric rhodium(I) and iridium(I) precursors occur easily affording the [MCl(COD)(trz)] (M = Rh, Ir, trz = 1-phenyl-2-methyl-4-tolyl-1,2,3-triazolin-5-ylidene) complexes in moderate yields. The iridium derivative was transformed into the dicarbonyl

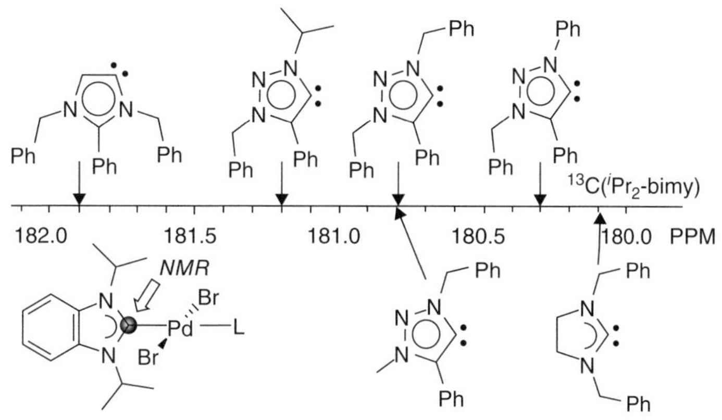

Figure 9.5
Comparison of donor strengths of selected NHCs on the ^{13}C NMR based scale.

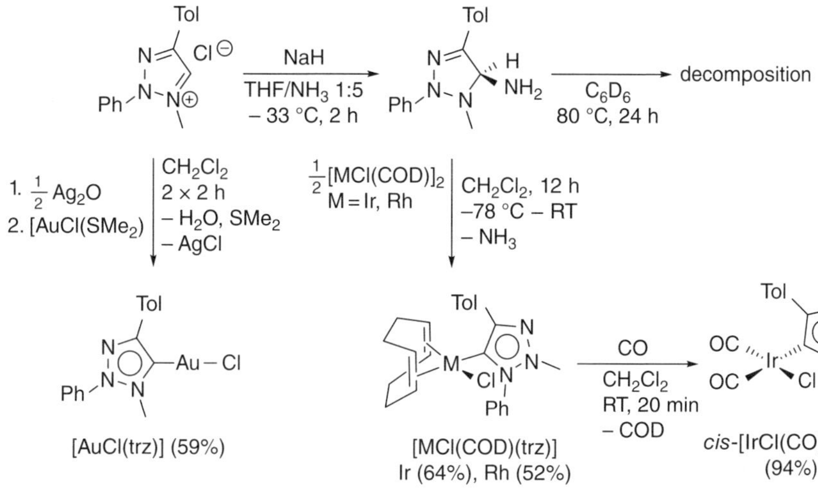

Scheme 9.18
Preparation of *normal* 1,2,3-triazolidin-5-ylidene (trz) complexes.

complex *cis*-[IrCl(CO)$_2$(trz)] for donor strength estimation of the *normal* 1,2,3-triazolin-5-ylidene. The carbonyl wavenumbers of 1975 and 2058 cm^{-1} {$\tilde{v}$(CO)$_{av}$ = 2017 cm^{-1}} indicate a similar donating power to their *mesoionic* cousins (Table 9.2). Other properties and spectroscopic features are equally comparable indicating that a change in the substitution pattern in triazolin-5-ylidenes has a small impact.

Transition metal complexes of such ligands are more easily prepared via the more routine silver-NHC transfer protocol [22]. Thus, treating the triazolium salt with silver(I) oxide and subsequently with chlorido-dimethylsulfidegold(I) afforded the monocarbene complex [AuCl(trz)] in a moderate yield.

The molecular structures of the first gold(I) and iridium(I) complexes bearing a normal 1,2,3-triazolin-5-ylidene ligand have also been established by single crystal X-ray diffraction and are depicted in Figure 9.6. The structural parameters indicate increased delocalization

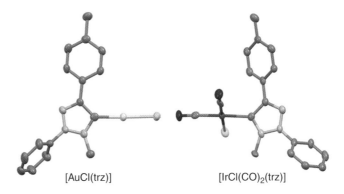

[AuCl(trz)]   [IrCl(CO)₂(trz)]

Figure 9.6
Molecular structures of the first normal 1,2,3-triazolin-5-ylidene complexes. Hydrogen atoms have been omitted for clarity.

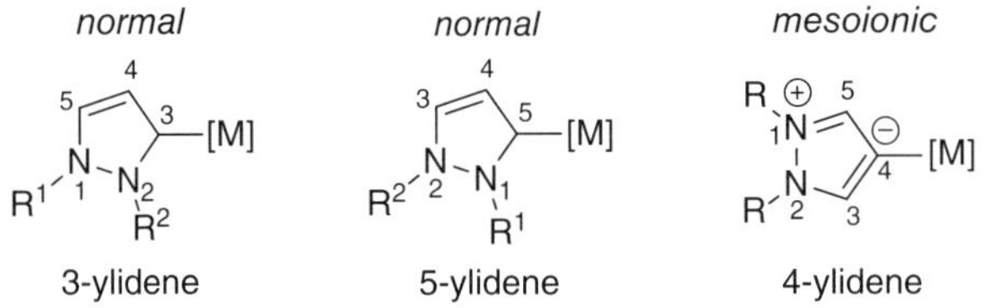

normal   *normal*   *mesoionic*

3-ylidene   5-ylidene   4-ylidene

Chart 9.2
Differentiation between *normal* versus *mesoionic* pyrazolinylidenes.

of the π-electrons within the five-membered heterocycle in comparison to its salt precursor. Again, these bond features are not significantly different from those of their mesoionic counterparts.

9.3   Pyrazolin-3/5-ylidenes (Pyry) and Indazolin-3-ylidenes (Indy)

Pyrazole-derived NHCs are isomers of the ubiquitous and well-known imidazolinylidenes. Similar to these, pyrazolinylidenes and their complexes can also exist in either *normal* or *mesoionic* form depending on the position of the carbene atom. Metallation at the C3 (R^1 of higher priority further away from donor) or C5 position (R^1 of higher priority closer to donor) leads to normal NHC complexes, while metallation at the more remote C4 carbon leads to a *mesoionic* and *remote* NHC complex (see Section 9.5). Chart 9.2 shows the three general types of pyrazolinylidenes.

9.3.1   Pyrazolin-3/5-ylidenes

As with most NHCs, pyrazolinylidenes are most commonly accessed via their respective salt precursors, that is, pyrazolium salts. Öfele and coworkers already reported the first examples of pyrazolin-3-ylidene (Pyry) complexes in 1976 [24]. In this work, the bis(1,2-dimethylpyrazolium)decacarbonyldimolybdate(-I) complex was heated in the solid state to its melting point, which results in a redox reaction between the azolium cations and dianionic carbonyl complex (Scheme 9.19). The metal centers are oxidized by one electron each from –I to a zero-valent oxidation state, while dihydrogen gas was released as a reduction product with the concurrent formation of the desired pentacarbonyl-monocarbene complex [Mo(CO)$_5$(Pyry)] (Pyry = 1,2-dimethylpyrazolin-3-ylidene). The complex was obtained in 40% yield as an air and moisture stable powder.

Comparison of its IR data with classical NHC analogs already revealed a stronger donating ability of pyrazolin-3-ylidenes because of the smaller number of direct nitrogen neighbors (Table 9.3). The reduced symmetry of the carbene ligand and its superior electron donation also induces a greater dipole moment in the respective pentacarbonyl complex compared to that of its classical counterparts.

Further heating of the monocarbene complex led to a ligand disproportionation affording the yellow bis(carbene) complex *cis*-[Mo(CO)$_4$(Pyry)$_2$] in 20% yield and homoleptic [Mo(CO)$_6$] among other decomposition products. In contrast to analogous complexes bearing classical NHC, the latter resists *cis* to *trans* isomerization upon further heating. This

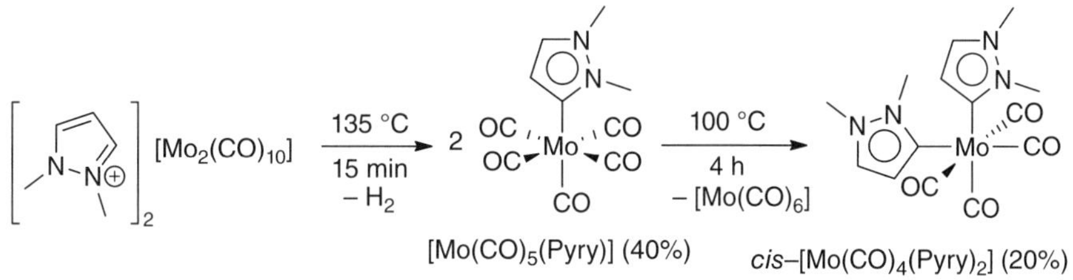

Scheme 9.19
Preparation of the first pyrazolin-3-ylidene complexes.

Table 9.3 IR spectroscopic data in hexane and dipole moments (μ) in benzene of [Mo(CO)$_5$(NHC)] complexes.

NHC	$\tilde{v}$(CO) [cm^{-1}]	$\tilde{v}$(CO)$_{av}$ [cm^{-1}]	μ [D]
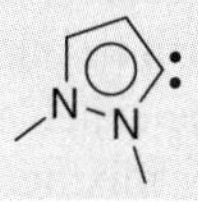	1919, 2054	1987	9.3
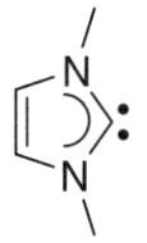	1930, 2062	1996	7.3
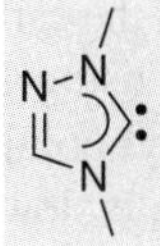	1931, 2066	1999	5.3
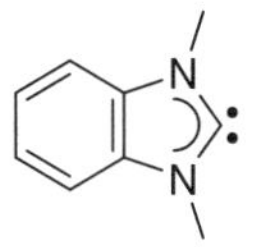	1934, 2066	2000	6.8

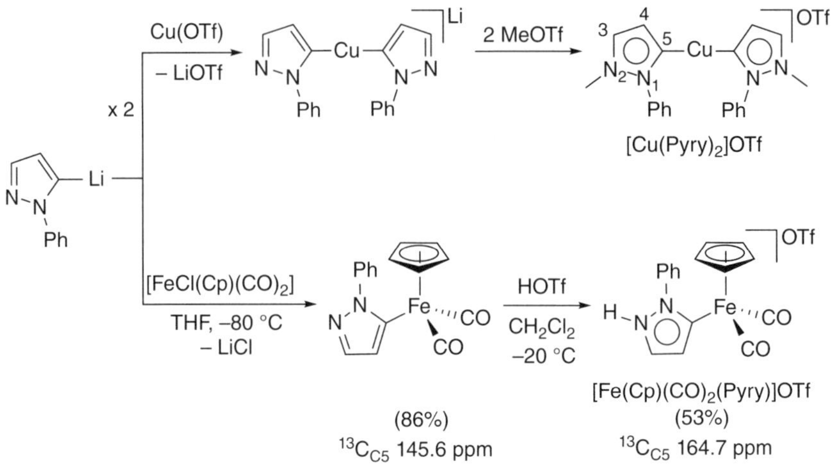

Scheme 9.20
Preparation pyrazolin-5-ylidene complexes using lithium pyrazolates.

observation provides further support for the greater donor strength of pyrazolin-3-ylidenes. In the *cis* form, the increased electron density can be most efficiently π-backdonated into the trans-standing carbonyl ligands, which would result in the most favorable overall bonding picture for the complex to prevent isomerization to the electronically less favorable *trans* complex.

Only 20 years later, Raubenheimer and coworkers reported the first pyrazolin-5-ylidene complexes by a very different route [25]. These were obtained in a two-step protocol from the reaction of lithium pyrazolates with copper(I) or iron(II) precursors under salt metathesis (Scheme 9.20). The intermediate pyrazolato complexes were then N-protonated or N-methylated with a strong acid or electrophile to furnish the pyrazolin-5-ylidene complexes.

Upon N-protonation, the ^{13}C NMR signal of the carbon donor in the iron(II) complex shifts downfield by ~19 ppm (CD_2Cl_2), which is indicative of carbene formation. However, such a downfield shift alone does not provide sufficient evidence for carbene formation.

The *in situ* acid-base reaction between a basic metal precursor and an azolium salt routinely used for the making of classical NHC complexes can also be adapted to pyrazolium salts as reported by Herrmann and coworkers in 1997 [26]. The suitable metal precursor [Rh(OEt)(COD)]$_2$ was also freshly prepared without isolation by mixing [RhCl(COD)]$_2$ and sodium hydride in ethanol. Introduction of 1,2-dimethylpyrazolium iodide to this mixture eventually led to the formation of the pyrazolin-3-ylidene complex [RhI(COD)(Pyry)] in 56% yield (Scheme 9.21). Compared to classical NHC analogs, this reaction proceeds very slowly, and a reaction time of 7 days was required even with an excess of pyrazolium salt. The low reactivity is attributable to the reduced acidity of the C3–H proton in the carbene precursor.

For donor strength comparison with classical NHCs using IR spectroscopy, the COD complex was quantitatively converted into the dicarbonyl complex *cis*-[RhI(CO)$_2$(Pyry)], which was noted to be stable in the solid state. However, it slowly decomposed when kept in solution for a few hours. In agreement with Öfele's study (Table 9.3), the 1,2-dimethyl-pyrazolin-3-ylidene ligand was found to be stronger donor compared to classical NHCs (Table 9.4) [27]. Despite the different topology, the chemical shift for the carbene carbon at 169.2 ppm was found to be very similar to the value of 168.9 ppm recorded for its direct isomer 1,3-dimethylimidazolin-2-ylidene.

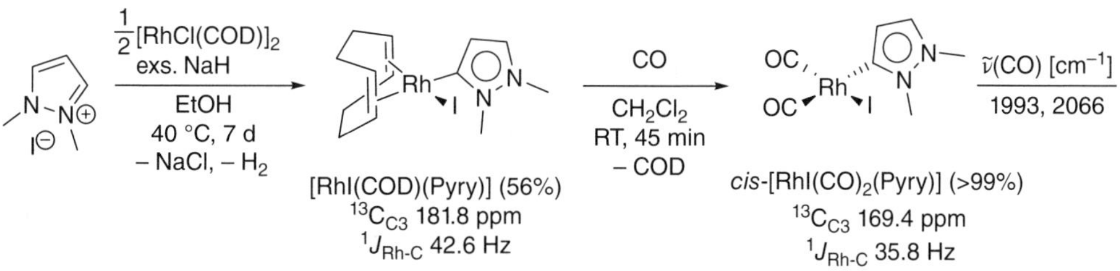

Scheme 9.21
Preparation of rhodium pyrazolin-3-ylidene complexes.

Table 9.4 Carbonyl wavenumbers in CH_2Cl_2, $^{13}C_{carbene}$ shifts and $^1J_{Rh\text{-}C}$ coupling constants in DMSO-d_6 of [RhI(CO)$_2$(NHC)] complexes.

NHC	$\tilde{v}(CO)$ [cm^{-1}]	$^{13}C_{carbene}$ [ppm]	$^1J_{Rh\text{-}C}$ [Hz]
	1993, 2066	169.2	35.8
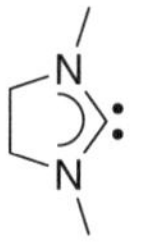	1999, 2072	195.2	n/a
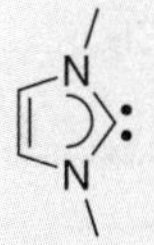	2000, 2073	168.9	40.8
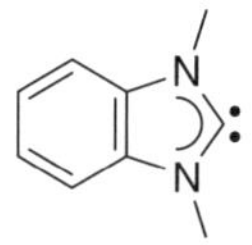	2001, 2075	180.8	40.5
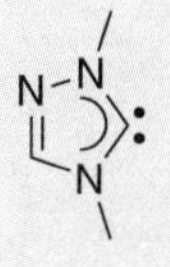	2006, 2078	174.2	42.6

The most efficient and versatile method for the preparation of pyrazolinylidene complexes to date is the silver-carbene transfer method. Peris and coworkers described the fact that the reaction of silver(I) oxide with 1,2-dimethylpyrazolium iodide under standard conditions afforded intermediate silver(I) carbene species; these were directly exposed to a chlorido-bridged ruthenium(II) dimer (Scheme 9.22) [28]. This led to the transmetallation of pyrazolin-3-ylidene ligand from silver to afford the respective ruthenium complexes. Depending on the ratio of the reactants, either the neutral, yellow monocarbene or the green cationic bis(carbene) complex were isolated in yields of 60% and 35%, respectively. The carbene signal of the cationic and thus more Lewis acidic bis(Pyry) complex [RuCl(cym)(Pyry)$_2$]PF$_6$ resonates at higher field compared to that of the neutral monocarbene species [RuCl$_2$(cym)(Pyry)]. In addition, the molecular structure of the latter could be

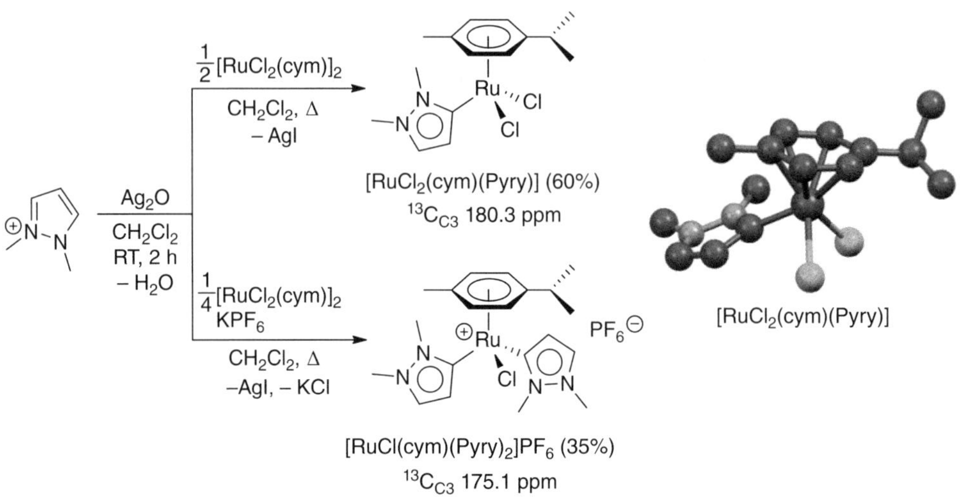

Scheme 9.22
Preparation of Ru^{II}-pyrazolin-3-ylidene complexes via Ag(I)-carbene transfer and molecular structure of [RuCl$_2$(cym)(Pyry)].

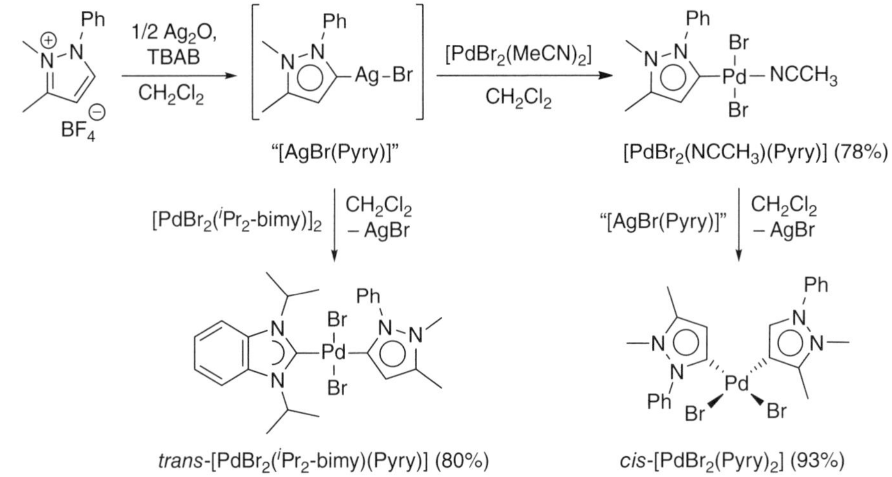

Scheme 9.23
Preparation of Pd^{II}-pyrazolin-5-ylidene complexes via Ag(I)-carbene transfer.

determined by single crystal X-ray diffraction providing evidence for the successful carbene transfer and the expected piano-stool geometry of the compound (Scheme 9.22). The complexes were tested in catalytic β-alkylation of secondary alcohols with primary alcohols, where particularly the cationic bis(Pyry) complex showed excellent activities.

Bernhammer and Huynh subsequently prepared palladium(II) pyrazolin-5-ylidene complexes using a similar strategy [29]. 1-phenyl-2,3-dimethylpyrazolium tetrafluoroborate was reacted with silver(I) oxide in the presence of one equivalent of tetrabutylammonium bromide (Scheme 9.23). The addition of the bromide is important as it increases the acidity of the pyrazolium salt via hydrogen bonding. Moreover, additional halide is provided for silver(I) precipitation, which drives the subsequent transfer of the carbene to palladium. The intermediate silver-carbene complex proved to be too unstable for isolation in substance and full characterization. It was therefore directly mixed with either bis(acetonitrile) dibromidopalladium(II) or the dimeric palladium(II)-benzimidazolin-2-ylidene complex [PdBr$_2$(^{i}Pr$_2$-bimy)]$_2$ to afford the pyrazolin-5-ylidene complexes *trans*-[PdBr$_2$(MeCN) (Pyry)] and *trans*-[PdBr$_2$(^{i}Pr$_2$-bimy)(Pyry)] in good yields. The influence of the 1-phenyl-2, 3-dimethylpyrazolin-5-ylidene ligand on the ^{13}C$_{carbene}$ NMR chemical shift of the ^{i}Pr$_2$-bimy reporter ligand in the latter hetero-bis(carbene) complex was used as a means to evaluate the Pyry donating ability in comparison to other carbenes. The observed value of 182.8 ppm for the reporter donor places this pyrazolin-5-ylidene above classical NHC, *mesoionic* triazolinylidenes and even imidazolin-4/5-ylidenes in terms of donor strength (cf. Figure 9.5).

The carbene chemical shift for the Pyry ligand in the electron-rich hetero-bis(carbene) complex was detected at 175.4 ppm in CDCl$_3$, which is substantially more downfield than that observed in the more Lewis acidic acetonitrile adduct, that is, 150.0 ppm.

By easy replacement of the labile acetonitrile ligand with triphenylphosphine, pyridine, and other classical NHCs, a small library of pyrazolin-5-ylidene palladium(II) complexes was prepared for catalytic screening in the direct arylation of pentafluorobenzene with aryl- and heteroaryl bromides. Its exposure to an additional equivalent of the silver complex resulted in the transfer of the second Pyry ligand, and the homo-bis(carbene) complex *cis*-[PdBr$_2$(Pyry)$_2$] was obtained. It is interesting to note that the direct reaction of excess silver-carbenes with palladium in a single step did not afford the same complex in decent yields. The *cis* complex was substantially less soluble in common organic solvents, which is attributed to its larger dipole moment compared to *trans* complexes. Thus, more polar solvents were required for acquisition of solution NMR data. The important chemical shift for the carbene carbon was noted downfield at 181.8 ppm in DMSO-d_6 pointing to a very electron-rich complex.

Single crystals for most of the complexes could be obtained for structural elucidation in the solid state. The molecular structures of the hetero-bis(NHC) and homo-bis(Pyry) complexes are depicted in Figure 9.7. They confirm the proposed bonding situation of the Pyry ligand. Also, the palladium–Pyry bonds were in the same range as those observed for analogous complexes of classical NHCs.

As with classical NHC complexes, Pyry complexes can also be prepared by oxidative addition/substitution using suitably substituted pyrazolium salts and low-valent metal

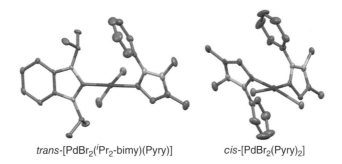

trans-[PdBr$_2$(^{i}Pr$_2$-bimy)(Pyry)]   *cis*-[PdBr$_2$(Pyry)$_2$]

Figure 9.7
Molecular structures of *trans*-[PdBr$_2$(^{i}Pr$_2$-bimy)(Pyry)] *cis*-[PdBr$_2$(Pyry)$_2$]. Hydrogen atoms are omitted for clarity.

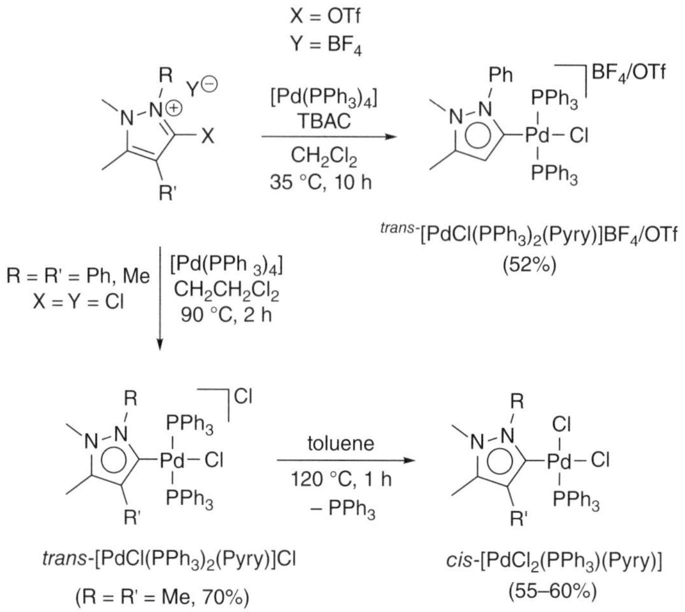

Scheme 9.24
Preparation of pyrazolin-5-ylidene complexes via oxidative addition.

precursors. In fact, the activation of 5-chloropyrazolium chlorides with the homoleptic phosphine complex [Pd(PPh$_3$)$_4$] in dichloroethane under reflux and in an inert atmosphere led to the first pyrazolinylidene complexes of palladium(II) (Scheme 9.24). The cationic monocarbene-bis(phosphine) complex of the type *trans*-[PdCl(PPh$_3$)$_2$(Pyry)]Cl could only be isolated for the tetramethylpyrazolin-5-ylidene ligand in 70% yield. The diphenyl-derivative, however, underwent rapid inner-outer sphere exchange under formation of the neutral dichlorido complex *cis*-[PdCl$_2$(PPh$_3$)(Pyry)]. This transformation can be effected more efficiently by further heating the intermediate cationic complexes in toluene to 120 °C for one hour, giving the desired neutral species in moderate yields. The carbene shifts for these complexes range from 161 to 164 ppm.

The isolation of the cationic bis(phosphine) complexes does not pose a problem when pyrazolium salts with non-coordinating anions are utilized. For example, metallation of the 5-triflatopyrazolium tetrafluoroborate salt in the presence of tetrabutylammonium chloride as a halide source gave the *trans*-[PdCl(PPh$_3$)$_2$(Pyry)]$^+$ complex cation in moderate yield. NMR spectroscopy revealed that both tetrafluoroborate and triflate anions were present in a 2:1 ratio. The ^{13}C NMR signal for the carbene donor was found slightly downfield at 165.6 ppm in CDCl$_3$.

An alternative way to prepare pyrazolinylidene complexes is through template-directed synthesis using either a Fischer carbene [30] or an allenylidene [31] complex of chromium(0). In both cases, the organometallic precursors are treated with a solution of N,N′-dimethylhydrazine, and hydrazinolosis occurs with cyclization to the pyrazolin-5-ylidene ligand (Scheme 9.25). At the same time, ethanol or dimethylamine is released, which can give rise to additional byproducts, hence making these methodologies less general and applicable. The need to use complicated precursor complexes is an additional drawback. Nevertheless, the chromium(0) pyrazolin-5-ylidene complex [Cr(CO)$_5$(Pyry)] can be used

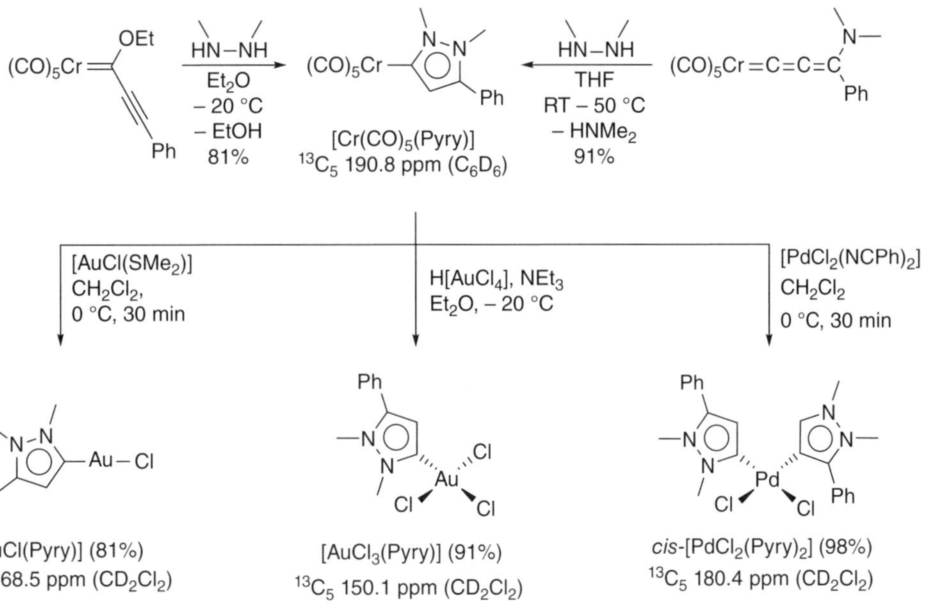

Scheme 9.25
Template-directed syntheses of a chromium(0) Pyry complex and its use as carbene transfer agent.

to gain access to monocarbene complexes of gold(I) and gold(III), and a bis(carbene) derivative of palladium(II) in high yields via transfer of the carbene ligand (Scheme 9.25). The transfer of the carbene ligand to gold(I) is straightforward and gave the linear [AuCl(Pyry)] complex, while that to gold(III) led to the formation of the hydrochloride adduct (not depicted) in addition to the desired neutral complex [AuCl₃(Pyry)] in a 1:1 ratio. Addition of triethylamine quantitatively transforms the former into the latter. The different pyrazolin-5-ylidene complexes exhibit carbene signals at 168.5, 150.1 and 180.4 ppm reflecting their Lewis acidity. The most downfield signal was detected for the *cis*-[PdCl₂(Pyry)₂] complex containing two strongly donating Pyry ligands, which leads to the least Lewis acidic complex.

9.3.2   Indazolin-3-ylidenes

Benzannulation involving two of the three carbon atoms of the parent five-membered pyrazole ring leads to the so-called indazole system. Carbenes derived from indazole are isomers of benzimidazolin-2-ylidenes, but contain the carbene atom inherently at the C3 position and are thus termed indazolin-3-ylidenes (Indy). Although, free indazolin-3-ylidenes have not been isolated and fully characterized, Schmidt and coworkers demonstrated that they can be generated *in situ* by thermal decarboxylation of indazolium-3-carboxylates and trapped using elemental sulfur, isocyanates, and isothiocyanates to give indazolin-3-thione, indazolium-3-amidates (X=O), and indazolium-3-thioamidates (X=S), respectively (Scheme 9.26) [32].

The first transition metal complex of an indazolin-3-ylidene was reported by Huynh and coworkers in 2009 [11]. The hetero-bis(NHC) complex *trans*-[PdBr₂(ⁱPr₂-bimy)

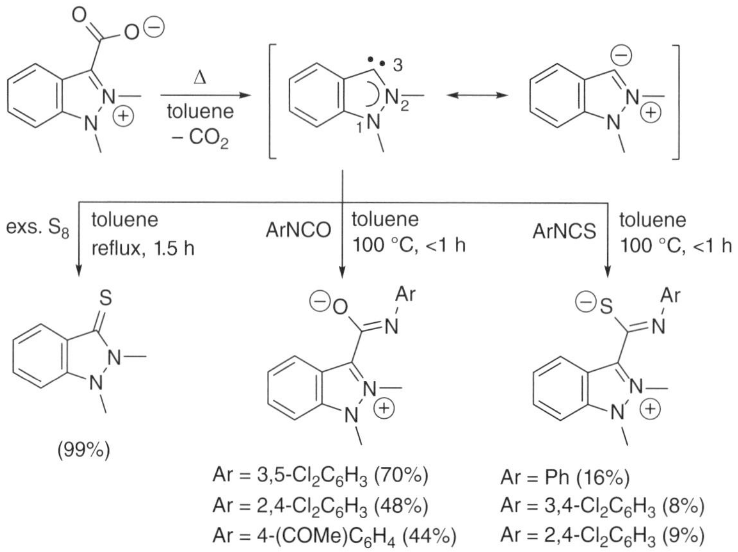

Scheme 9.26
In situ preparation and trapping of 1,2-dimethylindazolin-3-ylidene.

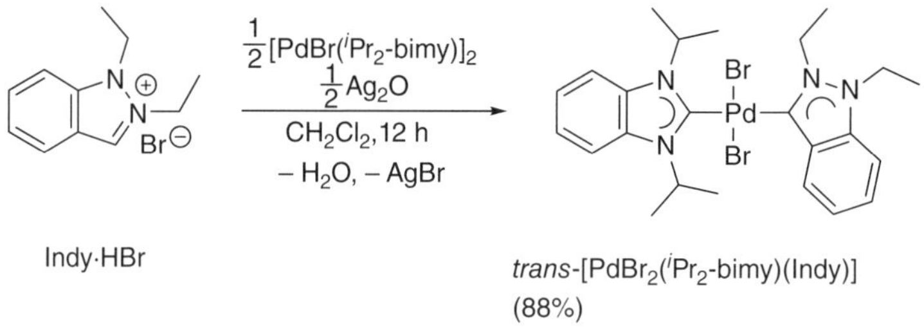

Scheme 9.27
Preparation of the first indazolin-3-ylidene complex.

(Indy)] containing the ^{i}Pr$_2$-bimy reporter and the 1,2-diethylindazolin-3-ylidene ligand was prepared in order to evaluate the electron donating ability of the latter by ^{13}C NMR spectroscopy. The one-pot synthetic procedure is straightforward and involves mixing and stirring all reactants in dichloromethane at ambient temperature for 12 hours (Scheme 9.27).

The air and moisture stable target complex was isolated as a yellow solid in 88% yield and exhibits signals for the two different carbenes at 180.6 (Indy) and 181.6 ppm (^{i}Pr$_2$-bimy) in CDCl$_3$, respectively. Comparison of the latter's chemical shift with the respective signal detected at 182.4 ppm for the 1,2-diethylpyrazolin-3-ylidene analog *trans*-[PdBr$_2$(^{i}Pr$_2$-bimy)(Pyry)], which was prepared in the same fashion using the respective pyrazolium salt, corroborates the view that benzannulation reduces the donor strength of NHCs. The same expected trend was also observed in the comparison between imidazolin-2-ylidenes and benzimidazolin-2-ylidenes.

Using the mild silver-carbene transfer method, a range of other gold(I), rhodium(I), and palladium(II) complexes have been prepared from both indazolium iodides and bromides (Scheme 9.28) [33]. The intermediate silver-indazolin-3-ylidene complexes were routinely

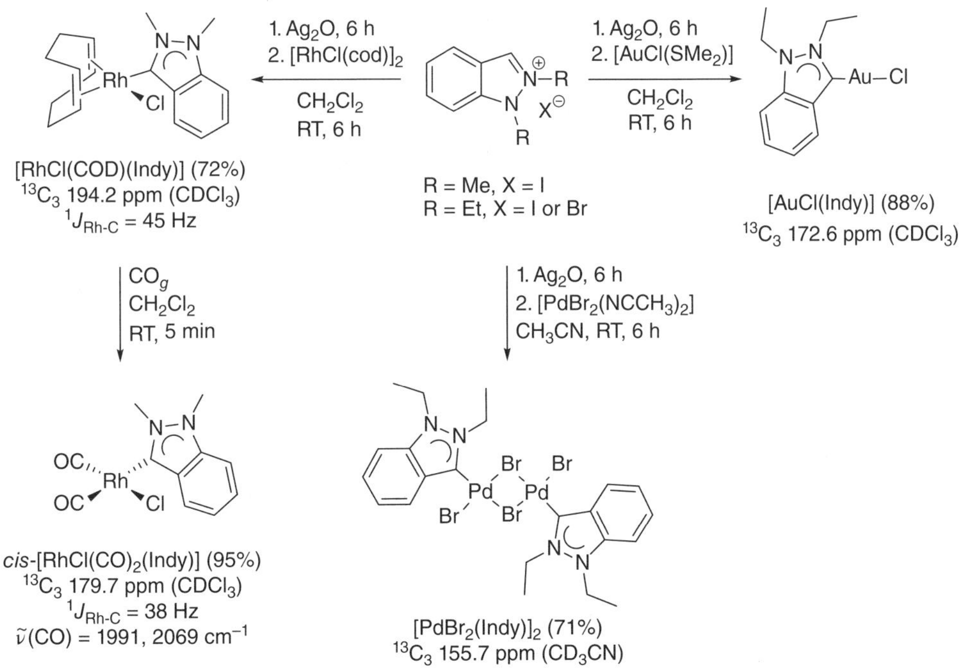

Scheme 9.28
Preparation of selected indazolin-3-ylidene complexes by carbene transfer.

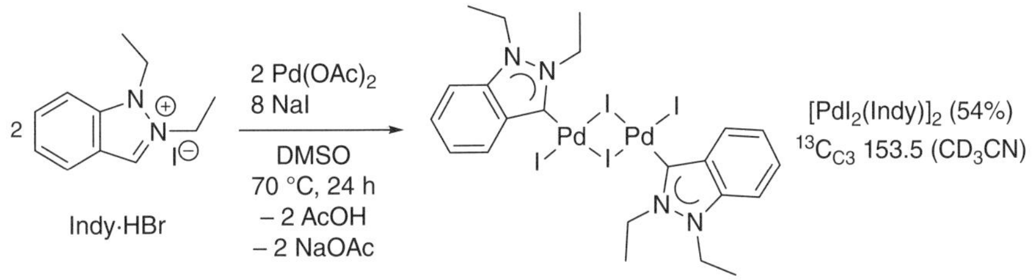

Scheme 9.29
Direct palladation of 1,2-diethylindazolium iodide with palladium(II) acetate.

prepared by reaction of the indazolium salt with silver oxide in dichloromethane under elimination of water. Subsequent addition of the transition metal complex precursors initiates the carbene transfer with the precipitation of silver halides. These reactions proceed under bridge-cleavage or displacement of labile ligands such as dimethylsulfide or acetonitrile, affording the desired complexes in generally good yields of >70% as stable compounds. Key spectroscopic data for these complexes are also summarized in Scheme 9.28.

The use of silver-carbene transfer agents is not always required, and the direct palladation of indazolium iodides using palladium(II) acetate is also possible. When this reaction is carried out in a 1:1 ratio with excess sodium iodide, the dimeric iodido-bridged palladium(II)-monocarbene complex $[PdI_2(Indy)]_2$ is obtained in moderate yield (Scheme 9.29). Exposure of this dimeric complex to additional donor ligands L leads to the formation of various heteroleptic, mononuclear Indy complexes $[PdI_2(Indy)L]$. Thus, the chemical shift observed

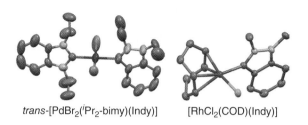

trans-[PdBr$_2$(^{i}Pr$_2$-bimy)(Indy)]   [RhCl$_2$(COD)(Indy)]

Figure 9.8
Molecular structures of *trans*-[PdBr$_2$(^{i}Pr$_2$-bimy)(Indy)] (left) and [RhCl(COD)(Indy)] (right). Hydrogen atoms are omitted for clarity.

at 153.5 ppm in CD$_3$CN for the carbene carbon is attributable to the carbon donor in the deutero-solvate *trans*-[PdI$_2$(Indy)(NCCD$_3$)]. Notably, the formation of neutral bis(Indy) palladium(II) complexes of the type [PdX$_2$(Indy)$_2$] was not observed in this direct palladation route even when an excess of indazolium salt was used.

Representative molecular structures of indazolin-3-ylidene complexes obtained by single crystal X-ray diffraction are depicted in Figure 9.8. These confirm the bonding of the non-classical NHC via the C3-position. Moreover, the detected metal–carbon bond distances are close to single bonds and essentially in the same range as those for their classical NHC counterparts. This fact underlines their similar coordination behavior.

9.4   Cyclic (Alkyl)(Amino)Carbenes (CAACs) Or Pyrrolidin-2-ylidenes

The formal replacement of one electronegative amino substituent in the classical and fully saturated imidazolidin-2-ylidene with an alkyl group gives rise to a parent pyrrolidin-2-ylidene, which in turn is formally derived from pyrrolidine by substitution of an α-CH$_2$ group with a carbene carbon (i.e. 2-ylidene, Chart 9.3). These carbenes also have a reduced heteroatom stabilization, and are therefore expected to be more basic than classical NHCs. Since they can be perfectly represented by a neutral resonance structure, they are *normal* carbenes. The increasingly popular cyclic (alkyl)(amino)carbenes (CAACs) [34] introduced by Bertrand and coworkers in 2005 are representatives of this pyrrolidin-2-ylidene family, where alkyl and aryl substiuents have been placed at the N1 and C3 positions [35].

The possibility of placing a quarternary, sp^3-hybridized carbon (i.e. C3 atom) adjacent to the carbene donor offers a uniquely different steric environment in Bertrand's CAACs compared to most other cyclic aminocarbenes. Calculations showed that the replacement of the σ-electron-withdrawing and π-donating N–H group with an only σ-donating CH$_2$ unit results in an increased energy of –5.0 eV (–482 kJ mol^{-1}) for the HOMO in the parent pyrrolidin-2-ylidene compared to that of the parent imidazolidin-2-ylidene of –5.2 eV (–502 kJ mol^{-1}). In addition, the singlet–triplet gap (S/T gap) was found to be 1.99 eV (192 kJ mol^{-1}), which is smaller than the 2.95 eV (285 kJ mol^{-1}) calculated for the fully saturated, classical NHC [34a]. Therefore, CAACs should be at the same time more σ-donating (nucleophilic), but also more π-accepting (electrophilic) than classical NHCs. This comparison is summarized in Figure 9.9 [36].

The preparation of the first free CAACs commences with a suitably substituted imine, which was formed by condensation of an aliphatic aldehyde with 2,6-diisopropylaniline

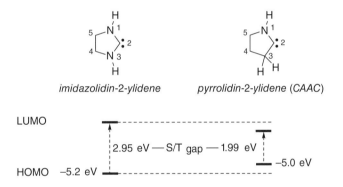

Chart 9.3
Relationship between imidazolidin-2-ylidene, pyrrolidin-2-ylidene, and CAACs.

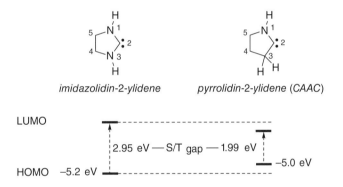

Figure 9.9
Comparison of calculated data for imidazolidin- and pyrrolidin-2-ylidenes [36].

(Scheme 9.30). Its deprotonation using lithium diisopropylamide (LDA) produces an aza-allyl anion, which readily affects ring-opening of 1,2-epoxy-2-methylpropane, affording the corresponding lithium alkoxide. Upon exposure to triflate anhydride the latter is transformed into a triflate derivative containing a better leaving group, and cyclization occurs on warming to room temperature giving the substituted 1-pyrrolinium triflate salt in 58% yield. Its deprotonation with LDA finally led to the quantitative formation of the first CAAC, which can be kept in solid state or in solution under inert atmosphere and at room temperature for at least two weeks.

Using the same approach, the authors also prepared another two sterically more elaborate, but yet flexible cyclohexyl-substituted carbenes demonstrating the versatility of the ligand design (Scheme 9.30, box). Notably, one of these could be obtained in an enantiopure form without the use of any additional chiral auxiliaries, but solely through steric control. Single crystals for this CAAC could be obtained to provide additional evidence for its identity and the arrangement of the cyclohexyl-type substituent, which results in an enantipure compound (Figure 9.10).

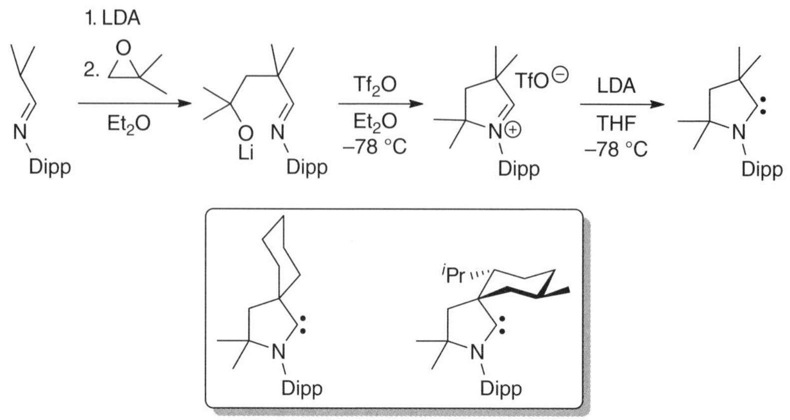

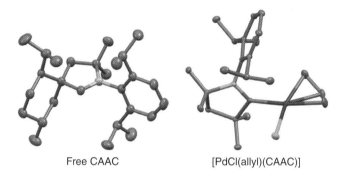

Scheme 9.30
Preparation of the first free CAACs.

Free CAAC   [PdCl(allyl)(CAAC)]

Figure 9.10
Molecular structures of a CAAC and a CAAC-palladium(II) complex.

Another striking feature is that the carbene carbon in CAACs is markedly more deshielded than their closest saturated classical NHC counterpart, despite having only one nitrogen atom in the ring. Thus, the $^{13}C_{carbene}$ signals of the first three CAACs were detected at 304.2, 309.4 and 319.0 ppm in tetrahydrofurane-d_8, respectively. Interestingly, one can note a downfield trend with increasing bulk at the quaternary C3 carbon.

The coordination chemistry of these new species was explored with iridium(I) and palladium(II) by simply mixing the free ligands with suitable dimeric metal precursors under enthalpically driven bridge-cleavage conditions (Scheme 9.31). The iridium(I)-COD complex was not characterized, but directly converted into the dicarbonyl complex *cis*-[IrCl(CO)$_2$(CAAC)] in an overall yield of 71% for donor strength estimation. The carbonyl stretches of of 1971 and 2055 cm^{-1} {$\tilde{v}(CO)_{av} = 2013$ cm^{-1}} indicate that they are, as expected, stronger donors compared to classical NHC and *mesoionic* triazolin-4-ylidenes {*cf.* $\tilde{v}(CO)_{av} = 2017$ cm^{-1}}. Maybe less intuitive, these values also suggests that their donating power is weaker than that of mesoionic imidazolin-4/5-ylidenes {*cf.* $\tilde{v}(CO)_{av} = 2003$ cm^{-1}}, although the latter contain an unsaturated heterocycle with two electronegative nitrogen atoms. Maybe, this is a consequence of the increased π-acceptor ability. Clearly, more detailed studies are required for a clearer picture of the electronic situation in CAACs.

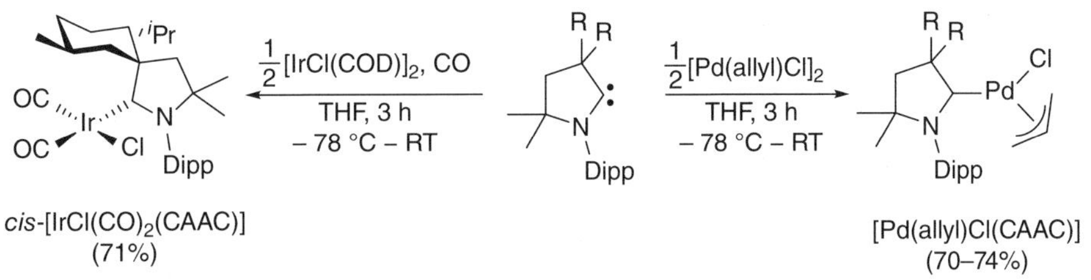

Scheme 9.31
Preparation of iridium(I) and palladium(II) CAAC complexes.

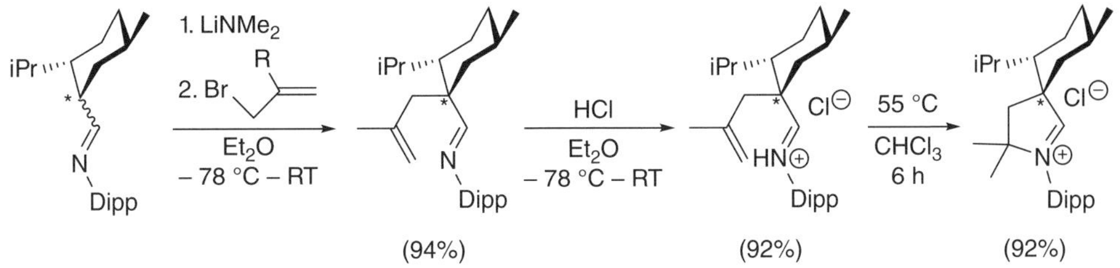

Scheme 9.32
"Hydroiminiumation" route to an enantiopure CAAC precursor.

Upon binding to iridium, a significant upfield shift of the carbene atom to 190.9 ppm (CDCl$_3$) was noted.

The cleavage of the allyl-chlorido-palladium(II) dimer with CAACs provides the respective monomeric [Pd(allyl)Cl(CAAC)] complexes in yields of 70–74% yields as colorless compounds. In contrast to the iridium(I) complex, a less pronounced upfield shift was observed for the carbon donor in the palladium complexes. Here, carbene chemical shifts of 267.4, 267.8 and 272.0 ppm were recorded for the three complexes in CDCl$_3$, which are among the most downfield signals recorded for palladium(II) NHC complexes. Moreover, these complexes gave rise to highly active catalysts for the α-arylation of ketones and even aldehydes, which underlines their suitability for catalysis.

Since the intial work on CAACs, many studies have been conducted on these new carbenes and their coordination chemistry. A prerequisite to these is an improved, easier, and cost effective synthetic route to the cyclic imminium salts, which are the direct precursors to CAACs. This has been realized by an "hydroiminiumation" strategy illustrated in Scheme 9.32 [37]. Introduction of an alkene is achieved by treatment of an chiral aza-allyl anion generated by deprotonation with lithium dimethylamide with 3-bromo-2-methylpropene. This reaction affords the enantiopure alkenylaldimine in 94% yield, which is subsequently N-protonated with hydrochloric acid in diethyl ether in 92%. Moderate heating of this salt in chloroform leads to intramolecular cycloaddition of the N–H moiety of the iminium cation to the olefin tether, furnishing the 1-pyrrolinium chloride salt in 92% yield.

With this strategy a series of CAAC precursors can be prepared, which increases the versatility and modularity of these intriguing new species. At the same time, this allows a more in depth study on their ligand properties, and one can envision more diversity in the applications of their respective complexes in the near future.

9.5 *Remote* N-Heterocyclic Carbenes (*r*NHC)

9.5.1 Pyridin-3/4-ylidenes

Carbenes derived from N-heterocycles, in which the carbene atom is not adjacent to any nitrogen atoms are termed *remote* N-heterocyclic carbenes (*r*NHCs). The descriptor "*remote*" refers to the positioning of the nitrogen atom(s) with respect to the carbene center. Thus, the *push-pull* stabilization usually exerted by α-heteroatoms in classical NHCs is essentially absent in *r*NHCs, and consequently, these species are generally more basic, but also less stable in the free form compared to their classical counterparts. As with many NHCs, both *normal* and *mesoionic* forms exist dependent on the location of the carbene atom within a given heterocycle.

The first *r*NHC complex was reported by Raubenheimer and coworkers in 2004 [38]. The authors explored two preparative pathways starting from 4-chloro-N-methylquinolinone. Its O-methylation with Meerwein's salt, that is, $[Me_3O]BF_4$, affords 4-chloro-2-methoxy-N-methylquinolinium tetrafluoroborate, which undergoes oxidative substitution with the zero-valent palladium precursor $[Pd(PPh_3)_4]$ under C–Cl activation to form the first *r*NHC palladium(II) complex *trans*-$[PdCl(PPh_3)_2(r\text{NHC})]BF_4$ in 95% yield (Scheme 9.33).

Alternatively, it is also possible to reverse the reaction steps by subjecting the neutral quinolinone first to oxidative substitution with palladium(0) followed by O-methylation of the resulting *trans*-$[PdCl(quin)(PPh_3)_2]$ complex with methyl triflate. In both cases, the same complex cation containing the *r*NHC ligand is obtained. Although the second route gives a lower yield, it allows for the direct spectroscopic comparison of the carbene complex and its direct neutral organometallic precursor. The most notable feature is the stark downfield shift of >30 ppm (CD_2Cl_2) that occurs upon going from anionic donor to a formally neutral *r*NHC ligand.

In subsequent works, it was demonstrated that similar *r*NHC complexes are easily accessible using simple halo-substituted pyridinium derivatives [39]. 2-, 3- and 4-chloropyridines were N-methylated using methyl triflate to give the respective pyridinium salts in 55 to 99% yields as suitable carbene precursors. Notably, the lowest yield was obtained for the

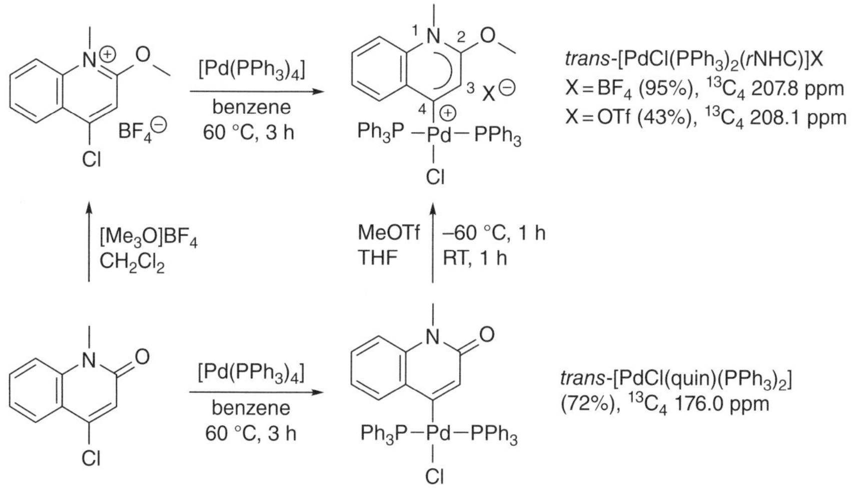

Scheme 9.33
Preparation of the first *r*NHC palladium(II) complex.

4-chloropyridinium salt, a result that can only be ascribed to electronic factors. Subsequent oxidative substitution with palladium(0) as described above affords three palladium(II) complexes bearing different carbenes in very good yields (Scheme 9.34).

The C–Cl activation of the *ortho*-derivative leads to a complex containing a pyridin-2-ylidene ligand, which does not qualify as a *remote* carbene since the carbene carbon is flanked by one nitrogen substituent. Moreover, a neutral carbene resonance structure exists for the ligand making it a *normal*, but still *reduced heteroatom-stabilized* carbene. The *meta*-pyridinium salt gives rise to a carbene ligand, which is both *mesoionic* and *remote*, while the *para*-substituted precursor affords a *normal* and *remote* carbene. Interestingly, comparison of their $^{13}C_{carbene}$ NMR resonances reveals that the *mesoionic* *r*NHC complex exhibits the most shielded carbene signal at 165.0 ppm. The two normal carbenes resonate at more downfield regions in their complexes and exhibit a smaller chemical shift difference. The most downfield carbene signal was observed for the *normal* *r*NHC complex at 197.7 ppm. Notably, the two *r*NHC complexes were found to be more active catalyst precursors than the pyridin-2-ylidene analog, with the normal *r*NHC complex being the best performer.

Representative molecular structures of selected quinoline- and pyridine-derived *r*NHC complexes obtained from single crystal analyses are depicted in Figure 9.11. The *"remote"*

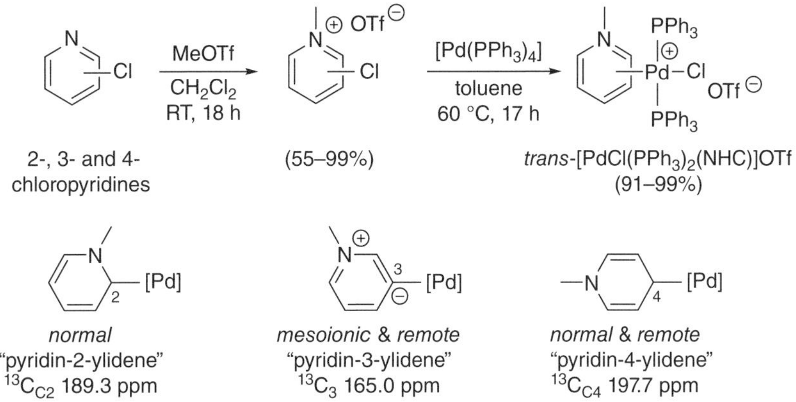

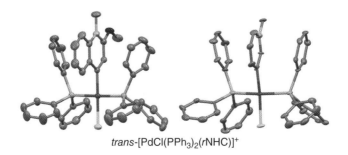

Scheme 9.34
Preparation of three different types of pyridinylidene Pd^{II} complexes. Only one mesomeric resonance structure is given for each carbene type.

trans-[PdCl(PPh$_3$)$_2$(*r*NHC)]$^+$

Figure 9.11
Representative molecular structures of *r*NHC complex cations of the type *trans*-[PdCl(PPh$_3$)$_2$(*r*NHC)]$^+$. Hydrogen atoms are omitted for clarity.

coordination site with respect to the nitrogen atom can be clearly seen in the essentially square planar palladium(II) complexes. The palladium–carbon separation in these complexes of 1.986(3) and 1.996(7)Å, respectively, are not exceptional, and fall comfortably within the normal 1.91–2.10Å range observed in a variety of classical NHC palladium(II) complexes.

9.5.2 Pyrazolin-4-ylidenes

Pyridine-derived *r*NHCs are six-membered ring systems. The first five-membered *r*NHC complexes were introduced by Han and Huynh in 2007 [40]. The *r*NHC ligand in these complexes was derived from pyrazole by blocking both C3 and C5 positions, which leaves only the C4 position of the heterocycle for metallation. Initial attempts to use simple 1,2,3,5-tetraalkylpyrazolium salts as carbene precursors were met with limited success due to the very low acidity of the C4-H proton as a result of the reduced polarization from the remote nitrogen atoms. Thus, metal coordination was achieved by the oxidative addition/substitution protocol using 4-iodopyrazolium salts and the zero-valent palladium complex precursor [$Pd_2(dba)_3$]. In general, the desired complexes are obtained by heating the ligand precursor with [$Pd_2(dba)_3$] in dichloromethane under reflux (Scheme 9.35) [41]. The use of pyrazolium iodide without any other donors yields a dimeric palladium(II)-*r*NHC complex with iodido bridges. With addition of pyridine or phosphine neutral species of the type *trans*-[$PdI_2(rNHC)(py)$] or *cis*-[$PdI_2(rNHC)(PPh_3)$] were obtained. The *cis* arrangement of the latter is due to *transphobia*, which is generally observed in mixed carbene/phosphine complexes. The use of triflate salts in combination with two equivalents of phosphine results in the cationic *trans*-[$PdI(rNHC)(PPh_3)_2$]OTf complex. All the complexes contain pyrazolin-4-ylidene ligands, which in contrast to *normal* pyrazolin-2/5-ylidenes are at the same time *mesoionic* and *remote*.

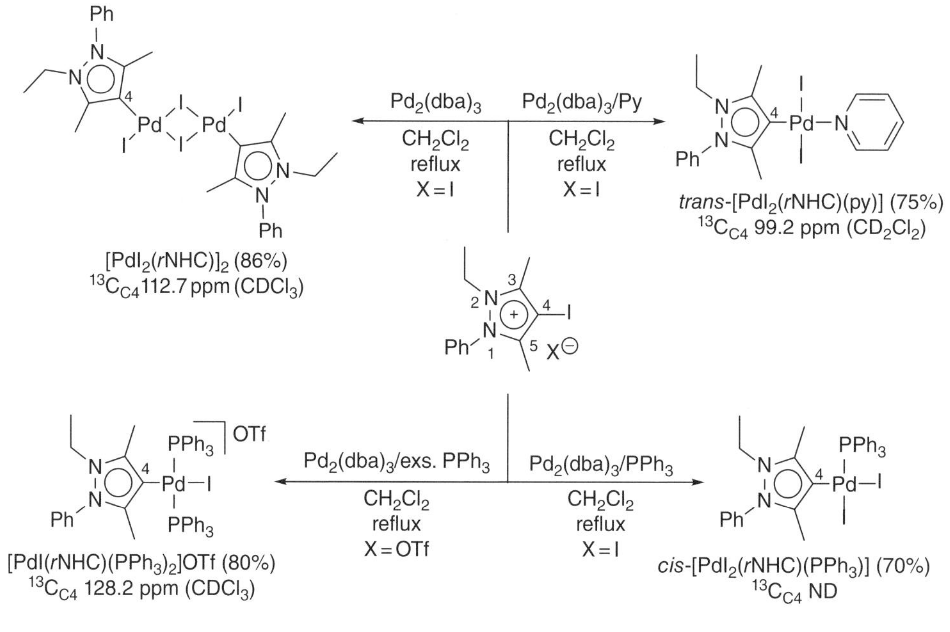

Scheme 9.35
Preparation of selected palladium(II) pyrazolin-4-ylidene complexes.

Notable spectroscopic features of these *remote* pyrazolin-4-ylidene complexes are their $^{13}C_{carbene}$ NMR resonances, which range from 99.2 to 128.2 ppm and are thus significantly shifted upfield compared to the carbene signals in their classical NHC counterparts. The reason for this is the absence of α-nitrogen atoms, which would otherwise exert a deshielding effect on the carbene donor. The chemical shifts are nevertheless substantially downfield shifted by ~26–55 ppm compared to the respective signals in their precursor salts.

Figure 9.12 shows two representative solid state molecular structures of such complexes obtained by single crystal X-ray diffraction. They adopt an essentially square planar geometry around the palladium center, and, as in classical NHC complexes, the carbene ring plane of the *r*NHC ligand is always orientated almost perpendicular to the respective coordination plane. The palladium–carbon bonds fall within a small range of 1.970(7)–2.033(7) Å, and are not distinctive for *normal* or *mesoionic* NHC complexes. All average bond lengths within the *r*NHC ring fall in the range of the respective aromatic bonds, and thus an "aromatic" Lewis structure was chosen for the five-membered pyrazole-derived ring to emphasize electron delocalization [42].

Attempts to isolate free pyrazolin-4-ylidenes with carbon-based 3,5-substituents (alkyl- or aryl) have not been successful thus far. However, by introducing exocyclic heteroatomic groups at these positions, Bertrand and coworkers were able to prepare the first examples of the free base, which they referred to as "*cyclic bent allenes.*" For this purpose, 1,2-diphenyl-3,5-diphenoxy-pyrazolium iodide was prepared and subsequently deprotonated with KN(SiMe$_3$)$_2$ in diethyl ether at –78 °C to afford the free pyrazolin-4-ylidene, the solid state molecular structure of which has also been obtained [43] (Scheme 9.36).

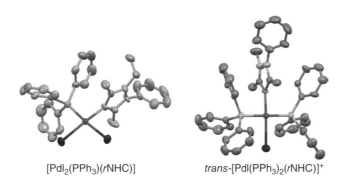

[PdI$_2$(PPh$_3$)(*r*NHC)]   *trans*-[PdI(PPh$_3$)$_2$(*r*NHC)]$^+$

Figure 9.12
Representative molecular structures of neutral and cationic palladium(II) pyrazolin-4-ylidene complexes. Hydrogen atoms and counter-anions are omitted for clarity.

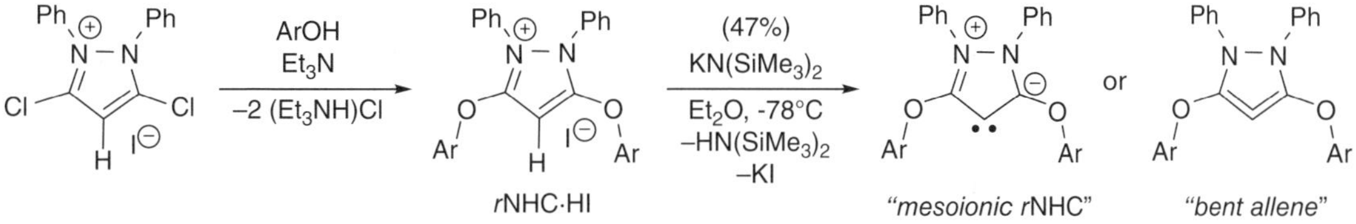

Scheme 9.36
Preparation of the first free pyrazolin-4-ylidene or "*cyclic bent allene.*"

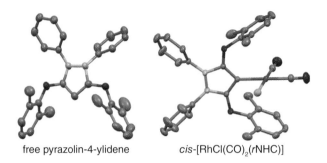

free pyrazolin-4-ylidene   *cis*-[RhCl(CO)₂(*r*NHC)]

Figure 9.13
Molecular structures of the free pyrazolin-4-ylidene and its rhodium(I) complex. Hydrogen atoms are omitted for clarity.

The 3,5-bis(aryloxy) substituents are capable of π-donation, and it has been reported that such 3,5-bis(π-donor) substituents lead to strong exocyclic delocalization and thus significantly reduce the ring-aromaticity in these heterocycles [44]. Exposure of the free ligand to the dirhodium species [RhCl(CO)₂]₂ in a 2:1 molar ratio straightforwardly leads to the mononuclear species *cis*-[RhCl(CO)₂(*r*NHC)] suitable for estimation of its ligand properties by IR spectroscopy. The low averaged CO stretching frequency of 2018 cm⁻¹ indicates that the pyrazolin-4-ylidene belongs to the strongest neutral carbon donors known to date.

The molecular structure of this complex and its free ligand in the solid state is depicted in Figure 9.13. The perpendicular orientation of this *r*NHC ligand with respect to the square planar coordination plane around the rhodium(I) centers indicates a similar coordination behavior to its purely carbon-substituted analog described above. Finally, it must be noted that the very nature of these unusual species has been subject to intensive debate [45].

References

[1] Recent review article: O. Schuster, L. Yang, H. G. Raubenheimer, M. Albrecht, *Chem. Rev.* **2009**, *109*, 3445.

[2] Recent review article: R. H. Crabtree, *Coord. Chem. Rev.* **2013**, *257*, 755.

[3] IUPAC. Compendium of Chemical Terminology, 2nd ed. (the "Gold Book"). Compiled by A. D. McNaught and A. Wilkinson. Blackwell Scientific Publications, Oxford (**1997**). XML on-line corrected version: http://goldbook.iupac.org (accessed October 7 2016) (2006-) created by M. Nic, J. Jirat, B. Kosata; updates compiled by A. Jenkins. ISBN 0-9678550-9-8. doi:10.1351/goldbook.

[4] S. Gruendemann, A. Kovacevic, M. Albrecht, J. W. Faller, R. H. Crabtree, *J. Am. Chem. Soc.* **2002**, *124*, 10473.

[5] H. Lebel, M. K. Janes, A. B. Charette, S. P. Nolan, *J. Am. Chem. Soc.* **2004**, *126*, 5046.

[6] B. Eguillor, M. A. Esteruelas, M. Oliván, M. Puerta, *Organometallics* **2008**, *27*, 445.

[7] G. Song, X. Wang, Y. Li, X. Li, *Organometallics* **2008**, *27*, 1187.

[8] D. Bacciu, K. J. Cavell, I. A. Fallis, L.-l. Ooi, *Angew. Chem. Int. Ed.* **2005**, *44*, 5282.

[9] A. R. Chianese, B. M. Zeglis, R. H. Crabtree, *Chem. Commun.* **2004**, 2176.

[10] A. R. Chianese, A. Kovacevic, B. M. Zeglis, J. W. Faller, R. H. Crabtree, *Organometallics* **2004**, *23*, 2461.

[11] H. V. Huynh, Y. Han, R. Jothibasu, J. A. Yang, *Organometallics* **2009**, *28*, 5395.

[12] M. Heckenroth, E. Kluser, A. Neels, M. Albrecht, *Angew. Chem. Int. Ed.* **2007**, *46*, 6293.

[13] E. Aldeco-Perez, A. J. Rosenthal, B. Donnadieu, P. Parameswaran, G. Frenking, G. Bertrand, *Science* **2009**, *326*, 556.

[14] E. Kluser, A. Neels, M. Albrecht, *Chem. Commun.* **2006**, 4495.

[15] E. Stander-Grobler, O. Schuster, C. E. Strasser, M. Albrecht, S. Cronje, H. G. Raubenheimer, *Polyhedron* **2011**, *30*, 2776.

[16] K. F. Donnelly, A. Petronilho, M. Albrecht, *Chem. Commun.* **2013**, *49*, 1145.

[17] P. Mathew, A. Neels, M. Albrecht, *J. Am. Chem. Soc.* **2008**, *130*, 13534.

[18] G. Guisado-Barrios, J. Bouffard, B. Donnadieu, G. Bertrand, *Angew. Chem. Int. Ed.* **2010**, *49*, 4759.

[19] A. Poulain, D. Canseco-Gonzalez, R. Hynes-Roche, H. Müller-Bunz, O. Schuster, H. Stoeckli-Evans, A. Neels, M. Albrecht, *Organometallics* **2011**, *30*, 1021.

[20] J. Bouffard, B. K. Keitz, R. Tonner, G. Guisado-Barrios, G. Frenking, R. H. Grubbs, G. Bertrand, *Organometallics* **2011**, *30*, 2617.

[21] D. Yuan, H. V. Huynh, *Organometallics* **2012**, *31*, 405.

[22] (a) L.-A. Schaper, K. Öfele, R. Kadyrov, B. Bechlars, M. Drees, M. Cokoja, W. A. Herrmann, F. E. Kühn, *Chem. Commun.* **2012**, *48*, 3857. (b) L.-A. Schaper, X. Wei, S. J. Hock, A. Pöthig, K. Öfele, M. Cokoja, W. A. Herrmann, F. E. Kühn, *Organometallics* **2013**, *32*, 3376.

[23] (a) D. Moderhack, M. Lorke, *Heterocycles* **1987**, *26*, 1751. (b) D. Moderhack, A. Daoud, *J. Heterocycl. Chem.* **2003**, *40*, 625. (c) D. Moderhack, *Liebigs Ann. Chem.* **1989**, 1271.

[24] K. Öfele, E. Roos, M. Herberhold, Z. Naturforsch. **1976**, *31b*, 1070.

[25] (a) H. G. Raubenheimer, M. Desmet, L. Lindeque, *J. Chem. Res., Synop.* **1995**, 184. (b) H. G. Raubenheimer, M. Desmet, P. Olivier, G. J. Kruger, *J. Chem. Soc., Dalton Trans.* **1996**, 4431.

[26] (a) C. Köcher, W. A. Herrmann, *J. Organomet. Chem.* **1997**, *532*, 261. (b) J. Schütz, E. Herdtweck, W. A. Herrmann, *Organometallics* **2004**, *23*, 6084.

[27] W. A. Herrmann, J. Schütz, G. D. Frey, E. Herdtweck, *Organometallics* **2006**, *25*, 2437.

[28] A. Prades, M. Viciano, M. Sanaú, E. Peris, *Organometallics* **2008**, *27*, 4254.

[29] J. C. Bernhammer, H. V. Huynh, *Organometallics* **2012**, *31*, 5121.

[30] R. Aumann, B. Jasper, R. Fröhlich, *Organometallics* **1996**, *14*, 2447.

[31] N. Szesni, C. Hohberger, G. G. Mohamed, N. Burzlaff, B. Weibert, H. Fischer, *J. Organomet. Chem.* **2006**, *691*, 5753.

[32] A. Schmidt, A. Beutler, T. Habeck, T. Mordhorst, B. Snovydovych, *Synthesis* **2006**, 1882.

[33] R. Jothibasu, H. V. Huynh, *Chem. Commun.* **2010**, *46*, 2986.

[34] Recent review articles: (a) M. Melaimi, M. Soleilhavoup, G. Bertrand, *Angew. Chem. Int. Ed.* **2010**, *49*, 8810. (b) D. Martin, M. Melaimi, M. Soleilhavoup, G. Bertrand, *Organometallics* **2011**, *30*, 5304.

[35] V. Lavallo, Y. Canac, C. Präsang, B. Donnadieu, G. Bertrand, *Angew. Chem. Int. Ed.* **2005**, *44*, 5705.

[36] *Note*: The singlet–triplet gap depicted in Figure 9.9 is simplified as the HOMO-LUMO gap.

[37] R. Jazzar, J.-B. Bourg, R. D. Dewhurst, B. Donnadieu, G. Bertrand, *J. Org. Chem.* **2007**, *72*, 3492.

[38] W. H. Meyer, M. Deetlefs, M. Pohlmann, R. Scholz, M. W. Esterhuysen, G. R. Julius, H. G. Raubenheimer, *Dalton Trans.* **2004**, 413.

[39] E. Stander-Grobler, O. Schuster, G. Heydenrych, S. Cronje, E. Tosh, M. Albrecht, G. Frenking, H. G. Raubenheimer, *Organometallics* **2010**, *29*, 5821.

[40] Y. Han, H. V. Huynh, *Chem. Commun.* **2007**, 1089.

[41] (a) Y. Han, H. V. Huynh, G. K. Tan, *Organometallics* **2007**, *26*, 6581. (b) Y. Han, H. V. Huynh, *Organometallics* **2009**, *28*, 2778. (c) Y. Han, D. Yuan, Q. Teng, H. V. Huynh, *Organometallics* **2011**, *30*, 1224.

[42] Y. Han, H. V. Huynh, *Dalton Trans.* **2011**, *40*, 2141.

[43] V. Lavallo, C. A. Dyker, B. Donnadieu, G. Bertrand, *Angew. Chem. Int. Ed.* **2008**, *47*, 5411.

[44] I. Fernández, C. A. Dyker, A. DeHope, B. Donnadieu, G. Frenking, G. Bertrand, *J. Am. Chem. Soc.* **2009**, *131*, 11875.

[45] (a) M. Christl, B. Engels, *Angew. Chem. Int. Ed.* **2009**, *48*, 1538. (b) M. M. Hanninen, A. Peuronen, H. M. Tuononen, *Chem. Eur. J.* **2009**, *15*, 7287. (c) H. V. Huynh, G. Frison, *J. Org. Chem.* **2013**, *78*, 328.

Index